Power Systems

Hebertt Sira-Ramírez and
Ramón Silva-Ortigoza

Control Design Techniques in Power Electronics Devices

With 202 Figures

Hebertt Sira-Ramírez, PhD
Ramón Silva-Ortigoza, PhD

Cinvestav-IPN
Avenida IPN
No. 2508
Departamento de Ingeniería Eléctrica
Sección de Mecatrónica
Colonia San Pedro Zacatenco AP 14740
07360 México D.F.
México

British Library Cataloguing in Publication Data
Sira-Ramirez, Hebertt
   Control design techniques in power electronics devices.
   (Power systems)
   1.Electronic control 2.Electronic controllers - Design and
   construction
   I.Title II.Silva-Ortigoza, Ramon
   629.8'9
ISBN-13: 9781846284588
ISBN-10: 1846284589

Library of Congress Control Number: 2006926892

Power Systems Series ISSN 1612-1287
ISBN-10: 1-84628-458-9   e-ISBN 1-84628-459-7   Printed on acid-free paper
ISBN-13: 978-1-84628-458-8

© Springer-Verlag London Limited 2006

MATLAB® and Simulink® are registered trademarks of The MathWorks, Inc., 3 Apple Hill Drive, Natick, MA 01760-2098, U.S.A. http://www.mathworks.com

National Instruments™ is a trademark of National Instruments Corporation, 11500 N. Mopac Expressway, Austin, Texas 78759-3504, USA.

Apart from any fair dealing for the purposes of research or private study, or criticism or review, as permitted under the Copyright, Designs and Patents Act 1988, this publication may only be reproduced, stored or transmitted, in any form or by any means, with the prior permission in writing of the publishers, or in the case of reprographic reproduction in accordance with the terms of licences issued by the Copyright Licensing Agency. Enquiries concerning reproduction outside those terms should be sent to the publishers.

The use of registered names, trademarks, etc. in this publication does not imply, even in the absence of a specific statement, that such names are exempt from the relevant laws and regulations and therefore free for general use.

The publisher makes no representation, express or implied, with regard to the accuracy of the information contained in this book and cannot accept any legal responsibility or liability for any errors or omissions that may be made.

Printed in Germany

9 8 7 6 5 4 3 2 1

Springer Science+Business Media
springer.com

To María Elena and Jessica.
To our families, and to the People of México.
To the "Bravo Pueblo" of Venezuela.

Preface

This book is intended for researchers, teachers, and students willing to explore conceptual bridges between the fields of Automatic Control and Power Electronics. The need to bring the two disciplines closer has been felt, for many years, both by Power Electronics specialists and by Automatic Control theorists, as a means of fruitful interaction between the two scientific communities. There have, certainly, been many steps given in that direction in the last decade as evidenced by the number of research articles in journals, special sessions in conferences, and summer courses throughout the world. This book hopes to become a small but positive contribution in the needed proximity of the two engineering fields. Automatic Control specialists are, generally speaking, not fully aware of the limitations, fundamental needs and nature of the technical problems in Power Electronics design. On the other hand, Power Electronics specialists are seldom Automatic Control theorists themselves, nor are they convinced of the advantageous viewpoint hidden in the, often rather complex, mathematical developments of Automatic Control theory. The net result has been a misunderstanding of the value of each others field with little interaction and a diminished chance for cross-fertilization of ideas, methods, visions and solutions.

Power electronics devices are physical devices that can be mathematically modelled as controlled dynamic systems and, hence, they are suitably conformed for the application of existing control theories. Specifically, control theory is mainly concerned in the design of the regulating subsystem in a power electronics device for enhancing its overall performance in accordance with the prescribed objective. Although difficult, the objectives behind the design of a certain power electronics device can usually be translated into a rather concrete "control objective" for which an arsenal of techniques exist nowadays. These facts makes Power Electronics an area of natural development and applications of Automatic Control while keeping its unique character and demands for enhanced reliability, reduced cost, and the need for experimental verification of mathematically founded claims. Automatic Control theories, on the other hand, offer a wide range of powerful techniques which can be immediately applied to problems in Power Electronics. An Automatic Control theorist dwelling in the Power Electronics area rapidly finds Power Electronics to be a challenging field with, sometimes, non clear cut, non direct mathematically tractable and, often, multiple and apparently contradictive objectives.

The basic aim of this book is to present some of the Automatic Control theories and techniques relevant to the design of feedback controllers in Power Electronics. Specifically, we deal with *switched* power electronics devices, mostly constituting DC to DC power converters, or power supplies, inverters and rectifiers of various kinds and of different topologies. Since most of switched power electronics devices may be cataloged as nonlinear (or, rather, bi-linear) control systems, we expose the reader to some few general, and rather traditional, methods for designing feedback controllers in nonlinear systems. The theoretical introduction to each one of the control design methods is accompanied by multiple case studies and examples showing the applicability of the explained theory. We concentrate in the following specific theoretical developments from the view point of Control Theory: Sliding mode control, feedback control by means of approximate linearization and nonlinear control design methods. These last are basically constituted by: exact feedback linearization methods, differential flatness and passivity based control. At the beginning of the book, we devote some space to the very important issue of modelling switched power electronics devices as controlled dynamic systems.

The material contained in this book has been used in teaching one semester graduate courses at Cinvestav (México City) in the topic of Control of Power Electronics Devices. The courses have been complemented by home-works, some of them including team work in the practical implementation of some of the control techniques, taught in the lectures, to a specific power electronics device. One of the most rewarding experiences in the teaching of these courses is to find out the ability of students to rapidly come up with a full implementation of the control ideas in their "proto-board" built electronic circuits and be able to reconfirm the values of the theoretical results. Some of the material has also been used in several motivating seminars and tutorials around the country to groups of engineering students, and teachers, interested in developing research application areas in their academic environments.

Many people have contributed to the development of this book, which arose from notes handed out to the Cinvestav master's and doctoral students. The students must be thanked first by their many pertinent observations and questions surrounding the clarity of the material. Among the people we would like to specially mention, and thank for their support and encouragement, is Dr. Gerardo Silva-Navarro of the Mechatronics section at Cinvestav. Gerardo helped us, in many ways, to make this book a reality. The assistance, in many ways, of Dr. Victor M. Hernández of the Autonoma University of Querétaro is gratefully acknowledged. The authors are indebted to Mr. Jesus Linares Flores, a PhD student at Cinvestav, for his involvement and help in the experimental implementations appearing in this book. The authors have also been positively influenced by the enthusiasm, in Power Electronics and Automatic Control theory related issues, of Dr. Jaime Arau-Roffiel of the Centro Nacional de Investigación y Desarrollo Tecnológico (Cenidet) at Cuernavaca, Morelos state. Both authors have benefited from invitations to Cuernavaca to deliver seminars, teach courses and participate in their many academic undertakings. The help and technical assistance of Dr. Mario Ponce-Silva and MSc. Rene Osorio of Cenidet is greatfully acknowledged. The advice of Dr. Gerardo Espinosa-Pérez of the Universidad Nacional Autónoma de México (UNAM) is also gratefully acknowledged. The friendly and informative discussions of topics, developments and trends, in many occasions, with Dr. Jesús Leyva-Ramos, and Dr. Gerardo Escobar of the Instituto Potosino de Investigación Científica y Tecnológica (IPICyT) in San Luis Potosí have been most fruitful and motivating. Their insight into Power Electronics has been particularly helpful to this undertaking. The first author has also immensely benefited from the capable experience and advice of Dr. Victor Cárdenas and Dr. Ciro Núñez, of the Universidad Autónoma de San Luis Potosí. The colleagues, former students, and friends at the Universidad de Los Andes (Mérida-Venezuela) have been particularly helpful and motivational. He would like to specially thank Professor Mario Spinnetti Rivera and Dr. Richard Márquez Contreras for their continuous support, generosity and their willingness to offer a helping hand with many practical as well as theoretical issues. Their good experience in directing practical implementations of many of the methods developed in these pages have been fundamental in our motivation to write this book. The generosity and advice of Dr. Joachim Rudolph of the Technical University in Dresden (Germany) is gratefully acknowledged. A one semester visit of R. Silva-Ortigoza to his laboratory was made possible thanks to the Consejo Nacional de Ciencia y Tecnología of México (CONACYT) and the Deutscher Akademischer Austauschdienst (DAAD).

The research work contained in this book has been primarily supported by the Centro de Investigación y Estudios Avanzados del Instituto Politécnico Nacional (Cinvestav-IPN) at México City and by the generous financial assistance of CONACYT, under Research Project 42231-Y, and a scholarship granted to the second author.

H. Sira-Ramírez dedicates his work in this book to his beloved wife, Maria Elena Gozaine, for the constant immense moral support and her kind understanding of the many demands implicit in the writing of this book. R. Silva-Ortigoza dedicates his work in this book, with all possible affection, to Jessica, to his parents and his family.

México City, *Hebertt Sira-Ramírez*
May 2006. *Ramón Silva-Ortigoza*

Contents

1 Introduction 1

Part I Modelling

2 Modelling of DC-to-DC Power Converters 11
   2.1 Introduction 11
   2.2 The Buck Converter 13
      2.2.1 Model of the Converter 14
      2.2.2 Normalization 15
      2.2.3 Equilibrium Point and Static Transfer Function 16
      2.2.4 A Buck Converter Prototype 18
   2.3 The Boost Converter 20
      2.3.1 Model of the Converter 22
      2.3.2 Normalization 23
      2.3.3 Equilibrium Point and Static Transfer Function 23
      2.3.4 Alternative Model of the Boost Converter 24
      2.3.5 A Boost Converter Prototype 25
   2.4 The Buck-Boost Converter 27
      2.4.1 Model of the Converter 27
      2.4.2 Normalization 28
      2.4.3 Equilibrium Point and Static Transfer Function 29
      2.4.4 A Buck-Boost Converter Prototype 30
   2.5 The Non-inverting Buck-Boost Converter 31
      2.5.1 Model of the Converter 31
      2.5.2 Normalization 32
      2.5.3 Equilibrium Point and Static Transfer Function 33
   2.6 The Cúk Converter 34
      2.6.1 Model of the Converter 35
      2.6.2 Normalization 36
      2.6.3 Equilibrium Point and Static Transfer Function 37

Contents

- 2.7 The Sepic Converter ... 38
  - 2.7.1 Model of the Converter ... 39
  - 2.7.2 Normalization ... 39
  - 2.7.3 Equilibrium Point and Static Transfer Function ... 40
- 2.8 The Zeta Converter ... 41
  - 2.8.1 Model of the Converter ... 41
  - 2.8.2 Normalization ... 43
  - 2.8.3 Equilibrium Point and Static Transfer Function ... 43
- 2.9 The Quadratic Buck Converter ... 44
  - 2.9.1 Model of the Converter ... 44
  - 2.9.2 Normalized Model ... 45
  - 2.9.3 Equilibrium Point ... 45
  - 2.9.4 Static Transfer Function ... 46
- 2.10 The Boost-Boost Converter ... 46
  - 2.10.1 Model of the Boost-Boost Converter ... 47
  - 2.10.2 Average Normalized Model ... 47
  - 2.10.3 Equilibrium Point and Static Transfer Function ... 47
  - 2.10.4 Alternative Model of the Boost-Boost Converter ... 49
  - 2.10.5 A Boost-Boost Converter Experimental Prototype ... 50
- 2.11 The Double Buck-Boost Converter ... 50
  - 2.11.1 Model of the Double Buck-Boost Converter ... 51
  - 2.11.2 Average Normalized Model ... 51
  - 2.11.3 Equilibrium Point and Static Transfer Function ... 51
- 2.12 Power Converter Models with Non-ideal Components ... 52
- 2.13 A General Mathematical Model for Power Electronics Devices ... 54
  - 2.13.1 Some Illustrative Examples of the General Model ... 56

Part II Controller Design Methods

3 Sliding Mode Control ... 61
- 3.1 Introduction ... 61
- 3.2 Variable Structure Systems ... 62
  - 3.2.1 Control of Single Switch Regulated Systems ... 62
  - 3.2.2 Sliding Surfaces ... 64
  - 3.2.3 Notation ... 65
  - 3.2.4 Equivalent Control and the Ideal Sliding Dynamics ... 65
  - 3.2.5 Accessibility of the Sliding Surface ... 67
  - 3.2.6 Invariance Conditions for Matched Perturbations ... 69
- 3.3 Control of the Boost Converter ... 71
  - 3.3.1 Direct Control ... 71
  - 3.3.2 Indirect Control ... 72
  - 3.3.3 Simulations ... 74
  - 3.3.4 Experimental Implementation ... 75
- 3.4 Control of the Buck-Boost Converter ... 78

		3.4.1	Direct Control 79
	3.5	Control of the Cúk Converter 82
		3.5.1	Direct Control 83
		3.5.2	Indirect Control................................... 84
		3.5.3	Simulations....................................... 86
	3.6	Control of the Zeta Converter 87
		3.6.1	Direct Control 88
		3.6.2	Indirect Control................................... 88
		3.6.3	Simulations....................................... 90
	3.7	Control of the Quadratic Buck Converter 91
		3.7.1	Direct Control 92
		3.7.2	Indirect Control................................... 93
		3.7.3	Simulations....................................... 95
	3.8	Multi-variable Case 95
		3.8.1	Sliding Surfaces 97
		3.8.2	Equivalent Control and Ideal Sliding Dynamics........ 99
		3.8.3	Invariance with Respect to Matched Perturbations..... 100
		3.8.4	Accessibility of the Sliding Surface................... 101
	3.9	Control of the Boost-Boost Converter 102
		3.9.1	Direct Control 103
		3.9.2	Indirect Control................................... 104
		3.9.3	Simulations....................................... 105
		3.9.4	Experimental Sliding Mode Control Implementation ... 105
	3.10	Control of the Double Buck-Boost Converter 108
		3.10.1	Direct Control 109
		3.10.2	Indirect Control................................... 110
		3.10.3	Simulations....................................... 111
	3.11	$\Sigma - \Delta$ Modulation 112
		3.11.1	$\Sigma - \Delta$-Modulators 113
		3.11.2	Average Feedbacks and $\Sigma - \Delta$ -Modulation........... 115
		3.11.3	A Hardware Realization of a $\Sigma - \Delta$-Modulator........ 118

4 Approximate Linearization in the Control of Power Electronics Devices 123
    4.1 Introduction 123
    4.2 Linear Feedback Control 124
          4.2.1 Pole Placement by Full State Feedback 124
          4.2.2 Pole Placement Based on Observer Design 126
          4.2.3 Reduced Order Observers 128
          4.2.4 Flatness.. 130
          4.2.5 Generalized Proportional Integral Controllers 133
          4.2.6 Passivity Based Control 136
          4.2.7 A Hamiltonian Systems Viewpoint.................... 139

4.3	The Buck Converter 142	
	4.3.1 Generalities about the Average Normalized Model 142	
	4.3.2 Controller Design by Pole Placement.................. 144	
	4.3.3 Proportional-Derivative Control via State Feedback.... 145	
	4.3.4 Trajectory Tracking 146	
	4.3.5 Fliess' Generalized Canonical Forms 150	
	4.3.6 State Feedback Control via Observer Design 152	
	4.3.7 GPI Controller Design 154	
	4.3.8 Passivity Based Control 156	
	4.3.9 The Hamiltonian Systems Viewpoint................. 159	
	4.3.10 Implementation of the Linear Passivity Based Control for the Buck Converter 162	
4.4	The Boost Converter...................................... 168	
	4.4.1 Generalities about the Average Normalized Model 168	
	4.4.2 Control via State Feedback 172	
	4.4.3 Proportional-Derivative State Feedback Control 174	
	4.4.4 Trajectory Tracking 176	
	4.4.5 Fliess' Generalized Canonical Form 181	
	4.4.6 State Feedback Control via Observer Design 182	
	4.4.7 GPI Controller Design 183	
	4.4.8 Passivity Based Control 185	
	4.4.9 The Hamiltonian Systems Viewpoint................. 187	
4.5	The Buck-Boost Converter 189	
	4.5.1 Generalities about the Model 189	
	4.5.2 State Feedback Controller Design.................... 193	
	4.5.3 Dynamic Proportional-Derivative State Feedback Control 195	
	4.5.4 Trajectory Tracking 198	
	4.5.5 Fliess' Generalized Canonical Forms 199	
	4.5.6 Control via Observer Design 200	
	4.5.7 GPI Controller Design 202	
	4.5.8 Passivity Based Control 204	
	4.5.9 The Hamiltonian Systems Viewpoint................. 205	
	4.5.10 Experimental Passivity based Control of the Buck-Boost Converter 207	
4.6	The Cúk Converter 210	
	4.6.1 Generalities about the Model 210	
	4.6.2 The Hamiltonian System Approach 213	
4.7	The Zeta Converter....................................... 214	
	4.7.1 Generalities about the Model 214	
	4.7.2 The Hamiltonian System Approach 218	
4.8	The Quadratic Buck Converter............................. 219	
	4.8.1 Generalities about the Model 219	
	4.8.2 State Feedback Controller Design.................... 223	
	4.8.3 The Hamiltonian System Approach 227	

	4.9	The Boost-Boost Converter 229
		4.9.1 Generalities about the Model 229
		4.9.2 The Hamiltonian System Approach 233

5 Nonlinear Methods in the Control of Power Electronics Devices 235

- 5.1 Introduction 235
- 5.2 Feedback Linearization 236
  - 5.2.1 Isidori's Canonical Form 236
  - 5.2.2 Input-Output Feedback Linearization 238
  - 5.2.3 State Feedback Linearization....................... 240
  - 5.2.4 The Boost Converter 243
  - 5.2.5 The Buck-Boost Converter 246
  - 5.2.6 The Cúk Converter 249
  - 5.2.7 The Sepic Converter 254
  - 5.2.8 The Zeta Converter................................. 258
  - 5.2.9 The Quadratic Buck Converter..................... 261
- 5.3 Passivity Based Control 261
  - 5.3.1 The Boost Converter 263
  - 5.3.2 The Buck-Boost Converter 266
  - 5.3.3 The Cúk Converter 269
  - 5.3.4 The Sepic Converter 272
  - 5.3.5 The Zeta Converter................................. 274
  - 5.3.6 The Quadratic Buck Converter..................... 279
- 5.4 Exact Error Dynamics Passive Output Feedback Control 282
  - 5.4.1 A General Result................................... 282
  - 5.4.2 The Boost Converter 286
  - 5.4.3 Experimental Implementation....................... 288
  - 5.4.4 The Buck-Boost Converter 291
  - 5.4.5 The Cúk Converter 293
  - 5.4.6 The Sepic Converter 294
  - 5.4.7 The Zeta Converter................................. 298
  - 5.4.8 The Quadratic Buck Converter..................... 301
  - 5.4.9 The Boost-Boost Converter 304
  - 5.4.10 The Double Buck-Boost Converter................... 306
- 5.5 Error Dynamics Passive Output Feedback 309
  - 5.5.1 The Boost Converter 312
  - 5.5.2 Experimental Results 315
- 5.6 Control via Fliess' Generalized Canonical Form............... 316
  - 5.6.1 The Boost Converter 317
  - 5.6.2 The Buck-Boost Converter 322
  - 5.6.3 The Quadratic Buck Converter..................... 326
- 5.7 Nonlinear Observers for Power Converters 331
  - 5.7.1 Full Order Observers 331
  - 5.7.2 The Boost Converter 333

		5.7.3	The Buck-Boost Converter 335

- 5.8 Reduced Order Observers 337
  - 5.8.1 The Boost Converter 337
  - 5.8.2 The Buck-Boost Converter 341
- 5.9 GPI Sliding Mode Control 343
  - 5.9.1 The Buck Converter 344
  - 5.9.2 The Boost Converter 350
  - 5.9.3 The Buck-Boost Converter 355

Part III Applications

6 DC-to-AC Power Conversion 361
- 6.1 Introduction 361
- 6.2 Nominal Trajectories in DC-to-AC Power Conversion 363
  - 6.2.1 The Buck Converter 363
  - 6.2.2 Two-Sided $\Sigma - \Delta$ Modulation 365
  - 6.2.3 The Boost Converter 366
  - 6.2.4 The Buck-Boost Converter 370
- 6.3 An Approximate Linearization Approach 371
  - 6.3.1 The Boost Converter 371
  - 6.3.2 The Buck-Boost Converter 373
- 6.4 A Flatness Based Approach 374
  - 6.4.1 The Double Bridge Buck Converter 374
  - 6.4.2 The Boost Converter 375
  - 6.4.3 The Buck-Boost Converter 376
- 6.5 A Sliding Mode Control Approach 378
  - 6.5.1 The Boost Converter 378
  - 6.5.2 A Feasible Indirect Input Current Tracking Approach .. 378
- 6.6 Exact Tracking Error Dynamics Passive Output Feedback Control 380
  - 6.6.1 The Double Bridge Buck Converter 380
  - 6.6.2 The Boost Converter 381
  - 6.6.3 The Buck-Boost Converter 383

7 AC Rectifiers 385
- 7.1 Introduction 385
- 7.2 Boost Unit Power Factor Rectifier 386
  - 7.2.1 Model of the Monophasic Boost Rectifier 386
  - 7.2.2 The Control Objectives 387
  - 7.2.3 Steady State Considerations 387
  - 7.2.4 Exact Open Loop Tracking Error Dynamics and Controller Design 388
  - 7.2.5 Simulations 389

		7.2.6	The Use of the Differential Flatness Property in the Passive Controller Design................................389

       7.2.6 The Use of the Differential Flatness Property in the
              Passive Controller Design........................... 389
       7.2.7 Simulations 392
  7.3 Three Phase Boost Rectifier 392
       7.3.1 The Three Phase Boost Rectifier Average Model 393
       7.3.2 A Static Passivity Based Controller 395
       7.3.3 Trajectory Planning 395
       7.3.4 Switched Implementation of the Average Design....... 398
       7.3.5 Simulations 399
  7.4 A Unit Power Factor Rectifier-DC Motor System 400
       7.4.1 The Combined Rectifier-DC Motor Model 400
       7.4.2 The Exact Tracking Error Dynamics Passive Output
              Feedback Controller 403
       7.4.3 Trajectory Generation 403
       7.4.4 Simulations 405
  7.5 A Three Phase Rectifier-DC Motor System 408
       7.5.1 The Combined Three Phase Rectifier DC Motor Model . 408
       7.5.2 The Exact Tracking Error Dynamics Passive Output
              Feedback Controller 409
       7.5.3 Trajectory Generation 410
       7.5.4 Simulations 412

References 415

Index 421

1
Introduction

This book contains a collection of Automatic Control techniques for the regulation of power electronics devices, such as: DC-to-DC power converters, DC-to-AC supplies (inverters), AC-to-AC conversion circuits (rectifiers) and some of its variants. As dynamical controlled systems, power electronics devices are prone to feedback controller design applications. In this respect, models of such devices may fall into one of two categories, *mono-variable* and *multi-variable*. The first class refers to the presence of only one control input variable, usually represented by the position of a switch. In the second class, we have multiple, independent switches. Typically, in this last category we find: cascaded arrangements of DC-to-DC power converters, three phase rectifiers etc. Since most of the devices treated in this book are of nonlinear nature, the control synthesis problem associated with the automatically regulated operation of these devices falls into the category of nonlinear control systems. We explore several feedback controller design methodologies. Namely; approximate linearization, exact feedback linearization and its versions: input-state and input-output linearization, passivity based control in its various forms (dynamic and static), observer design and Generalized Proportional Integral (GPI) control. In order to make this feedback controller design techniques relevant and applicable to switched power electronics devices, we need to use *average* models of the switched dynamics describing these devices. Such average models are usually obtained under the assumption of ideal *infinite* switching frequency operation. The infinite frequency idealization is, of course, never verified in practise. Nevertheless, as a necessary step in the controller design procedure, its use sidesteps cumbersome exact discretization of the dynamic models and the mathematically involved, complex, form of the resulting sampled data controllers. Also, the infinite frequency idealization has the enormous advantage of rather accurately predicting the actual behavior of the implemented finite switching frequency controller. The finite frequency implementation of the average feedback controller design is here tackled in a manner which is rather different from the traditional approach. We specifically resort to $\Sigma - \Delta$ modulation as a means of synthesizing a binary valued input to the actual switch command subsystem. Realistic $\Sigma - \Delta$ modulation can also be suitably proposed to provide a finite frequency logic input to the switching arrangement while preserving the most relevant qualitative features of the closed loop average based feedback controller design.

In this book, we advocate the use of *normalization* as a systematic time scale and state variable model transformation that offers several advantageous features to the designer. In the first place, normalization simplifies the mathematical description of the system by eliminating the presence of superfluous parameters and exhibiting only those parameters which are responsible for important qualitative changes in the system response and behavior. Normalization, therefore, enormously simplifies the algebraic manipulation of the model equations at the controller design stage and allows for qualitative insight into the form of the proposed control solutions. It also portrays the relevance, and implications, of the partial analytic results derived from the model, such as: equilibrium points, steady state behavior, control amplitude restrictions and the like. Simulation runs on normalized models considerably facilitates the mathematical processor operations and it result in fast, accurate, reliable, computations devoid of the traditional numerical "stiffness" present in most power electronics devices models arising from small capacitance and small inductance values (i.e., exceedingly large right hand sides of the involved differential equations). Finally, reverting to non-normalized variables amounts to the multiplication by constant factors of the state variables magnitudes and of the simulation time scale.

Most chapters in this book include the description of a laboratory implementation of at least one of the feedback controller design options explained in that chapter. We include circuit layouts and details that will allow the interested reader to obtain a physical realization of some of the studied controllers, thus creating the opportunity to synthesize and try out some other controllers of his (her) interest.

Chapter 2 deals with the modelling of DC-to-DC power Converters. Even though an Euler-Lagrange modelling approach could have been undertaken in this part. The authors feel that this modelling technique has been sufficiently explained, and illustrated, in the book by Ortega *et al.* [48]. A closely related modelling technique, with many interesting implications in modern electronics circuits, is that of "Port Controlled Hamiltonian" systems (see Escobar [11] and Escobar *et al.* [13]). Here, we prefer the more direct approach of using Kirkchoff's voltage and Kirkchoff's current laws on each constitutive part of the system, obtained from each possible commanded switch position, and then combine the obtained dynamical equations. Throughout, we hypothesize the presence of an ideal switch, characterized by a switch position function taking values in the discrete set: $\{0, 1\}$. In some special instances, we advocate the use of discrete sets of the form: $\{-1, 1\}$, or $\{-1, 0, 1\}$.

In this chapter, we undertake the detailed modelling of the several DC-to-DC power converters. In particular we examine the derivation of the models of the "Buck", the "Boost", the "Buck-Boost", the "non-inverting Buck-Boost", the "Cúk ", the "Sepic", the "Zeta" and the "quadratic Buck" converter. These are all *mono-variable* converters, i.e., the control action is constituted by a *single* switch acting as a control input. A more interesting and versatile class of DC-to-DC power converters is constituted by the cascade arrangement of several converters. These, in general, constitute the *multi-variable* converters. Two examples of this class of converters are introduced and modelled in detail in this chapter. They are the "Boost-Boost" converter and the "double Buck-Boost" converter.

The models analyzed and derived in Chapter 2 include *ideal switch* models. This assumption is found to be quite unrealistic when one attempts actual laboratory implementation of certain feedback controllers. Switches are often realized by suitable arrangements of diodes and transistors. These electronic components include non-ideal components, such as: parasitic resistances and offset voltages. For this reason, the chapter concludes by examining some more refined models of standard DC-to-DC power converters (see the article by Kazimierczuk and Czarkowski [36]) for further details about non-ideal models of switches. Here, we simply adopt models which include inductor resistances, diode internal resistors and parasitic voltages as well as transistor resistances. These models, which are slightly more involved, have been found useful in several realistic simulations leading to actual controller designs.

In Chapter 3, we present a tutorial introduction to sliding mode control of switch regulated systems. The sliding mode control technique is perhaps the simplest control technique that may be applied to switched controller converters. The reasons being, the simplicity of generating a meaningful *sliding surface*, the relative ease for implementation of the derived control law and the fact that the analysis of the *ideal sliding dynamics* and steady state characteristics of the closed loop system is relatively straightforward and easy to understand. The reader is advised to read the fundamental work in this area by Professor V. Utkin in his books [75] and [76]. In Utkin *et al.* [77] the reader may find applications of sliding mode control to DC-to-DC power conversion. The chapter deals with the essential elements of sliding mode control in rather general single-input and multiple-input systems (i.e., in mono-variable and multi-variable switched systems). We demonstrate, in a theoretical manner, the traditional foundation of the robustness claims usually advertised in sliding mode control as a nonlinear, discontinuous, control technique. The robustness characteristic of sliding mode control refers to the annihilation of *matched* perturbations. We explore the implications of a sliding mode control approach in a wide variety of DC-to-DC power converters. The several examples will provide the reader with a systematic view towards sliding mode controller design, thus revealing the conceptual advantages and the simplicity of the approach. In order to avoid some of the consequences of *unmatched perturbations* and, more importantly, to be able to effectively use linear as well as nonlinear traditional feedback controller design techniques in the realm of switched controlled systems, we propose an alternative to sliding mode implementation, known as the $\Sigma - \Delta$ modulation approach. In fact, this implementation technique will be present throughout this book. $\Sigma - \Delta$ modulation is quite well known, and popular, in areas other than control theory (mainly: communications, signal processing, analog to binary conversion, etc.). The idea is to have a *block* capable of translating continuous valued, bounded, signals into high frequency (ideally, infinite frequency) switched signals with one important property: that the *average* or *ideal equivalent* behavior of the output signal coincides, exactly, with the continuous input to the modulator. Δ modulation and its variants: $\Sigma - \Delta$ modulation, Double Σ modulation and two sided $\Sigma - \Delta$ modulation, etc., was used in the early days of voice transmission in space flights, and many other areas of signal digitalization (see the books by Steele and, most notably, that by Norsworthy *et al.* [47]). Its use in sliding mode controller design has been more recent (see Sira-Ramírez [62] and [61] although initial developments may be traced back to [56]).

Chapter 4 revisits the most popular feedback control technique used in the area of DC-to-DC power electronics. It deals with the *approximate linearization* based feedback control. The new feature we advocate is the $\Sigma - \Delta$ modulation implementation of the derived average feedback controllers. We explore several controller design techniques: Linear static state feedback control, Linear dynamic state feedback control, Generalized PI control, and linear passivity based control. As expected, in this chapter we deal with linearized average models of DC-to-DC power converters. The linearized average models are computed via standard first order Taylor series approximations of the average nonlinear dynamics around desired average equilibrium points. We first assume the availability of the entire linearized state of the system. Application of linear state feedback control is then quite natural and direct. Linear state feedback control is quite well known in the control literature and only a tutorial introduction is presented in this chapter which explains how to achieve closed loop desired pole placement and stabilization of the incremental state trajectory. The relation with traditional Proportional Derivative (PD) control is immediate, as PD control can be reinterpreted as a state feedback control technique. A second proposal is that of using Fliess' Generalized Canonical forms (GCF) of the various converters models in order to obtain a *dynamic* state feedback controller. GCF were introduced by Fliess in [15] as a generalization of Kalman state representations of linear and nonlinear dynamic systems. In fact, these canonical forms correspond to input-output descriptions of the average linearized model of the dynamic plant. We also examine the implications of this approach, within the nonlinear setting, in Chapter 5. The limitation of this design method is evidently related to the *non-minimum phase* phenomenon present in some of those circuits. In such cases, the output exhibits a transfer function characterized by an unstable numerator complex variable polynomial.

If, as it may usually be the case, the state of the system is not available for measurement and feedback, then, one traditionally resorts to *asymptotic observer* design. Although this last statement is certainly true in most of control problems, in power electronics one may always sidestep the need with an extra measurement of a needed current or voltage. However, extra measurements tend to complicate the circuit design, introduces more hardware into the circuit, it somewhat increases the costs, and it results in a decrease of the feedback controller reliability. The fundamental property to be tested is then the *observability* of the system as this property is related to the possibilities of obtaining states from inputs and outputs (and a finite number of its time derivatives). Once the system is determined to be observable, an asymptotic observer of the *Luenberger* type may be proposed to asymptotically obtain the unavailable, or unmeasured, linearized states needed in the designed feedback control law. However, the use of observers in Power Electronics is not popular due also to circuit cost increase. In fact, a dynamic observer system has to be synthesized, via analog electronics, or software, creating a system of the same order, or of reduced order depending on the number of outputs, than that of the observed system. For this reason, we prefer to advocate a different option that integrates observer and controller in a single design. This technique has become known as GPI. This input-output feedback control technique is based on the idea of avoiding traditional observer design by using only *structural estimates* or *integral reconstructors* of the unmeasured states as estimates for such variables. One of the attractive features of the GPI control method lies in the fact that it is based only on measurements of inputs and outputs, and linear combinations of finite numbers of iterated integrals of these available signals. These estimates are computed modulo initial conditions and modulo the influence of classical perturbations (such as steps, ramps, parabolas, etc.). As a result, the integral reconstructors differ from the actual signals in errors that can be described by finite order time polynomials signals. The structural errors thus being fundamentally *unstable*. The superposition principle is then invoked to complement, at the feedback controller design stage, the reconstructor based feedback with a suitable finite number of iterated integrals of the stabilization output error, so as to counteract the destabilizing effect of the structural estimator. As a result, one obtains a higher order controller which effectively stabilizes the closed loop system using nothing more than a simple pole placement technique on an increased order characteristic polynomial. GPI control has only been recently introduced in the control literature by Fliess and his coworkers [21] and its application to the control of power electronics devices has been advocated in [69] and in [72].

Within the framework of approximate state linearization, one may also obtain, in a relatively simple fashion, a *Generalized Hamiltonian* model of the linear incremental dynamic model (see [58] and also [59]). Such models are also known as "Port controlled Hamiltonian systems" (see Escobar *et al.* [13]). Using this special Hamiltonian form of the linearized incremental model, we develop a simple controller design procedure, which is based on *static passivity* considerations. The technique allows for the specification of a stabilizing incremental state feedback controller provided a certain *dissipation matching* condition is satisfied. Fortunately, most of the popular power electronics devices for DC-to-DC power conversion do satisfy such a matching condition. Incidentally, this static passivity based controller design procedure has an interesting implication and generalization in the nonlinear framework, as explained in Chapter 5.

Chapter 5 is devoted to the relevance of nonlinear feedback controller design in power electronics devices of the DC-to-DC power conversion type. We explore, and illustrate, several feedback controller design methods: Feedback linearization, passivity based control through the traditional method of "energy shaping and damping injection" advocated in Ortega *et al.* [48]. Static and dynamic input-output linearization and static nonlinear passivity based control. We start by revisiting the exact feedback linearization technique in the context of stabilization problems. The technique proves to be tractable only in the simplest of cases. The geometric theory of nonlinear control systems, as nicely described in the book by Isidori [31], is used for testing the linearizability of the average models of the most popular DC-to-DC power converters. The input-output static feedback linearization is also examined through the use of Isidori's canonical form. As a result, the minimum and non-minimum phase nature of the corresponding *zero* dynamics is presented for all the illustrative case studies treated in that chapter. Through exact models of the open loop tracking error dynamics, we also explore the possibilities of static passivity based control. This feedback technique invariably results in linear, time invariant, state feedback controllers for stabilization problems and in linear, time-varying, state feedback controllers for trajectory tracking problems. A *dissipation matching* condition, satisfied by most of the traditional average models of DC-to-DC converters topologies, guarantees the semi-global asymptotic stability of the state tracking error equilibrium point, located at the origin. The linear controller requires, nevertheless, of the nominal state and nominal input trajectories. A most useful property that can be exploited in off-line determining these trajectories, with great ease, is that of *differential flatness*; a technique introduced 14 years ago by Fliess and his colleagues (see Fliess *et al.* [18] and, also, Sira-Ramírez and Agrawal [63]). The dynamic input-output linearization of average converters is examined through Fliess' generalized canonical forms [15]. This chapter also explores the use of nonlinear asymptotic observers for the determination of unmeasured states from input and output data. Special attention is devoted to a new class of observers whose state reconstruction error dynamics explicitly depends, in a "non-harming manner" upon the control input. Thanks to the energy managing structure of the system, this control input dependance is proven to be irrelevant in the stability properties of the average reconstruction error. As a means to avoid the use of observers, we also present an extension of the Generalized Proportional Integral control design technique to some of the better known nonlinear average DC-to-DC power converter topologies.

In Chapter 6, we explore the nonlinear feedback controller design problem for DC-to-AC power conversion schemes, using some of the traditional DC-to-DC power conversion topologies. Our approach is to treat the controller design problem as an output trajectory tracking controller design problem for the given average converter plant. The problem of specifying nominal state and input trajectories compatible with the desired output voltage is found to be particularly, and surprisingly, challenging in the cases of plants whose output signal variable happens to be a *non-minimum phase* output, such as in the *Boost* and the *Buck-Boost* topologies. An effective approximate solution scheme, based on functional iterations involving an unbounded "operator", are presented. Emphasis in this chapter is devoted to the underlying nominal trajectory generation, or trajectory planning, problem. Several relevant feedback controller design techniques are briefly summarized.

Chapter 7, is devoted to the regulation of several AC-to-DC switched converters of various kinds, also widely known as *rectifiers*. This topic has been extensively treated in Escobar [11] and, in fact, an abundant literature exists on the problem and its solutions (see, for instance, Escobar *et al.* [12], Wu [80] and also Blasko and Kaura [3] and the many references therein). Here, we concentrate mainly on the static passivity based control of the exact average tracking error model dynamics. We explore the implications of this controller design technique, invariably leading to quite simple linear, time-varying, feedback controllers in the *mono-phasic* and the *three-phasic* type of rectifiers. The emphasis is placed on simultaneously achieving unit power factor along with rectified output voltage command. Two applications of the proposed control technique are explained in detail which deal with the more complex cases of the angular velocity regulation of a DC motor fed by either a monophasic unit power factor rectifier or a three phase *Boost* rectifier.

Part I

Modelling

2
Modelling of DC-to-DC Power Converters

2.1 Introduction

In this chapter, we derive the dynamic models of DC-to-DC power converters. The most elementary structures of these converters are broadly classified into *second order converters* and *fourth order converters*. In attention to the number of independent switches they are classed into two groups: *mono-variable*, or *Single Input Single Output* (SISO), and *multi-variable*, or *Multiple Input Multiple Outputs* (MIMO). The most commonly used converters correspond to the SISO second order converters. The advantages and difficulties of the MIMO converters is just beginning to be fully understood. We remark that there are converters with *multiple dependent* switches. These may still be SISO or MIMO. The second order converters that we study in this book are: the Buck converter, the Boost converter, the Buck-Boost converter and the non-inverting Buck-Boost converter. The fourth order converters are: the Cúk converter, the Sepic converter, the Zeta converter and the quadratic Buck converter. Some multi-variable converters can be obtained by a simple cascade arrangement of the basic SISO converter topologies while considering the switch in each stage as being completely independent of the other switches present in the arrangement. Many books in the Power Electronics literature present derivations of the power converters models. For a rather thorough presentation of the Euler-Lagrange modelling technique in DC-to-DC power converters, the reader is referred to the book by Ortega *et al.* [48]. The authors find the pioneering book by Severns and Bloom [54] quite accessible and direct. The thoughtful book by Kassakian *et al.* [35] contains also detailed derivations of the most popular DC-to-DC power converters topologies. Standard reference textbooks, which do contain models of DC-to-DC power converters but with a special emphasis on the steady state PWM switched behavior, are those of Bose [4], Czaki *et al.* [8], Rashid [50], Mohan *et al.* [44], Wood [79] and Batarseh [1].

We extensively use, in the derivation of the dynamic controlled models of the several converters, the fundamental Kirchoff's current and Kirchoff's voltage laws. The methodology for the derivation of the models is, therefore, quite straightforward. We fix the position of the switch, or switches, and derive the differential equations of the circuit model. We then combine the derived models into a single one parameterized by the switch position function whose value must coincide, for each possible case, with the numerical values of either "zero" or "one". In other words, the numerical values ascribed to the switch position function is the binary set $\{0, 1\}$. The obtained switched model is then interpreted as an *average model* by letting the switch position function take values on the closed interval of the real line $[0, 1]$. This state averaging procedure has been extensively justified in the literature since the early days of power electronics and, therefore, we do not dwell into the theoretical justifications of such averaging procedure. The consequences of this idealization will not be counterproductive in the controller design procedure, nor in its actual implementation through Pulse Width Modulated (PWM) "electronic actuators" or its corresponding sliding mode counterparts. In order to simplify the exposition, we make no distinction between the average model variables and the switched model variables. At the beginning, we shall only distinguish between these models by using u_{av} for the control input variable in the average model and by using u for the switched model. In later chapters, we shall also lift this distinction. It will be clear from the context whether we are referring to the average or to the switched model.

After the derivation of the average model of each converter, we systematically proceed to *normalize* the controlled differential equations constituting the dynamic model of each one of the studied converters. This normalization procedure has a definite advantage in the simulation of the converters and their derived controllers, aside from producing a rather simplified model of the system with as few parameters as possible. We point out that DC-to-DC power converters are somewhat difficult to simulate in computer packages, such as Simnon®, or MATLAB®, when considered in their traditional physical circuit form equations. This is due to the small values of inductances and capacitances which multiply the left hand sides of the involved differential equations. This fact produces quite large right hand sides thus making the model numerically "stiff" for computer simulations. The required numerical precision may then be achieved only at the cost of extremely small integration steps thus requiring longer simulation periods with a definite negative consequence in the trust placed on the obtained numerical precision. We, thus, evade these difficulties by resorting to normalization. We clarify that the normalization procedure not only refers to an appropriate scaling of the magnitudes of currents and voltages, but also to a re-scaling of the time variable to dimensionless units thus considerably simplifying the right hand side of the associated differential equations with an effective "acceleration" of the simulation time. This advantageous stand is achieved without sacrificing the required numerical precision. It is also quite straightforward to revert the normalization procedure, back to original variables magnitudes and time magnitudes, with a simple multiplication operation on the trajectories and time spans obtained for the normalized variables. Naturally, as long as actual laboratory implementation goes, the normalization considerably simplifies the controller design but the obtain design cannot be directly implemented. The actual gain values and expressions in the derived controllers have to be naturally "de-normalized" (i.e., placed in original physical units) before the implementation. We believe such an extra effort is worth the pain.

In the exposition about each converter, average models are utilized in establishing the average values of the equilibrium points. We usually parameterize the derived equilibrium points in terms of the desired average normalized value of the output voltage. Other parameterizations are still possible and, in fact, the normalized model equations allow us to carry them out with relative ease. The nature of the parametrization of the equilibrium points usually determines the fundamental characteristic of the converter in the sense that its static features define the amplifying, attenuating, or even both, features present in a specific converter. We refer to the static average normalized input-output relation as the *static transfer function*. This quantity is readily obtained from the average input value parametrization of the desired equilibrium output voltage.

2.2 The Buck Converter

The circuit diagram of the Buck converter is shown in Figure 2.1. In this figure, we actually depict the circuit schematic with the transistor-diode symbols. These arrangements constitute the actual synthesis, or realization, of the *switching* element. In Figure 2.2 however we show the ideal switch representation of the same converter circuit. In any of the two cases, the presented topological arrangement is addressed as the Buck converter. The Buck converter belongs to the class of "chopper" circuits, or attenuation circuits. It actually multiplies the constant input voltage E by a scalar factor, smaller than unity, at the output.

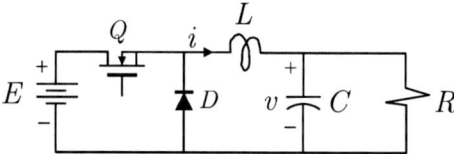

Fig. 2.1. Semiconductor realization of the Buck converter.

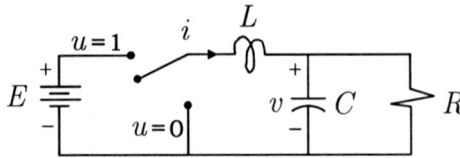

Fig. 2.2. Ideal switch representation of the Buck converter.

2.2.1 Model of the Converter

To obtain the differential equations describing the Buck converter, we consider the ideal topology shown in Figure 2.2. The system of differential equations describing the dynamics of the Buck converter is obtained through the direct application of Kirchoff's current and Kirchoff's voltage laws for each one of the possible circuit topologies arising from the assumed particular switch position function value. Thus, when the switch position function exhibits the value $u = 1$, we obtain the topology corresponding to the *non-conducting mode* for the diode. Alternatively, when the switch position exhibits the value $u = 0$ we obtain the second possible circuit topology corresponding to the *conducting mode* for the diode.

We first let the switch position function to be $u = 1$, and proceed to apply Kirchoff's current and Kirchoff's voltages laws to the resulting circuit (see Figure 2.3(a)). We obtain then the following system of differential equations:

$$L\frac{di}{dt} = -v + E$$
$$C\frac{dv}{dt} = i - \frac{v}{R} \quad (2.1)$$

When the diode is in the non-conducting mode, i.e., when the switch position function is: $u = 0$ (see Figure 2.3(b)), the dynamics of the system is described by the following differential equations:

$$L\frac{di}{dt} = -v$$
$$C\frac{dv}{dt} = i - \frac{v}{R} \quad (2.2)$$

By comparing the obtained particular dynamic systems descriptions, we immediately obtain the following *unified* dynamic system model. This results in:

$$L\frac{di}{dt} = -v + uE$$
$$C\frac{dv}{dt} = i - \frac{v}{R} \quad (2.3)$$

Indeed, when $u = 1$ or $u = 0$, the model (2.3) recovers the system models (2.1) and (2.2), respectively. The Buck converter model is then represented by Equation 2.3. We usually refer to this model as the *switched model*, and, sometimes, we make emphasis on the binary valued nature of the switch position function u by using the set theoretic relation $u \in \{0, 1\}$.

The *average* converter model would be represented exactly by the same mathematical model (2.3), possibly by renaming the state variables with different symbols and by redefining the control variable u as a sufficiently smooth function taking values in the compact interval of the real line $[0, 1]$. In order to simplify the exposition, we shall refer to the model (2.3), with u replaced by u_{av}, as the *average model* and use it to derive average feedback control laws, for the average (continuous) input variable u_{av}. We shall however distinguish between the *average control input*, denoted by u_{av} and the *switched control input*, denoted by u.

The only feature distinguishing the average model from the switched model will then be the control input. This will surely make things unequivocal.

The average model of the Buck converter is then described by

$$L\frac{di}{dt} = -v + u_{av}E$$
$$C\frac{dv}{dt} = i - \frac{v}{R} \quad (2.4)$$

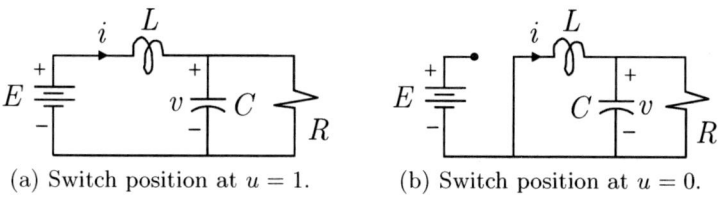

(a) Switch position at $u = 1$.   (b) Switch position at $u = 0$.

Fig. 2.3. Circuit topologies in the Buck converter.

2.2.2 Normalization

Once the average model of the converter is obtained, we proceed to make some convenient changes in the scales measuring the magnitudes of the state variables and the time variable.

We define the new set of variables for the normalized system as follows:

$$\begin{pmatrix} x_1 \\ x_2 \end{pmatrix} = \begin{pmatrix} \frac{1}{E}\sqrt{\frac{L}{C}} & 0 \\ 0 & \frac{1}{E} \end{pmatrix} \begin{pmatrix} i \\ v \end{pmatrix}, \quad \tau = \frac{t}{\sqrt{LC}} \quad (2.5)$$

Note that the voltages in the system are being divided by the constant value of the external source voltage E. The normalized voltage of the source is therefore represented by 1. Using this state and input coordinate transformation on the average system (2.4) we readily obtain the *average normalized model* of the Buck converter

$$\frac{dx_1}{d\tau} = -x_2 + u_{av}$$
$$\frac{dx_2}{d\tau} = x_1 - \frac{x_2}{Q} \tag{2.6}$$

where now the derivations in the left hand sides represent differentiations with respect to normalized (dimensionless) time τ. The variable x_1 is the normalized average current in the inductor L, x_2 is the normalized average output voltage and u_{av} represents the average switch position function, necessarily restricted to continuously take values on the set $[0, 1]$. The parameter Q is the inverse of the *quality factor of the circuit* and it is related to the resistance of the load by means of the relation: $Q = R\sqrt{C/L}$.

2.2.3 Equilibrium Point and Static Transfer Function

The control objective will be, most often, to regulate, the output voltage of the converter towards the desired average output voltage equilibrium value, taken as a constant reference signal. This is to be achieved by the application of an appropriate feedback control law u which will command the switch position in reference to an average value (this average value is most often interpreted as a *duty ratio* in a PWM scheme, but it may also be interpreted as an *equivalent control* in a sliding mode control scheme).

In general, it is desirable to relate the average values of the system variables, in steady state equilibrium, with the corresponding constant average value of the control input. These relations, also addressed as: *steady state relations* are useful in establishing the main *static* features of the converter.

In equilibrium, the time derivatives of the normalized average currents and voltages is set to zero while letting the average control input u_{av} to adopt a constant value $u_{av} = U$. As a result, we obtain a simple linear system of equations for the steady state equilibrium values of the average normalized state variables. Using the normalized average state representation (2.6) we obtain, denoting the average equilibrium values of the current and the output voltage as $\bar{x}_1$ and $\bar{x}_2$, the following relation:

$$\begin{pmatrix} 0 & 1 \\ 1 & -\frac{1}{Q} \end{pmatrix} \begin{pmatrix} \bar{x}_1 \\ \bar{x}_2 \end{pmatrix} = \begin{pmatrix} U \\ 0 \end{pmatrix} \tag{2.7}$$

Solving the system of equations for the unknowns $\bar{x}_1$ and $\bar{x}_2$, we obtain the equilibrium state of the system as:

$$\bar{x}_1 = \frac{1}{Q}U, \quad \bar{x}_2 = U \tag{2.8}$$

This average control input parametrization of the equilibrium point is useful in establishing the "attenuation" characteristics of the converter. Indeed the average normalized output voltage exhibits the numerical value of the average control input U which is restricted to the interval $[0, 1]$. The average steady state output voltage is then restricted to such an interval. Since the normalized input voltage is fixed to the value of 1, this means that the output voltage will be only a fraction of the input voltage. The converter cannot "amplify" the input voltage.

The equilibrium state (2.8) can also be conveniently parameterized in terms of the desired equilibrium value of the output voltage. Suppose such a desired voltage is represented by V_d. We would then have, $\bar{x}_2 = V_d$, and thus,

$$\bar{x}_1 = \frac{1}{Q}V_d, \qquad \bar{x}_2 = V_d \qquad (2.9)$$

We define the *normalized static transfer function* of the converter (also known as the *normalized converter gain*) as the normalized steady state normalized output voltage $\bar{x}_2$, written in terms of the constant average input U. We denote this quantity by $\mathcal{H}$ and, since it will be parameterized by the average input value U, we denote it by $\mathcal{H}(U)$. In the Buck converter case, it is given by the simple relation

$$\bar{x}_2 = \mathcal{H}(U) = U \qquad (2.10)$$

In original coordinates, we can readily write the corresponding steady state relation by using (2.5). We define the *non-normalized static transfer function*, or simply the *static transfer function*, as the ratio of the steady state output voltage $\bar{v}$ and the constant input voltage E. We have

$$\frac{\bar{v}}{E} = \frac{UE}{E} = U = \mathcal{H}(U) \qquad (2.11)$$

Clearly, the normalized and non-normalized static transfer functions are equivalent. Also, the maximum value of the gain is seen to be 1. For this reason, the Buck converter is sometimes addressed as the *voltage chopper*, or the *down converter*. The characteristic curve, corresponding to the value of the gain under feasible variations of the average control input equilibrium value U, is represented by a straight line as it is illustrated in Figure 2.4.

The equilibrium point for the actual state variables is obtained by inverting the transformation used in the normalization. This yields:

$$\bar{i} = \frac{1}{R}\bar{v}, \qquad \bar{v} = \bar{u}E \qquad (2.12)$$

18   2 Modelling of DC-to-DC Power Converters

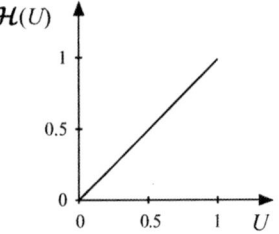

Fig. 2.4. Static transfer function of the Buck DC-to-DC power converter.

2.2.4 A Buck Converter Prototype

The Buck converter circuit, shown in Figure 2.1, was synthesized according to the diagram shown in Figure 2.5.

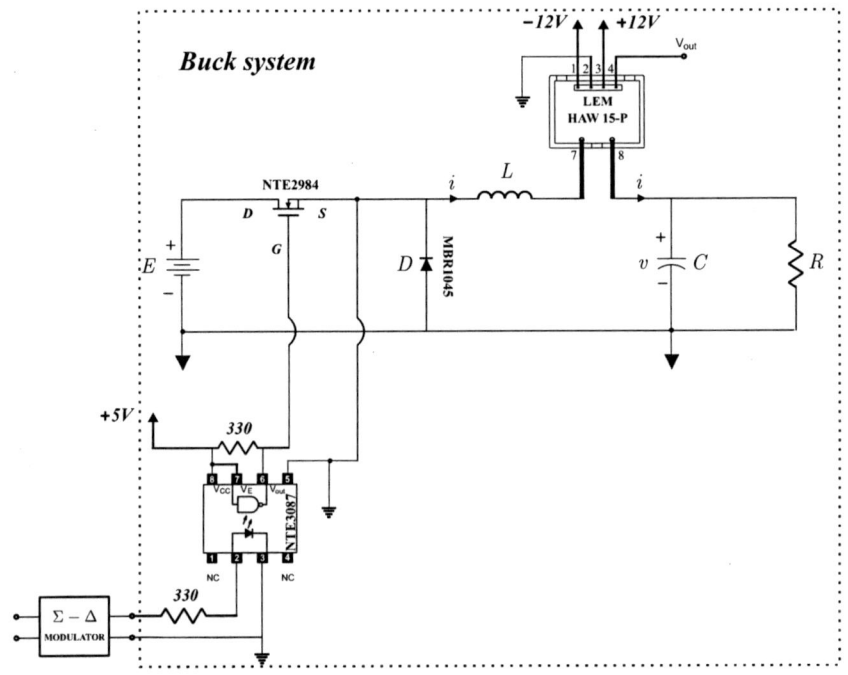

Fig. 2.5. Circuit diagram of the *Buck converter prototype*.

In this prototype circuit we will be implementing a feedback control law designed for the regulation of this system. The following parameters characterize the experimental test bed:

$$L = 15.91 \text{ mH}, \quad C = 50 \text{ }\mu\text{F}, \quad R = 25 \text{ }\Omega, \quad E = 24 \text{ V}$$

2.2 The Buck Converter

The Buck converter was designed for an operation frequency of 45 kHz. The circuit diagram shown in Figure 2.5 illustrates the corresponding parts of the *Buck system*, while Figure 2.6 depicts a photograph of the actual circuit.

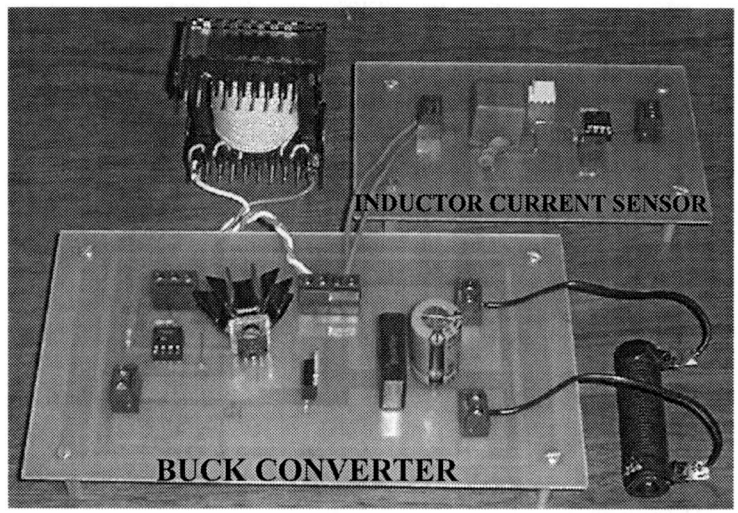

Fig. 2.6. Picture of the experimental Buck system.

According to Figure 2.5 we remark that the experimental *Buck converter prototype* consists of the following parts: *Buck system* (which includes the Buck converter, the *inductor current sensor*, the *capacitor voltage sensor* and the *driver*) and the *actuator*, represented by a $\Sigma - \Delta$-*modulator* or its corresponding *sliding mode* (SM) counterpart.

- *Buck system*: The core of this block is the Buck converter. The *inductor current sensor* consist of a LEM HAW 15-P sensor which operate under *Hall effect* principle. Additionally, the sensor bestows galvanic isolation between the Buck converter and the corresponding *control circuit*. The output voltage of the sensor is proportional to the inductor current i i.e., of the form ki. To obtain a relation 1 A : 1 V, we propose the circuit diagram shown in Figure 2.7. The *capacitor voltage sensor* let us obtain a measurement of the output voltage v. It consist of a voltage divisor so that we can reduce the amplitude of this signal in such a way that its final value is always in the 0-9 V interval. Figure 2.8 shows the voltage sensor. On the other hand, the *driver* is made up of the NTE3087 integrated circuit (IC). This circuit provides optical isolation between the *actuator* and the Buck converter. It also provides a suitably switching pulsed signal with amplitudes of 0 and 5 V, programmed to have a sampling rate of 45

kHz. The provided output signal allows to command the gate of a Mosfet (NTE2984) acting as a switch.
- $\Sigma - \Delta$-*modulator*: In this block the average synthesized control strategies are appropriately implemented in a switched manner. $\Sigma - \Delta$ modulation is a *sliding mode* based implementation technique which will be extremely useful in the actual realization of feedback control laws designed on the basis of average models. We present, at the end of Chapter 3 a detailed theoretical treatment of $\Sigma - \Delta$ modulation, and we also provide a proposal for the practical implementation of the $\Sigma - \Delta$-*modulator* which allows to limit the commutation frequency of the circuit to a finite value.

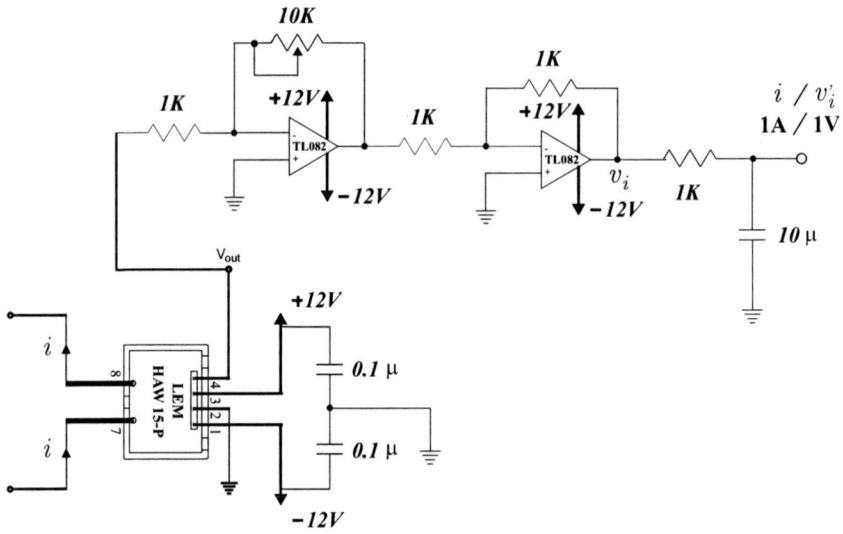

Fig. 2.7. Conditioning circuit for the inductor current measurement.

2.3 The Boost Converter

The electronic circuit of the Boost converter, also known as the *up converter*, is shown in Figure 2.9. We assume that the semi-conductors are ideal, i.e., the transistor Q has an infinitely fast response while the diode D has a threshold value equal to zero. This allows that the *conduction state* and the *blocking* states are activated with no loss of time whatsoever. From the preceding, we have the following behavior: when the transistor Q is in the ON state, the diode D is inversely polarized. As a consequence, there is no connection between the source voltage E and the system load R. This can be seen from

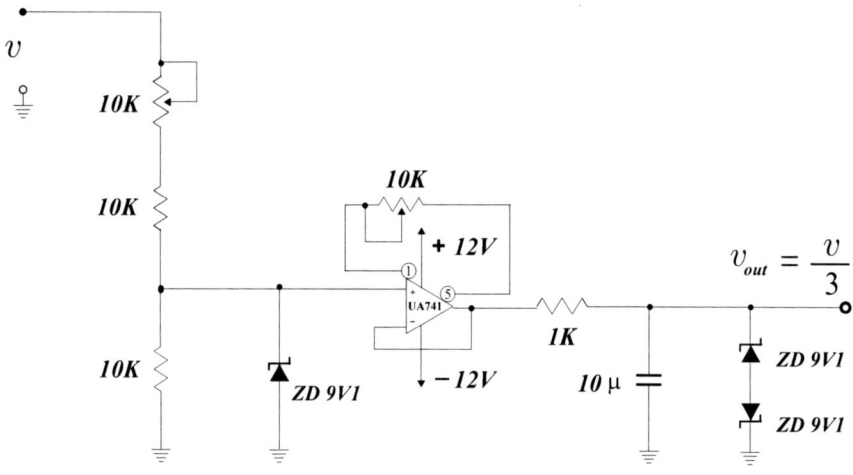

Fig. 2.8. Conditioning circuit for the capacitor voltage measurement.

the Figure 2.10(a). On the other hand, when the transistor Q is in the OFF state, the diode D is directly polarized, or D is conducting. This allows the flow of energy between the voltage source E and the load of the system R, as illustrated in Figure 2.10(b).

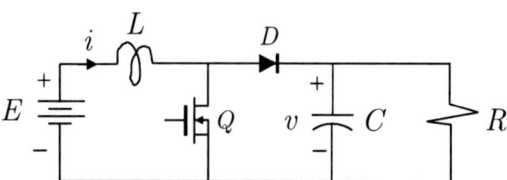

Fig. 2.9. Switched DC-to-DC power converter Boost using semi-conductor devices.

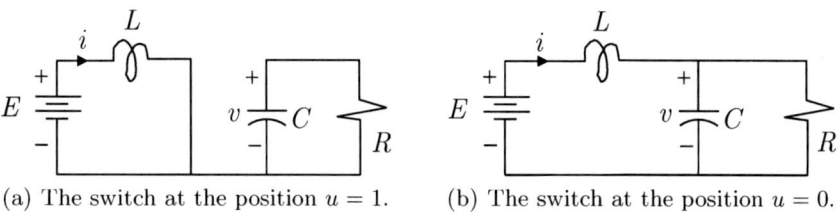

(a) The switch at the position $u = 1$. (b) The switch at the position $u = 0$.

Fig. 2.10. Circuit topologies involved in the Boost converter.

22   2 Modelling of DC-to-DC Power Converters

The two circuit topologies associated with the Boost converter (see Figure 2.10) may be combined into a single circuit diagram by means of the introduction of an ideal switch as shown in Figure 2.11.

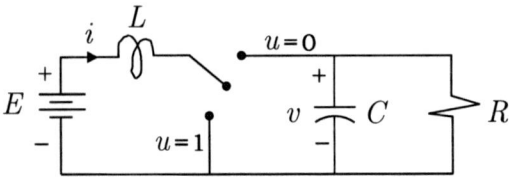

Fig. 2.11. DC-to-DC Boost converter with an ideal switch.

2.3.1 Model of the Converter

To obtain the dynamics of the Boost converter, we may apply Kirchoff's laws in each one of the circuit topologies arising as a consequence of the two switch positions. The first circuit topology is obtained when the switch position function is set to adopt the numerical value $u = 1$, and the second circuit topology is obtained when the switch position function takes the value $u = 0$. The two circuit topologies are shown in Figure 2.10.

When the switch position function is set to $u = 1$, we obtain, using Kirchoff's voltage and Kirchoff's current laws, the dynamics described by the following set of equations,

$$L\frac{di}{dt} = E$$
$$C\frac{dv}{dt} = -\frac{v}{R} \qquad (2.13)$$

When the switch position function is set to $u = 0$, we obtain the dynamics described by the equations,

$$L\frac{di}{dt} = -v + E$$
$$C\frac{dv}{dt} = i - \frac{v}{R} \qquad (2.14)$$

The Boost converter dynamics is then described by the following *bilinear* [1] type of system:

---

[1] We say that a system is *bilinear* when it is, independently, linear in the control u and linear in the state variables x, but not in both. In other words, the dynamics contains as nonlinearities, only the products of the form $x_i u$.

$$L\frac{di}{dt} = -(1-u)v + E$$
$$C\frac{dv}{dt} = (1-u)i - \frac{v}{R} \qquad (2.15)$$

2.3.2 Normalization

The *normalization* of the Boost converter system equations is carried out by redefining the state variables and the time variable as follows:

$$\begin{pmatrix} x_1 \\ x_2 \end{pmatrix} = \begin{pmatrix} \frac{1}{E}\sqrt{\frac{L}{C}} & 0 \\ 0 & \frac{1}{E} \end{pmatrix} \begin{pmatrix} i \\ v \end{pmatrix}, \qquad \tau = \frac{t}{\sqrt{LC}} \qquad (2.16)$$

We obtain the following *average normalized model* for the Boost converter,

$$\frac{dx_1}{d\tau} = -(1-u_{av})x_2 + 1$$
$$\frac{dx_2}{d\tau} = (1-u_{av})x_1 - \frac{x_2}{Q} \qquad (2.17)$$

where the parameter Q, representing the inverse of a circuit quality factor, is obtained by the relation: $Q = R\sqrt{C/L}$. The variable x_1 is the normalized inductor current while x_2 represents the normalized output voltage. The switch position function is invariant with respect to the normalization process.

2.3.3 Equilibrium Point and Static Transfer Function

One of the control objectives, which we desire to achieve when using or designing a DC-to-DC power converter, is to regulate the output voltage so as to stabilize it to a constant value or to track a given reference signal. In the case of stabilization it becomes quite important to understand the steady state behavior of the circuit.

In the steady state regime, corresponding to constant equilibrium values, all time derivatives of the state variables in the description of the system are set to zero. Thus, the control input must also remain constant, i.e., $u_{av} = U = constant$. This condition results in a set of simultaneous equations whose solutions describe the equilibrium points of the system.

The normalized average model of the Boost converter corresponding to a constant value of the control input $u_{av} = U$, generates the following system of equations for the equilibrium states:

$$\begin{pmatrix} 0 & (1-U) \\ (1-U) & -\frac{1}{Q} \end{pmatrix} \begin{pmatrix} \overline{x}_1 \\ \overline{x}_2 \end{pmatrix} = \begin{pmatrix} 1 \\ 0 \end{pmatrix} \qquad (2.18)$$

The solution of this system of equations for the steady state equilibrium values: $\overline{x}_1$ and $\overline{x}_2$ is given by

$$\bar{x}_1 = \frac{1}{Q}\frac{1}{(1-U)^2}, \quad \bar{x}_2 = \frac{1}{(1-U)} \quad (2.19)$$

A different parametrization is obtained by expressing the equilibrium value in terms of the desired average output voltage of the converter, denoted by $\bar{x}_2 = V_d$:

$$\bar{x}_1 = \frac{1}{Q}V_d^2, \quad \bar{x}_2 = V_d, \quad U = \frac{V_d - 1}{V_d} \quad (2.20)$$

In this manner, from the relation (2.19), it is clear that the *static normalized transfer function* of the Boost converter is given by:

$$\mathcal{H}(U) = \bar{x}_2 = \frac{1}{(1-U)} \quad (2.21)$$

It is clear that the gain of the converter circuit is always larger than 1. For this reason, this converter is addressed as the *up converter* or the *Boost converter*. The *characteristic curve* of the *static transfer function* for the Boost converter is depicted in Figure 2.12. It is clear that through the variation of the *duty cycle* or *average control input* U, we can read the actual steady state output voltage of desired value $\bar{v}$ provided it is larger than 1.

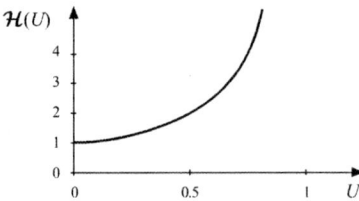

Fig. 2.12. Characteristic curve of the static transfer function of the Boost DC-to-DC power converter.

The equilibrium values of the non-normalized Boost power converter state variables are obtained as:

$$\bar{i} = \frac{1}{R}\frac{\bar{v}^2}{E}, \quad \bar{v} = \frac{E}{(1-U)} \quad (2.22)$$

2.3.4 Alternative Model of the Boost Converter

It is important to notice that the switch position function values $u = 1$ and $u = 0$ of the Buck and Boost power converters can be realized in a non-unique form, i.e., any of the ideal switch positions can be made to correspond with, say, the value $u = 1$. In general, it is more convenient to assign the position function value $u = 1$ to that corresponding to the *conducting mode* for the power transistor Q. As a consequence, the other position function value, $u = 0$, corresponds to the power transistor Q being in the *non-conducting mode*. This is a natural convention that will be adopted in this chapter. However, in the following chapters, we will also use the alternative model for some of the converters, in particular for the *Boost converter*. The alternative models may be of significant help in the algebraic manipulations for the derivation of the various controller schemes. Thus, the alternative model for the normalized average model of the Boost converter (2.17) is given by:

$$\frac{dx_1}{d\tau} = -\mathbf{u}_{av}x_2 + 1$$
$$\frac{dx_2}{d\tau} = \mathbf{u}_{av}x_1 - \frac{1}{Q}x_2 \qquad (2.23)$$

where clearly, the new control input $\mathbf{u}_{av}$ satisfies,

$$\mathbf{u}_{av} = 1 - u_{av} \qquad (2.24)$$

2.3.5 A Boost Converter Prototype

Figure 2.13 shows the circuit diagram of a *Boost converter prototype*. The circuit parameters adopted for this experimental system are given by:

$$L = 15.91 \text{ mH}, \qquad C = 50 \text{ }\mu\text{F}, \qquad R = 52 \text{ }\Omega, \qquad E = 12 \text{ V}$$

where the inductor L was designed to operate at 45 kHz. Figure 2.14 depicts a picture of the actual *Boost system*. Similar to the *Buck converter prototype*, the *Boost converter prototype* is made up of two blocks: *Boost system* and a $\Sigma - \Delta$-*modulator*, or its corresponding SM counterpart, acting as an *actuator*.

- *Boost system*: It is made up of four circuits: 1) a Boost converter, 2) an *inductor current sensor*, 3) a *capacitor voltage sensor* and 4) a *driver*. The *inductor current sensor* was chosen to be a LEM HAW 15-P. The conditioning circuits for the inductor current and capacitor voltage measurements are the same that we employed for the *Buck system*. These circuits are shown in Figure 2.7 and Figure 2.8, respectively. On the other hand, the *driver* is made up of an NTE3087 IC. The NTE3087 IC provides optical isolation between the $\Sigma - \Delta$-*modulator* and the Boost converter circuit. It provides a suitably switching pulsed signal with amplitudes restricted to 0 and 5 V. It is programmed to sustain a sampling rate of 45 kHz. The provided output signal allows to command the gate of the Mosfet NTE2984, which acts as the switch (see Figure 2.13).
- $\Sigma - \Delta$-*modulator*: In this block, the control strategies, designed on the basis of average models, are appropriately implemented.

26   2 Modelling of DC-to-DC Power Converters

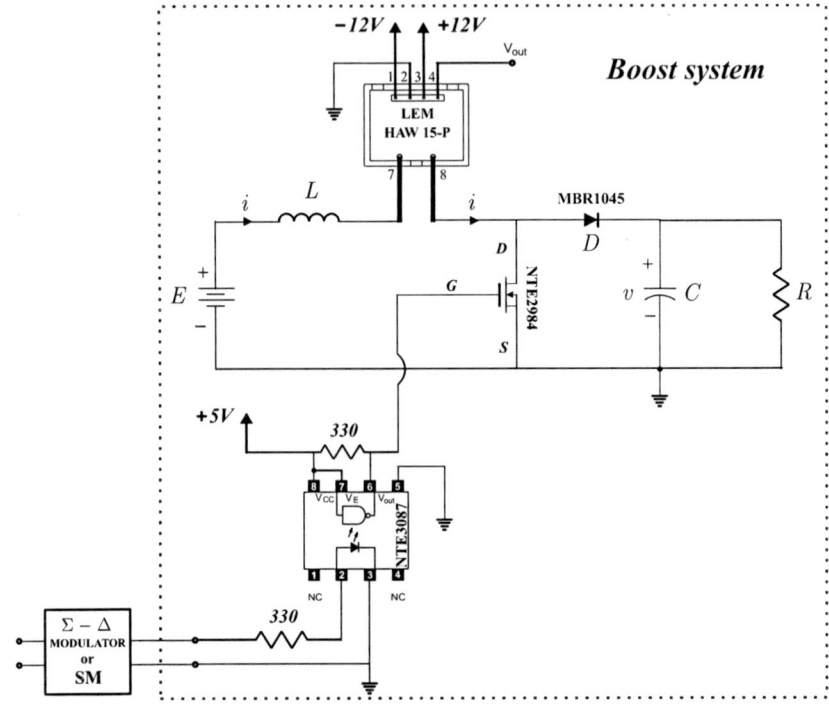

Fig. 2.13. Circuit diagram for the experimental *Boost converter prototype*.

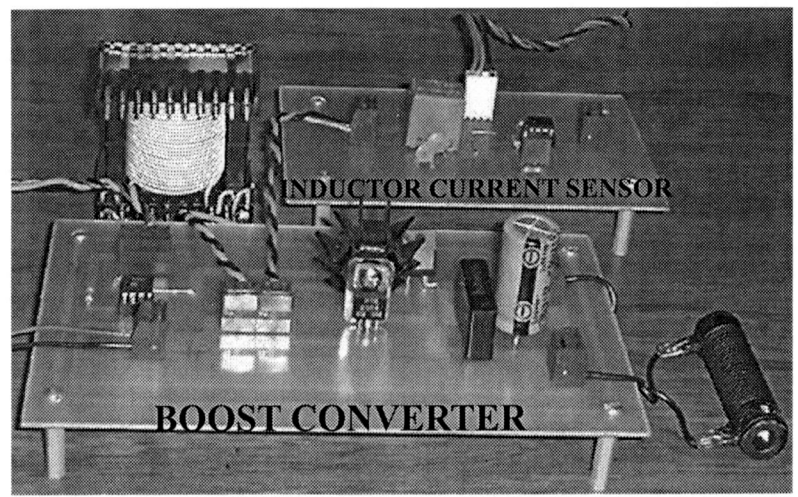

Fig. 2.14. Photograph of the experimental Boost system.

2.4 The Buck-Boost Converter

Another possible arrangement of the semiconductor switches, gives room to a third type of DC-to-DC power converter known as the Buck-Boost converter. In fact, this new converter is obtained by interchanging the diode D and the inductor L of the Buck converter. The circuit is shown in Figure 2.15. This converter is also known as the *chopper-amplifier converter*. In this type of converter, the circuit gain may be higher or lower than 1 modulo a polarity change. The fundamental difference of this class of converter with the Buck and the Boost converters is that the output voltage is of *opposite sign* to that of the constant source E.

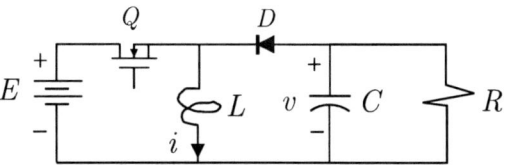

Fig. 2.15. Buck-Boost converter with semiconductor switch realization.

Assuming that the Buck-Boost circuit components are ideal, the resulting circuit is the one shown in Figure 2.16, where the semiconductors (Q, D) have been substituted by an ideal switch.

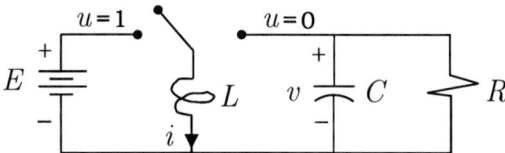

Fig. 2.16. Ideal switch representation of the Buck-Boost DC-to-DC converter.

2.4.1 Model of the Converter

The operation of this system is as follows: when the transistor is switched to the ON state (conduction state), the diode is inversely polarized generating a circuit topology which is shown in Figure 2.17(a). During this period, the inductor current is generated from the source voltage E. While the diode remains inversely polarized we say the circuit is operating in the "charging period". When the transistor is switched OFF, the diode is directly polarized generating the circuit topology shown in Figure 2.17(b). This second period is known as the "discharging period" due to the fact that the stored energy in the inductor L is transferred to the system load R.

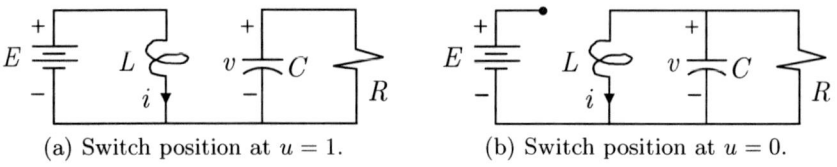

Fig. 2.17. Circuits topologies associated with the Buck-Boost converter.

When the Kirchoff's voltage and current laws are applied to the two circuit topologies of Figure 2.17, and the obtained models are combined into a single dynamic model, the resulting system of differential equations describing the Buck-Boost converter is the following:

$$L\frac{di}{dt} = (1-u)\,v + uE$$
$$C\frac{dv}{dt} = -(1-u)\,i - \frac{v}{R} \qquad (2.25)$$

2.4.2 Normalization

The *average normalized model* of the Buck-Boost converter is given by

$$\frac{dx_1}{d\tau} = (1-u_{av})\,x_2 + u_{av}$$
$$\frac{dx_2}{d\tau} = -(1-u_{av})\,x_1 - \frac{x_2}{Q} \qquad (2.26)$$

where the variable x_1 represents the normalized inductor current, x_2 is the normalized output voltage and u_{av}, represents, as before, the average control variable.

Clearly the underlying transformation is, just as before, given by

$$\begin{pmatrix} x_1 \\ x_2 \end{pmatrix} = \begin{pmatrix} \frac{1}{E}\sqrt{\frac{L}{C}} & 0 \\ 0 & \frac{1}{E} \end{pmatrix} \begin{pmatrix} i \\ v \end{pmatrix}, \qquad Q = R\sqrt{C/L}, \qquad \tau = \frac{t}{\sqrt{LC}} \qquad (2.27)$$

Similarly as in the Boost converter case, we can justify the use of the following alternative model for the Buck-Boost converter,

$$\frac{dx_1}{d\tau} = u_{av}x_2 + (1-u_{av})$$
$$\frac{dx_2}{d\tau} = -u_{av}x_1 - \frac{1}{Q}x_2 \qquad (2.28)$$

where
$$\mathbf{u}_{av} = 1 - u_{av} \tag{2.29}$$

2.4.3 Equilibrium Point and Static Transfer Function

The equilibrium point of the Buck-Boost corresponding to a constant value of the average control input is obtained by letting the right hand side of the state equations (2.26) to be zero while the control variable is set to be $u_{av} = U = constant$. We thus obtain a system of equations for $\overline{x}_1$ and $\overline{x}_2$ given by

$$\begin{pmatrix} 0 & (1-U) \\ -(1-U) & -\frac{1}{Q} \end{pmatrix} \begin{pmatrix} \overline{x}_1 \\ \overline{x}_2 \end{pmatrix} = \begin{pmatrix} -U \\ 0 \end{pmatrix} \tag{2.30}$$

The equilibrium point of the Buck-Boost converter parameterized in terms of the constant value U of the control input is then given by,

$$\overline{x}_1 = \frac{1}{Q}\frac{U}{(1-U)^2}, \qquad \overline{x}_2 = -\frac{U}{(1-U)} \tag{2.31}$$

The equilibrium point, parameterized now in terms of the desired constant normalized average output voltage $\overline{x}_2 = V_d$, is given by the following relations:

$$\overline{x}_1 = (V_d - 1)\frac{V_d}{Q}, \qquad \overline{x}_2 = V_d, \qquad U = \frac{V_d}{V_d - 1} \tag{2.32}$$

On the other hand, the actual (or de-normalized) steady state variables corresponding to the equilibrium point (2.32) are obtained when we introduce the redefining state variables (2.27) into (2.32), generating:

$$\overline{i} = \left(\frac{\overline{v}}{E} - 1\right)\frac{\overline{v}}{R}, \qquad \overline{v} = -\left(\frac{U}{1-U}\right)E \tag{2.33}$$

The *normalized static transfer function* of the Buck-Boost converter is immediately obtained from Equation 2.31 as:

$$\mathcal{H}(U) = -\frac{U}{(1-U)} \tag{2.34}$$

The graph in Figure 2.18 depicts the *static transfer function* of the Buck-Boost converter. It is also clear that we may read the steady state output voltage of the system $\overline{v}$, in correspondence with the average control input equilibrium value U. It is also clear from the characteristic curve of the Buck-Boost converter that this circuit may either amplify, or reduce, the constant input voltage but with the output voltage polarity being opposite to that of the system constant input voltage source E.

30   2 Modelling of DC-to-DC Power Converters

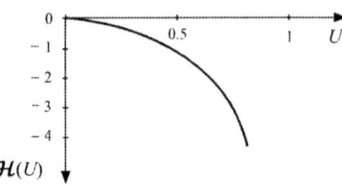

Fig. 2.18. Characteristic curve of the Buck-Boost DC-to-DC power converter.

2.4.4 A Buck-Boost Converter Prototype

The circuit schematics, the actual circuit, as well as a picture of the prototype for the *Buck-Boost system* are shown in Figures 2.15, 2.19, and 2.20, respectively. The values of the components for this system were set to be:

$$L = 15.91 \text{ mH}, \quad C = 470 \text{ } \mu\text{F}, \quad R = 52 \text{ } \Omega, \quad E = 12 \text{ V}$$

The switching frequency for this converter, as in the previous converters, is 45 kHz.

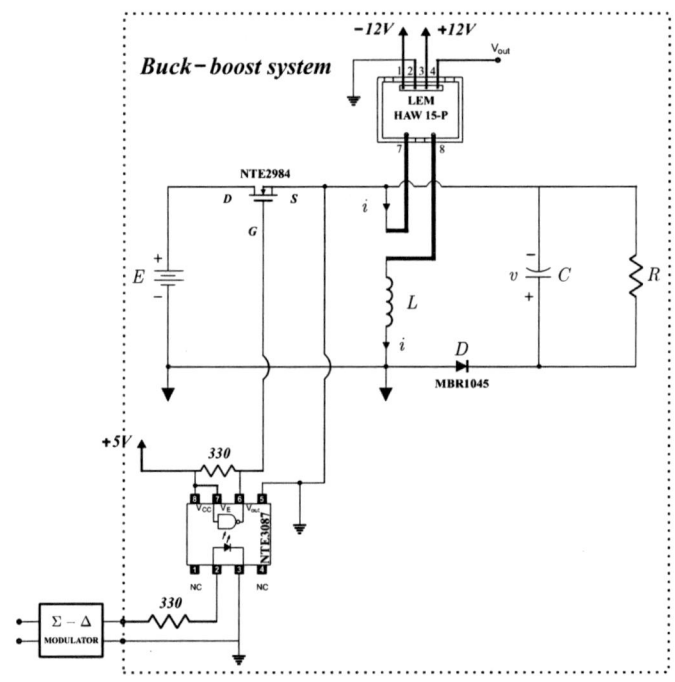

Fig. 2.19. Circuit of the *Buck-Boost converter prototype*.

Fig. 2.20. Hardware implementation of the Buck-Boost system.

2.5 The Non-inverting Buck-Boost Converter

Our main objective in the previous section was centered around the study of the DC-to-DC Buck-Boost converter, which exhibited the particular property of delivering an output voltage of opposite polarity with respect to that of the input voltage source E along with the possibility of amplifying or scaling down this value. Following the same presentation scheme used in the previous section, we shall now deal with the non-inverting Buck-Boost converter. This converter also has the capability of scaling and of amplifying the constant input voltage source value E at the output. The fundamental difference is that this new converter does not change the polarity of the input voltage source E at the output.

2.5.1 Model of the Converter

The configuration of the non-inverting Buck-Boost converter, assuming that the circuit components are all ideal, is shown in Figure 2.21.

If we consider the ideal switch version of the converter, shown in Figure 2.21, the model of the system may be directly obtained using the same procedure as in the previous examples.

The dynamics describing to the non-inverting Buck-Boost converter is found to have the following state representation:

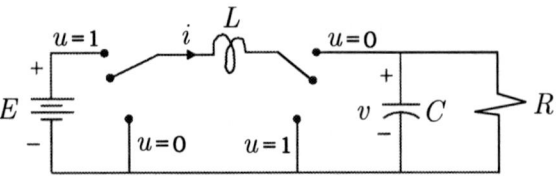

Fig. 2.21. Simplified non-inverting Buck-Boost converter.

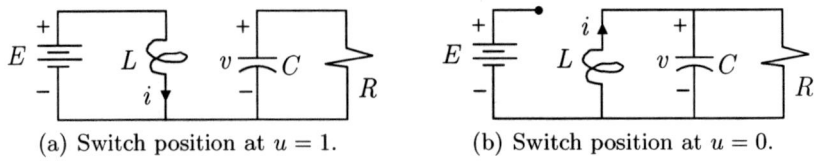

(a) Switch position at $u = 1$.   (b) Switch position at $u = 0$.

Fig. 2.22. Circuit topologies involved in the non-inverting Buck-Boost converter.

$$L\frac{di}{dt} = -(1-u)v + uE$$
$$C\frac{dv}{dt} = (1-u)i - \frac{v}{R} \qquad (2.35)$$

where i and v are, respectively, the current through the inductor L and the voltage across the terminals of the capacitor C. The external source voltage E has a constant value. The variable u is the control input, which represents the switch position, restricted to take values in the discrete set $\{0,1\}$. The control objective consists in regulating the output voltage v around a desired equilibrium point.

2.5.2 Normalization

Using precisely the same state coordinate transformation and time scaling used in the three previous cases, the normalized model of the non-inverting Buck-Boost converter is written as,

$$\dot{x}_1 = -(1-u)x_2 + u$$
$$\dot{x}_2 = (1-u)x_1 - \frac{x_2}{Q} \qquad (2.36)$$

With some abuse of notation we use " $\cdot$ " to denote the derivative with respect to the dimensionless time τ. The normalized variables x_1 y x_2 represent, respectively, the normalized current and the normalized output voltage. The switch position function is still represented by u. The only remaining parameter in the normalized model, Q, is expressed as,

$$Q = R\sqrt{\frac{C}{L}} \qquad (2.37)$$

2.5.3 Equilibrium Point and Static Transfer Function

The equilibrium point of the average non-inverting Buck-Boost converter is obtained from the solution of the corresponding set of algebraic equations when u_{av} is set to be the constant value U and the time derivatives of the normalized state variables is set to zero. We obtain,

$$\overline{x}_1 = \frac{1}{Q}\frac{U}{(1-U)^2}, \qquad \overline{x}_2 = \frac{U}{(1-U)} \qquad (2.38)$$

Using these expressions, we may rewrite the normalized value of the equilibrium inductor current $\overline{x}_1$ in terms of the normalized output voltage $\overline{x}_2$ as follows:

$$\overline{x}_1 = (\overline{x}_2 + 1)\frac{\overline{x}_2}{Q} \qquad (2.39)$$

Thus, if we wish to regulate the normalized output voltage x_2 towards a desired equilibrium value $\overline{x}_2$, then, this may be achieved in an indirect fashion by regulating the variable x_1 towards its corresponding equilibrium value given by (2.39).

According to (2.38) the normalized *gain* for the non-inverting Buck-Boost converter is given by:

$$\mathcal{H}(U) = \overline{x}_2 = \frac{U}{(1-U)} \qquad (2.40)$$

From here, we may confirm that, in steady state, the non-inverting Buck-Boost converter may either *attenuate*, or *rise*, at the output terminals, the constant *input voltage* E. Moreover, this is achieved without polarity inversion with respect to the input source E.

The graph of the corresponding characteristic curve for this converter system is shown in Figure 2.23.

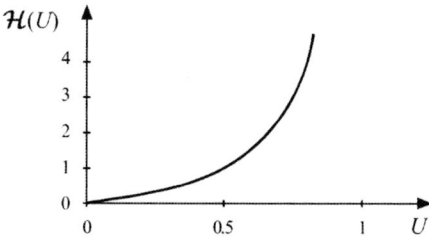

Fig. 2.23. Characteristic curve for the non-inverting Buck-Boost converter.

2.6 The Cúk Converter

By suitable combination of some of the basic converter topologies representing the Buck, the Boost and the Buck-Boost converters, one may obtain some other useful DC-to-DC power converters. A typical example is the cascade connection of the Boost and the Buck converter which produces the well known Cúk converter. This converter is shown in Figure 2.24. The input circuit in the Cúk converter is, clearly, a Boost, converter and the output circuit is seen to be a Buck converter. Thus, we may also think of the Cúk converter as a *"Boost-Buck"* converter. In contradistinction to the basic topologies, the Cúk converter requires two (dependent) switches instead of one as well as two inductors L_1, L_2, and two capacitors; one for storing the energy and the second one to transfer the energy from the input circuit towards the output circuit load. This results in a higher complexity for the analysis and construction of the converter. The *static transfer function* of the Cúk converter exhibits the same characteristic curve we obtained for the Buck-Boost converter, as it will be shown in this section.

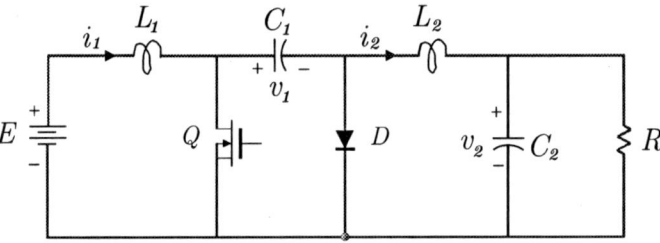

Fig. 2.24. Practical Cúk converter realization.

The ideal switch (and ideal components) circuit diagram for the Cúk converter is shown in Figure 2.24. This idealization may be contrasted against a practical realization of the Cúk converter, shown in Figure 2.25. This simplified circuit representation will allow us to obtain, rather directly, the dynamic model of the converter.

The Cúk converter exhibits two different modes of operation. The first mode is obtained when the transistor is ON and instantaneously, the diode D is inversely polarized generating an circuit topology shown in Figure 2.26(a). During this period, the current through the inductor L_1 is drawn from the voltage source E. This mode represents the *charging* mode. The second mode of operation starts when the transistor is OFF and the diode D is directly polarized generating the circuit topology shown in Figure 2.26(b). This stage or mode of operation is known as the *discharging* mode since all the energy stored in L_1 is now transferred to the load R.

2.6 The Cúk Converter   35

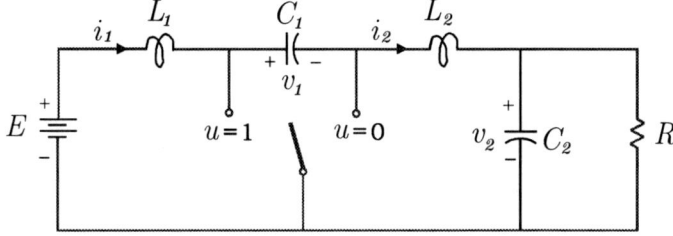

Fig. 2.25. Ideal switch representation of the Cúk converter.

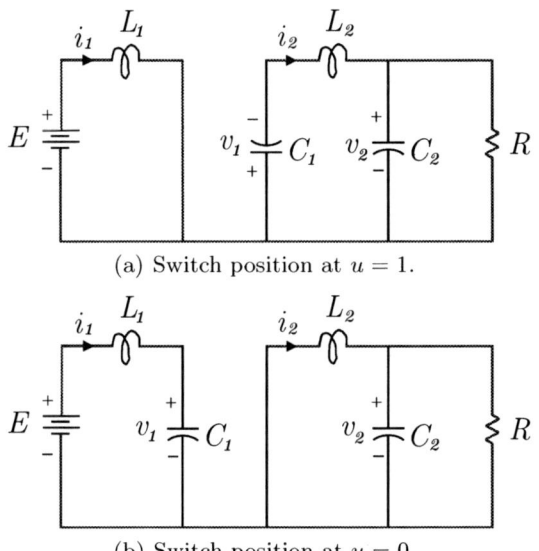

Fig. 2.26. Equivalent circuits of the Cúk converter.

2.6.1 Model of the Converter

The derivation of the dynamics of the *Cúk* converter is carried out in the same manner in which we analyzed the topologies of the previous basic DC-to-DC power converters.

When $u = 1$, we obtain the following equations for i_1 and i_2 in the obtained circuit topology,

$$L_1 \frac{di_1}{dt} = E$$
$$L_2 \frac{di_2}{dt} = -v_1 - v_2 \qquad (2.41)$$

and the following equations for the capacitor voltages v_1 and v_2,

$$C_1 \frac{dv_1}{dt} = i_2$$
$$C_2 \frac{dv_2}{dt} = i_2 - \frac{v_2}{R} \qquad (2.42)$$

When $u = 0$, we obtain the following equations for i_1 and i_2,

$$L_1 \frac{di_1}{dt} = -v_1 + E$$
$$L_2 \frac{di_2}{dt} = -v_2 \qquad (2.43)$$

The capacitor voltages v_1 and v_2 are described by,

$$C_1 \frac{dv_1}{dt} = i_1$$
$$C_2 \frac{dv_2}{dt} = i_2 - \frac{v_2}{R} \qquad (2.44)$$

The *Cúk* converter dynamics is then described by combining the previous partial models. We obtain the following system of differential equations:

$$L_1 \frac{di_1}{dt} = -(1-u)v_1 + E$$
$$C_1 \frac{dv_1}{dt} = (1-u)i_1 + ui_2$$
$$L_2 \frac{di_2}{dt} = -uv_1 - v_2$$
$$C_2 \frac{dv_2}{dt} = i_2 - \frac{v_2}{R} \qquad (2.45)$$

where v_1 and i_1 are, respectively, the voltage across the capacitor C_1 and the current in the inductor L_1, while v_2 and i_2 are, respectively, the voltage across the parallel branches formed by the capacitor C_2 and the load R, and the current through the inductor L_2. As usual, the external voltage source E has a constant value. The variable u is the control input, which represents the switch position restricted to take values in the discrete set $\{0, 1\}$. It is assumed that the converter operates in the *continuous conduction mode*, i.e., neither of the inductor currents are identically zero on an open interval of time.

2.6.2 Normalization

Once we have obtained the *Cúk* converter model, we proceed to perform the normalizing transformations of the state variable and time coordinates.

The state coordinates transformation and time scaling:

$$\begin{pmatrix} x_1 \\ x_2 \\ x_3 \\ x_4 \end{pmatrix} = \begin{pmatrix} \frac{1}{E}\sqrt{\frac{L_1}{C_1}} & 0 & 0 & 0 \\ 0 & \frac{1}{E} & 0 & 0 \\ 0 & 0 & \frac{1}{E}\sqrt{\frac{L_1}{C_1}} & 0 \\ 0 & 0 & 0 & \frac{1}{E} \end{pmatrix} \begin{pmatrix} i_1 \\ v_1 \\ i_2 \\ v_2 \end{pmatrix}, \quad \tau = \frac{t}{\sqrt{L_1 C_1}} \quad (2.46)$$

yields the following normalized model for the Cúk converter:

$$\begin{aligned} \dot{x}_1 &= -(1-u)x_2 + 1 \\ \dot{x}_2 &= (1-u)x_1 + ux_3 \\ \alpha_1 \dot{x}_3 &= -ux_2 - x_4 \\ \alpha_2 \dot{x}_4 &= x_3 - \frac{x_4}{Q} \end{aligned} \quad (2.47)$$

where the symbol: " $\cdot$ " again represents (abusively) the derivative with respect to the dimensionless time coordinate τ. The variables x_1, x_3 and x_2, x_4 represent, respectively, the currents and the voltages of the normalized system, while u represents the switch position function. The parameter Q is defined as $Q = R\sqrt{C_1/L_1}$, while the constants α_1 and α_2 are defined by the quotients

$$\alpha_1 = L_2/L_1, \qquad \alpha_2 = C_2/C_1 \quad (2.48)$$

2.6.3 Equilibrium Point and Static Transfer Function

Setting to zero the right hand sides of the average normalized model (2.47) with $u_{av} = U = constant$, yields the following system of equations:

$$\begin{pmatrix} 0 & -(1-U) & 0 & 0 \\ 1-U & 0 & U & 0 \\ 0 & -U & 0 & -1 \\ 0 & 0 & 1 & -\frac{1}{Q} \end{pmatrix} \begin{pmatrix} \overline{x}_1 \\ \overline{x}_2 \\ \overline{x}_3 \\ \overline{x}_4 \end{pmatrix} = \begin{pmatrix} -1 \\ 0 \\ 0 \\ 0 \end{pmatrix} \quad (2.49)$$

whose solution is found to be given by:

$$\overline{x}_1 = \frac{1}{Q}\frac{U^2}{(1-U)^2}, \quad \overline{x}_2 = \frac{1}{1-U}, \quad \overline{x}_3 = -\frac{1}{Q}\frac{U}{(1-U)}, \quad \overline{x}_4 = -\frac{U}{1-U} \quad (2.50)$$

A different parametrization of the equilibria is obtained by using a constant desired value of the output voltage $\overline{x}_4$. Such a parametrization is given by:

$$\overline{x}_1 = \frac{\overline{x}_4^2}{Q}, \quad \overline{x}_2 = 1 - \overline{x}_4, \quad \overline{x}_3 = \frac{\overline{x}_4}{Q}, \quad U = \frac{\overline{x}_4}{\overline{x}_4 - 1} \quad (2.51)$$

From the relation existing between the constant output voltage of the converter, $\overline{x}_4$ and the corresponding value of the average control input U, determined in (2.50), the *static normalized transfer function* of the Cúk converter is readily found to be

38   2 Modelling of DC-to-DC Power Converters

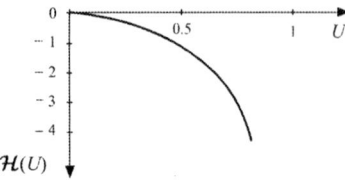

Fig. 2.27. Characteristic curve of the static transfer function of the Cúk DC-to-DC power converter.

$$\mathcal{H}(U) = \overline{x}_4 = -\frac{U}{(1-U)} \qquad (2.52)$$

The characteristic curve of the *static transfer function* of the Cúk converter is shown in Figure 2.27. As it was previously stated, the characteristic curve of the voltage gain of the Cúk converter is the same as that of the Buck-Boost converter.

2.7 The Sepic Converter

Figure 2.28 shows the Sepic DC-to-DC converter circuit with switches realized by means of semiconductor devices(Q, D). These operate in a complementary fashion i.e., when the transistor Q is in the conducting mode then the diode D is inversely polarized and viceversa.

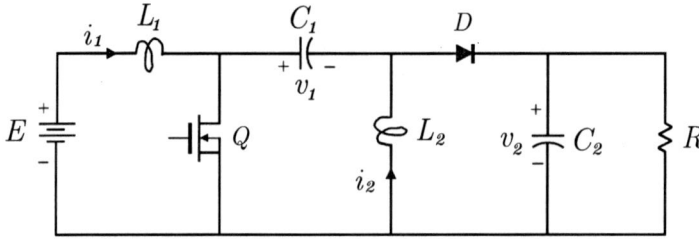

Fig. 2.28. The Sepic DC-to-DC power converter.

Figure 2.29 depicts the ideal switch realization of the Sepic converter. The equivalent circuits, corresponding to the switch position function values, $u = 1$ and $u = 0$, are shown in Figure 2.30

2.7 The Sepic Converter

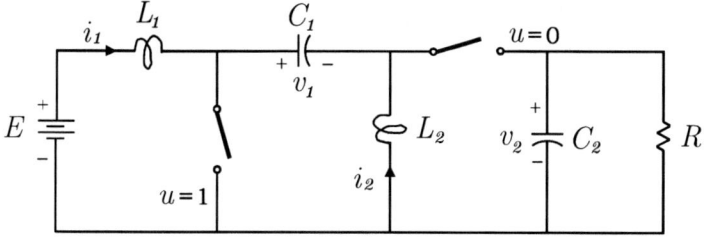

Fig. 2.29. Sepic converter realization with ideal switches.

2.7.1 Model of the Converter

The model of the converter is derived to be:

$$L_1 \frac{di_1}{dt} = -(1-u)(v_1 + v_2) + E$$
$$C_1 \frac{dv_1}{dt} = (1-u)i_1 - ui_2$$
$$L_2 \frac{di_2}{dt} = uv_1 - (1-u)v_2$$
$$C_2 \frac{dv_2}{dt} = (1-u)(i_1 + i_2) - \frac{v_2}{R} \quad (2.53)$$

where v_1 and i_1 are, respectively, the voltage across capacitor C_1 and the current through the inductor L_1, v_2 and i_2 are, respectively, the voltage across the capacitor C_2 and the load R, and the inductor current L_2. The source voltage E is constant. The control input u is the switch position function taking values in $\{0, 1\}$.

2.7.2 Normalization

The following state coordinate transformation and time scaling,

$$\begin{pmatrix} x_1 \\ x_2 \\ x_3 \\ x_4 \end{pmatrix} = \begin{pmatrix} \frac{1}{E}\sqrt{\frac{L_1}{C_1}} & 0 & 0 & 0 \\ 0 & \frac{1}{E} & 0 & 0 \\ 0 & 0 & \frac{1}{E}\sqrt{\frac{L_1}{C_1}} & 0 \\ 0 & 0 & 0 & \frac{1}{E} \end{pmatrix} \begin{pmatrix} i_1 \\ v_1 \\ i_2 \\ v_2 \end{pmatrix}, \quad \tau = \frac{t}{\sqrt{L_1 C_1}} \quad (2.54)$$

yields the following normalized model of the Sepic converter:

$$\dot{x}_1 = -(1-u)(x_2 + x_4) + 1$$
$$\dot{x}_2 = (1-u)x_1 - ux_3$$
$$\alpha_1 \dot{x}_3 = ux_2 - (1-u)x_4$$
$$\alpha_2 \dot{x}_4 = (1-u)(x_1 + x_3) - \frac{x_4}{Q} \quad (2.55)$$

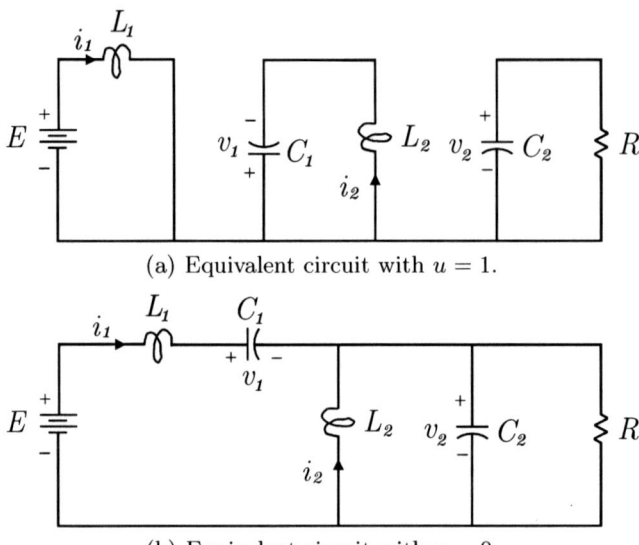

Fig. 2.30. Circuit topologies associated with the Sepic converter.

The constants α_1, α_2 and Q are defined by:

$$\alpha_1 = L_2/L_1, \qquad \alpha_2 = C_2/C_1, \qquad Q = R\sqrt{C_1/L_1} \qquad (2.56)$$

2.7.3 Equilibrium Point and Static Transfer Function

The normalized average state equilibrium point, parameterized in terms of the constant average input U, is readily obtained as

$$\bar{x}_1 = \frac{1}{Q}\frac{U^2}{(1-U)^2}, \qquad \bar{x}_2 = 1, \qquad \bar{x}_3 = \frac{1}{Q}\frac{U}{(1-U)}, \qquad \bar{x}_4 = \frac{U}{(1-U)} \qquad (2.57)$$

Parameterizing the equilibrium values in terms of the constant output voltage $\bar{x}_4$ leads to:

$$\bar{x}_1 = \frac{\bar{x}_4^2}{Q}, \qquad \bar{x}_2 = 1, \qquad \bar{x}_3 = \frac{\bar{x}_4}{Q}, \qquad U = \frac{\bar{x}_4}{\bar{x}_4+1} \qquad (2.58)$$

The *static transfer function* is obtained to be:

$$\mathcal{H}(U) = \bar{x}_4 = \frac{U}{(1-U)} \qquad (2.59)$$

which points to the fact that the Sepic converter can *reduce* or *amplify*, in steady state, the constant input source voltage value. Clearly, the output voltage is of the same polarity as that of the source input E. Figure 2.31 shows the corresponding characteristic curve.

The equilibrium point for the actual state variables is obtained by inverting the transformation used in the normalization. This yields:

$$\overline{i}_1 = \frac{1}{R}\frac{\overline{v}_2^2}{E}, \quad \overline{v}_1 = E, \quad \overline{i}_2 = \frac{\overline{v}_2}{R}, \quad \overline{v}_2 = \frac{U}{(1-U)}E \qquad (2.60)$$

where,

$$U = \frac{\overline{v}_2}{\overline{v}_2 + E} \qquad (2.61)$$

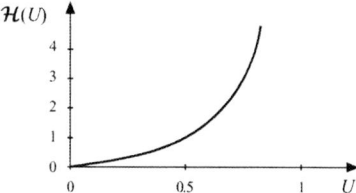

Fig. 2.31. Characteristic curve for the Sepic DC-to-DC power converter.

2.8 The Zeta Converter

Similarly to the Cúk and the Sepic converters, the Zeta converter may be represented by a fourth order nonlinear (bilinear) system. The reason being is that it includes two capacitors and two inductors as dynamic storage elements. The Zeta converter can both amplify and reduce, without polarity inversions, the value of the input source voltage E. We briefly summarize next the most important features involved in the modelling of the Zeta converter.

Figure 2.32 depicts a semiconductor realization of a Zeta DC-to-DC power converter. The ideal switch based realization of the Zeta converter is depicted in Figure 2.33.

2.8.1 Model of the Converter

The Zeta converter exhibits two different modes of operation. The first mode is obtained when the transistor is ON and instantaneously, the diode D is inversely polarized generating an equivalent circuit shown in Figure 2.34(a). During this period, the current through the inductor L_1 and L_2 are drawn from the voltage source E. This mode is the *charging* mode. The second mode of operation starts when the transistor is OFF and the diode D is directly

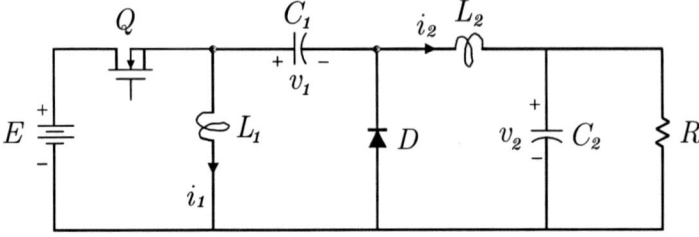

Fig. 2.32. A Zeta converter using a semiconductor realization of the switches.

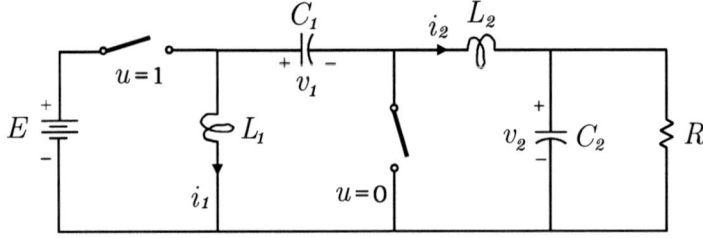

Fig. 2.33. The Zeta converter with ideal switches.

polarized generating the equivalent circuit shown in Figure 2.34(b). This stage or mode of operation is known as the *discharging* mode since all the energy stored in L_2 is now transferred to the load R.

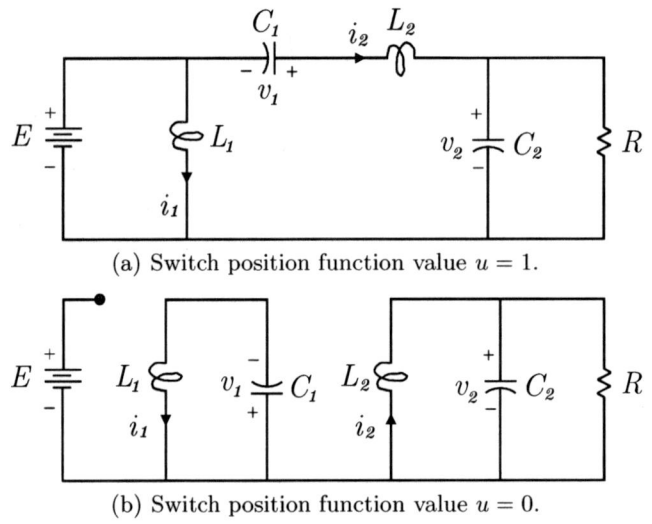

Fig. 2.34. Circuit topologies associated with the Zeta converter.

The dynamical model of the Zeta converter is found to be

$$L_1 \frac{di_1}{dt} = -(1-u)v_1 + uE$$

$$C_1 \frac{dv_1}{dt} = (1-u)i_1 - ui_2$$

$$L_2 \frac{di_2}{dt} = uv_1 - v_2 + uE$$

$$C_2 \frac{dv_2}{dt} = i_2 - \frac{v_2}{R} \quad (2.62)$$

2.8.2 Normalization

After the required change of state and time variables one obtains the following normalized model for the converter:

$$\dot{x}_1 = -(1-u)x_2 + u$$

$$\dot{x}_2 = (1-u)x_1 - ux_3$$

$$\alpha_1 \dot{x}_3 = ux_2 - x_4 + u$$

$$\alpha_2 \dot{x}_4 = x_3 - \frac{x_4}{Q} \quad (2.63)$$

with

$$\alpha_1 = L_2/L_1, \qquad \alpha_2 = C_2/C_1, \qquad Q = R\sqrt{C_1/L_1} \quad (2.64)$$

2.8.3 Equilibrium Point and Static Transfer Function

The equilibrium equations are given by

$$\begin{pmatrix} 0 & -(1-U) & 0 & 0 \\ 1-U & 0 & -U & 0 \\ 0 & U & 0 & -1 \\ 0 & 0 & 1 & -\frac{1}{Q} \end{pmatrix} \begin{pmatrix} \bar{x}_1 \\ \bar{x}_2 \\ \bar{x}_3 \\ \bar{x}_4 \end{pmatrix} = \begin{pmatrix} -U \\ 0 \\ -U \\ 0 \end{pmatrix} \quad (2.65)$$

The average normalized equilibrium point, parameterized in terms of $u_{av} = U$ is found to be given by

$$\bar{x}_1 = \frac{1}{Q} \frac{U^2}{(1-U)^2}, \quad \bar{x}_2 = \frac{U}{1-U}, \quad \bar{x}_3 = \frac{1}{Q} \frac{U}{(1-U)}, \quad \bar{x}_4 = \frac{U}{1-U} \quad (2.66)$$

A parametrization in terms of the desired output equilibrium voltage $\bar{x}_4$ is found by elimination of the parameter U, yielding:

$$\bar{x}_1 = \frac{\bar{x}_4^2}{Q}, \quad \bar{x}_2 = \bar{x}_4, \quad \bar{x}_3 = \frac{\bar{x}_4}{Q}, \quad U = \frac{\bar{x}_4}{\bar{x}_4 + 1} \quad (2.67)$$

The static transfer function is hence given by:

2 Modelling of DC-to-DC Power Converters

$$\mathcal{H}(U) = \overline{x}_4 = \frac{U}{(1-U)} \qquad (2.68)$$

which confirms the basic features of the Zeta converter as a possible *scaling* or *amplifying* converter.

The characteristic curve of the static transfer function is shown in Figure 2.35.

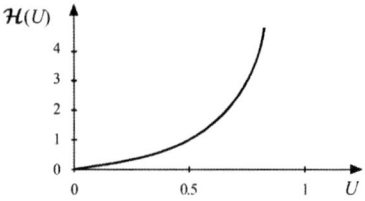

Fig. 2.35. Characteristic curve of the static transfer function for the Zeta DC-to-DC power converter.

2.9 The Quadratic Buck Converter

The quadratic Buck converter owes its name to the quadratic nature of the static transfer function in terms of a constant average control input value. This quadratic feature enhances the adjustment properties of the steady state equilibrium when the input is found to be close to the saturation limits. Here, we summarize the modelling features of a quadratic Buck converter shown in Figure 2.36.

2.9.1 Model of the Converter

$$L_1 \frac{di_1}{dt} = -v_1 + uE$$
$$C_1 \frac{dv_1}{dt} = i_1 - ui_2$$
$$L_2 \frac{di_2}{dt} = uv_1 - v_2$$
$$C_2 \frac{dv_2}{dt} = i_2 - \frac{v_2}{R} \qquad (2.69)$$

2.9 The Quadratic Buck Converter

Circuit Diagram

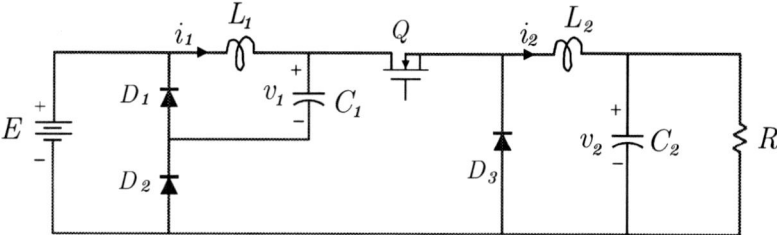

Fig. 2.36. The quadratic Buck converter with semiconductor realizations of the switches.

Ideal Circuit Realization

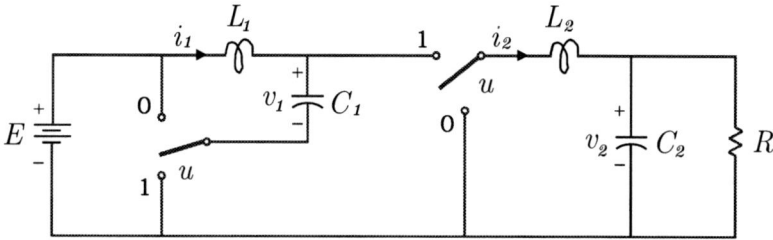

Fig. 2.37. Ideal switches realization of the quadratic Buck converter.

2.9.2 Normalized Model

$$\dot{x}_1 = -x_2 + u$$
$$\dot{x}_2 = x_1 - ux_3$$
$$\alpha_1 \dot{x}_3 = ux_2 - x_4$$
$$\alpha_2 \dot{x}_4 = x_3 - \frac{x_4}{Q} \quad (2.70)$$

with,

$$\alpha_1 = L_2/L_1, \quad \alpha_2 = C_2/C_1, \quad Q = R\sqrt{C_1/L_1} \quad (2.71)$$

2.9.3 Equilibrium Point

The equilibrium points, parameterized in terms of a constant value U of the average control input, are found to be:

$$\bar{x}_1 = \frac{1}{Q}U^3, \quad \bar{x}_2 = U, \quad \bar{x}_3 = \frac{1}{Q}U^2, \quad \bar{x}_4 = U^2 \quad (2.72)$$

These equilibrium points, parameterized now in terms of constant output voltage $\overline{x}_4$, are written as:

$$\overline{x}_1 = \frac{1}{Q}(\overline{x}_4)^{3/2}, \qquad \overline{x}_2 = (\overline{x}_4)^{1/2}, \qquad \overline{x}_3 = \frac{1}{Q}\overline{x}_4 \qquad (2.73)$$

2.9.4 Static Transfer Function

The static transfer function of the quadratic Buck converter is obtained from (2.72) as:

$$\mathcal{H}(U) = \overline{x}_4 = U^2 \qquad (2.74)$$

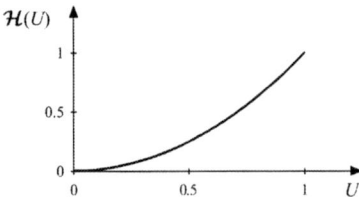

Fig. 2.38. Characteristic curve of the static transfer function for the quadratic Buck DC-to-DC power converter.

2.10 The Boost-Boost Converter

A tandem connection of two Boost converters, while preserving the independence of the control switches, results in a multi-variable DC-to-DC power converter. This converter has interest in applications where two loads need to be independently controlled with a single converter device. With some limitations, this is possible using the described combination of Boost converters.

Figure 2.39 depicts the Boost-Boost circuit with the switches realized by suitable arrangement of diodes and transistors.

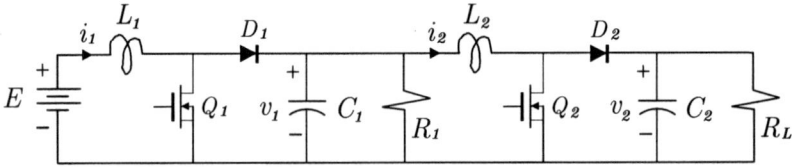

Fig. 2.39. The Boost-Boost circuit.

A realization of the circuit, entitling ideal switches, is shown in Figure 2.40.

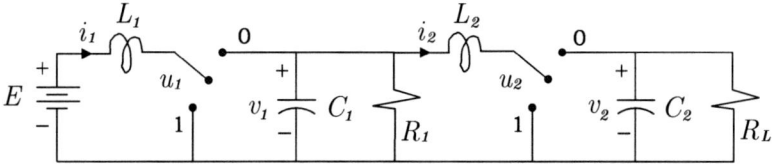

Fig. 2.40. Ideal switch realization of the Boost-Boost converter.

2.10.1 Model of the Boost-Boost Converter

$$L_1 \frac{di_1}{dt} = -(1 - u_1) v_1 + E$$
$$C_1 \frac{dv_1}{dt} = (1 - u_1) i_1 - \frac{v_1}{R_1} - i_2$$
$$L_2 \frac{di_2}{dt} = v_1 - (1 - u_2) v_2$$
$$C_2 \frac{dv_2}{dt} = (1 - u_2) i_2 - \frac{v_2}{R_L} \quad (2.75)$$

2.10.2 Average Normalized Model

$$\frac{dx_1}{d\tau} = -(1 - u_{1av}) x_2 + 1$$
$$\frac{dx_2}{d\tau} = (1 - u_{1av}) x_1 - \frac{1}{Q_1} x_2 - x_3$$
$$\alpha_1 \frac{dx_3}{d\tau} = x_2 - (1 - u_{2av}) x_4$$
$$\alpha_2 \frac{dx_4}{d\tau} = (1 - u_{2av}) x_3 - \frac{1}{Q_L} x_4 \quad (2.76)$$

with,

$$\alpha_1 = \frac{L_2}{L_1}, \quad \alpha_2 = \frac{C_2}{C_1}, \quad Q_1 = R_1 \sqrt{\frac{C_1}{L_1}}, \quad Q_L = R_L \sqrt{\frac{C_1}{L_1}} \quad (2.77)$$

2.10.3 Equilibrium Point and Static Transfer Function

Under the assumption of constant average control inputs U_1 and U_2 we find the following equations for the state equilibrium point,

$$\begin{pmatrix} 0 & -(1-U_1) & 0 & 0 \\ (1-U_1) & -\frac{1}{Q_1} & -1 & 0 \\ 0 & 1 & 0 & -(1-U_2) \\ 0 & 0 & (1-U_2) & -\frac{1}{Q_L} \end{pmatrix} \begin{pmatrix} \overline{x}_1 \\ \overline{x}_2 \\ \overline{x}_3 \\ \overline{x}_4 \end{pmatrix} = \begin{pmatrix} -1 \\ 0 \\ 0 \\ 0 \end{pmatrix} \qquad (2.78)$$

The solution of the above equations, parameterized by the constant values of the average control inputs is given by

$$\begin{pmatrix} \overline{x}_1 \\ \overline{x}_2 \\ \overline{x}_3 \\ \overline{x}_4 \end{pmatrix} = \begin{pmatrix} \frac{1}{Q_1 Q_L} \frac{Q_1 + Q_L(1-U_2)^2}{(1-U_1)^2(1-U_2)^2} \\ \frac{1}{1-U_1} \\ \frac{1}{Q_L} \frac{1}{(1-U_1)(1-U_2)^2} \\ \frac{1}{(1-U_1)(1-U_2)} \end{pmatrix} \qquad (2.79)$$

The output variables of the Boost-Boost converter are considered to be the voltage variables, x_2 and x_4. A parametrization of the equilibrium point in terms of the steady state output voltages: $\overline{x}_2 = V_{2d}$ and $\overline{x}_4 = V_{4d}$, is obtained as

$$\overline{x}_1 = \frac{V_{2d}^2}{Q_1} + \frac{V_{4d}^2}{Q_L}, \qquad \overline{x}_2 = V_{2d}, \qquad \overline{x}_3 = \frac{V_{4d}^2}{Q_L V_{2d}}, \qquad \overline{x}_4 = V_{4d} \qquad (2.80)$$

where:

$$U_1 = \frac{V_{2d} - 1}{V_{2d}}, \qquad U_2 = \frac{V_{4d} - V_{2d}}{V_{4d}} \qquad (2.81)$$

Such an equilibrium point, in original state variables, is found to be:

$$\overline{i}_1 = \frac{1}{R_1} \frac{\overline{v}_1^2}{E} + \frac{1}{R_L} \frac{\overline{v}_2^2}{E}, \qquad \overline{i}_2 = \frac{1}{R_L} \frac{\overline{v}_2^2}{\overline{v}_1} \qquad (2.82)$$

and

$$\overline{v}_1 = \frac{1}{(1-U_1)} E, \qquad \overline{v}_2 = \frac{1}{(1-U_1)(1-U_2)} E \qquad (2.83)$$

Therefore

$$U_1 = 1 - \frac{E}{\overline{v}_1}, \qquad U_2 = 1 - \frac{\overline{v}_1}{\overline{v}_2} \qquad (2.84)$$

The *normalized static transfer matrix* is now defined as a *row vector* relating the two output voltages with the scalar constant input voltage. Evidently, the entries of such a matrix are now functions of U_1 and U_2. This matrix is constituted by the steady state values of the output voltages. We have

$$\mathcal{H}(U_1, U_2) = \begin{bmatrix} \overline{x}_2(U_1, U_2) & \overline{x}_4(U_1, U_2) \end{bmatrix} = \begin{bmatrix} \frac{1}{1-U_1} & \frac{1}{(1-U_1)(1-U_2)} \end{bmatrix} \qquad (2.85)$$

2.10.4 Alternative Model of the Boost-Boost Converter

In order to simplify the manipulations involved in the controller design for the multi-variable Boost-Boost converter, we present the alternative average model of the system in the same spirit as that presented for the mono-variable case.

$$\frac{dx_1}{d\tau} = -\mathbf{u}_{1av}x_2 + 1$$
$$\frac{dx_2}{d\tau} = \mathbf{u}_{1av}x_1 - \frac{1}{Q_1}x_2 - x_3$$
$$\alpha_1 \frac{dx_3}{d\tau} = x_2 - \mathbf{u}_{2av}x_4$$
$$\alpha_2 \frac{dx_4}{d\tau} = \mathbf{u}_{2av}x_3 - \frac{1}{Q_L}x_4 \qquad (2.86)$$

Clearly, we have defined the following new control inputs:

$$\mathbf{u}_{1av} = (1 - u_{1av}) \qquad (2.87)$$
$$\mathbf{u}_{2av} = (1 - u_{2av}) \qquad (2.88)$$

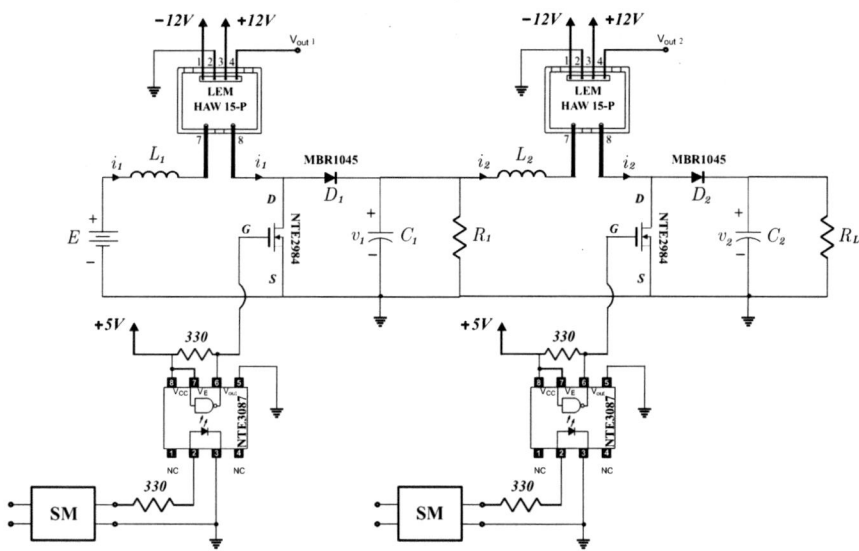

Fig. 2.41. Circuit diagram of the *Boost-Boost converter prototype*.

2.10.5 A Boost-Boost Converter Experimental Prototype

The actual circuit and a picture of the *Boost-Boost system* are shown in Figures 2.41, and 2.42, respectively. We have set the following parameter values:

$$L_1 = 15.91 \text{ mH}, \quad C_1 = 48 \ \mu\text{F}, \quad L_2 = 40 \text{ mH}, \quad C_2 = 107 \ \mu\text{F},$$

$$R_1 = 52 \ \Omega, \quad R_L = 52 \ \Omega, \quad E = 12 \text{ V}$$

The switching frequency for the Boost-Boost converter, as in the previous converters, is 45 kHz. Thus, the inductors L_1 and L_2 were designed for this operation frequency.

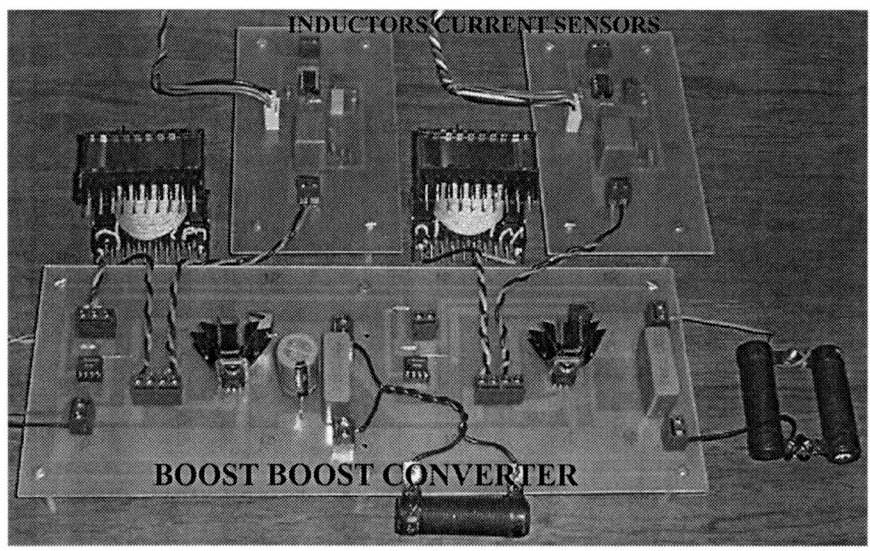

Fig. 2.42. Photograph of the *Boost-Boost system*.

2.11 The Double Buck-Boost Converter

Similarly to the Boost-Boost system, we propose a double Buck-Boost converter as a multi-variable DC-to-DC power converter with interesting, but limited, independence features for handling two loads with different steady state, or tracking, requirements. As in the previous case we limit ourselves to a summary of the relevant equations of the model and the steady state features.

Figure 2.43 depicts the tandem connection of two Buck-Boost DC-to-DC power converters.

2.11 The Double Buck-Boost Converter

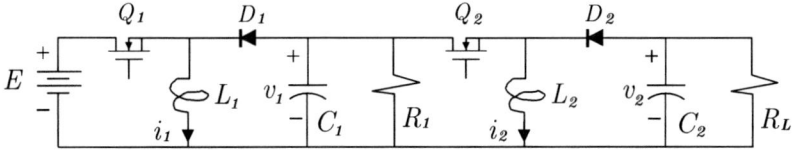

Fig. 2.43. Double Buck-Boost converter circuit.

2.11.1 Model of the Double Buck-Boost Converter

$$L_1 \frac{di_1}{dt} = (1 - u_1) v_1 + u_1 E$$
$$C_1 \frac{dv_1}{dt} = -(1 - u_1) i_1 - \frac{v_1}{R_1} - u_2 i_2$$
$$L_2 \frac{di_2}{dt} = u_2 v_1 + (1 - u_2) v_2$$
$$C_2 \frac{dv_2}{dt} = -(1 - u_2) i_2 - \frac{v_2}{R_L} \qquad (2.89)$$

2.11.2 Average Normalized Model

$$\frac{dx_1}{d\tau} = (1 - u_{1av}) x_2 + u_{1av}$$
$$\frac{dx_2}{d\tau} = -(1 - u_{1av}) x_1 - \frac{1}{Q_1} x_2 - u_{2av} x_3$$
$$\alpha_1 \frac{dx_3}{d\tau} = u_{2av} x_2 + (1 - u_{2av}) x_4$$
$$\alpha_2 \frac{dx_4}{d\tau} = -(1 - u_{2av}) x_3 - \frac{1}{Q_L} x_4 \qquad (2.90)$$

with,

$$\alpha_1 = \frac{L_2}{L_1}, \quad \alpha_2 = \frac{C_2}{C_1}, \quad Q_1 = R_1 \sqrt{\frac{C_1}{L_1}}, \quad Q_L = R_L \sqrt{\frac{C_1}{L_1}} \qquad (2.91)$$

2.11.3 Equilibrium Point and Static Transfer Function

The parametrization of the equilibrium point in terms of U_1 and U_2 is found to be

$$\begin{pmatrix} \overline{x}_1 \\ \overline{x}_2 \\ \overline{x}_3 \\ \overline{x}_4 \end{pmatrix} = \begin{pmatrix} \frac{U_1}{Q_1 Q_L} \frac{Q_1 U_2^2 + Q_L (1-U_2)^2 +}{(1-U_1)^2 (1-U_2)^2} \\ -\frac{U_1}{1-U_1} \\ -\frac{1}{Q_L} \frac{U_1 U_2}{(1-U_1)(1-U_2)^2} \\ \frac{U_1 U_2}{(1-U_1)(1-U_2)} \end{pmatrix} \qquad (2.92)$$

The parametrization in terms of the desired output voltages $\bar{x}_2 = V_{2d}$ and $\bar{x}_4 = V_{4d}$, is found to be

$$\bar{x}_1 = -\left(\frac{V_{2d}^2}{Q_1} + \frac{V_{4d}^2}{Q_L}\right)\left(\frac{1-V_{2d}}{V_{2d}}\right),\ \bar{x}_2 = V_{2d},\ \bar{x}_3 = \frac{V_{4d}}{Q_L}\left(\frac{V_{4d}}{V_{2d}} - 1\right),\ \bar{x}_4 = V_{4d} \tag{2.93}$$

where

$$U_1 = -\frac{V_{2d}}{1 - V_{2d}},\quad U_2 = \frac{V_{4d}}{V_{4d} - V_{2d}} \tag{2.94}$$

The normalized static transfer matrix is, as before, given by a row vector whose entries are functions of U_1 and U_2,

$$\mathcal{H}(U_1, U_2) = \left[-\frac{U_1}{1-U_1}\ \frac{U_1 U_2}{(1-U_1)(1-U_2)}\right] \tag{2.95}$$

2.12 Power Converter Models with Non-ideal Components

In this chapter, we have emphasized ideal components in the constitution of dynamical average DC-to-DC power converter models. For instance, we have not considered resistances in series with the inductors, we have neglected parallel conductances in combination with capacitors and have also assumed that the switches have no imperfections attached. In real life, inductors do exhibit associated resistances, capacitors must be considered along with parallel conductances, or equivalent series resistances, and switches are synthesized by means of physical transistor and diode arrangements which exhibit important resistances and parasitic voltage sources.

A more realistic model of a DC-to-DC power converter must, therefore, include a number of "parasitic" resistances and voltage sources in connection with the diode-transistor arrangements synthesizing the regulating switch. Here we will only present one such example, concerning the "Boost" converter model, in order to highlight the important differences with the already derived ideal model.

Figure 2.44 shows a Boost converter model including parasitic components surrounding the ideal switch. The model of such a converter, using the Kirchoff's voltage and current laws as in the several previous examples presented in this chapter, results in the following set of differential equations (see also Ortega et al. [48])

$$\dot{x}_1 = -\frac{r}{L}x_1 - u\frac{R}{(r_C + R)L}x_2 + \frac{E}{L} - u\frac{V_F}{L}$$

$$\dot{x}_2 = u\frac{R}{(r_C + R)C}x_1 - \frac{1}{(r_C + R)C}x_2$$

$$x_0 = \frac{R}{r_C + R}x_2 + u\frac{r_C R}{r_C + R}x_1$$

where $(r_C||R)$ stands for the resistance associated with the parallel arrangement of resistors of values: r_C and R. The resistance r is given by

$$r = [r_L + (1-u)r_{DS} + u(R_F + (r_C||R))]$$

The variable x_0 is the output voltage which, in this instance, is not directly constituted by the capacitor voltage x_2 and directly receives the influence of x_1 and of the control input variable u. Note that if the values of the parasitic resistances described by r_C, r_L and r_{DS} are all set to zero, we recover the ideal model of the "Boost" converter.

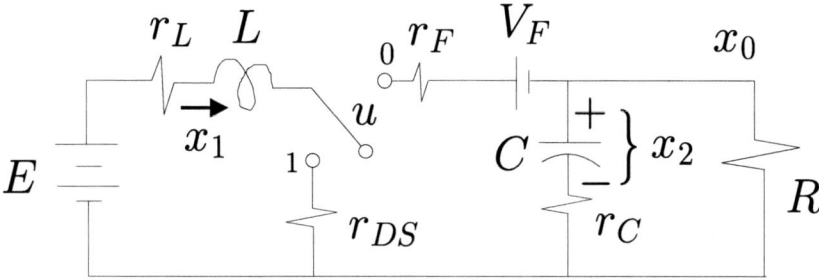

Fig. 2.44. A realistic model of the Boost converter.

A rigorous mathematical analysis aimed at the derivation of control laws for a DC-to-DC power converter should entitle a realistic model of the converter of the same nature as that described in the previous paragraphs for the Boost converter. However, experience tells that the complexity of the derivations associated with such models does not substantially improve the quality of the overall converter performance, as far as the closed loop operation is concerned, when compared with that achieved by a controller derived in terms of the idealized models. Naturally, the fundamental discrepancies refer to *constant* steady state stabilization or tracking errors, and transients quality. These features may, in principle, be remedied by automatic adjustment features (say, of the integral tracking error type) bestowed on the simplified controller. Nevertheless, in order to justify the assertion we have just made, we state that *all* the laboratory tests of controller performance, carried out on the actual prototypes described in this book, involved controller designs entirely based on the ideal models.

2.13 A General Mathematical Model for Power Electronics Devices

Most of power electronics devices, such as: DC-to-DC power converters, DC-to-AC converters, ac rectifiers, inverters, and combinations of these devices with dc and ac electric motors are accurately described by nonlinear system models of varying complexity. A closer examination of such models reveals that even though these systems are nonlinear, they belong, generally speaking, to the class of *bilinear* systems. As such, it is also revealing to study the "energy dissipation structure" of most of these dynamic models.

In a state space framework viewpoint electric circuits are considered as *lumped parameter* dynamical systems. this is the case of electric and electronic circuits in general, as opposed to *distributed* i.e., infinite dimensional circuits. Lumped parameter circuits are described, in general, by ordinary nonlinear vector differential equations. The right hand side of these systems of equations are vectors which depend nonlinearly on the state vector of the system, These nonlinear vector functions are generally addressed as *vector fields*. The presence of control inputs in the right hand side of the describing differential equations is termed *controlled vector fields*. A nonlinear dynamic system is then generally described by the set of equations

$$\dot{x} = f(x, u) \tag{2.96}$$

where x is a vector in R^n and the control input $u \in R^m$. The nonlinear map f is assumed to be either *smooth*, i.e., it admits an infinite number of derivatives with respect to its arguments, of *analytic* i.e., at any point in the (x, u) space we can obtain a infinite convergent Taylor series expansion of the function $f(x,u)$. In the description of power electronics devices, the control input functions u represent, in general, *switch position functions*, i.e., each component of the vector u independently takes values in a *discrete set*. In some instances such set is represented by a binary set with two numerical values, say $\{0, 1\}$. In other instances, such binary set is of the form $\{-W, +W\}$ where W is a constant real number. We denote the set where the switched control input vector takes values by a "power" of the binary set where each component takes values: i.e., $\{0,1\}^m$, or $\{-W, W\}^m$. Such powers evidently mean, cartesian products of the base set a finite number of times defined by m.

Power electronics models are thus, generally speaking, *Variable Structure Systems*. This class of systems is naturally characterized by the possibility of sudden changes in their differential equation description. This immediately translates, in the circuit description arena, to a sudden change in the *circuit topology*. Such topological changes may be made 1) "occasionally" in a non-periodic fashion, 2) periodically, with a significant frequency or, even, at an ideally *infinite* frequency rate. This last idealization, most often, proves to be useful in rescuing the main qualitative features of high-frequency topological

2.13 A General Mathematical Model for Power Electronics Devices

changes (such as those provided, say, by a PWM train of pulses triggering such topology changes), without resorting to exact or approximate discretizations of the nonlinear set of differential equations describing the power device. In general, we term such ideally infinite frequency models as *average* models. These average models are most suitable for feedback control considerations. With an abuse of notation, we customarily mathematically describe such average models with the same symbols for state variables as in the switched model, except for a distinction in the control input. We write,

$$\dot{x} = f(x, u_{av}) \tag{2.97}$$

where now x is addressed as the average state vector, still taking values in R^n and u_{av} is now a *continuous* valued m-dimensional function. The average control input functions u_{av} are now *bounded* functions continuously taking values over compact sets in R^m. If the original control input function components $u_i, i = 1, ..., m$ take values, independently, on, say, the binary set $\{0, 1\}$, then the components of the average control input vector u_{av} will take values on the closed interval $[0, 1]$ of the real line. Hence, the vector u_{av} takes values on $[0, 1]^m$.

The most general model one can think of, for power electronics devices, clearly exhibits the following decomposition of the "energy managing" structure: 1) An input dependent conservative vector field, characterized by the product of a skew symmetric matrix with the state vector. The skew symmetric matrix is, generally speaking, an affine function of the input vector components and its most important property is that it does not intervene in the system stability considerations. 2) A dissipative vector field characterized by the product of a constant symmetric, positive semi-definite, matrix with the state vector. This term accounts for the dissipation forces in the system due to resistances and frictions. 3) The control input channels which entitle a constant matrix multiplying the input vector. The input vector is constituted by switching functions representing the discontinuous control inputs to the system. 4) A time varying or, alternatively constant, vector field representing the external input sources. Such sources are of fixed nature, i.e., their amplitude values and frequencies are not subject to our command. The sum of this various fields produces the rate of change of the state of the system. Such a general model is summarized in the following n-dimensional differential equation:

$$\mathcal{A}\dot{x} = \mathcal{J}(u)x - \mathcal{R}x + \mathcal{B}u + \mathcal{E}(t) \tag{2.98}$$

where: x is an n-dimensional vector, $\mathcal{A}$ is a symmetric, positive definite, constant, matrix, $\mathcal{J}$ is a skew symmetric matrix of the form:

$$\mathcal{J} = \mathcal{J}_0 + \sum_{i=1}^{m} \mathcal{J}_i u_{i\,av} \tag{2.99}$$

for all u_i, where $\mathcal{J}_0$ is constant and skew symmetric and $\mathcal{J}_i$ is also constant skew symmetric for all i. $\mathcal{R}$ is a symmetric, positive semi-definite constant matrix. $\mathcal{B}$ is a constant $n \times m$ matrix and, hence y is an m dimensional output vector. In terms of its n dimensional column vectors, the matrix $\mathcal{B}$ is given by $\mathcal{B} = [b_1, b_2, \cdots, b_m]$. The vector u_{av} is the average control input vector assumed to be m-dimensional, with each component $u_{i\,av}$ taking values either in the closed set $[0, 1]$, or in the closed set $[-1, 1]$, of the real line. In any case, $u_{i\,av}$ represents a *bounded* average control input function. $\mathcal{E}(t)$ is a n-dimensional smooth vector function of t or, sometimes, a vector of constant entries.

Note that the matrix $\mathcal{R}$ represents the *dissipative field* of the system while $\mathcal{J}(u)$ represents the, possibly control input dependent, *conservative field* of the system. The *control input channels* are represented by the constant matrix $\mathcal{B}$ while $\mathcal{E}(t)$ represent *external input sources*, such as batteries or ac line voltages.

We will be using this rather general *bilinear system* model, rather extensively, in later chapters in connection with linear feedback controller designs for power electronics devices. The average models, or infinite frequency models, corresponding to this general mathematical model will be described by the following set of controlled differential equations:

$$\mathcal{A}\dot{x} = \mathcal{J}(u_{av})x - \mathcal{R}x + \mathcal{B}u_{av} + \mathcal{E}(t) \qquad (2.100)$$

The general model (2.100) has a number of interesting properties which enormously ease the feedback controller design task. We defer the study of the mathematical properties of this model to Chapter 5, where we address the study of nonlinear methods for controller design.

2.13.1 Some Illustrative Examples of the General Model

Here, we present just a couple of examples illustrating the validity of the general mathematical model, proposed in this section, in relation to some DC-to-DC power converter models. In Chapter 7 we shall also use this general model for monophasic and three phase rectifiers. The general model is also valid for power converters loaded by DC motors as well as mono and three -phase rectifiers loaded with DC motors.

The Boost-Boost Converter

Consider the following average normalized model of the Boost-Boost converter.

$$\frac{dx_1}{d\tau} = -u_{1av}x_2 + 1, \qquad \frac{dx_2}{d\tau} = u_{1av}x_1 - \frac{1}{Q_1}x_2 - x_3$$
$$\alpha_1 \frac{dx_3}{d\tau} = x_2 - u_{2av}x_4, \qquad \alpha_2 \frac{dx_4}{d\tau} = u_{2av}x_3 - \frac{1}{Q_L}x_4 \qquad (2.101)$$

The mathematical model of the average normalized Boost-Boost converter may be written, in matrix form, as:

2.13 A General Mathematical Model for Power Electronics Devices

$$\begin{bmatrix} 1 & 0 & 0 & 0 \\ 0 & 1 & 0 & 0 \\ 0 & 0 & \alpha_1 & 0 \\ 0 & 0 & 0 & \alpha_2 \end{bmatrix} \begin{bmatrix} \dot{x}_1 \\ \dot{x}_2 \\ \dot{x}_3 \\ \dot{x}_4 \end{bmatrix} = \begin{bmatrix} 0 & -u_{1av} & 0 & 0 \\ u_{1av} & 0 & -1 & 0 \\ 0 & 1 & 0 & -u_{2av} \\ 0 & 0 & u_{2av} & 0 \end{bmatrix} \begin{bmatrix} x_1 \\ x_2 \\ x_3 \\ x_4 \end{bmatrix}$$

$$- \begin{bmatrix} 0 & 0 & 0 & 0 \\ 0 & \frac{1}{Q_1} & 0 & 0 \\ 0 & 0 & 0 & 0 \\ 0 & 0 & 0 & \frac{1}{Q_L} \end{bmatrix} \begin{bmatrix} x_1 \\ x_2 \\ x_3 \\ x_4 \end{bmatrix} + \begin{bmatrix} 1 \\ 0 \\ 0 \\ 0 \end{bmatrix}$$

Hence, in reference to the proposed general model (2.100) we have

$$\mathcal{A} = \begin{bmatrix} 1 & 0 & 0 & 0 \\ 0 & 1 & 0 & 0 \\ 0 & 0 & \alpha_1 & 0 \\ 0 & 0 & 0 & \alpha_2 \end{bmatrix}, \quad \mathcal{R} = \begin{bmatrix} 0 & 0 & 0 & 0 \\ 0 & \frac{1}{Q_1} & 0 & 0 \\ 0 & 0 & 0 & 0 \\ 0 & 0 & 0 & \frac{1}{Q_L} \end{bmatrix}, \quad \mathcal{B} = b = 0, \quad \mathcal{E} = \begin{bmatrix} 1 \\ 0 \\ 0 \\ 0 \end{bmatrix}$$

$$\mathcal{J}(u_{av}) = \mathcal{J}_0 + \mathcal{J}_1 u_{1av} + \mathcal{J}_2 u_{2av}$$

$$= \begin{bmatrix} 0 & 0 & 0 & 0 \\ 0 & 0 & -1 & 0 \\ 0 & 1 & 0 & 0 \\ 0 & 0 & 0 & 0 \end{bmatrix} + \begin{bmatrix} 0 & -1 & 0 & 0 \\ 1 & 0 & 0 & 0 \\ 0 & 0 & 0 & 0 \\ 0 & 0 & 0 & 0 \end{bmatrix} u_{1av} + \begin{bmatrix} 0 & 0 & 0 & 0 \\ 0 & 0 & 0 & 0 \\ 0 & 0 & 0 & -1 \\ 0 & 0 & 1 & 0 \end{bmatrix} u_{2av}$$

The Sepic Converter

Consider now, the following average normalized model of the Sepic converter:

$$\begin{aligned} \dot{x}_1 &= -(1 - u_{av})(x_2 + x_4) + 1 \\ \dot{x}_2 &= (1 - u_{av}) x_1 - u_{av} x_3 \\ \alpha_1 \dot{x}_3 &= u_{av} x_2 - (1 - u_{av}) x_4 \\ \alpha_2 \dot{x}_4 &= (1 - u_{av})(x_1 + x_3) - \frac{x_4}{Q} \end{aligned} \quad (2.102)$$

Writing the model in matrix form we have:

$$\begin{bmatrix} 1 & 0 & 0 & 0 \\ 0 & 1 & 0 & 0 \\ 0 & 0 & \alpha_1 & 0 \\ 0 & 0 & 0 & \alpha_2 \end{bmatrix} \begin{bmatrix} \dot{x}_1 \\ \dot{x}_2 \\ \dot{x}_3 \\ \dot{x}_4 \end{bmatrix} = \begin{bmatrix} 0 & -(1 - u_{av}) & 0 & -(1 - u_{av}) \\ 1 - u_{av} & 0 & -u_{av} & 0 \\ 0 & u_{av} & 0 & -(1 - u_{av}) \\ 1 - u_{av} & 0 & 1 - u_{av} & 0 \end{bmatrix} \begin{bmatrix} x_1 \\ x_2 \\ x_3 \\ x_4 \end{bmatrix}$$

$$- \begin{bmatrix} 0 & 0 & 0 & 0 \\ 0 & 0 & 0 & 0 \\ 0 & 0 & 0 & 0 \\ 0 & 0 & 0 & \frac{1}{Q} \end{bmatrix} \begin{bmatrix} x_1 \\ x_2 \\ x_3 \\ x_4 \end{bmatrix} + \begin{bmatrix} 1 \\ 0 \\ 0 \\ 0 \end{bmatrix}$$

In reference to the proposed general model (2.100), we have:

$$\mathcal{A} = \begin{bmatrix} 1 & 0 & 0 & 0 \\ 0 & 1 & 0 & 0 \\ 0 & 0 & \alpha_1 & 0 \\ 0 & 0 & 0 & \alpha_2 \end{bmatrix}, \quad \mathcal{R} = \begin{bmatrix} 0 & 0 & 0 & 0 \\ 0 & 0 & 0 & 0 \\ 0 & 0 & 0 & 0 \\ 0 & 0 & 0 & \frac{1}{Q} \end{bmatrix}, \quad \mathcal{B} = b = 0, \quad \mathcal{E} = \begin{bmatrix} 1 \\ 0 \\ 0 \\ 0 \end{bmatrix}$$

$$\mathcal{J}(u_{av}) = \mathcal{J}_0 + \mathcal{J}_1 u_{av} = \begin{bmatrix} 0 & -1 & 0 & -1 \\ 1 & 0 & 0 & 0 \\ 0 & 0 & 0 & -1 \\ 1 & 0 & 1 & 0 \end{bmatrix} + \begin{bmatrix} 0 & 1 & 0 & 1 \\ -1 & 0 & -1 & 0 \\ 0 & 1 & 0 & 1 \\ -1 & 0 & -1 & 0 \end{bmatrix} u_{av}$$

The proposed mathematical model is also valid for original (i.e., non-normalized) switched DC-to-DC power converter models when they are suitably rewritten in matrix form. We leave it to the reader to verify that anyone of the converter models presented in this chapter, except for the Boost converter exhibiting realistic parasitic components, conform to the general mathematical model (2.100).

When parasitic components are considered, the symmetric dissipative matrix $\mathcal{R}$ may be an affine function of the control input, i.e., we must consider control dependent dissipations, $\mathcal{R}(u) = \mathcal{R}_0 + \mathcal{R}_1 u$.

Part II

Controller Design Methods

3
Sliding Mode Control

3.1 Introduction

This chapter is devoted to an exposition of sliding mode control of switch-regulated non linear systems and its implications in the feedback regulation of DC-to-DC power converters exhibiting one or multiple, independent, switches.

Sliding mode control is a well known discontinuous feedback control technique which has been exhaustively explored in many books and journal articles by many authors. The technique is naturally suited for the regulation of switched controlled systems, such as power electronics devices, in general, and DC-to-DC power converters, in particular. Sliding mode control was studied primarily by Russian scientists in the former Soviet Union. A complete account of the history and fundamental results of sliding modes, or *sliding regimes*, is found in books, such as that by Emelyanov [10], Utkin [75], [76] and Utkin *et al.* [77]. In [77], the discontinuous feedback control of a rather complete collection of physical electro-mechanical systems is addressed along with remarkable laboratory implementation results. In that book, there is some detailed attention devoted to the control and stabilization of DC-to-DC power converters. A recent book, mainly devoted to the area of linear systems, with a terse and very clear exposition of the topics along with some interesting laboratory and industry applications, is that of Edwards and Spurgeon [9]. Well documented books, containing chapters on sliding mode control, are those of Slotine and Li [73], Kwatny and Blankenship [38], Sastry [53], Żak [84], among many others. The state of the art has been recently summarized in the edited books by Sabanovic *et al.* [52], Perruquetti and Barbot [49] and that by Young and Özgüner [83].

In this chapter, we provide an introduction to the sliding mode control of switch-regulated nonlinear systems. We formulate the sliding mode creation problem in a rather general set up, using the language of elementary *differential geometry*. We revisit both the single switch case and the multiple switch case (i.e., the SISO and the MIMO cases). We examine the most salient features and theoretical elements of sliding mode control: the sliding surface accessibility, or reachability, problem, the definition of the equivalent control and its corresponding ideal sliding dynamics and, finally, the robustness of closed loop sliding mode responses with respect to additive perturbation fields satisfying the so called *perturbation matching condition*. The approach naturally allows one to relate these important features with well known concepts of nonlinear geometric control such as: invariance, zero dynamics, minimum and non-minimum phase outputs, projection operators (over tangent subspaces along, or parallel to, the span of given input vector fields, or input matrix spans) and local stability in the sense of Lyapunov. After the theoretical introductions to SISO and MIMO sliding mode control cases, we then center our attention on the sliding mode control of the most popular DC-to-DC power converters written, for ease of treatment, in normalized form. Incidentally, normalization is an invariant throughout our book. This practise not only greatly facilitates the algebraic manipulation in the controller design process and the computer simulation tasks, but it is also a good guide and handy check possibility for actual implementation of feedback controllers in many areas of Power Electronics. We also present, at the end of the chapter, a detailed treatment of $\Sigma - \Delta$ modulation, a sliding mode technique which will be extremely useful in the actual implementation of feedback control laws designed on the basis of average models.

3.2 Variable Structure Systems

A variable structure system is a system in which the current dynamic model heavily depends on the region of the state space where the operation of the system is circumstantially found. The discontinuous nature of the model is characteristic and these abrupt changes occur due to either a voluntary action on the part of the operator, due to the automatic activation of one or more switches present in the system, or due to the change of the temporary values of certain system parameters.

The class of variable structure systems is quite wide for its detailed study and, moreover, its interest in Power Electronics is limited. For this reason, we shall study variable structure systems regulated by one or several switches. The position of the switches constitute our only set of available control inputs.

Additionally we restrict ourselves to the class of systems where such descriptions, or structures, have in common the dimension of the resulting system as well as the nature of the describing state of the system.

3.2.1 Control of Single Switch Regulated Systems

We study the control of systems represented by nonlinear state space models of the following form:

$$\dot{x} = f(x) + g(x)u, \qquad y = h(x) \tag{3.1}$$

where $x \in R^n$, $u \in \{0, 1\}$, $y \in R$. The vector functions $f(x)$ and $g(x)$ represent smooth *vector fields*, i.e., infinitely differentiable vector fields, defined over the tangent space to R^n. The output function, $h(x)$, is a smooth scalar function of x taking values in the real line R. We refer to x as the *state* of the system. The variable u is addressed as the *control input*, or simply as *the control*. The variable y is the *output* of the system. We usually refer to $f(x)$ as the *drift vector field* and to $g(x)$ as the *control input field*.

The main feature of the systems we consider is the *binary* valued nature of the control input variable. Without loss of generality, we assume that the control input takes values on the discrete set $\{0, 1\}$. Note that if the set of possible values for the scalar control input u, were the discrete set $\{W_1, W_2\}$ with $W_i \in R, i = 1, 2$, then the following invertible input coordinate transformation: $v = (u - W_2)/(W_1 - W_2)$, $u = W_2 + v(W_1 - W_2)$ makes the new control input v a binary valued control input function taking values in the set $\{0, 1\}$

Example 3.1. The following circuit represents a DC-to-DC power converter, known as the "Boost" converter, controlled by a single switch.

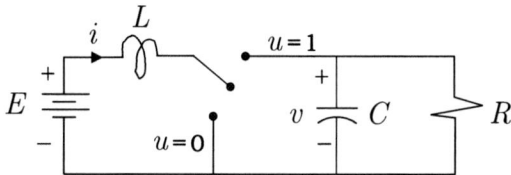

Fig. 3.1. Alternative DC-to-DC Boost converter with an ideal switch.

The controlled differential equations describing the system are given by

$$L\frac{di}{dt} = -uv + E$$
$$C\frac{dv}{dt} = ui - \frac{1}{R}v$$

where i is the input inductor current, v is the output voltage and u is the switch position function satisfying, $(u \in \{0, 1\})$.

In matrix terms, the mathematical description of the "Boost" converter is given by:

$$\frac{d}{dt}\begin{bmatrix} i \\ v \end{bmatrix} = \begin{bmatrix} 0 & 0 \\ 0 & -\frac{1}{RC} \end{bmatrix}\begin{bmatrix} i \\ v \end{bmatrix} + \begin{bmatrix} -\frac{v}{L} \\ \frac{i}{L} \end{bmatrix}u + \begin{bmatrix} \frac{E}{L} \\ 0 \end{bmatrix}$$

Letting $x = [x_1 \ x_2]^T = [i \ v]^T$, we have

$$f(x) = \begin{bmatrix} 0 & 0 \\ 0 & -\frac{1}{RC} \end{bmatrix}x + \begin{bmatrix} \frac{E}{L} \\ 0 \end{bmatrix} = \begin{bmatrix} \frac{E}{L} \\ -\frac{x_2}{RC} \end{bmatrix}$$

and
$$g(x) = \begin{bmatrix} -x_2/L \\ x_1/C \end{bmatrix}$$

3.2.2 Sliding Surfaces

In the context of single switch, n-dimensional, systems, a sliding surface, denoted by $\mathcal{S}$, is represented by the set of state vectors in R^n where the algebraic restriction, $h(x) = 0$, is satisfied, where $h : R^n \to R$ is a smooth scalar output function of the system. We define

$$\mathcal{S} = \{x \in R^n \mid h(x) = 0\} \quad (3.2)$$

The set $\mathcal{S}$ represents a *smooth variety* or *smooth manifold* of dimension $n-1$ in R^n.

The main assumption is the following:

There exist feedback control actions $u(x)$, possibly of discontinuous nature, that render the restriction: $h(x) = 0$ to be locally satisfied by the state trajectory $x(t)$. The motions of the system state, x, on the smooth surface $\mathcal{S}$ ideally produces an overall desired, local, behavior for the state of the controlled system. The constrained evolution of the state is accomplished thanks to appropriate control input actions satisfying: $u \in \{0, 1\}$.

One of the primordial features in the design of feedback control laws for switch-regulated systems is represented by the fact that the specification of the smooth scalar function $h(x)$ is *part of the design problem*. The choice of the output function $y = h(x)$ and, hence, of the smooth manifold $\mathcal{S}$ depends entirely of our will in correspondence with the specified control objective for the system.

Example 3.2. In the previous "Boost" converter example, a sliding surface may be proposed as specified by the output function:

$$h(x) = v - \bar{v} = x_2 - V_d$$

where $\bar{v} = V_d$ is the average desired output equilibrium voltage. If one succeeds in forcing $h(x)$ to be zero, even if only locally, along the controlled trajectories of the system, then, the output voltage locally ideally coincides with the desired voltage.

Another sliding surface that one may consider, in this particular case, is given by,

$$h(x) = i - \bar{i} = x_1 - I_d$$

where $\bar{i} = I_d = V_d^2/(RE)$ represents an average equilibrium input current which corresponds, with the desired average output equilibrium voltage V_d.

Even though both sliding surfaces above represent a desired behavior of the output, only one of them is actually feasible due to internal stability considerations.

3.2.3 Notation

Let $f(x)$ and $g(x)$ be smooth vector fields, locally defined on the tangent space to R^n. Let $h(x)$ be a scalar function taking values on R.

We denote the *directional derivative* of $h(x)$ in the direction of $f(x)$ as the scalar quantity $\frac{\partial h}{\partial x^T} f(x)$ and we denote it by means of $L_f h(x)$. Similarly, we refer to $L_g h(x)$ as the directional derivative of $h(x)$ in the direction of $g(x)$.

In local coordinates we have:

$$\frac{\partial h}{\partial x^T} = \left(\frac{\partial h}{\partial x_1} \; \frac{\partial h}{\partial x_2} \; \cdots \; \frac{\partial h}{\partial x_n} \right), \qquad f(x) = \begin{bmatrix} f_1(x) \\ f_2(x) \\ \vdots \\ f_n(x) \end{bmatrix}$$

and

$$L_f h(x) = \sum_{i=1}^{n} \frac{\partial h}{\partial x_i} f_i(x)$$

3.2.4 Equivalent Control and the Ideal Sliding Dynamics

Let us assume that thanks to the use of an appropriate switching law $u \in \{0, 1\}$, we manage to make the state x of the system locally evolve restricted to the smooth manifold, $\mathcal{S}$. While the condition $x \in \mathcal{S}$ is satisfied, it is assumed that we are complying with a specific control objective. In other words, assume we can achieve the *invariance* of $\mathcal{S}$ with respect to the trajectories of the state of the system by means of appropriate control input commutations taking place in the set $\{0, 1\}$, regardless how fast these commutations should be performed as needed. It is not difficult to realize that when the state trajectories collide obliquely with the sliding surface, then the control input commutations have to be, necessarily, of *infinite frequency*, for finite frequency switchings may make the trajectory temporarily stray away from the surface. The evolution of the state along $\mathcal{S}$ takes place then as if it had been produced by a smooth control input, rather than a switched control input. This equivalence between an infinite frequency switched control input and a smooth feedback control is known as the *equivalent control* concept.

We define the equivalent control as the smooth feedback control law, denoted by $u_{eq}(x)$ which locally sustains the evolution of the state trajectory ideally restricted to the smooth manifold $\mathcal{S}$ when the initial state of the system $x(t_0) = x_0$ is located precisely on the manifold $\mathcal{S}$, i.e., when $h(x_0) = 0$.

The coordinate function $h(x)$ satisfies then the following invariance condition,

$$\dot{h}(x) = \frac{\partial h}{\partial x}(f(x) + g(x)u_{eq}(x)) = 0 \qquad (3.3)$$

In other words,

$$L_f h(x) + [L_g h(x)]u_{eq}(x) = 0$$

and therefore, the equivalent control is expressed, in a unique fashion, as the quotient:

$$u_{eq}(x) = -\frac{L_f h(x)}{L_g h(x)} \qquad (3.4)$$

The controlled vector field, $f(x) + g(x)u_{eq}(x)$, and the corresponding evolution over the smooth manifold $\mathcal{S}$ of the state trajectories of the system, is expressed as,

$$\dot{x} = f(x) - g(x)\frac{L_f h(x)}{L_g h(x)} \qquad (3.5)$$

Note that for any other initial condition which is not over the smooth manifold $\mathcal{S}$ evolves, under the actions of $u_{eq}(x)$, in such a manner that the function $h(x)$ remains constant since $\dot{y}$ is identically, locally, zero. Such a constant value of $y = h(x)$ only adopts the value of zero when the initial state x_0 is located on $\mathcal{S}$. The closed loop system, fed back by the equivalent control may be alternatively described as follows:

$$\dot{x} = \left\{ I - \frac{1}{L_g h(x)} g(x) \frac{\partial h}{\partial x} \right\} f(x) = \mathcal{M}(x) f(x) \qquad (3.6)$$

Proposition 3.3. *The square $n \times n$ matrix $\mathcal{M}(x)$, is a **projection operator**, over the tangent space to $\mathcal{S}$, along the span $g(x)$. The operator $\mathcal{M}(x)$ projects any smooth vector field defined in the tangent space of R^n over the tangent subspace to the manifold $\mathcal{S}$ in a parallel fashion to the $\{\text{span } g(x)\}$ or in the direction of the control input field $g(x)$.*

Indeed, let v be a vector field in the tangent space to R^n such that $v \in$ span $g(x)$ i.e., $v(x)$ may be expressed as $v(x) = g(x)\alpha(x)$ where $\alpha(x)$ is a smooth scalar function. We then have,

$$\begin{aligned}
\mathcal{M}(x)v(x) &= \left\{ I - \frac{1}{L_g h(x)} g(x) \frac{\partial h}{\partial x} \right\} g(x)\alpha(x) \\
&= \left\{ g(x) - \frac{1}{L_g h(x)} g(x) \frac{\partial h}{\partial x} g(x) \right\} \alpha(x) \\
&= \left\{ g(x) - \frac{1}{L_g h(x)} g(x) L_g h(x) \right\} \alpha(x) \\
&= [g(x) - g(x)] \alpha(x) = 0
\end{aligned}$$

Additionally, the n-th dimensional row vector: $\partial h/\partial x^T$, is orthogonal to the image under $\mathcal{M}(x)$ of the vector fields lying in the tangent space of R^n. For this, it is enough to show that any 1-form in the span of $\frac{\partial h}{\partial x^T}$ annihilates all the column vectors of $\mathcal{M}(x)$.

A 1-form in the span of $\partial h/\partial x^T$ is written as: $\xi(x) \left(\partial h/\partial x^T \right)$, where $\xi(x)$ is a completely arbitrary nonzero scalar function. Indeed:

$$\xi(x)\frac{\partial h}{\partial x^T}\mathcal{M}(x) = \xi(x)\frac{\partial h}{\partial x^T}\left\{I - \frac{1}{L_g h(x)}g(x)\frac{\partial h}{\partial x^T}\right\}$$
$$= \xi(x)\left[\frac{\partial h}{\partial x^T} - L_g h(x)[L_g h(x)]^{-1}\frac{\partial h}{\partial x^T}\right]$$
$$= \xi(x)\left[\frac{\partial h}{\partial x^T} - \frac{\partial h}{\partial x^T}\right] = 0$$

The image, under $\mathcal{M}(x)$, of any vector field in the tangent space to R^n are in the null space of $\partial h/\partial x^T$. In other words, they are in the tangent subspace to the manifold $\mathcal{S}$.

Clearly, $\mathcal{M}^2(x) = \mathcal{M}(x)$ given that $\mathcal{M}(x)G(x) = 0$.

3.2.5 Accessibility of the Sliding Surface

Let x be a representative point of a state trajectory, located in an open neighborhood of the manifold $\mathcal{S}$ (This neighborhood strictly contains its intersection with the sliding manifold). Assume, without loss of generality that at this point the surface coordinate function $h(x)$ of the manifold $\mathcal{S}$ is strictly positive, i.e., $h(x) > 0$. We may, conventionally, say that we are located above the surface $\mathcal{S}$. Our objective is to prescribe an appropriate control action which guarantees that the trajectory of the system reaches and crosses the manifold $\mathcal{S}$. The time derivative of $h(x)$ at the point x is given by

$$\frac{d}{dt}h(x) = \frac{\partial h}{\partial x}(f(x) + g(x)u) = L_f h(x) + [L_g h(x)]u$$

If we assume that $L_g h(x) > 0$ in a neighborhood of $\mathcal{S}$ (i.e., $L_g h(x)$ is strictly positive, "above" and "below" $\mathcal{S}$ in the vicinity of this surface), then we require that the time derivative of $h(x)$ be strictly negative at the point x.

Since by assumption $L_g h(x) > 0$, we must choose the control that annihilates the positive incremental effect that this term has over the derivative of h,. We must then let $u = 0$. The time derivative of $h(x)$ for this control input entirely coincides with the directional derivative $L_f h(x)$. It follows that being $L_g h > 0$ in an open neighborhood of $\mathcal{S}$, it is necessary that $L_f h(x)$ be strictly negative in a neighborhood of $\mathcal{S}$.

If we now assume that the point x is located "below" the surface, i.e., $h(x) < 0$ then, it is easy to see that for the trajectories to reach and cross the sliding manifold $\mathcal{S}$, the time derivative of $h(x)$ must be strictly positive. In other words, $L_f h(x) + [L_g h(x)]u > 0$. Since $L_g(x) > 0$ and $L_f h(x) < 0$, we must choose $u = 1$ so as to magnify the positive incremental effect of $L_g h(x)$ over the time derivative of $h(x)$, but, besides, it is necessary that this positive term be of such magnitude that it may overcome the effect of the negative increment represented by $L_f h(x)$ over the time derivative.

We conclude that, assuming $L_g h(x) > 0$, in an open neighborhood of $\mathcal{S}$, the necessary condition for the existence of a sliding regime on $\mathcal{S}$ is that

$L_g h(x) > -L_f h(x) > 0$. In other words, dividing this inequality by the strictly positive quantity $L_g h(x)$, it is necessary that:

$$1 > -\frac{L_f h(x)}{L_g h(x)} > 0$$

Note that this inequality must be valid in an open neighborhood of R^n exhibiting a non-empty intersection with $\mathcal{S}$. In particular, if this inequality is locally valid for $x \in \mathcal{S}$, then it is also valid in an open neighborhood of $\mathcal{S}$ in R^n, given the smooth character of the involved vector fields and of the surface coordinate function $h(x)$.

Under the assumption that $L_g h(x) > 0$ around $\mathcal{S}$, it is easy to see that the previously discussed existence condition is also sufficient.

Indeed, if the representative point is located, say, above the sliding manifold $\mathcal{S}$, the inequality tells us that $L_f h(x) < 0$ and then it suffices to take $u = 0$ since then $\dot{h}(x) < 0$ in any open neighborhood of $\mathcal{S}$. The state trajectory thus approaches, and crosses, the manifold $\mathcal{S}$ from any neighboring point located above the surface. If the representative point is located below $\mathcal{S}$ then, the inequality establishes that $L_f(x) + L_g h(x) > 0$ and, therefore, the choice, $u = 1$, forces the condition: $\dot{h}(x) > 0$ for any point in an open neighborhood of $\mathcal{S}$. This says that the state trajectory approaches the manifold $\mathcal{S}$.

Note that if we locally had $L_g h(x) < 0$, we then should have $L_f h(x) > 0$ in any neighborhood of $\mathcal{S}$. The changes in the previous arguments for surface reachability are referred only to the choice of u in each case. In this case, we would choose $u = 1$ when x is located above $\mathcal{S}$ and we should set $u = 0$ when we are below the sliding surface.

Nevertheless, and in order to avoid confusion, we note that if locally, $L_g h(x) < 0$, we may always redefine $\mathcal{S}$ taking as a sliding surface coordinate function, $-h(x)$ instead of $h(x)$, and now all the previous analysis becomes valid.

The condition $L_g h(x) > 0$ is particularly important and it determines the switching policy that locally achieves a sliding regime over the sliding manifold $\mathcal{S}$. We address this condition as the *transversal condition* of the control input field $g(x)$ in relation to the sliding manifold $\mathcal{S}$. Note that if $L_g h(x) = 0$ on an open set around the sliding manifold, the system is not controllable and the quantity $\dot{h}(x)$ cannot be made to change its sign in such a vicinity of $\mathcal{S}$. Therefore, the transversal condition is a necessary condition for the local existence of a sliding regime.

By virtue of the fact that the quantity $-L_f h(x)/L_g h(x)$ coincides with the equivalent control we conclude that the following theorem is valid.

Theorem 3.4. *The necessary and sufficient condition for the local existence of a sliding regime over the smooth manifold* $\mathcal{S} = \{\, x \mid h(x) = 0\,\}$ *is that the equivalent control satisfies:*

$$0 < u_{eq}(x) < 1, \qquad x \in \mathcal{S}.$$

The transversal condition $L_g h(x) > 0$, or, more generally: $L_g h(x) \neq 0$, tells us that if the sliding surface coordinate function $h(x)$, is considered as a system output function, $y = h(x)$, then, this function must be, necessarily, locally relative degree equals to 1, around the value $y = 0$. Note that for $y = 0$ the *zero dynamics* entirely coincides with the *ideal sliding dynamics* given by,

$$\dot{x} = f(x) - g(x)\frac{L_f h(x)}{L_g h(x)} = f(x) + g(x)u_{eq}(x)$$

Under the assumption that the transversal condition adopts the form:

$$L_g h(x) > 0$$

in a sufficiently large open neighborhood of the sliding surface $\mathcal{S}$, the control law, that locally forces the state trajectories to reach the sliding surface and thus acquire the possibility of "crossing" this surface, is given by

$$u = \begin{cases} 1 \text{ if } h(x) < 0 \\ 0 \text{ if } h(x) > 0 \end{cases}, \quad u = \frac{1}{2}[1 - \operatorname{sign} h(x)]$$

Evidently, any incipient incursion of the state trajectory to the "other side" of the sliding manifold causes an immediate control reaction commanding the switch to change its position to the other only available value. As a consequence, the trajectory is forced to return towards the surface possibly crossing it again with the a corresponding new change in the switch control position. The resulting motion taking place around an arbitrarily small neighborhood of the sliding surface is characterized by a "zig-zag" motion whose frequency is, theoretically speaking, infinitely large and known as a sliding regime or a sliding motion.

3.2.6 Invariance Conditions for Matched Perturbations

One of the main features of sliding regimes, or sliding mode control, is their robustness with respect to certain external perturbation inputs affecting the system behavior. In this section, we explore what type of conditions should be satisfied by the perturbation for them to be automatically rejected from the description of the ideal sliding dynamics.

Consider the nonlinear additively perturbed system:

$$\dot{x} = f(x) + g(x)u + \xi(x)$$

controlled by a single switch and, moreover, let $\mathcal{S}$ be a smooth sliding surface over which we may create a local sliding regime in spite of the presence of the perturbation. The perturbation field $\xi(x)$ is assumed to be an unknown smooth function of the state x and it is assumed that its values are bounded.

Suppose then that it is possible to create a sliding regime over the sliding surface $\mathcal{S}$ in spite of the presence of the perturbation field ξ. The existence of such a sliding regime implies the existence of an equivalent control, u_{eq}, which ideally, and possibly locally, sustains the state trajectories on the smooth manifold $\mathcal{S}$. The equivalent control is, necessarily, a function of the unknown perturbation field ξ and it is given by

$$u_{eq}(x) = -\frac{L_f h(x) + L_\xi h(x)}{L_g h(x)}$$

The ideal sliding dynamics, with $x \in \mathcal{S}$, is then obtained to be,

$$\begin{aligned}\dot{x} &= f(x) - g(x)\frac{L_f h(x) + L_\xi h(x)}{L_g h(x)} + \xi(x) \\ &= \left[I - \frac{1}{L_g h(x)}g(x)\frac{\partial h}{\partial x^T}\right]f(x) \\ &\quad + \left[I - \frac{1}{L_g h(x)}g(x)\frac{\partial h}{\partial x^T}\right]\xi(x)\end{aligned}$$

The projection operator $M(x)$ over the tangent space to $\mathcal{S}$, along the span of $g(x)$, acts over the addition of the vector fields: $f(x) + \xi(x)$, in the creation of the local sliding regime on $\mathcal{S}$.

Clearly, the ideal sliding dynamics is totally independent of the influence of the perturbation vector $\xi(x)$ if and only if the vector field $\xi(x)$ is in the null space of $M(x)$, i.e.,

$$\left[I - \frac{1}{L_g h(x)}g(x)\frac{\partial h}{\partial x^T}\right]\xi(x) = 0$$

In other words, the sliding motions are invariant with respect to the perturbation if and only if the vector field $\xi(x)$ is in the span of $g(x)$, i.e., there exists a non-zero scalar function $\alpha(x)$ such that

$$\xi(x) = \alpha(x)g(x)$$

The perturbation field $\xi(x)$ is thus aligned with the control vector field $g(x)$. Such perturbations receive the name of *matched perturbations* and the condition

$$\xi \in span\ \{g\}$$

is known as the *perturbation matching condition*.

3.3 Control of the Boost Converter

We now present the control of the sliding mode of the Boost converter, with the converter model appropriately normalized, as explained in Chapter 2.

$$\dot{x}_1 = -ux_2 + 1$$
$$\dot{x}_2 = ux_1 - \frac{1}{Q}x_2$$

In the context of the previously defined notation we have,

$$f(x) = \begin{bmatrix} 1 \\ -\frac{1}{Q}x_2 \end{bmatrix}, \quad g(x) = \begin{bmatrix} -x_2 \\ x_1 \end{bmatrix}$$

The control objective is to drive the normalized average voltage x_2 to a desired equilibrium value $\bar{x}_2$. We propose first a direct control approach in which the output variable x_2 is used to synthesize a suitable sliding surface representing the desired objective.

3.3.1 Direct Control

Consider the following sliding surface coordinate function,

$$h(x) = x_2 - \bar{x}_2$$

Driving the output function $h(x)$ to zero by means of discontinuous control means that the output voltage coincides with the desired average equilibrium output voltage. Nevertheless, we wish to establish the nature and the stability of the corresponding remaining ideal sliding dynamics, or zero dynamics. In our case, we have

$$L_f h(x) = \frac{\partial h}{\partial x^T} f(x) = -\frac{1}{Q}x_2$$
$$L_g h(x) = \frac{\partial h}{\partial x^T} g(x) = x_1$$

The equivalent control is found to be

$$u_{eq}(x) = -\frac{L_f h(x)}{L_g h(x)} = \frac{1}{Q}\left(\frac{x_2}{x_1}\right)$$

The ideal sliding dynamics occurs when $u_{eq}(x)$ acts on the system as a feedback function while the system is ideally satisfying the condition $x_2 = \bar{x}_2$. We then have,

$$\dot{x}_1 = -\frac{1}{Q}\left(\frac{\bar{x}_2^2}{x_1}\right) + 1$$

It is not difficult to see that this dynamics exhibits an unstable equilibrium point. We may establish this fact via several approaches.

An Approximate Linearization Approach

This technique will provide us with the local nature of the stability around the equilibrium point of the zero dynamics, $\bar{x}_1$, corresponding with $h(x) = 0$. The incremental model (or tangent linearization model) of the normalized inductor current is given by:

$$\frac{d}{dt}x_{1\delta} = \frac{1}{Q}\left(\frac{\bar{x}_2}{\bar{x}_1}\right)^2 x_{1\delta}$$

where $x_{1\delta} = x_1 - \bar{x}_1$ and $\bar{x}_1 = \frac{\bar{x}_2^2}{Q}$.

The equilibrium point $\bar{x}_1$ is clearly unstable in view of the fact that the linearized zero dynamics exhibits a characteristic polynomial with a zero in the right half part of the complex plane.

A Lyapunov Stability Theory Approach

We rewrite the zero dynamics corresponding to $h(x) = 0$ as:

$$\frac{dx_1}{d\tau} = \frac{1}{x_1}\left(x_1 - \frac{\bar{x}_2^2}{Q}\right)$$

Consider the following Lyapunov function candidate in the x_1 variable space

$$V(x_1) = \frac{1}{2}\left(x_1 - \frac{\bar{x}_2^2}{Q}\right)^2$$

The derivative of this function, taking into account that $x_1 > 0$ is given by

$$\dot{V}(x_1) = \frac{1}{x_1}\left(x_1 - \frac{\bar{x}_2^2}{Q}\right)^2 \geq 0$$

Thus, the zero dynamics is unstable.

A Phase Diagram Approach

Figure 3.2 depicts the phase diagram of the zero dynamics associated with the motions of the controlled system on the sliding surface $\mathcal{S} = x_2 - V_d = 0$. Clearly the nature of the equilibrium point depicted on the diagram is unstable.

3.3.2 Indirect Control

The alternative is then to use, as a sliding surface coordinate function, a function that, when set to zero, reproduces the desired equilibrium value of the input inductor current, in correspondence with the desired output equilibrium voltage. We may propose:

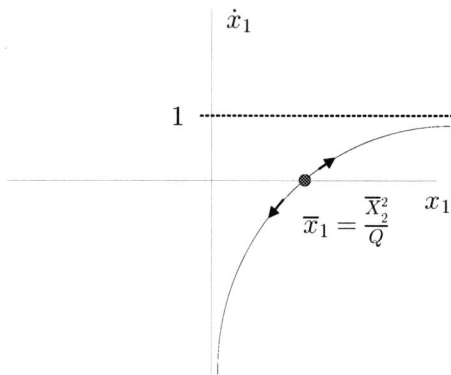

Fig. 3.2. Phase diagram of the Boost converter ideal sliding dynamics (direct control).

$$h_1(x) = x_1 - \overline{x}_1$$

To specify this function, in terms of the desired output voltage, we compute the equilibrium point of the system under ideal sliding conditions. We write the equilibrium value of the current in terms of the equilibrium value of the output voltage as,

$$\overline{x}_1 = \frac{1}{Q}\overline{x}_2^2$$

We now have,

$$L_f h(x) = 1, \qquad L_g h(x) = -x_2$$

The equivalent control is then given by

$$u_{eq}(x) = \frac{1}{x_2}$$

The ideal sliding dynamics corresponding to $h(x) = 0$, i.e., $x_1 = \overline{x}_1$, is given by:

$$\dot{x}_2 = \frac{\overline{x}_2^2}{Q x_2} - \frac{x_2}{Q}$$

It is easy to see that the unique equilibrium point of the zero dynamics is an asymptotically stable equilibrium point.

Indeed, consider the following Lyapunov function candidate, defined in the x_2 space describing the ideal sliding dynamics, or zero dynamics,

$$V(x_2) = \frac{1}{2}(x_2 - \overline{x}_2)^2 \qquad (3.7)$$

The time derivative of this candidate function is given by

$$\dot{V}(x_2) = -\frac{1}{Qx_2}(x_2 - \bar{x}_2)(x_2^2 - \bar{x}_2^2) = -\frac{1}{Qx_2}(x_2 - \bar{x}_2)^2(x_2 + \bar{x}_2)$$

Evidently, the last expression is negative definite around the equilibrium point $\bar{x}_2$, given that $x_2 > 0$ around the equilibrium. The ideal sliding dynamics exhibits an asymptotically stable equilibrium point given by the desired voltage.

According to the developed theory, the sliding surface is reachable, or accessible, by means of the following switching policy,

$$u = \begin{cases} 1 \text{ if } (x_1 - \bar{x}_1) > 0 \\ 0 \text{ if } (x_1 - \bar{x}_1) < 0 \end{cases} \quad (3.8)$$

In other words, the control policy given by:

$$u = \frac{1}{2}[1 + \text{sign}(x_1 - \bar{x}_1)]$$

yields the desired regulation with an internally stable system. In non normalized form this expression is given by:

$$u = \frac{1}{2}[1 + \text{sign}(i - I_{ref})] \quad (3.9)$$

where i is the real inductor current and $I_{ref} = \bar{i}$.

The fact that in a Boost converter, the zeroing of the output voltage $y = x_2$ entitles an unstable corresponding zero dynamics is addressed, in the control systems literature, by stating that the output voltage is a non-minimum phase output. On the contrary, the inductor current, regarded as an output of the converter, is said to be a minimum phase output.

3.3.3 Simulations

We take a Boost converter with the following parameter values

$$L = 15.91 \text{ mH}, \quad C = 50 \text{ }\mu\text{F}, \quad R = 52 \text{ }\Omega, \quad E = 12 \text{ V}$$

It is desired to regulate the output voltage to the average equilibrium value

$$\bar{v} = 24 \text{ V}$$

The equilibrium value of the corresponding average non-normalized inductor current is given by

$$\bar{i} = 0.923 \text{ A}$$

Figure 3.3 depicts the simulated state trajectories of the sliding mode controlled Boost converter.

3.3 Control of the Boost Converter   75

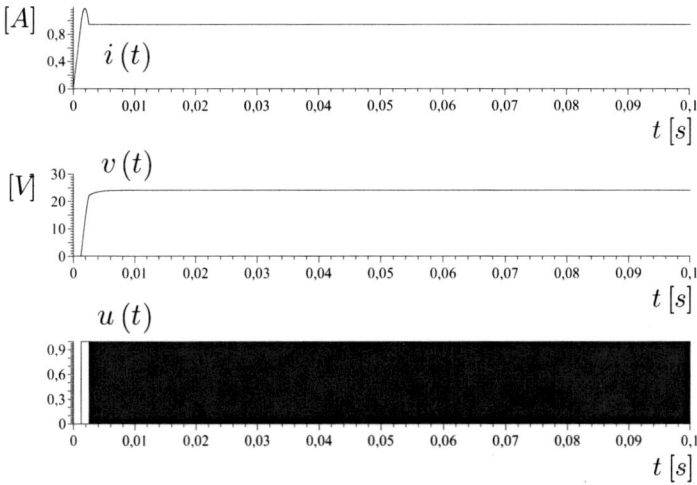

Fig. 3.3. Sliding mode controlled responses of a Boost converter.

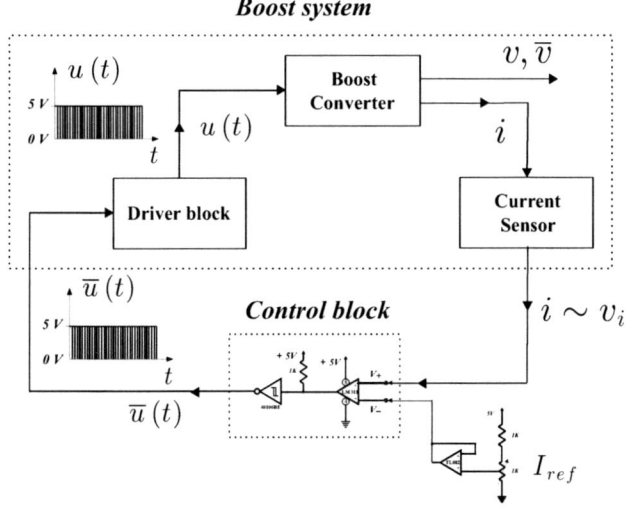

Fig. 3.4. Block diagram of the Boost converter with sliding mode controller.

3.3.4 Experimental Implementation

We now present the experimental sliding mode controller performance results, obtained on the basis of the experimental prototype of the Boost converter described in Chapter 2.

Figure 3.4 depicts a block diagram of the Boost system, along with the corresponding sliding mode control block.

- *Control block.* In this block, the designed sliding mode control strategy determined by (3.9) is implemented using analog electronics. The core of this block is a *comparator circuit*.

Comparator Circuit

A comparator circuit is a device which compares two voltages, or two currents signals, and determines which one is greater. In Figure 3.5 we shown a general op-amp configured as a comparator.

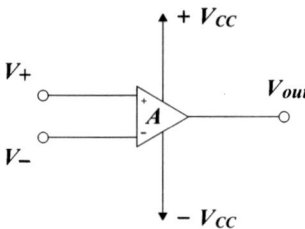

Fig. 3.5. Comparator circuit.

The operation principle for the comparator circuit is the following:

- When the input voltage (V_+) is higher than the input (V_-), the high gain of the op-amp, $\mathbf{A}$, causes the op-amp's output to be approximately saturated to the positive supply voltage (i.e., $+V_{CC}$). For this case we can represent the output voltage by the following expression:

$$V_{out} = \mathbf{A}\,(V_+ - V_-) \approx +V_{CC}$$

- On the other hand, when the input (V_-) is greater than the input (V_+), the high gain $\mathbf{A}$ of the op-amp causes that the op-amp's output be also approximately saturated to the negative supply voltage (i.e., to $-V_{CC}$). We have,

$$V_{out} = \mathbf{A}\,(V_+ - V_-) \simeq -V_{CC}$$

The above expressions are unified by the following equation:

$$V_{out} = V_{CC}\,\text{sign}(V_+ - V_-) \tag{3.10}$$

The actual control block circuit is shown in Figure 3.6. This figure shows how the comparator circuit is connected to produce the sliding mode controller circuit. The output of the comparator is an ON OFF digital signal. The input (V_+) of the comparator is connected to the inductor current signal i (which was transformed into a proportional voltage signal v_i, via the LEM HAW 15-P current sensor, see Figure 2.7) coming from the Boost system, while the

3.3 Control of the Boost Converter

comparator's input (V_-) is feeded by a reference current signal I_{ref}, which is the desired current for the Boost converter. This current is determined by the corresponding desired equilibrium value of the output voltage of the converter, denoted by $\bar{v}$.

The control block circuit was designed using the voltage comparator LM311 IC. The control circuit provides a digital output signal, with amplitudes of 0V or +5V, so it can be interfaced to a TTL logic circuit. The LM311 IC achieves this feature with its open collector output arrangement, a *pull-up resistor* $(R = 1 \text{ k}\Omega)$ and a 0 V to +5 V power supply. The resulting signal is directly interfaced with the TTL compatible NTE3087 IC, via the Hex Inverting Schmidt Trigger 74HC14 IC, which commands the gate of the Mosfet NTE2984 acting as a switch. We can see in Figure 3.4 the entire block diagram.

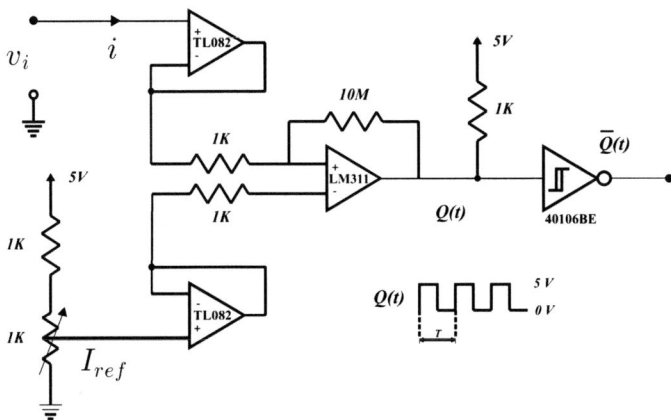

Fig. 3.6. Control circuit diagram for the sliding mode controller.

Experimental Results

Figure 3.7 illustrates the experimental performance of the indirect output voltage sliding mode regulation scheme for a typical "Boost" converter circuit, characterized by the parameters:

$$L = 15.91 \text{ mH}, \quad C = 50 \text{ }\mu\text{F}, \quad R = 52 \text{ }\Omega, \quad E = 12 \text{ V}$$

We set an actual desired output voltage of $\bar{v} = 24$ V. This corresponding to an actual current $I_{ref} = \bar{i} = 0.923$ A.

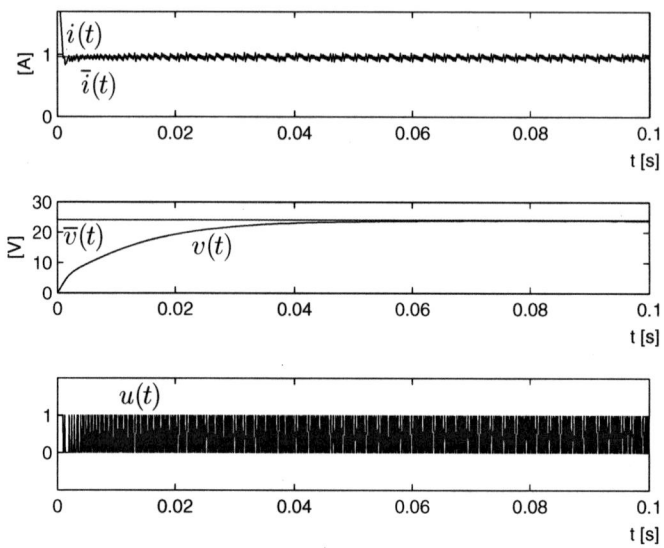

Fig. 3.7. Experimental sliding mode closed loop response of the Boost converter.

3.4 Control of the Buck-Boost Converter

The circuit shown in Figure 3.8 represents a DC-to-DC power converter controlled by a switch. This system is better known as the "Buck-Boost" converter.

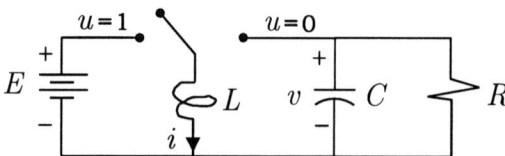

Fig. 3.8. The Buck-Boost DC-to-DC power converter.

As it was shown in Chapter 2, the normalized model of the Buck-Boost DC-to-DC power converter is given by

$$\frac{dx_1}{d\tau} = (1-u)x_2 + u$$
$$\frac{dx_2}{d\tau} = -(1-u)x_1 - \frac{1}{Q}x_2$$

In the vector field notation, introduced earlier, we specifically have,

3.4 Control of the Buck-Boost Converter

$$f(x) = \begin{bmatrix} x_2 \\ -x_1 - \frac{1}{Q}x_2 \end{bmatrix}, \quad g(x) = \begin{bmatrix} 1 - x_2 \\ x_1 \end{bmatrix}$$

The control objective is to have the normalized average voltage x_2 to converge towards the desired equilibrium value $\bar{x}_2$.

3.4.1 Direct Control

We try first with the following sliding surface coordinate function

$$h(x) = x_2 - \bar{x}_2$$

Clearly, if $h(x)$ is forced to be zero, the output capacitor voltage coincides with the desired value. As before, we must establish the stability features of the corresponding internal, or zero, dynamics of this output function.

In our case we have,

$$L_f h(x) = \frac{\partial h}{\partial x^T} f(x) = -x_1 - \frac{1}{Q} x_2$$

$$L_g h(x) = \frac{\partial h}{\partial x^T} g(x) = x_1$$

and the equivalent control is then given by

$$u_{eq}(x) = -\frac{L_f h(x)}{L_g h(x)} = 1 + \frac{1}{Q}\left(\frac{x_2}{x_1}\right)$$

The ideal sliding dynamics, corresponding with the equivalent control, taking place on the sliding surface is given by:

$$\dot{x}_1 = \frac{(1-\bar{x}_2)\bar{x}_2}{Q}\left(\frac{1}{x_1}\right) + 1$$

This dynamics has a unique equilibrium point, $\bar{x}_1 = (\bar{x}_2 - 1)\frac{\bar{x}_2}{Q}$, which is unstable. We show this fact via two approaches.

An Approximate Linearization Approach

The linear incremental model (or the tangent linearization model) of the normalized average current is given, after defining the incremental variable as: $x_{1\delta} = x_1 - \bar{x}_1$, by:

$$\frac{d}{d\tau} x_{1\delta} = \frac{Q}{(\bar{x}_2 - 1)\bar{x}_2} x_{1\delta}$$

This linear dynamics has the origin as an unstable equilibrium point due to the fact that the characteristic polynomial of the linearized dynamics exhibits a zero in the right hand side of the complex plane. This is established from the fact that $\bar{x}_2 < 0$.

A Lyapunov Stability Theory Approach

We rewrite the zero dynamics corresponding to $h(x) = 0$ as:

$$\frac{dx_1}{d\tau} = \frac{1}{x_1}\left(x_1 - (\overline{x}_2 - 1)\frac{\overline{x}_2}{Q}\right)$$

and consider the positive definite Lyapunov function in the x_1 space

$$V(x_1) = \frac{1}{2}\left(x_1 - (\overline{x}_2 - 1)\frac{\overline{x}_2}{Q}\right)^2$$

By virtue of the fact that $x_1 > 0$, the time derivative of this function is seen to be positive semi-definite. Indeed,

$$\dot{V}(x_1) = \frac{1}{x_1}\left(x_1 - (\overline{x}_2 - 1)\frac{\overline{x}_2}{Q}\right)^2 \geq 0$$

The zero dynamics is thus unstable.

3.4.2 Indirect Control

The alternative is then to use, as the sliding surface coordinate function, a function which reproduces for the variable x_1 the desired equilibrium current in correspondence with the desired average normalized voltage. We set

$$h(x) = x_1 - \overline{x}_1$$

Writing the equilibrium current in terms of the average equilibrium normalized output voltage we get,

$$\overline{x}_1 = -(1 - \overline{x}_2)\frac{\overline{x}_2}{Q}$$

We have, in our directional derivative notation,

$$L_f h(x) = x_2, \quad L_g h(x) = 1 - x_2$$

The equivalent control is therefore given by

$$u_{eq}(x) = -\frac{x_2}{1 - x_2}$$

The ideal sliding dynamics corresponding to the zero value of the output function $h(x)$, yielding $x_1 = \overline{x}_1$, is, after some algebraic manipulations, given by,

$$\dot{x}_2 = -\frac{1}{Q}\left(1 + \frac{\overline{x}_2}{x_2 - 1}\right)(x_2 - \overline{x}_2)$$

3.4 Control of the Buck-Boost Converter

Note that the factor $1+\frac{\bar{x}_2}{x_2-1}$ is strictly positive due to the fact that $x_2 < 0$ and $\bar{x}_2 < 0$. It is easy to verify that the unique equilibrium point of this zero dynamics, or ideal sliding dynamics, is asymptotically stable.

Indeed take as a Lyapunov function candidate the following function,

$$V(x_2) = \frac{1}{2}(x_2 - \bar{x}_2)^2$$

which is globally strictly positive in the x_2 space, except at $x_2 = \bar{x}_2$ where it is zero. The time derivative of this function, along the trajectories of the zero dynamics is given by

$$\dot{V}(x_2) = -\frac{1}{Q}\left(1 + \frac{\bar{x}_2}{x_2-1}\right)(x_2 - \bar{x}_2)^2$$

This quantity is zero at $x_2 = \bar{x}_2$ and strictly negative in the operating region of the converter $x_2 < 0$. The equilibrium point $x_2 = \bar{x}_2$ is hence asymptotically stable.

According to the developed theory, the sliding surface is reachable or accessible and the sliding motion is feasible due to internal stability considerations. The switching policy, which allows the state trajectory to reach the sliding surface and it is capable of sustaining the sliding motion on the sliding manifold is given by,

$$u = \begin{cases} 1 \text{ if } (x_1 - \bar{x}_1) < 0 \\ 0 \text{ if } (x_1 - \bar{x}_1) > 0 \end{cases}$$

3.4.3 Simulations

We take, as the Buck-Boost converter parameters the following ones:

$$L = 20 \text{ mH}, \quad C = 20 \text{ μF}, \quad R = 30 \text{ Ω}, \quad E = 15 \text{ V}$$

It is desired to control the output voltage to the following desired equilibrium value

$$\bar{v} = -22.5 \text{ V}$$

The corresponding equilibrium current is just found to be,

$$\bar{i} = 1.875 \text{ A}$$

Figure 3.9 depicts the simulated state trajectories of the sliding mode controlled Buck-Boost converter.

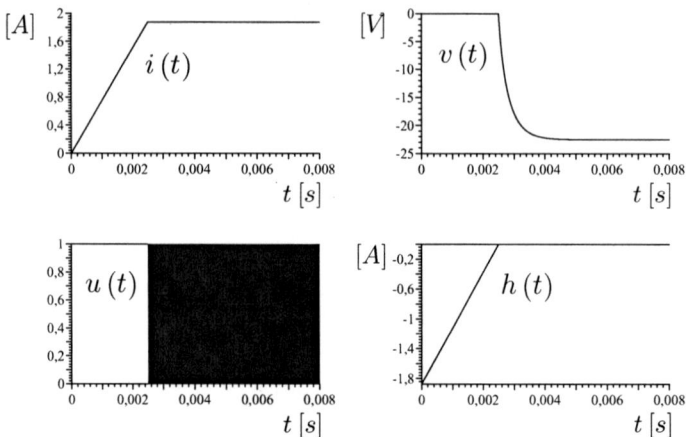

Fig. 3.9. Simulated responses of sliding mode controlled Buck-Boost converter.

3.5 Control of the Cúk Converter

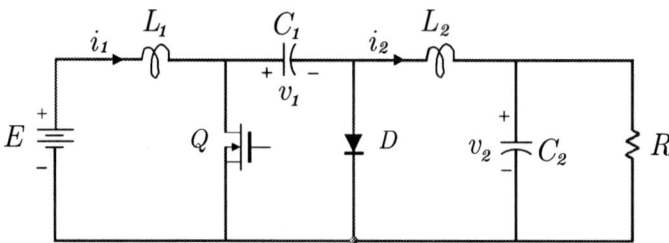

Fig. 3.10. The Cúk DC-to-DC power converter.

The Cúk converter, shown in Figure 3.10, is described by the following normalized set of equations.

$$\frac{dx_1}{d\tau} = -(1-u)x_2 + 1$$

$$\frac{dx_2}{d\tau} = (1-u)x_1 + ux_3$$

$$\alpha_1 \frac{dx_3}{d\tau} = -ux_2 - x_4$$

$$\alpha_2 \frac{dx_4}{d\tau} = x_3 - \frac{1}{Q}x_4$$

Using the notation previously established, we identify the vector fields defining the system as,

$$f(x) = \begin{bmatrix} 1 - x_2 \\ x_1 \\ -\frac{1}{\alpha_1} x_4 \\ \frac{1}{\alpha_2}\left(x_3 - \frac{x_4}{Q}\right) \end{bmatrix}, \quad g(x) = \begin{bmatrix} x_2 \\ -x_1 + x_3 \\ -\frac{1}{\alpha_1} x_2 \\ 0 \end{bmatrix}$$

Similarly as in the previous examples, the control objective is to drive the normalized output voltage, represented in this case by the state variable x_4, towards the desired equilibrium value $\bar{x}_4 = V_d$.

3.5.1 Direct Control

The following sliding surface coordinate function:

$$h(x) = x_4 - \bar{x}_4 = x_4 - V_d$$

appears to be natural and simple enough. As already discussed, forcing the sliding surface coordinate function $h(x)$ to be zero means that the output voltage coincides with the desired equilibrium voltage V_d.

In this case, we have

$$L_f h(x) = \frac{\partial h(x)}{\partial x^T} f(x) = \frac{1}{\alpha_2}\left(x_3 - \frac{x_4}{Q}\right)$$

$$L_g h(x) = \frac{\partial h(x)}{\partial x^T} g(x) = 0$$

The proposed sliding surface coordinate function $h(x)$ has relative degree greater than one and, hence, the equivalent control cannot be defined. We must therefore propose a sliding surface coordinate function of the form,

$$h(x) = \dot{x}_4 + \lambda(x_4 - V_d) = \frac{1}{\alpha_2}\left(x_3 - \frac{x_4}{Q}\right) + \lambda(x_4 - V_d)$$

where λ is a strictly positive constant. This function is relative degree one since

$$L_g h(x) = \frac{\partial h(x)}{\partial x^T} g(x) = -\frac{1}{\alpha_1 \alpha_2} x_2 \neq 0$$

If $h(x) = 0$ the dynamics corresponding to x_4 produces trajectories with exponential convergence towards the desired equilibrium point $x_4 = V_d$. We now evaluate the zero dynamics corresponding to this steady state behavior.

The equivalent control corresponding to the ideal sliding dynamics is given by $u_{eq} = -V_d/x_2$. From the last equation it follows that the ideal behavior of the x_3 variable corresponds itself to a constant value, i.e., $x_3 = \bar{x}_3 = \frac{V_d}{Q}$. We therefore have that the zero dynamics is characterized by

$$\frac{dx_1}{d\tau} = -(x_2 + V_d) + 1$$

$$\frac{dx_2}{d\tau} = \left(1 + \frac{V_d}{x_2}\right) x_1 - \frac{V_d}{x_2}\bar{x}_3$$

The linearized zero dynamics around the equilibrium point

$$(\bar{x}_1, \bar{x}_2) = \left(V_d^2/Q, 1 - V_d\right)$$

described by the incremental variables

$$x_{1\delta} = x_1 - \frac{V_d^2}{Q}, \qquad x_{2\delta} = x_2 - (1 - V_d)$$

is given by

$$\begin{bmatrix} \dot{x}_{1\delta} \\ \dot{x}_{2\delta} \end{bmatrix} = \begin{bmatrix} 0 & -1 \\ \frac{1}{(1-V_d)} & \frac{V_d^2}{Q(1-V_d)} \end{bmatrix} \begin{bmatrix} x_{1\delta} \\ x_{2\delta} \end{bmatrix}$$

whose characteristic polynomial is just obtained as: $s^2 - \frac{1}{Q}\frac{V_d^2}{(1-V_d)}s + \frac{1}{(1-V_d)}$, which clearly has at least one unstable root in the complex plane.

The proposed sliding surface is therefore not viable since the zero dynamics corresponding to its zero level set is unstable. i.e., the system, along with the proposed sliding surface coordinate function viewed as an output, exhibits a non-minimum phase output.

3.5.2 Indirect Control

The alternative is then to use, as a sliding surface coordinate function, a function involving the a desired average equilibrium behavior for the input inductor current. We set then,

$$h(x) = x_1 - \bar{x}_1 = x_1 - \frac{V_d^2}{Q}$$

In this case we have, corresponding to this value of $h(x)$, the following quantities:

$$L_f h(x) = \frac{\partial h(x)}{\partial x^T} f(x) = 1 - x_2$$

$$L_g h(x) = \frac{\partial h(x)}{\partial x^T} g(x) = x_2$$

The equivalent control is then well defined and given by

$$u_{eq}(x) = 1 - \frac{1}{x_2}$$

which under, non-saturated operating conditions, satisfies

$$0 < u_{eq}(x) < 1$$

and, hence $x_2 \in (1, \infty)$.

3.5 Control of the Cúk Converter

The ideal sliding dynamics, or the zero dynamics, corresponding to $x_1 = \frac{V_d^2}{Q}$, is given by:

$$\frac{dx_2}{d\tau} = \bar{x}_1 \frac{1}{x_2} + \left(1 - \frac{1}{x_2}\right) x_3$$

$$\alpha_1 \frac{dx_3}{d\tau} = 1 - x_2 - x_4$$

$$\alpha_2 \frac{dx_4}{d\tau} = x_3 - \frac{1}{Q} x_4 \qquad (3.11)$$

The equilibrium point of the ideal sliding dynamics is clearly given by

$$\bar{x}_2 = 1 - V_d, \qquad \bar{x}_3 = \frac{V_d}{Q}, \qquad \bar{x}_4 = V_d$$

In order to assess the stability of the zero dynamics (3.11), we propose the following candidate Lyapunov function

$$V(x_2, x_3, x_4) = \frac{1}{2}\left[(x_2 - \bar{x}_2)^2 + \alpha_1 (x_3 - \bar{x}_3)^2 + \alpha_2 (x_4 - \bar{x}_4)^2\right] + \gamma$$
$$+ \int_0^\tau \frac{[x_2(\sigma) - \bar{x}_2][x_3(\sigma) - \bar{x}_3]}{x_2(\sigma)} d\sigma$$

with γ being a strictly positive constant parameter, which is assumed to be sufficiently large so that V is strictly positive, and $x_2 \in (1, \infty)$. The time derivative of V, along the solution of the system of differential equation yields, after quite straightforward but tedious algebraic manipulations, the following expression:

$$\dot{V}(x_2, x_3, x_4) = -\frac{1}{Q}(x_4 - \bar{x}_4)^2 + \bar{x}_3 \frac{(x_2 - \bar{x}_2)^2}{x_2} \leq 0$$

The inequality follows from the fact that, both, Q and x_2 are strictly positive quantities, while that $\bar{x}_3$ is a strictly negative quantity. By LaSalle's theorem, the trajectories: $x_4 = \bar{x}_4$ and $x_2 = \bar{x}_2$, which produce the relation $\dot{V} = 0$, should not be incompatible with the equilibrium trajectory, for the zero dynamics system to be asymptotically stable to its equilibrium point.

From the third equation of the zero dynamics system (3.11), we have that: $\bar{x}_3 = \frac{\bar{x}_4}{Q}$, which is true at the equilibrium point. From the second equation in (3.11), we have that $x_2 = 1 - \bar{x}_4$, which implies $x_2 = 1 - \bar{x}_4 = \bar{x}_2$. Finally, from the first equation in (3.11), we obtain: $\bar{x}_1 = (1 - \bar{x}_2) \bar{x}_3 = \bar{x}_3 \bar{x}_4 = \frac{\bar{x}_4^2}{Q}$. Hence, the only trajectory for which $\dot{V}(x_2, x_3, x_4) = 0$ is the one represented by the equilibrium point itself. Also, since $\dot{V} \leq 0$ outside the equilibrium point, then V is bounded and, in particular, the integral quantity found in the expression for V is bounded. Hence, a constant $\gamma > 0$ exists which bounds this integral quantity.

It is then clear that the average normalized input inductor current x_1, taken as a system output, is a minimum phase output. We, thus, attempt an indirect regulation of the converter average normalized output voltage, x_4, towards the desired value $\bar{x}_4 = V_d$. This is accomplished by primarily regulating the inductor current x_1 towards its corresponding average equilibrium value, $\bar{x}_1 = \frac{V_d^2}{Q}$. The demonstrated asymptotic stability of the zero dynamics corresponding to x_1 takes care of the complete internal stabilization of the controlled system. To achieve the stabilization goal, we propose a sliding mode control strategy.

According to the developed theory, the sliding surface is reachable, or accessible, by means of the following switching policy:

$$u = \begin{cases} 1 \text{ if } (x_1 - \bar{x}_1) < 0 \\ 0 \text{ if } (x_1 - \bar{x}_1) > 0 \end{cases}$$

with an asymptotically stable ideal sliding dynamics.

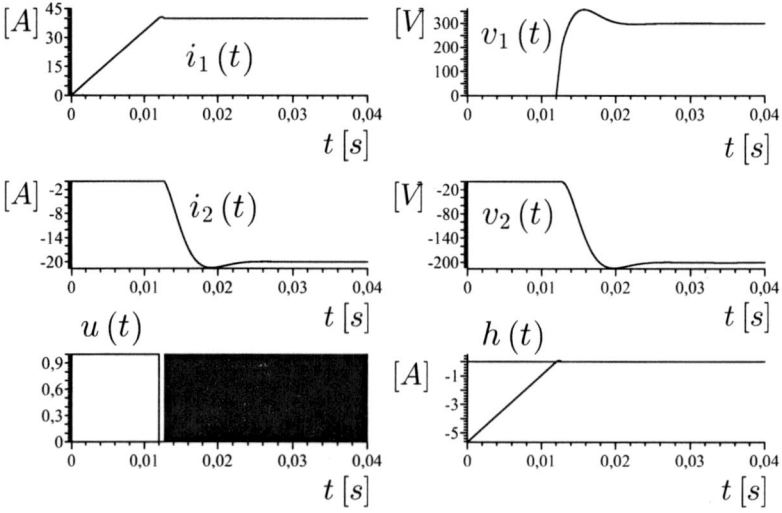

Fig. 3.11. Simulated responses of sliding mode controlled Cúk converter.

3.5.3 Simulations

We take a Cúk converter with the following parameter values

$$L_1 = 30 \text{ mH}, \quad C_1 = 150 \text{ }\mu\text{F}, \quad L_2 = 30 \text{ mH}, \quad C_2 = 50 \text{ }\mu\text{F},$$

$$R = 10 \text{ }\Omega, \quad E = 100 \text{ V}$$

It is desired to regulate the output voltage to the average equilibrium value $\bar{v}_2 = -200$ V. The actual equilibrium values of the corresponding currents and the internal capacitor voltage are given by

$$\bar{i}_1 = 40 \text{ A}, \qquad \bar{v}_1 = 300 \text{ V}, \qquad \bar{i}_2 = -20 \text{ A}$$

Figure 3.11 shows the simulated behavior of the sliding mode controlled Cúk converter.

3.6 Control of the Zeta Converter

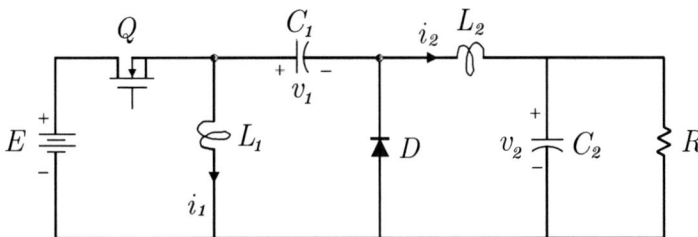

Fig. 3.12. The Zeta converter circuit.

The Zeta converter, shown in Figure 3.12, is described by the following normalized set of differential equations, including switched control inputs.

$$\frac{dx_1}{d\tau} = -(1-u)x_2 + u$$

$$\frac{dx_2}{d\tau} = (1-u)x_1 - ux_3$$

$$\alpha_1 \frac{dx_3}{d\tau} = ux_2 - x_4 + u$$

$$\alpha_2 \frac{dx_4}{d\tau} = x_3 - \frac{1}{Q}x_4$$

Using the notation previously established, we identify the vector fields defining the system,

$$f(x) = \begin{bmatrix} -x_2 \\ x_1 \\ -\frac{1}{\alpha_1}x_4 \\ \frac{1}{\alpha_2}\left(x_3 - \frac{x_4}{Q}\right) \end{bmatrix}, \qquad g(x) = \begin{bmatrix} 1 + x_2 \\ -x_1 - x_3 \\ \frac{1}{\alpha_1}(1 + x_2) \\ 0 \end{bmatrix}$$

The control objective is to drive the normalized output voltage x_4 towards the desired equilibrium value $\bar{x}_4 = V_d$.

3.6.1 Direct Control

We propose a sliding surface coordinate function of the form,

$$h(x) = \dot{x}_4 + \lambda(x_4 - V_d) = \frac{1}{\alpha_2}\left(x_3 - \frac{x_4}{Q}\right) + \lambda(x_4 - V_d)$$

where λ is a strictly positive constant. This function is clearly relative degree one since

$$L_g h(x) = \frac{\partial h(x)}{\partial x^T} g(x) = \frac{1}{\alpha_1 \alpha_2}(1 + x_2) \neq 0$$

If $h(x) = 0$ the dynamics corresponding to x_4 produces trajectories with exponential convergence towards the desired equilibrium point $x_4 = V_d$. We now evaluate the zero dynamics corresponding to this steady state behavior.

The equivalent control corresponding to the ideal sliding dynamics is given by $u_{eq} = V_d/(1 + x_2)$. It follows that the ideal behavior of the x_3 variable corresponds itself to a constant value, i.e., $x_3 = \bar{x}_3 = \frac{V_d}{Q}$. We therefore have that the zero dynamics is characterized by

$$\frac{dx_1}{d\tau} = -x_2 + V_d$$

$$\frac{dx_2}{d\tau} = x_1 - \frac{V_d}{(1+x_2)}(x_1 + \bar{x}_3)$$

The linearized zero dynamics around the equilibrium point $(\bar{x}_1, \bar{x}_2) = (V_d^2/Q, V_d)$, described by the incremental variables $x_{1\delta} = x_1 - \frac{V_d^2}{Q}, x_{2\delta} = x_2 - V_d$, is given by

$$\dot{x}_\delta = \begin{bmatrix} 0 & -1 \\ \frac{1}{1+V_d} & \frac{V_d^2}{Q(1+V_d)} \end{bmatrix} x_\delta$$

whose characteristic polynomial is just obtained as: $s^2 - \frac{1}{Q}\frac{V_d^2}{1+V_d}s + \frac{1}{1+V_d}$. Clearly this polynomial has at least one unstable root in the complex plane.

The sliding surface is therefore not viable since the zero dynamics corresponding to the zero level set of the sliding surface coordinate function is unstable. i.e., the system along with the proposed sliding surface coordinate function, viewed as an output, exhibits a non-minimum phase output.

3.6.2 Indirect Control

The alternative is then to use, as a sliding surface coordinate function, a function involving the a desired average equilibrium behavior for the input inductor current. We set then,

$$h(x) = x_1 - \bar{x}_1 = x_1 - \frac{V_d^2}{Q}$$

3.6 Control of the Zeta Converter

In this case we have, corresponding to this value of $h(x)$, the following directional derivatives:

$$L_f h(x) = \frac{\partial h(x)}{\partial x^T} f(x) = -x_2$$

$$L_g h(x) = \frac{\partial h(x)}{\partial x^T} g(x) = 1 + x_2$$

The equivalent control is then given by

$$u_{eq}(x) = \frac{x_2}{1 + x_2}$$

which under, non-saturated operating conditions, satisfies

$$0 < u_{eq}(x) < 1 \Rightarrow x_2 > 0 \quad (3.12)$$

The ideal sliding dynamics, or the zero dynamics, corresponding to $x_1 = \frac{V_d^2}{Q}$, is given by:

$$\frac{dx_2}{d\tau} = \left(\frac{1}{1+x_2}\right)(\overline{x}_1 - x_2 x_3)$$

$$\alpha_1 \frac{dx_3}{d\tau} = x_2 - x_4$$

$$\alpha_2 \frac{dx_4}{d\tau} = x_3 - \frac{1}{Q} x_4 \quad (3.13)$$

The equilibrium point of the ideal sliding dynamics is clearly given by

$$\overline{x}_2 = V_d, \quad \overline{x}_3 = \frac{V_d}{Q}, \quad \overline{x}_4 = V_d$$

In order to assess the stability of the zero dynamics (3.13), we propose the following candidate Lyapunov function

$$V(x_2, x_3, x_4) = \frac{1}{2}\left[(x_2 - \overline{x}_2)^2 + \alpha_1(x_3 - \overline{x}_3)^2 + \alpha_2(x_4 - \overline{x}_4)^2\right] + \gamma$$
$$- \int_0^\tau \frac{[x_2(\sigma) - \overline{x}_2][x_3(\sigma) - \overline{x}_3]}{[1 + x_2(\sigma)]} d\sigma$$

with γ being a strictly positive constant parameter, assumed to be sufficiently large so that V is strictly positive, with $x_2 > 0$ by (3.12). The time derivative of V, along the solution of the system of differential equation yields, after quite straightforward but tedious algebraic manipulations, the following expression:

$$\dot{V}(x_2, x_3, x_4) = -\frac{1}{Q}(x_4 - \overline{x}_4)^2 - \overline{x}_3 \frac{(x_2 - \overline{x}_2)^2}{(1 + x_2)} \leq 0$$

The inequality follows from the fact that, Q, x_2 and $\overline{x}_3$ are strictly positive quantities. By LaSalle's theorem, the trajectories: $x_4 = \overline{x}_4$ and $x_2 = \overline{x}_2$, which make $\dot{V} = 0$, should not be incompatible with the equilibrium trajectory, for the zero dynamics system to be asymptotically stable to that equilibrium. From the , third equation of the zero dynamics set of equations (3.13) we have that: $\overline{x}_3 = \frac{\overline{x}_4}{Q}$, which is true at the equilibrium point. From the second equation in (3.13), we have that $x_2 = \overline{x}_4$, which implies $x_2 = \overline{x}_4 = \overline{x}_2$. Finally, from the first equation in (3.13), $\overline{x}_1 = \overline{x}_2 \overline{x}_3 = \frac{\overline{x}_4^2}{Q}$. Hence, the only trajectory for which $\dot{V}(x_2, x_3, x_4) = 0$ is the one represented by the equilibrium point itself. Also, since $\dot{V} \leq 0$ outside the equilibrium point, then V is bounded and, in particular, the integral quantity found in the expression for V is bounded. Hence, a constant $\gamma > 0$ exists which bounds this integral quantity.

It is then clear that the average normalized input inductor current x_1, taken as a system output, is a minimum phase output. We, thus, attempt an indirect regulation of the converter average normalized output voltage, x_4, towards the desired value $\overline{x}_4 = V_d$. This is accomplished by primarily regulating the inductor current x_1 towards its corresponding average equilibrium value, $\overline{x}_1 = \frac{V_d^2}{Q}$. The demonstrated asymptotic stability of the zero dynamics corresponding to x_1 takes care of the complete internal stabilization of the controlled system. To achieve the stabilization goal, we propose an sliding mode control.

Thus, according to the developed theory, the sliding surface is reachable, or accessible, by means of the following switching policy:

$$u = \begin{cases} 1 \text{ if } (x_1 - \overline{x}_1) < 0 \\ 0 \text{ if } (x_1 - \overline{x}_1) > 0 \end{cases}$$

The sliding motions achieved by this control law yield an asymptotically stable ideal sliding dynamics.

3.6.3 Simulations

Taking as the converter parameters the following ones,

$$L_1 = 600 \ \mu H, \quad C_1 = 15 \ \mu F, \quad L_2 = 1.3 \ mH, \quad C_2 = 12 \ \mu F,$$

$$R = 25 \ \Omega, \quad E = 120 \ V$$

It is desired to control the average output voltage to the following desired equilibrium value

$$\overline{v}_2 = 60 \ V$$

The corresponding steady state equilibrium for the rest of the state variables are given by:

$$\overline{i}_1 = 1.2 \ A, \quad \overline{v}_1 = 60 \ V, \quad \overline{i}_2 = 2.4 \ A$$

Figure 3.13 shows the simulated behavior of the sliding mode controlled Zeta converter.

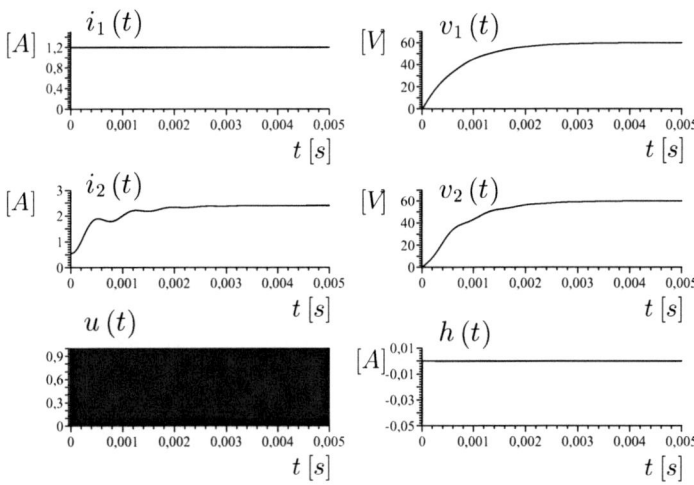

Fig. 3.13. Simulated sliding mode controlled responses for the Zeta converter.

3.7 Control of the Quadratic Buck Converter

Let us now consider the quadratic Buck converter, shown in Figure 3.14. The normalized differential equations describing this system are given by:

$$\frac{dx_1}{d\tau} = -x_2 + u$$

$$\frac{dx_2}{d\tau} = x_1 - ux_3$$

$$\alpha_1 \frac{dx_3}{d\tau} = ux_2 - x_4$$

$$\alpha_2 \frac{dx_4}{d\tau} = x_3 - \frac{1}{Q}x_4$$

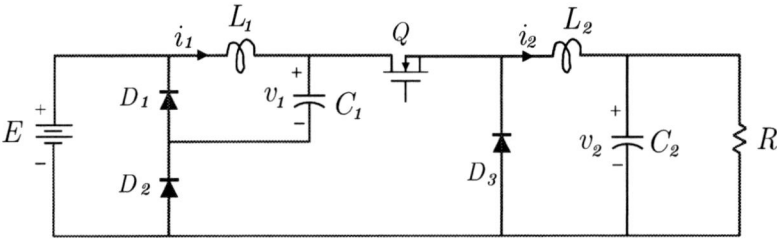

Fig. 3.14. Switch regulated DC-to-DC quadratic Buck power converter.

The equilibrium point of the quadratic Buck converter, parameterized in terms of the desired output voltage $\bar{x}_4 = V_d$, is computed to be:

$$\bar{x}_1 = \frac{(V_d)^{\frac{3}{2}}}{Q}, \quad \bar{x}_2 = \sqrt{V_d}, \quad \bar{x}_3 = \frac{V_d}{Q}, \quad U = \sqrt{V_d} \qquad (3.14)$$

Using the notation previously established, we identify the vector fields defining the system,

$$f(x) = \begin{bmatrix} -x_2 \\ x_1 \\ -\frac{1}{\alpha_1} x_4 \\ \frac{1}{\alpha_2}\left(x_3 - \frac{x_4}{Q}\right) \end{bmatrix}, \quad g(x) = \begin{bmatrix} 1 \\ -x_3 \\ \frac{1}{\alpha_1} x_2 \\ 0 \end{bmatrix}$$

3.7.1 Direct Control

The control objective is to drive the normalized output voltage x_4 towards the desired equilibrium value $\bar{x}_4 = V_d$.

We first try with the following sliding surface coordinate function:

$$h(x) = x_4 - \bar{x}_4 = x_4 - V_d$$

Forcing the sliding surface coordinate function $h(x)$ to be zero means that the output voltage coincides with the desired equilibrium voltage.

In this case we have

$$L_f h(x) = \frac{\partial h(x)}{\partial x^T} f(x) = \frac{1}{\alpha_2}\left(x_3 - \frac{x_4}{Q}\right)$$

$$L_g h(x) = \frac{\partial h(x)}{\partial x^T} g(x) = 0$$

From the preceding we conclude that the proposed sliding surface coordinate function has relative degree equals greater than one and hence we must propose a sliding surface coordinate function of the form,

$$h(x) = \dot{x}_4 + \lambda(x_4 - V_d) = \frac{1}{\alpha_2}\left(x_3 - \frac{x_4}{Q}\right) + \lambda(x_4 - V_d)$$

where λ is a strictly positive constant. This function is relative degree one since

$$L_g h(x) = \frac{\partial h(x)}{\partial x^T} g(x) = \frac{1}{\alpha_1 \alpha_2} x_2 \neq 0$$

If $h(x) = 0$ the dynamics corresponding to x_4 produces trajectories with exponential convergence towards the desired equilibrium point $x_4 = V_d$. We now evaluate the zero dynamics corresponding to this steady state behavior.

The equivalent control corresponding to the ideal sliding dynamics is given by $u_{eq} = \frac{V_d}{x_2}$. From the last equation it follows that the ideal behavior of the x_3 variable corresponds itself to a constant value, i.e., $x_3 = \overline{x}_3 = \frac{V_d}{Q}$. We therefore have that the zero dynamics is characterized by

$$\frac{dx_1}{d\tau} = -x_2 + \frac{V_d}{x_2}$$

$$\frac{dx_2}{d\tau} = x_1 - \frac{V_d}{x_2}\overline{x}_3$$

The linearized zero dynamics around the equilibrium point:

$$\overline{x}_1 = (V_d)^{\frac{3}{2}}/Q, \qquad \overline{x}_2 = \sqrt{V_d}$$

described by the incremental variables

$$x_{1\delta} = x_1 - \frac{(V_d)^{\frac{3}{2}}}{Q}, \qquad x_{2\delta} = x_2 - \sqrt{V_d}$$

is given by

$$\dot{x}_{1\delta} = -2x_{2\delta}$$
$$\dot{x}_{2\delta} = x_{1\delta} + \frac{V_d}{Q}x_{2\delta}$$

whose characteristic polynomial is just obtained as: $s^2 - \frac{V_d}{Q}s + 2$, which, clearly, has at least one unstable root in the complex plane.

The sliding surface is therefore not viable since the zero dynamics corresponding to the zero level set of the sliding surface coordinate function is unstable. i.e., the system along with the proposed sliding surface coordinate function, viewed as an output, is a non-minimum phase system.

3.7.2 Indirect Control

The alternative is then to use as a sliding surface coordinate function one involving the a desired average equilibrium behavior for the input inductor current. We set then,

$$h(x) = x_1 - \overline{x}_1 = x_1 - \frac{(V_d)^{\frac{3}{2}}}{Q}$$

In this case we have, corresponding to this value of $h(x)$, the following quantities:

$$L_f h(x) = \frac{\partial h(x)}{\partial x^T} f(x) = -x_2$$

$$L_g h(x) = \frac{\partial h(x)}{\partial x^T} g(x) = 1$$

3 Sliding Mode Control

The equivalent control is then given by

$$u_{eq}(x) = x_2$$

which, under non-saturated operating conditions, satisfies

$$0 < x_2 < 1$$

The ideal sliding dynamics, or the zero dynamics, corresponding to $x_1 = \bar{x}_1 = \frac{(V_d)^{\frac{3}{2}}}{Q}$, is given by:

$$\frac{dx_2}{d\tau} = \bar{x}_1 - x_2 x_3$$
$$\alpha_1 \frac{dx_3}{d\tau} = x_2^2 - x_4$$
$$\alpha_2 \frac{dx_4}{d\tau} = x_3 - \frac{1}{Q} x_4 \qquad (3.15)$$

The equilibrium point of the ideal sliding dynamics is clearly given by

$$\bar{x}_2 = \sqrt{V_d}, \qquad \bar{x}_3 = \frac{V_d}{Q}, \qquad \bar{x}_4 = V_d$$

In order to assess the stability of the zero dynamics (3.15), we propose the following candidate Lyapunov function

$$V(x_2, x_3, x_4) = \frac{1}{2}\left[(x_2 - \bar{x}_2)^2 + \alpha_1(x_3 - \bar{x}_3)^2 + \alpha_2(x_4 - \bar{x}_4)^2\right] + \gamma$$
$$- \bar{x}_2 \int_0^\tau [x_2(\sigma) - \bar{x}_2][x_3(\sigma) - \bar{x}_3]\, d\sigma$$

with γ being a strictly positive constant parameter, which is assumed to be sufficiently large so that V is strictly positive, and $\bar{x}_2 = U \in (0,1)$. The time derivative of V, along the solution of the system of differential equation yields, after quite straightforward but tedious algebraic manipulations, the following expression:

$$\dot{V}(x_2, x_3, x_4) = -\frac{1}{Q}(x_4 - \bar{x}_4)^2 - \bar{x}_3(x_2 - \bar{x}_2)^2 \leq 0$$

The inequality follows from the fact that, both, Q and $\bar{x}_3$ are strictly positive quantities. By LaSalle's theorem, the trajectories: $x_4 = \bar{x}_4$ and $x_2 = \bar{x}_2$, which force the relation $\dot{V} = 0$, should not be incompatible with the equilibrium trajectory (3.14), for the zero dynamics system to be asymptotically stable to that equilibrium. From the zero dynamics equations (3.15), third equation, we have that: $\bar{x}_3 = \frac{\bar{x}_4}{Q}$, which is true at the equilibrium point. From the second equation in (3.15), we have that $x_2^2 = \bar{x}_4$, which implies $x_2 = \sqrt{\bar{x}_4} = \bar{x}_2$.

Finally, from the first equation in (3.15), $\overline{x}_2\overline{x}_3 = (\overline{x}_4)^{\frac{3}{2}}/Q = \overline{x}_1$. Hence, the only trajectory for which $\dot{V}(x_2, x_3, x_4) = 0$ is the one represented by the equilibrium point itself. Also, since $\dot{V} \leq 0$ outside the equilibrium point, then V is bounded and, in particular, the integral quantity found in the expression for V is bounded. Hence, a constant $\gamma > 0$ exists which bounds this integral quantity.

It is then clear that the average normalized input inductor current x_1, taken as a system output, is a minimum phase output. We, thus, attempt an indirect regulation of the converter average normalized output voltage, x_4, towards the desired value $\overline{x}_4 = V_d$. This is accomplished by primarily regulating the inductor current x_1 towards its corresponding average equilibrium value, $\overline{x}_1 = (V_d)^{\frac{3}{2}}/Q$. The demonstrated asymptotic stability of the zero dynamics corresponding to x_1 takes care of the complete internal stabilization of the controlled system. To achieve the stabilization goal, we propose an sliding mode control.

Thus, according to the developed theory, the sliding surface is reachable, or accessible, by means of the following switching policy:

$$u = \begin{cases} 1 \text{ if } (x_1 - \overline{x}_1) < 0 \\ 0 \text{ if } (x_1 - \overline{x}_1) > 0 \end{cases}$$

3.7.3 Simulations

We consider a quadratic Buck converter with the following parameters:

$$L_1 = 600 \ \mu H, \quad C_1 = 10 \ \mu F, \quad L_2 = 600 \ \mu H, \quad C_2 = 10 \ \mu F,$$

$$R = 40 \ \Omega, \quad E = 100 \text{ V}$$

It is desired to regulate the output capacitor voltage to the equilibrium value

$$\overline{v}_2 = 25 \text{ V}$$

The equilibrium values of the currents and the internal capacitor voltage are given by

$$\overline{i}_1 = 0.3125 \text{ A}, \quad \overline{v}_1 = 50 \text{ V}, \quad \overline{i}_2 = 0.625 \text{ A}$$

Figure 3.15 shows the behavior of the sliding mode controlled quadratic Buck converter.

3.8 Multi-variable Case

The general description of systems controlled by multiple independent switches corresponds, within the framework of the state space representation, to the following form:

$$\dot{x} = f(x) + G(x)u, \quad y = h(x)$$

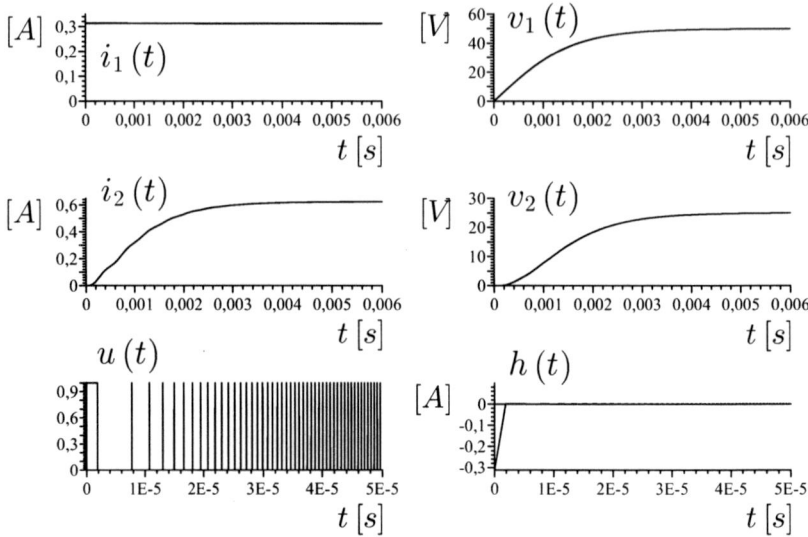

Fig. 3.15. Sliding mode controlled responses of a quadratic Buck converter.

where $x \in R^n$, $u \in \{0,1\}^m$, and $y \in R^m$. The function $f(x)$ is a smooth vector field defined over the tangent space to R^n and usually addressed as the drift vector field. $G(x)$ is a matrix whose entries are smooth functions of the state x of the system and its dimensions are $n \times m$, i.e., n rows and m columns. The columns of $G(x)$, denoted by means of $g_i(x), i = 1, 2, \ldots, m$ also represent smooth vector fields. The matrix $G(x)$ is called the input matrix. The output function $h(x)$ is a smooth map taking values in R^m. We refer to the point x as the state vector of the system, while u is the input vector and y is the output vector.

Example 3.5. Figure 3.16 represents a MIMO DC-to-DC power converter controlled by two switches and known as the Boost-Boost converter. This converter consists of two stages, each one independently controlled by means of a switch position function,

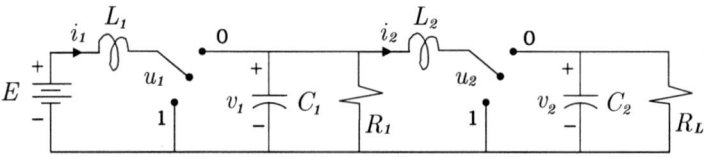

Fig. 3.16. Ideal switch realization of the Boost-Boost converter.

The differential equations describing the system are the following:

$$L_1 \frac{di_1}{dt} = -(1 - u_1) v_1 + E$$

$$C_1 \frac{dv_1}{dt} = (1 - u_1) i_1 - \frac{v_1}{R_1} - i_2$$

$$L_2 \frac{di_2}{dt} = v_1 - (1 - u_2) v_2$$

$$C_2 \frac{dv_2}{dt} = (1 - u_2) i_2 - \frac{v_2}{R_L} \quad (3.16)$$

where i_1 is the input current, v_1 is the output voltage of the first stage, i_2 is the input current to the second stage and v_2 represents the output voltage of the second stage.

In matrix terms, the mathematical description of the system is given by:

$$\frac{d}{dt} \begin{bmatrix} i_1 \\ v_1 \\ i_2 \\ v_2 \end{bmatrix} = \begin{bmatrix} 0 & -\frac{1}{L_1} & 0 & 0 \\ \frac{1}{C_1} & -\frac{1}{R_1 C_1} & -\frac{1}{C_1} & 0 \\ 0 & \frac{1}{L_2} & 0 & -\frac{1}{L_2} \\ 0 & 0 & \frac{1}{C_2} & -\frac{1}{R_L C_2} \end{bmatrix} \begin{bmatrix} i_1 \\ v_1 \\ i_2 \\ v_2 \end{bmatrix} + \begin{bmatrix} \frac{v_1}{L_1} & 0 \\ -\frac{i_1}{C_1} & 0 \\ 0 & \frac{v_2}{L_2} \\ 0 & -\frac{i_2}{C_2} \end{bmatrix} \begin{bmatrix} u_1 \\ u_2 \end{bmatrix} + \begin{bmatrix} \frac{E}{L_1} \\ 0 \\ 0 \\ 0 \end{bmatrix}$$

Here, evidently, letting $x = [i_1 \; v_1 \; i_2 \; v_2]^T$, yields the following expressions for the drift vector field $f(x)$ and the input matrix $G(x)$,

$$f(x) = \begin{bmatrix} 0 & -\frac{1}{L_1} & 0 & 0 \\ \frac{1}{C_1} & -\frac{1}{R_1 C_1} & -\frac{1}{C_1} & 0 \\ 0 & \frac{1}{L_2} & 0 & -\frac{1}{L_2} \\ 0 & 0 & \frac{1}{C_2} & -\frac{1}{R_L C_2} \end{bmatrix} x + \begin{bmatrix} \frac{E}{L_1} \\ 0 \\ 0 \\ 0 \end{bmatrix}$$

and

$$G(x) = \begin{bmatrix} \frac{x_2}{L_1} & 0 \\ -\frac{x_1}{C_1} & 0 \\ 0 & \frac{x_4}{L_2} \\ 0 & -\frac{x_3}{C_2} \end{bmatrix}, \quad u = \begin{bmatrix} u_1 \\ u_2 \end{bmatrix}$$

3.8.1 Sliding Surfaces

In the context of n dimensional controlled systems regulated by m independent switches and where m sliding surface coordinate functions are defined as system outputs, a sliding surface is represented by the simultaneous satisfaction of m smooth, independent, algebraic restrictions summarized in the equation: $h(x) = 0$.

The fundamental assumption is the following: The simultaneous verification of m restrictions, $h(x) = 0$, produces, ideally, a desired behavior for the system's state trajectory, $x(t)$, of the controlled system. These restrictions are represented by a smooth variety, $\mathcal{S}$, locally of dimension $n - m$. The condition $x \in \mathcal{S}$, happens thanks to the control actions, which in turn are restricted by: $u \in \{0, 1\}^m$, i.e., $u_i \in \{0, 1\}$ for $i = 1, 2, \cdots, m$.

The smooth algebraic restrictions, $h_i(x) = 0, i = 1, \ldots, m$ define m smooth manifolds, or surfaces, in R^n, each one of dimension $n - 1$. We denote each smooth manifold by $\mathcal{S}_i$. We have,

$$\mathcal{S}_i = \{x \in R^n \mid h_i(x) = 0\}$$

The intersection of the m smooth manifolds $\mathcal{S}_i$ is denoted by $\mathcal{S}$ and it is defined as follows:

$$\mathcal{S} = \{x \in R^n \mid x \in \mathcal{S}_i, \; i = 1, 2, \cdots, m\}$$

One of the primordial facets of the design of feedback control laws for multi-variable switch regulated systems is given by the fact that the m smooth functions, $h_i(x)$, constitute a part of the control design problem. The choice of the outputs and, therefore, of the restrictions $h_i(x) = 0$, i.e., of $\mathcal{S} = \bigcap \mathcal{S}_i$, depend entirely on the control objectives.

We assume that the surfaces $\mathcal{S}_i$ are locally functionally independent, this means that at any given point x of their intersection, the set of surface gradients, expressed by the row vectors: $\partial h_i / \partial x^T$, $i = 1, 2, ..., m$ are locally linearly independent. This means that the $m \times n$ matrix, $[\partial h / \partial x^T]$, is locally full rank around the point x.

Notation

Let $f(x)$ be a smooth vector field, defined on the tangent space to R^n and let $G(x)$ be a smooth matrix constituted by m columns representing smooth vector fields, $g_i(x), i = 1, 2, \cdots, m$. We assume that $dim \; \{\text{span } G(x)\} = m$, i.e., that span $G(x)$ is a proper m-dimensional subspace of the tangent space to R^n. Also, we say that the range of $G(x)$ is m. Let $h(x)$ be a smooth m-dimensional map, i.e., one taking values in R^m.

From our assumptions about the local functional independence of the conditions $h_i(x) = 0$ and the local full rank condition on the matrix $G(x)$, it follows that the $m \times m$ matrix, $\partial h / \partial x^T G(x)$, is locally full rank m and, hence, invertible.

We define the directional derivative of a smooth map $h(x)$, along the direction of the vector field $f(x)$, as the vector quantity: $\frac{\partial h}{\partial x^T} f(x)$. We denote this m dimensional vector as, $L_f h$. Similarly, we represent by means of $L_G h(x)$ the invertible matrix:

$$\frac{\partial h}{\partial x^T} G(x) = \frac{\partial h}{\partial x^T} [g_1(x), \cdots, g_m(x)] = [L_{g_1} h(x), L_{g_2} h(x), \cdots, L_{g_m} h(x)]$$

Let σ be an m-dimensional vector. We denote by "SIGN(σ)" a vector, also of dimension m, whose i-th component is just $sign(\sigma_i)$.

3.8.2 Equivalent Control and Ideal Sliding Dynamics

Let us assume that through the application of suitable smooth feedback control laws, $u(x)$, that define the position of the m switches in the system, $\dot{x} = f(x) + G(x)u$, as a function of the state vector, we manage to locally make the condition: $x \in \mathcal{S}$ valid. In other words, assume we may force the state x of the system to evolve on the intersection of the the smooth manifolds $\mathcal{S}_i$, that represent the desired algebraic restrictions which, in turn, allow the system to satisfy the specified control objectives.

We define the equivalent control as the smooth feedback control law, denoted by $u_{eq}(x)$, which, ideally, locally sustains the state evolution on the smooth variety $\mathcal{S}$ when the initial state of the system happens to be located, precisely, on $\mathcal{S}$.

The sliding surface coordinate functions, $h_i(x)$, satisfy then, simultaneously, the following invariance condition:

$$\dot{h}(x) = \frac{\partial h}{\partial x^T}(f(x) + G(x)u_{eq}(x)) = 0$$

i.e.,

$$L_f h(x) + [L_G h(x)] u_{eq}(x) = 0$$

and, therefore, the equivalent control is expressed, in a unique fashion, as:

$$u_{eq}(x) = -[L_G h(x)]^{-1} L_f h(x)$$

The closed loop controlled vector field evolving on the manifold $\mathcal{S}$ is expressed as:

$$\dot{x} = f(x) - G(x)[L_G h(x)]^{-1} L_f h(x)$$

Note that for any other initial condition which is not located on the smooth manifold $\mathcal{S}$, the state of the system, governed by $u_{eq}(x)$, evolves in such a manner that $h(x)$ remains constant. Clearly, this constant value adopts the value 0 only when the initial state x_0 satisfies $x_0 \in \mathcal{S}$. The closed loop system, virtually controlled by the equivalent control, may be alternatively written as:

$$\dot{x} = \left\{ I - G(x)[L_G h(x)]^{-1} \frac{\partial h}{\partial x^T} \right\} f(x) = \mathcal{M}(x) f(x)$$

Proposition 3.6. *The square $n \times n$, matrix $\mathcal{M}(x)$, is a **projection operator**, over the tangent space to $\mathcal{S}$, whose null space is represented by the span $G(x)$. In other words, $\mathcal{M}(x)$ projects any smooth vector field lying in the tangent space to R^n over the tangent subspace to $\mathcal{S}$ in a parallel fashion to the span of $G(x)$.*

Indeed, let v be a vector field defined in the tangent space to R^n such that $v \in \text{span } G(x)$ i.e., v may be expressed as $v(x) = G(x)\alpha(x)$ for a certain m-dimensional smooth vector field $\alpha(x)$. Then,

$$\mathcal{M}(x)v(x) = \left\{ I - G(x)[L_G h(x)]^{-1}\frac{\partial h}{\partial x} \right\} G(x)\alpha(x)$$
$$= \left\{ G(x) - G(x)[L_G h(x)]^{-1}\frac{\partial h}{\partial x}G(x) \right\} \alpha(x)$$
$$= \left\{ G(x) - G(x)[L_G h(x)]^{-1} L_G h(x) \right\} \alpha(x)$$
$$= [G(x) - G(x)]\,\alpha(x) = 0$$

Additionally, the n-dimensional row vectors of the matrix $\partial h/\partial x^T$, are all orthogonal to the images under $\mathcal{M}(x)$ of the vector fields lying in the tangent space to R^n. To see this, it is enough to demonstrate that any 1-form lying in the span of the matrix, $\partial h/\partial x^T$, annihilates all the (column) vector fields constituting the matrix, $\mathcal{M}(x)$.

A 1-form in the span of $\frac{\partial h}{\partial x^T}$ is written as: $\xi^T(x)\frac{\partial h}{\partial x^T}$, where $\xi^T(x)$ is a non-zero, completely arbitrary, m-dimensional row vector.

Indeed:

$$\xi^T(x)\frac{\partial h}{\partial x^T}\mathcal{M}(x) = \xi^T(x)\frac{\partial h}{\partial x^T}\left\{ I - G(x)[L_G h(x)]^{-1}\frac{\partial h}{\partial x^T} \right\}$$
$$= \xi^T(x)\left[\frac{\partial h}{\partial x^T} - L_G h(x)[L_G h(x)]^{-1}\frac{\partial h}{\partial x^T} \right]$$
$$= \xi^T(x)\left[\frac{\partial h}{\partial x^T} - \frac{\partial h}{\partial x^T} \right] = 0$$

The image, under $\mathcal{M}(x)$, of any vector lying in the tangent space of R^n is found in the null space of $\partial h/\partial x^T$. In other words, they belong to the tangent subspace of $\mathcal{S}$.

It is clear that $\mathcal{M}^2(x) = \mathcal{M}(x)$, given that $\mathcal{M}(x)G(x) = 0$.

3.8.3 Invariance with Respect to Matched Perturbations

Consider the multi-variable nonlinear system, additively perturbed by an unknown, possibly state dependent, vector field, of unknown nature denoted by, $\xi(x)$, affecting the system as follows: $\dot{x} = f(x) + G(x)u + \xi(x)$. The system is assumed to be controlled by m independent switches. Let $\mathcal{S}$ be a sliding surface, obtained as the intersection of m smooth manifolds represented by the algebraic conditions: $h_i(x) = 0$ for $i = 1, 2, \cdots, m$. Over this sliding surface, $\mathcal{S}$, we want to locally induce a forced trajectory of the system state as that obtained through the creation of a sliding regime. The perturbation field $\xi(x)$ is assumed to be a bounded function of the state of the system.

Assume we may create a sliding motion on the sliding surface: $\mathcal{S}$ in spite of the presence of the perturbation field $\xi(x)$. The existence of such a sliding regime implies the existence of a smooth control, the perturbed equivalent control, still denoted by: $u_{eq}(x)$, which, in an ideal fashion, would maintain the trajectories of the system constrained to the manifold $\mathcal{S}$.

Necessarily, the equivalent control is, in this case, a function of the unknown vector field $\xi(x)$. It would be given by:

$$u_{eq}(x) = -[L_G h(x)]^{-1} (L_f h(x) + L_\xi h(x))$$

The corresponding ideal sliding dynamics is then written as,

$$\dot{x} = f(x) - G(x)[L_G h(x)]^{-1} (L_f h(x) + L_\xi h(x))$$
$$= \left[I - G(x)[L_G h(x)]^{-1} \frac{\partial h}{\partial x^T} \right] f(x) + \left[I - G(x)[L_G h(x)]^{-1} \frac{\partial h}{\partial x^T} \right] \xi(x)$$

The projection operator $\mathcal{M}(x)$ over the tangent space to $\mathcal{S}$, parallel to the span of $G(x)$, acts over the sum of vector fields $f(x) + \xi(x)$, in the creation of a sliding regime on $\mathcal{S}$.

Clearly, the ideal sliding dynamics is totally independent of the perturbation input vector $\xi(x)$, if, and only if, the vector field $\xi(x)$ lies in the null space of $\mathcal{M}(x)$, i.e.,

$$\left[I - G(x)[L_G h(x)]^{-1} \frac{\partial h}{\partial x^T} \right] \xi(x) = 0$$

According to our previous result, the ideal sliding dynamics is therefore invariant with respect to the perturbation field if, and only if, the vector $\xi(x)$ belongs to the span of $G(x)$. There exists then a non-zero vector function, taking values in R^m and denoted by $\alpha(x)$, such that,

$$\xi(x) = G(x)\alpha(x)$$

The perturbation field $\xi(x)$ is contained in the span of the columns of $G(x)$. Such perturbations receive the name of matched perturbations and the previous condition has been addressed as the perturbation matching condition.

3.8.4 Accessibility of the Sliding Surface

Consider the scalar quantity:

$$V(y) = \frac{1}{2} y^T y = \frac{1}{2} h^T(x) h(x) \geq 0$$

This quantity represents a sort of instantaneous sliding surface "output error energy" measuring the distance from the representative point x in the state space to the smooth manifold $\mathcal{S}$. The quantity $V(y)$ is identically zero precisely over the manifold $\mathcal{S}$ and it represents a positive semi-definite function of the sliding surface coordinate function y.

Therefore, a plausible strategy to reach the sliding surface from a neighborhood of the manifold $\mathcal{S}$ which allows us to satisfy the desired restriction $h(x) = 0$, is to exercise control actions $u \in \{0, 1\}^m$ that result in a strict decrease of the quantity $V(h(x))$.

This is achieved influencing the system in such a manner that the velocity of variation of $V(h(x))$ be strictly negative. This means,

$$\frac{d}{dt}(V(h(x))) = \frac{1}{2}\frac{d}{dt}\left(h^T(x)h(x)\right) = h^T(x)\dot{h}(x) < 0$$

Using the relation, $\dot{h}(x) = L_f h(x) + L_G h(x)u$ and realizing that $L_f h(x) + L_G h(x)u_{eq} = 0$ for any $x \notin \mathcal{S}$ and further adding and subtracting the quantity: $L_G h(x)u_{eq}$ to the first order time derivative of $h(x)$ in the previous expression, we have the following relations:

$$h^T(L_f h(x) + L_G h(x)u) = h^T\left(L_f h(x) + L_G h(x)(u - u_{eq}) + L_G h(x)u_{eq}\right)$$
$$= h^T L_G h(x)(u - u_{eq}) < 0$$

This inequality may be expressed in the following manner:

$$h^T[L_{g_1}h]u_1 + h^T[L_{g_2}h]u_2 + \cdots h^T[L_{g_m}h]u_m <$$
$$h^T[L_{g_1}h]u_{1eq} + h^T[L_{g_2}h]u_{2eq} + \cdots h^T[L_{g_m}h]u_{meq}$$

A sufficient condition to achieve this last inequality is to apply one of the two possible values for $u_j, j = 1, \ldots, m$, according to the sign of the factor multiplying the control input u_j represented by $h^T L_{g_j} h$. We use then,

$$u_j = \begin{cases} 1 \text{ if } h^T L_{g_j} h(x) < 0 \\ 0 \text{ if } h^T L_{g_j} h(x) > 0 \end{cases}$$

In other words,

$$u_j = \frac{1}{2}\left[1 - \text{sign}\left(h^T L_{g_j} h(x)\right)\right]$$

If we denote $\mathbf{1}_m$ an m dimensional column vector constituted by 1 in each entry, the suggested control law is written as follows:

$$u = \frac{1}{2}\left[\mathbf{1}_m - \text{SIGN}\left(h^T L_G h(x)\right)^T\right]$$

3.9 Control of the Boost-Boost Converter

Consider the normalized model of the Boost-Boost system.

$$\dot{x}_1 = -(1 - u_1)x_2 + 1$$
$$\dot{x}_2 = (1 - u_1)x_1 - \frac{1}{Q_1}x_2 - x_3$$
$$\alpha_1 \dot{x}_3 = x_2 - (1 - u_2)x_4$$
$$\alpha_2 \dot{x}_4 = (1 - u_2)x_3 - \frac{1}{Q_L}x_4 \tag{3.17}$$

Using the previously introduced notation, we have:

$$f(x) = \begin{bmatrix} 1 - x_2 \\ x_1 - \frac{1}{Q_1}x_2 - x_3 \\ \frac{1}{\alpha_1}(x_2 - x_4) \\ \frac{1}{\alpha_2}\left(x_3 - \frac{1}{Q_L}x_4\right) \end{bmatrix}, \quad G(x) = \begin{bmatrix} x_2 & 0 \\ -x_1 & 0 \\ 0 & \frac{1}{\alpha_1}x_4 \\ 0 & -\frac{1}{\alpha_2}x_3 \end{bmatrix}$$

The control objective is to have the normalized average capacitor voltages x_2 and x_4 to adopt the following constant desired equilibrium values: $\bar{x}_2 = V_{1d}$, $\bar{x}_4 = V_{2d}$, respectively.

3.9.1 Direct Control

We try the following sliding surface coordinate functions:

$$h_1(x) = x_2 - \bar{x}_2, \quad h_2(x) = x_4 - \bar{x}_4$$

Clearly, forcing to zero the vector of sliding surface coordinate functions means that the capacitor voltages reach the desired equilibrium values. We must nevertheless establish the nature and stability of the corresponding zero dynamics, or ideal sliding dynamics.

In our case we have

$$L_f h(x) = \frac{\partial h}{\partial x^T} f(x) = \begin{bmatrix} x_1 - \frac{1}{Q_1}x_2 - x_3 \\ \frac{1}{\alpha_2}\left(x_3 - \frac{1}{Q_L}x_4\right) \end{bmatrix}$$

$$L_G h(x) = \frac{\partial h}{\partial x^T} G(x) = \begin{bmatrix} -x_1 & 0 \\ 0 & -\frac{1}{\alpha_2}x_3 \end{bmatrix}$$

and the equivalent control is given by

$$u_{eq}(x) = -[L_G h(x)]^{-1} L_f h(x) = \begin{bmatrix} 1 - \frac{(1/Q_1)x_2 + x_3}{x_1} \\ 1 - \frac{1}{Q_L}\left(\frac{x_4}{x_3}\right) \end{bmatrix}$$

The ideal sliding dynamics occurs when $u_{eq}(x)$ acts over the system and this satisfies the conditions: $x_2 = \bar{x}_2$ and $x_4 = \bar{x}_4$. We then have

$$\dot{x}_1 = -\left(\frac{(1/Q_1)\bar{x}_2 + x_3}{x_1}\right)\bar{x}_2 + 1$$

$$\alpha_1 \dot{x}_3 = -\frac{1}{Q_L}\left(\frac{\bar{x}_4^2}{x_3}\right) + \bar{x}_2$$

It is not difficult to see that these set of dynamics is unstable around the desired equilibrium point.

3.9.2 Indirect Control

The alternative is then to consider, as coordinate functions of the sliding surfaces, other functions which stably reproduce the desired output voltages when forced to be zero. These alternative functions are represented by the stabilization errors of the input inductor currents.

$$h_1(x) = x_1 - \overline{x}_1, \qquad h_2(x) = x_3 - \overline{x}_3$$

To specify these functions we compute the state and input equilibrium points in terms of the desired average output equilibrium voltages, which expressed in terms of the output voltages $\overline{x}_2 = V_{2d}$ and $\overline{x}_4 = V_{4d}$ are given by

$$\overline{x}_1 = \frac{V_{2d}^2}{Q_1} + \frac{V_{4d}^2}{Q_L}, \qquad \overline{x}_2 = V_{2d}, \qquad \overline{x}_3 = \frac{V_{4d}^2}{Q_L V_{2d}}, \qquad \overline{x}_4 = V_{4d} \qquad (3.18)$$

We now have:

$$L_f h(x) = \begin{bmatrix} 1 - x_2 \\ \frac{1}{\alpha_1}(x_2 - x_4) \end{bmatrix}, \quad L_G h(x) = \begin{bmatrix} x_2 & 0 \\ 0 & \frac{1}{\alpha_1} x_4 \end{bmatrix}, \quad u_{eq}(x) = \begin{bmatrix} 1 - \frac{1}{x_2} \\ 1 - \frac{x_2}{x_4} \end{bmatrix}$$

Thus, the ideal sliding dynamics corresponding to $x_1 = \overline{x}_1$, $x_3 = \overline{x}_3$ is given by:

$$\dot{x}_2 = \frac{1}{x_2}\overline{x}_1 - \frac{1}{Q_1}x_2 - \overline{x}_3$$

$$\alpha_2 \dot{x}_4 = \frac{x_2}{x_4}\overline{x}_3 - \frac{1}{Q_L}x_4 \qquad (3.19)$$

It is not difficult to verify, with the help of 3.18, that the obtained zero dynamics system 3.19 has the desired average output equilibrium voltages as its asymptotically stable equilibrium point.

According to the developed theory, the intersection of the sliding surfaces is reachable by means of the following switching policy:

$$u_1 = \begin{cases} 1 \text{ if } (x_1 - \overline{x}_1)x_2 < 0 \\ 0 \text{ if } (x_1 - \overline{x}_1)x_2 > 0 \end{cases}, \qquad u_2 = \begin{cases} 1 \text{ if } \frac{1}{\alpha_1}(x_3 - \overline{x}_3)x_4 < 0 \\ 0 \text{ if } \frac{1}{\alpha_1}(x_3 - \overline{x}_3)x_4 > 0 \end{cases}$$

In other words, the control policy are given by:

$$u_1 = \frac{1}{2}\left[1 - \text{sign}\left(h^T L_{g_1} h(x)\right)\right] = \frac{1}{2}\left[1 - \text{sign}\left((x_1 - \overline{x}_1)x_2\right)\right]$$

$$u_2 = \frac{1}{2}\left[1 - \text{sign}\left(h^T L_{g_2} h(x)\right)\right] = \frac{1}{2}\left[1 - \text{sign}\left((x_3 - \overline{x}_3)x_4\right)\right] \qquad (3.20)$$

since $\alpha_1 > 0$.

Since E, L_1 and C_1 are all positives and given that $v_1 \geq 0$ and $v_2 \geq 0$ (because $x_2 \geq 0$ and $x_4 \geq 0$) around the equilibrium point of the system. We can rewrite (3.20) in non-normalized form as:

$$u_1 = \frac{1}{2}\left[1 - \text{sign}\left(i_1 - \bar{i}_1\right)\right]$$

$$u_2 = \frac{1}{2}\left[1 - \text{sign}\left(i_2 - \bar{i}_2\right)\right] \tag{3.21}$$

where $\bar{i}_1$ and $\bar{i}_2$ are given by:

$$\bar{i}_1 = \frac{1}{R_1}\frac{\bar{v}_1^2}{E} + \frac{1}{R_L}\frac{\bar{v}_2^2}{E}$$

$$\bar{i}_2 = \frac{1}{R_L}\frac{\bar{v}_2^2}{\bar{v}_1}$$

3.9.3 Simulations

We take a typical converter with the following parameters

$L_1 = 15.91$ mH, $\quad C_1 = 48\ \mu$F, $\quad L_2 = 40$ mH, $\quad C_2 = 107\ \mu$F,

$R_1 = 52\ \Omega, \quad R_L = 52\ \Omega, \quad E = 12$ V

It is desired to control the capacitor voltages to the values:

$$\bar{v}_1 = 15 \text{ V}, \quad \bar{v}_2 = 24 \text{ V}$$

The equilibrium values of the average input inductor currents to each stage correspond approximately to the following values:

$$\bar{i}_1 = 1.28 \text{ A}, \quad \bar{i}_2 = 0.738 \text{ A}$$

Figure 3.17 shows the behavior of the sliding mode controlled Boost-Boost converter.

3.9.4 Experimental Sliding Mode Control Implementation

The experimental prototype of the Boost-Boost converter, described in Chapter 2, was used for the implementation of the software implementation of the multi-variable sliding mode feedback controller (3.21). It was implemented on a National Instruments™ PCI-6025E data acquisition board with MATLAB® Simulink®

Figure 3.18 depicts the entire block diagram of the Boost-Boost converter prototype including the multi-variable sliding mode control.

In Figure 3.19 the main block diagram used for code generation is shown, and in Figure 3.20 the multi-variable sliding control is shown.

106   3 Sliding Mode Control

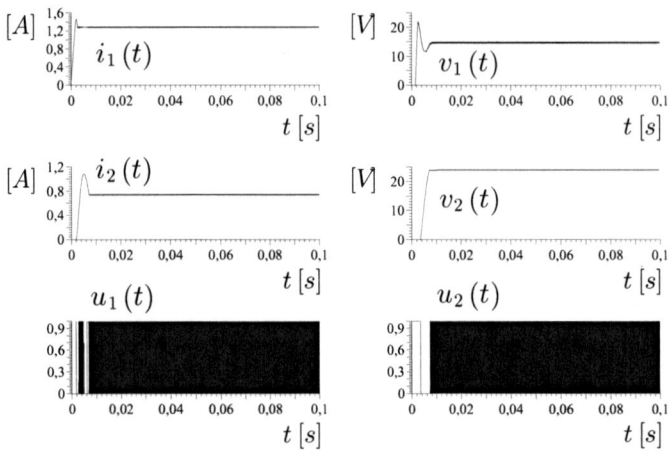

Fig. 3.17. Closed loop responses of the Boost-Boost converter to a multi-input sliding mode controller.

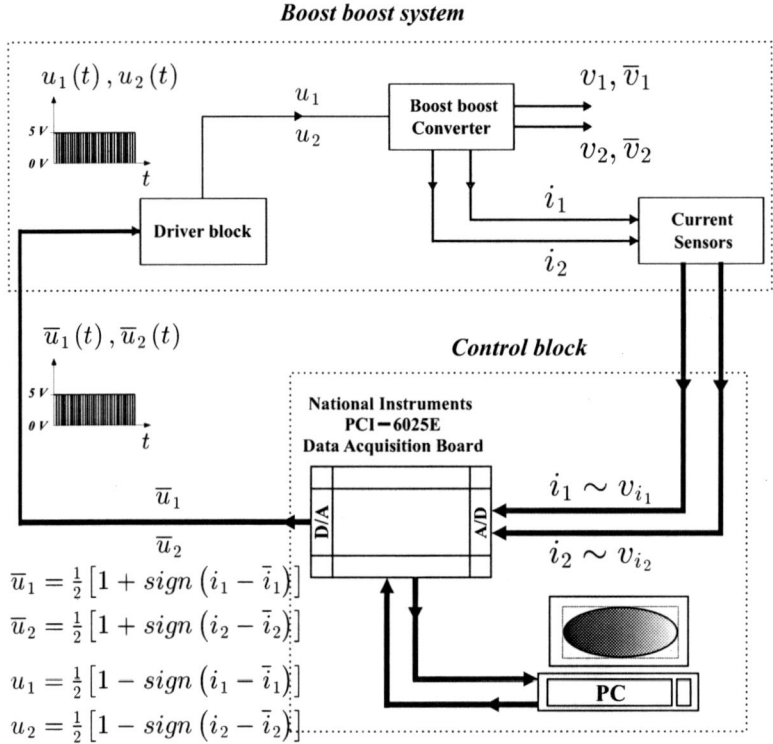

Fig. 3.18. Block diagram of experimental setup.

3.9 Control of the Boost-Boost Converter   107

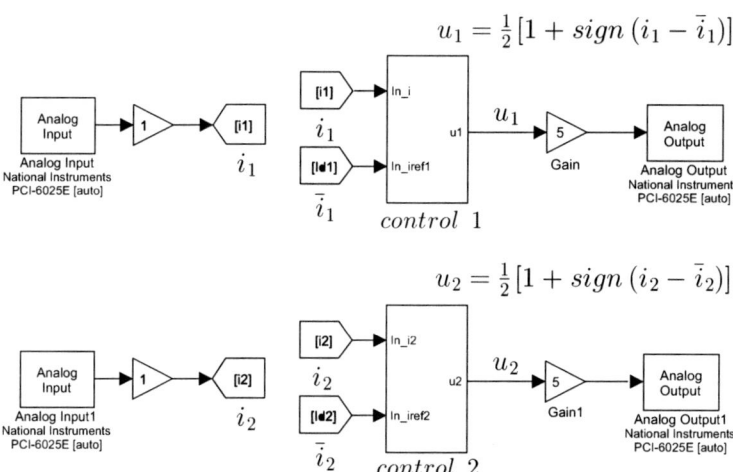

Fig. 3.19. Main block diagram for C-code generation from Simulink®

Figure 3.20 depicts the software realization of the multi-variable sliding mode control law, based on the indirect regulation of the output voltage vector for the Boost-Boost converter. It consists of two controls (control 1, and control 2). Each one of these controls accepts, as inputs, two signals: the inductor current signal and the desired current reference signal. The output of the control block is constituted by two pulsed signals with amplitudes of 0 V and 1 V, which are amplified by 5, see Figure 3.19. These two signals command, respectively, the gate of the two Mosfet NT2984 IC's acting as switches.

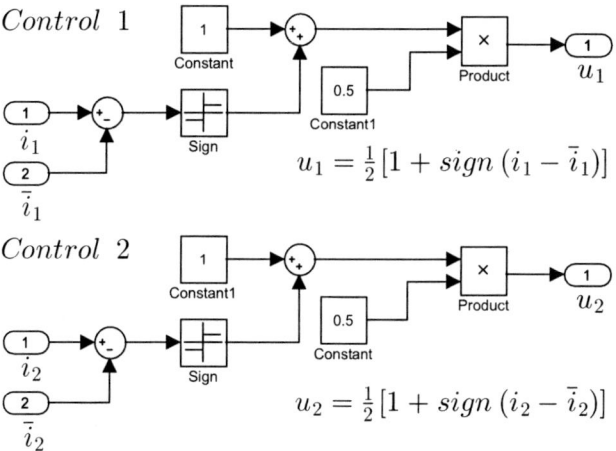

Fig. 3.20. Simulink® block diagram for the multi-variable control.

Experimental Results

Figure 3.21 presents the experimental results portraying the closed loop response of the Boost-Boost controlled by the designed multi-variable sliding mode controller. The controller and the system parameters were chosen to be exactly the same as in the simulation results presented in the previous section.

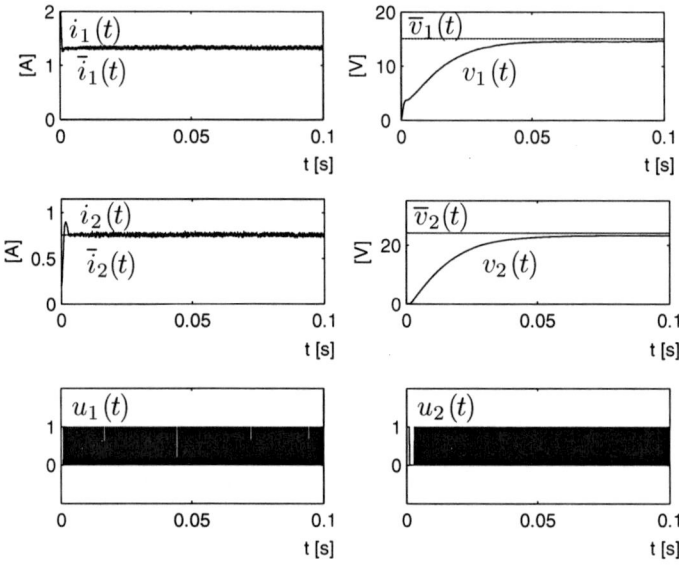

Fig. 3.21. Experimental closed loop responses of the Boost-Boost converter variables to a multi-input sliding mode controller.

3.10 Control of the Double Buck-Boost Converter

Consider the composite converter constituted by the cascade connection of two stages of the Buck-Boost converter, which we address as the double Buck-Boost converter. This circuit is shown in Figure 3.22. Clearly this system represents a multi-input converter regulated by two independent switches.

The set of normalized differential equations describing the converter dynamics was already obtained in Chapter 2. These are given by,

3.10 Control of the Double Buck-Boost Converter

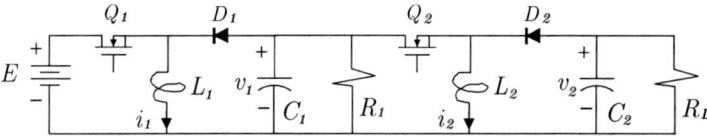

Fig. 3.22. Switch regulated DC-to-DC double Buck-Boost power converter.

$$\frac{dx_1}{d\tau} = (1 - u_1)x_2 + u_1$$
$$\frac{dx_2}{d\tau} = -(1 - u_1)x_1 - \frac{1}{Q_1}x_2 - u_2 x_3$$
$$\alpha_1 \frac{dx_3}{d\tau} = u_2 x_2 + (1 - u_2)x_4$$
$$\alpha_2 \frac{dx_4}{d\tau} = -(1 - u_2)x_3 - \frac{1}{Q_L}x_4 \quad (3.22)$$

In terms of vector fields and input matrices, we clearly have the following identifications:

$$f(x) = \begin{bmatrix} x_2 \\ -x_1 - \frac{1}{Q_1}x_2 \\ \frac{1}{\alpha_1}x_4 \\ -\frac{1}{\alpha_2}\left(x_3 + \frac{1}{Q_L}x_4\right) \end{bmatrix}, \quad G(x) = \begin{bmatrix} 1 - x_2 & 0 \\ x_1 & -x_3 \\ 0 & \frac{1}{\alpha_1}(x_2 - x_4) \\ 0 & \frac{1}{\alpha_2}x_3 \end{bmatrix}$$

The control objective consists in stably regulating the average normalized output voltages, x_2 and x_4, towards the desired equilibrium values: $\overline{x}_2 = V_{2d}$ and $\overline{x}_4 = V_{4d}$, respectively.

3.10.1 Direct Control

Consider the following sliding surface coordinate functions:

$$h_1(x) = x_2 - V_{2d}, \quad h_2(x) = x_4 - V_{4d}$$

Forcing these functions to zero, evidently, means that the output capacitor voltages coincide with the desired values. We must establish the nature of the stability of the corresponding zero dynamics.

For this system we have,

$$L_f h(x) = \frac{\partial h}{\partial x^T} f(x) = \begin{bmatrix} -x_1 - \frac{1}{Q_1}x_2 \\ -\frac{1}{\alpha_2}\left(x_3 + \frac{1}{Q_L}x_4\right) \end{bmatrix}$$

$$L_G h(x) = \frac{\partial h}{\partial x^T} G(x) = \begin{bmatrix} x_1 & -x_3 \\ 0 & \frac{1}{\alpha_2}x_3 \end{bmatrix}$$

The equivalent control is then given by

$$u_{eq}(x) = -[L_G h(x)]^{-1} L_f h(x) = \begin{bmatrix} 1 + \left(\frac{1}{Q_1}x_2 + x_3 + \frac{1}{Q_L}x_4\right)\frac{1}{x_1} \\ 1 + \frac{1}{Q_L}\frac{x_4}{x_3} \end{bmatrix}$$

The ideal sliding dynamics is readily obtained to be given by,

$$\frac{dx_1}{d\tau} = (1 - V_{2d})\left(\frac{V_{2d}}{Q_1} + \frac{V_{4d}}{Q_L} + x_3\right)\frac{1}{x_1} + 1$$

$$\alpha_1 \frac{dx_3}{d\tau} = -\frac{V_{4d}}{Q_L}(V_{4d} - V_{2d})\frac{1}{x_3} + V_{2d}$$

It is not difficult to see that this dynamics is unstable around the equilibrium point. We show this fact below by means of approximate linearization.

The ideal sliding dynamics, or the zero dynamics, for the state variable x_3 represents a decoupled system whose equilibrium point is given by

$$\overline{x}_3 = \frac{V_{4d}}{Q_L}\left(\frac{V_{4d} - V_{2d}}{V_{2d}}\right)$$

The approximate linearization model, of the nonlinear dynamics for the variable x_3 is found to be,

$$\alpha_1 \dot{x}_{3\delta} = \frac{Q_L V_{2d}^2}{V_{4d}(V_{4d} - V_{2d})} x_{3\delta}$$

where $x_{3\delta} = x_3 - \overline{x}_3$. The linearized system is, evidently, unstable for its characteristic polynomial exhibits a zero in the right hand of the complex plane. The zero dynamics is therefore unstable, regardless of the stability characteristics of the variable x_1.

3.10.2 Indirect Control

Consider now the following indirect approach represented by the sliding surfaces:

$$h_1 = x_1 - \overline{x}_1, \qquad h_2 = x_3 - \overline{x}_3$$

The equilibrium point of the system under ideal sliding conditions is given by

$$\overline{x}_1 = -\left(\frac{V_{2d}^2}{Q_1} + \frac{V_{4d}^2}{Q_L}\right)\left(\frac{1 - V_{2d}}{V_{2d}}\right), \qquad \overline{x}_3 = \frac{V_{4d}}{Q_L}\left(\frac{V_{4d} - V_{2d}}{V_{2d}}\right)$$

In this case, we have

$$L_f h(x) = \frac{\partial h}{\partial x^T} f(x) = \begin{bmatrix} x_2 \\ \frac{1}{\alpha_1} x_4 \end{bmatrix}$$

$$L_G h(x) = \frac{\partial h}{\partial x^T} G(x) = \begin{bmatrix} 1 - x_2 & 0 \\ 0 & \frac{1}{\alpha_1}(x_2 - x_4) \end{bmatrix}$$

3.10 Control of the Double Buck-Boost Converter

The equivalent control is then given by,

$$u_{eq}(x) = -[L_G h(x)]^{-1} L_f h(x) = \begin{bmatrix} \frac{x_2}{x_2-1} \\ \frac{x_4}{x_4-x_2} \end{bmatrix}$$

The ideal sliding dynamics corresponding to $x_1 = \bar{x}_1, x_3 = \bar{x}_3$ is given by:

$$\frac{dx_2}{d\tau} = -\left(\frac{1}{1-x_2}\right)\bar{x}_1 - \frac{1}{Q_1}x_2 - \left(\frac{x_4}{x_4-x_2}\right)\bar{x}_3$$

$$\alpha_2 \frac{dx_4}{d\tau} = \left(\frac{x_2}{x_4-x_2}\right)\bar{x}_3 - \frac{1}{Q_L}x_4$$

It is easy to verify that the equilibrium point of this zero dynamics is an asymptotically stable equilibrium point.

According to the developed theory, the intersection of the sliding surfaces is reachable by means of the following switching policy

$$u_1 = \begin{cases} 1 \text{ if } (x_1-\bar{x}_1)(1-x_2) < 0 \\ 0 \text{ if } (x_1-\bar{x}_1)(1-x_2) > 0 \end{cases}, \quad u_2 = \begin{cases} 1 \text{ if } \frac{1}{\alpha_1}(x_3-\bar{x}_3)(x_2-x_4) < 0 \\ 0 \text{ if } \frac{1}{\alpha_1}(x_3-\bar{x}_3)(x_2-x_4) > 0 \end{cases}$$

In other words, the control policy given by:

$$u_1 = \frac{1}{2}\left[1 - \text{sign}\left(h^T L_{g_1} h(x)\right)\right] = \frac{1}{2}\left[1 - \text{sign}\left((x_1-\bar{x}_1)(1-x_2)\right)\right]$$

$$u_2 = \frac{1}{2}\left[1 - \text{sign}\left(h^T L_{g_2} h(x)\right)\right] = \frac{1}{2}\left[1 - \text{sign}\left((x_3-\bar{x}_3)(x_2-x_4)\right)\right]$$

since $\alpha_1 > 0$, yields the desired control objective with an internally stable ideal sliding dynamics.

3.10.3 Simulations

Simulations were carried out with the following design parameter values:

$$L_1 = 20 \text{ mH}, \quad C_1 = 20 \text{ }\mu\text{F}, \quad L_2 = 20 \text{ mH}, \quad C_2 = 20 \text{ }\mu\text{F},$$

$$R_1 = 30 \text{ }\Omega, \quad R_L = 30 \text{ }\Omega, \quad E = 15 \text{ V}$$

The prescribed control objectives are set to regulate the voltage variables to the values

$$\bar{v}_1 = -22.5 \text{ V}, \quad \bar{v}_2 = 22.5 \text{ V}$$

The corresponding equilibrium currents are given by

$$\bar{i}_1 = 3.75 \text{ A}, \quad \bar{i}_2 = -1.5 \text{ A}$$

Figure 3.23 shows the behavior of the sliding mode controlled double Buck-Boost converter.

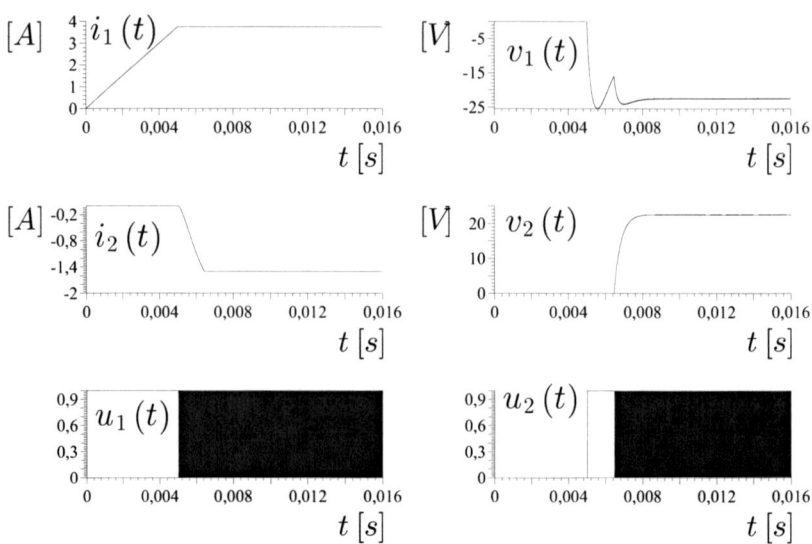

Fig. 3.23. Closed loop responses of the double Buck-Boost converter variables to a multi-input sliding mode controller.

3.11 $\Sigma - \Delta$ Modulation

As was seen in previous sections, the sliding mode control technique is, fundamentally, a state space-based discontinuous feedback control technique. The lack of complete knowledge of the state vector components forces the designer to use asymptotic state observers, of the Luenberger, or of the sliding mode type [77], or perhaps to resort to direct output feedback control schemes (see [9]). Unfortunately, the first approach is not robust with respect to unforseen exogenous perturbation inputs, even if they happen to be of the "classical type" (by this we mean: steps, ramps, parabolas, etc.). The second approach is quite limited in nature and it is not directly applicable in a host of non-minimum phase systems. Generically speaking, state space based sliding mode techniques fall into the unmatched perturbation input case, while output feedback control techniques do not suffer such realistic drawback.

In this section, we propose a rather practical approach for the synthesis of sliding mode feedback control schemes in nonlinear switch controlled systems, in general, and in DC-to-DC power converters and other switch controlled Power Electronics devices, in particular. We carry out a sliding mode implementation of average feedback controller design schemes, based on an analog version of $\Sigma - \Delta$ modulators. Although we could use average state feedback controllers, we concentrate on dynamical output feedback control schemes as the desired paradigm of control laws. This points to the issue that states are not really needed for sliding mode control. We show that the use of analog

$\Sigma - \Delta$-modulators [1] allows for the switched synthesis of any feedback controller which has been synthesized from an average viewpoint (i.e., assuming that the control input continuously takes values on a closed subset of the real line, usually restricted to be the closed interval $[0, 1]$). We show that a $\Sigma - \Delta$-modulator can be used to translate such a continuous average design into a discontinuous one with the property that the "equivalent output" signal of the modulator, in an ideal sliding mode sense, precisely matches the modulator's input signal generated by the continuous average feedback controller.

When we combine $\Sigma - \Delta$-modulation with standard state or output feedback control, the result is that the induced sliding motion retains all the desirable essential features of the average devised controller (robustness, adaptability, perturbation rejection properties etc). This technique is, therefore, most interesting in the case of dynamical output feedback control. The corresponding induced sliding mode control does not require, in that case, of the state of the plant. A limitation which has been imposed on sliding mode control for a number of years.

The $\Sigma - \Delta$ modulation approach for the sliding mode implementation provides a systematic approach to average based controller design for switched systems. In this respect, the proposed approach is quite different from that found in Yeung et al. [82] where the sliding surface is synthesized in terms of (filtered) differential polynomials acting on inputs and outputs (see also [62] for a yet different perspective). As an additional outcome, the scheme here presented requires no matching conditions whatsoever.

3.11.1 $\Sigma - \Delta$-Modulators

Consider the basic block diagram of Figure 3.24, reminiscent of a traditional $\Sigma - \Delta$-modulator block used in early communications systems theory and analog to digital conversion schemes, but with a binary valued forward nonlinearity, taking values in the discrete set $\{0, 1\}$. The following theorem summarizes the relation of the considered modulator with sliding mode control while establishing the basic features of its input-output performance.

Theorem 3.7. *Consider the $\Sigma - \Delta$-modulator of Figure 3.24. Given a sufficiently smooth, bounded, signal $\mu(t)$, then the integral error signal, $e(t)$, converges to zero in a finite time, t_h. Moreover, from any arbitrary initial value, $e(t_0)$, a sliding motion exists on the perfect encoding condition surface, represented by $e = 0$, for all $t > t_h$, provided the following encoding condition is satisfied for all t,*

$$0 < \mu(t) < 1 \tag{3.23}$$

---

[1] A complete account of Δ-modulators, and their modification: $\Sigma - \Delta$ modulators, extensively used in analog signal encoding, which never benefited from the theoretical basis of sliding mode control, is found in the classical book by Steele [74] and in the excellent book by Norsworthy et al. [47].

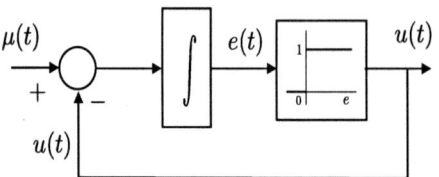

Fig. 3.24. $\Sigma - \Delta$-modulator.

Proof. From Figure 3.24, the variables in the $\Sigma - \Delta$-modulator satisfy the following relations:

$$\dot{e} = \mu(t) - u, \quad u = \frac{1}{2}[1 + \text{sign}(e)] \tag{3.24}$$

The quantity $e\dot{e}$ is given by

$$e\dot{e} = e\left[\mu - \frac{1}{2}(1 + \text{sign}(e))\right] = -|e|\left[\frac{1}{2}(1 + \text{sign}(e)) - \mu\text{sign}(e)\right]$$

For $e > 0$ we have $e\dot{e} = -e(1 - \mu)$, which, according with the assumption in (3.23) leads to $e\dot{e} < 0$. On the other hand, when $e < 0$, we have $e\dot{e} = -|e|\mu < 0$. A sliding regime exists then on $e = 0$ for all time t after the hitting time t_h (see [75]). Under ideal sliding, or *ideal encoding*, conditions, $e = 0, \dot{e} = 0$, we have that the, so called, equivalent value of the switched output signal, u, denoted by $u_{eq}(t)$ satisfies $u_{eq}(t) = \mu(t)$.

An estimate of the hitting time t_h is obtained by examining the modulator system equations with the worst possible bound for the input signal μ in each of the two conditions: $e > 0$ and $e < 0$, along with the corresponding value of u. Consider then $e(0) > 0$ at time $t = 0$. We have for all $0 < t \leq t_h$,

$$e(t) = e(0) + \int_0^t (\mu(\sigma) - u(\sigma))d\sigma \leq e(0) + t\left[\sup_{t \in [0,t]} \mu(t) - 1\right]$$

$$< e(0) + t_h\left[\sup_t \mu(t) - 1\right]. \tag{3.25}$$

Since $e(t_h) = 0$, we have:

$$t_h \leq \frac{e(0)}{1 - \sup_t \mu(t)} \tag{3.26}$$

$\square$

The average $\Sigma - \Delta$-modulator output u_{eq}, ideally yields the modulator's input signal $\mu(t)$ in an equivalent control sense (see [75]).

To illustrate, by means of simulations, the feature just stated about $\Sigma - \Delta$ modulation, we let $\mu(t) = 0.5(1 + A\sin(\omega t))$ with $A = 0.8$, $\omega = 3$ rad/s. At the output of the modulator we put a second order low pass filter of the form

$$y = \frac{\omega_n^2}{s^2 + 2\zeta\omega_n s + \omega_n^2}$$

with $\zeta = 0.81$ and $\omega_n = 30$. We may compare the filter output y with the input signal $\mu(t)$: modulo a small delay and the second order filter transient from zero initial conditions, the filtering of the switched output signal, $u(t)$, of the modulator, represented by the variable $y(t)$, reproduces, quite accurately, the sinusoidal input to the modulator. Figure 3.25 depicts the results.

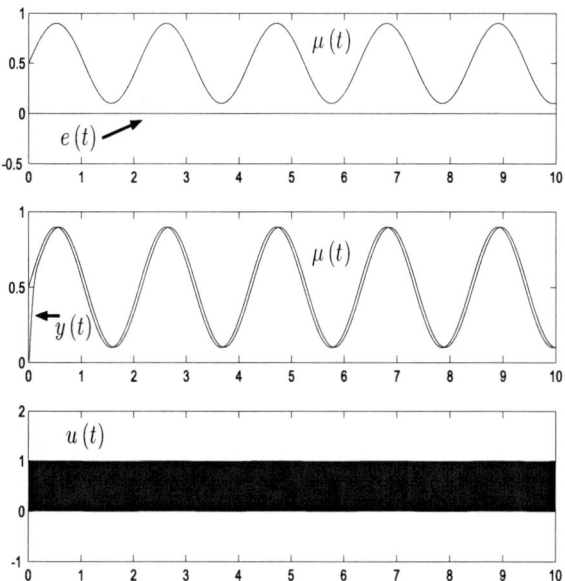

Fig. 3.25. Performance of $\Sigma - \Delta$ modulator and tracking properties of the low pass filtered switched output.

3.11.2 Average Feedbacks and $\Sigma - \Delta$ -Modulation

Suppose we have a smooth nonlinear system of the form $\dot{x} = f(x) + ug(x)$ with u being a (continuous) control input signal that, due to some physical limitations, requires to be bounded by the closed interval $[0, 1]$. Suppose, furthermore, that we have been able to specify a dynamic output feedback controller of the form $u = -\kappa(y, \zeta)$, $\dot{\zeta} = \varphi(y, \zeta)$, with desirable closed loop performance features. Assume, furthermore, that for some reasonable set of initial states of the system (and of the dynamic controller), the values of the generated feedback signal function, $u(t)$, are uniformly strictly bounded by the closed interval $[0, 1]$.

3 Sliding Mode Control

If an additional implementation requirement entitles now that the control input u of the system is no longer allowed to continuously take values within the interval $[0,1]$, but that it may only take values in the discrete set, $\{0,1\}$, the natural question is: how can we now implement the previously derived continuous controller, so that we can recover, possibly in an average sense, the desirable features of the derived dynamic output feedback controller design in view of the newly imposed actuator restriction?

The answer is clearly given by the average reproducing features of the input signal in the previously considered $\Sigma - \Delta$-modulator. Recall, incidentally, that the output signal of such a modulator is restricted to take values, precisely, in the discrete set $\{0,1\}$. Thus, if the output of the designed continuous controller, call it $u_{av}(t)$, is feeded into the proposed $\Sigma - \Delta$ modulator, the output signal of the modulator reproduces, on the average, the required control input signal $u_{av}(t)$. Figure 3.26 shows the switch based implementation of an average designed output feedback controller, through a $\Sigma - \Delta$-modulator, which reproduces, in an average sense, the features of a designed continuous controller.

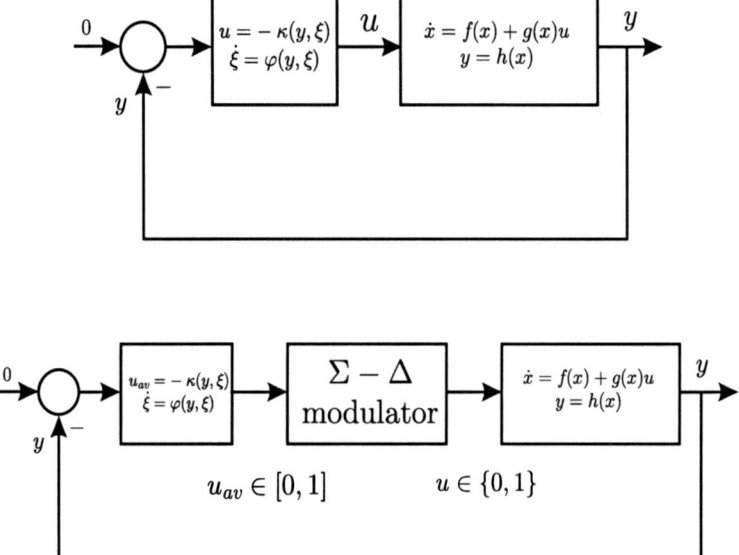

Fig. 3.26. Sliding mode implementation of a designed continuous output feedback controller through a $\Sigma - \Delta$-modulator.

In view of the previous result, we have the following general result concerning the control of nonlinear systems through sliding modes synthesized on the basis of an average feedback controller and a $\Sigma - \Delta$-modulator. We only deal with the dynamic output feedback controller case for the stabilization problem around an equilibrium. The result, however, can also be extended to be valid for trajectory tracking problems.

Theorem 3.8. *Consider the following n-dimensional smooth, nonlinear, single input, system: $\dot{x} = f(x) + ug(x)$, with the smooth scalar output map, $y = h(x)$. Assume the dynamic smooth output feedback controller $u = -\kappa(y, \xi), \dot{\xi} = \varphi(y, \xi)$, with $\xi \in R^p$, locally (globally, semi-globally) asymptotically stabilizes the system towards a desired constant equilibrium state, represented by X. Assume, furthermore, that the control signal, u, is uniformly strictly bounded by the closed interval $[0, 1]$ of the real line. Then the closed loop system:*

$$\dot{x} = f(x) + ug(x)$$
$$y = h(x)$$
$$u_{av}(y, \xi) = -\kappa(y, \xi, X)$$
$$\dot{\xi} = \varphi(y, \xi, X)$$
$$u = \frac{1}{2}[1 + \text{sign } e]$$
$$\dot{e} = u_{av}(y, \xi) - u$$

exhibits an ideal sliding dynamics which is locally (globally, semi-globally) asymptotically stable to the same constant state equilibrium point, X, of the system.

Proof. The proof of this theorem is immediate upon realizing that under the hypothesis made on the average control input, u_{av}, the previous theorem establishes that a sliding regime exists on the manifold $e = 0$. Under the invariance conditions: $e = 0, \dot{e} = 0$, which characterize ideal sliding motions (see Sira-Ramírez [57]), the corresponding equivalent control, u_{eq}, associated with the system satisfies: $u_{eq}(t) = u_{av}(t)$. The ideal sliding dynamics is then represented by

$$\dot{x} = f(x) + u_{av}g(x)$$
$$y = h(x)$$
$$u_{av}(y, \xi) = -\kappa(y, \xi, X)$$
$$\dot{\xi} = \varphi(y, \xi, X)$$

which is assumed to be locally (globally, semi-globally) asymptotically stable towards the desired equilibrium point.

Remark 3.9. Note that the Σ-Δ modulator state, e, can be initialized at the value $e(t_0) = 0$. This implies that the induced sliding regime exists uniformly for all times after t_0. Hence, no reaching time of the sliding surface, $e = 0$, is required. This practical feature is adopted throughout this book.

118   3 Sliding Mode Control

3.11.3 A Hardware Realization of a $\Sigma - \Delta$-Modulator

Ideal sliding motions imply infinite frequency in the switched signals. Such a demand is impossible to achieve in practise. A practical solution for the realization of a $\Sigma - \Delta$-modulator, characterized by high, but finite switching frequency, would not markedly affect the key results. For this, we propose a Pulse Width Modulator (PWM) based $\Sigma - \Delta$-modulator, which allows a finite switching frequency to be selected by the user. The block diagram of the proposed $\Sigma - \Delta$-modulator circuit is shown in Figure 3.27, while Figure 3.28 shows the actual $\Sigma - \Delta$-modulator realization. This practical realization arises from replacing the switching block of the $\Sigma - \Delta$-modulator by a classical (PWM) block.

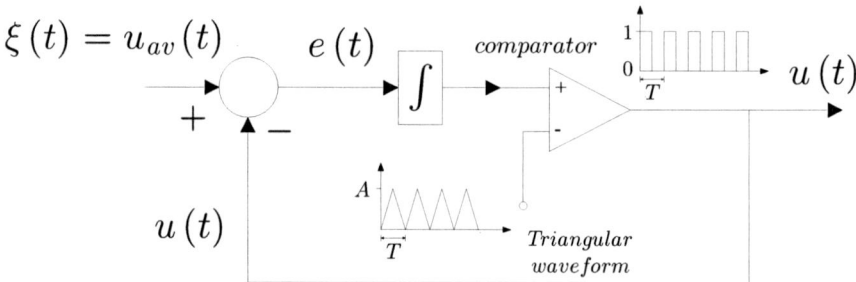

Fig. 3.27. Block diagram for the realization of a Σ-Δ-modulator.

In reference to Figure 3.27 (or Figure 3.28) realizing the $\Sigma - \Delta$ modulator, we distinguish a *subtractor* block, an *integrator* block, a *triangular-wave generator system* and a *classical PWM circuit*.

The *subtractor* block, shown in Figure 3.29, will produce an output which represents the voltage difference between the input signals V_1 and V_2. For arbitrary values of R_1, R_2, R_3 and R_4, the difference amplifier circuit produces the following output:

$$V_{out} = \left(\frac{(R_3 + R_1) R_4}{(R_4 + R_2) R_1}\right) V_2 - \left(\frac{R_3}{R_1}\right) V_1$$

Letting $R_1 = R_2$ and $R_3 = R_4$, one obtains the output signal V_{out} given by,

$$V_{out} = \left(\frac{R_3}{R_1}\right)(V_2 - V_1)$$

Naturally, for $R_1 = R_3$ and $R_2 = R_4$, the output V_{out} of the difference amplifier is now,

$$V_{out} = V_2 - V_1$$

3.11 $\Sigma - \Delta$ Modulation   119

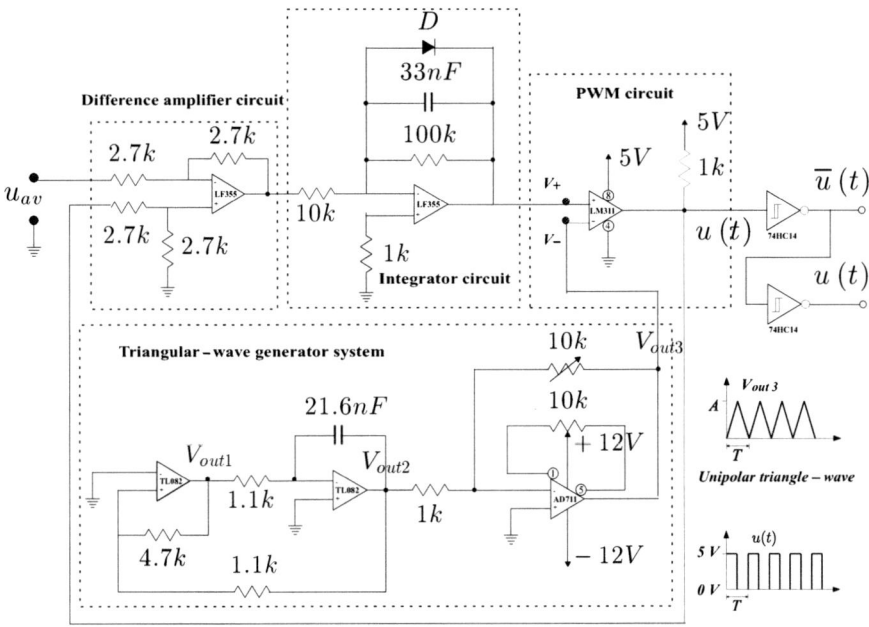

Fig. 3.28. Circuit diagram of an experimental $\Sigma - \Delta$-modulator.

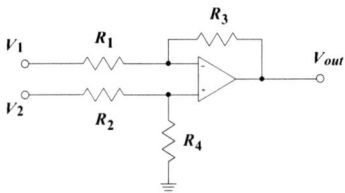

Fig. 3.29. Difference amplifier circuit.

In the realization we are proposing of the $\Sigma - \Delta$-modulator, we used the last choice of resistor values with all the resistors value taken to be, $R = 2.7$ kΩ. In our control oriented applications, we set, $V_1 = u_{av}$ and $V_2 = u$.

The basic *integrator* circuit is shown in Figure 3.30. It integrates, over time, the signal V_{out}, properly inverted.

$$V_{out} = -\frac{1}{RC}\int_0^t V_{in}(\sigma)d\sigma + V_{0\,initial}$$

where $V_{0\,initial}$ is the output voltage of the integrator at time $t = 0$.

The *triangular-wave generator system* uses an triangular-wave generator circuit and an inverting amplifier circuit with offset null configuration. These circuits are shown in Figure 3.31 and Figure 3.32, respectively.

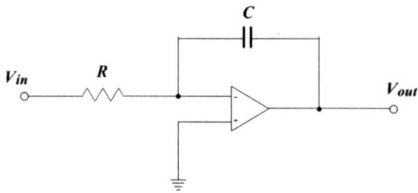

Fig. 3.30. Basic integrator circuit.

In reference to Figure 3.31, the triangular-wave generator circuit was synthesized using two operational amplifiers (Op-Amp) identified as the IC-TL082 Op-Amp. The first op-amp works as a Schmidt circuit and the second op-amp works as an integration circuit. At the output, V_{out1}, of the Schmidt circuit a square wave is obtained. The output V_{out1} is the input to the integration circuit which generates the triangular wave, V_{out2}. The IC-TL082 require 12 Volt supply. Additionally, the condition $R_2 > R_3$ is necessary, to work in the oscillation mode of the circuit. Finally, if the absolute magnitudes of $+V_{sat}$ and $-V_{sat}$ are equal, the frequency of oscillation can be calculated by the following formula:

$$f = \frac{1}{4R_1C}\left(\frac{R_2}{R_3}\right)$$

for the component values of the circuit diagram, $R_1 = R_3 = 1.1$ kΩ, $R_2 = 4.7$ kΩ and $C = 21.6$ nF, the frequency of oscillation is approximately of 45 kHz.

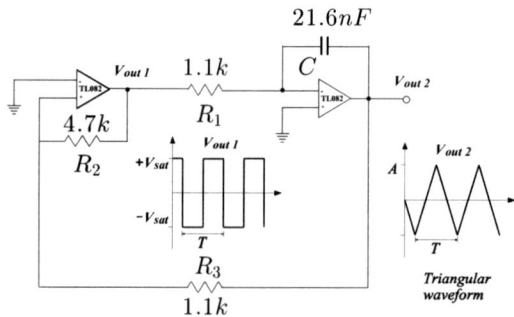

Fig. 3.31. Triangular-wave generator circuit.

On the other hand, the inverting amplifier circuit with offset null configuration generates an unipolar triangular-wave, which is necessary for the synthesis of the classical PWM block placed inside the $\Sigma - \Delta$-modulator

3.11 $\Sigma - \Delta$ Modulation

structure. The circuit diagram for this constitutive block is shown in Figure 3.31. The unipolar triangle-wave circuit was implemented via the op-amp IC-AD711. The output of the integration circuit, V_{out2} is the input to the inverting amplifier circuit, which additionally has offset null configuration. The output of the inverting amplifier circuit, denoted by V_{out3}, becomes the unipolar triangular-wave represented in Figure 3.32.

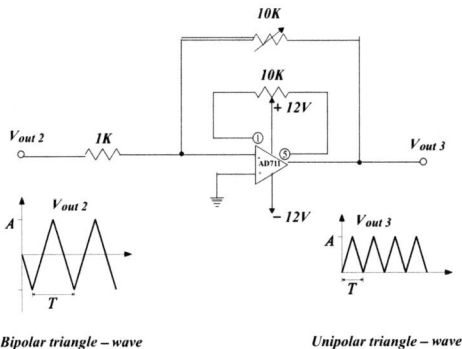

Fig. 3.32. Inverting amplifier circuit with offset null configuration.

A classical PWM circuit is synthesized with the help of a comparator circuit, which has, as inputs, a unipolar triangle-wave and a signal for comparison which is the output of the integrator circuit. The actual classical PWM circuit is shown in Figure 3.33.

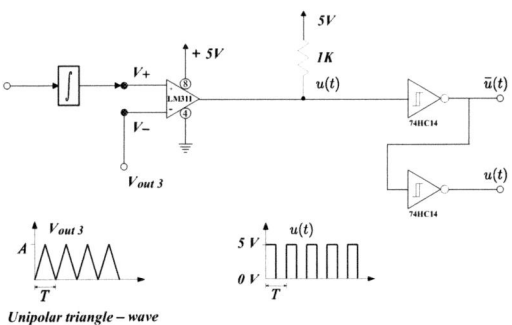

Fig. 3.33. Classical PWM circuit.

The PWM circuit was designed using the voltage comparator IC LM311, programmed to have a sampling rate of 45 kHz. The PWM circuit provides a digital output signal, with amplitudes of $0V$ or $+5V$, so can be interfaced to

TTL logic. The IC LM311 achieves this with their open collector output arrangement, a pull-up resistor ($R = 1\text{k}\Omega$) and a $0V$ to $+5V$ power supply. The resulting signal is directly interfaced with the TTL compatible IC NTE3087, via the Hex Inverting Schmidt Trigger IC 74HC14, which commands the gate of the Mosfet NTE2984 acting as a switch.

Figure 3.34 depicts a picture of the implemented $\Sigma - \Delta$-modulator.

Fig. 3.34. Hardware implementation of the experimental $\Sigma - \Delta$-modulator.

4

Approximate Linearization in the Control of Power Electronics Devices

4.1 Introduction

In this chapter, we undertake the simplest, most studied and most applied, controller design method for nonlinear systems, in general, and for dc to dc power converters in particular: the approximate linearization method. Many books have been devoted to linear control design techniques. The list is immense and a detailed survey of the available literature seems to be out of place in this book. We recommend a few fundamental references which have been particularly useful to the authors in understanding the issues and grasping the feeling for controller design. The book by Brockett [5] is a classical reference for a solid understanding of linear systems theory in general. An excellent reference textbook is that of Kailath [33] which is highly recommended to students and people unfamiliar with linear systems theory. A most readable and clearly written book is that of Furuta *et al.* [26]. The book by Rugh [51] is also highly recommended. A recent contribution with many interesting topics and a clear, didactic, approach in the exposition is the book by Glad and Ljung [29].

The philosophy of the approximate linearization method in the control of switched power electronics devices is quite simple: by designing a state, or output, feedback controller on the basis of the average tangent linearization model around a desired equilibrium, or reference trajectory, the use of the derived incremental average feedback controller will also prove to be successful on the full nonlinear system. The desired equilibrium, or the reference trajectory, will be achieved, or tracked, provided the controlled system (initial) state is sufficiently close to the desired control objective. Nevertheless, average dc-to-dc power converters enjoy a unique property which will be fully explored in the next chapter: The average approximate linearization based controller design is successful in semi-globally controlling the nonlinear system itself. This is true for stabilization and trajectory tracking tasks.

To ease the controller design computations, we resort, as in previous chapters, to system normalization. For the most known converter topologies, the incremental average feedback controller is here designed by resorting to several *linear feedback control* techniques primarily designed on the basis of the approximately linearized, normalized, average converter model. There exist a host of recent linear controller design techniques that we do not exploit, such as H_∞, matrix inequalities, L_2, etc. These will be left for the interested reader to explore and develop in the context of power electronics.

4.2 Some Linear Feedback Control Design Methods

In this section, we first briefly revisit some of the most traditional linear feedback control design methods and its applications in the regulation of dc-to-dc power converters. The basic road map is as follows: We start by obtaining, or presenting, the average normalized model of switched converters, next we compute the physically meaningful constant average equilibrium points of the system. Here we make an emphasis in parameterizing such equilibria in terms of the desired average normalized output voltage of the converter. We proceed to design a feedback controller by means of one of the following methods: pole placement via state feedback, pole placement via observer design, flatness, Generalized Proportional Integral (GPI) control, passivity based control by means of the energy shaping plus damping injection method and, finally, static passivity based control via a Hamiltonian viewpoint using feedback of the passive output error associated with the linearized exact tracking, or stabilization, error dynamics.

4.2.1 Pole Placement by Full State Feedback

The average linearized model of the normalized average converter circuit model, assumed controllable, is placed in the traditional state space form:

$$\dot{x}_\delta = A x_\delta + b u_{av,\delta} \quad (4.1)$$

where x_δ is the average *incremental state* defined as the difference, $x - \overline{x}$, with $\overline{x} = X$ being an average constant equilibrium state corresponding to the constant average equilibrium input, $\overline{u}_{av} = U$. The incremental average control input is then, $u_{av,\delta} = u_{av} - U$. The matrix A is a constant square $n \times n$ matrix and b is a constant, n-dimensional column vector. The controllability of the system is expressed by the fact,

$$\operatorname{rank} \mathcal{C} = \operatorname{rank} \begin{bmatrix} b, Ab, \cdots, A^{n-1}b \end{bmatrix} = n \quad (4.2)$$

The matrix $\mathcal{C}$ is known as the *controllability* matrix of the system (4.1).

Under such conditions, a linear incremental average feedback controller of the form:

$$u_{av,\delta} = -k^T x_\delta$$

exists which guarantees exponential stability of the origin of the average incremental state coordinates.

The trajectories of the closed loop linearized system,

$$\dot{x}_\delta = (A - bk^T)x_\delta$$

arising from initial incremental states $x_{0,\delta}$ are guaranteed to converge to the origin thanks to appropriate placement of the closed loop eigenvalues of the matrix $A - bk^T$ in the stable region of the complex plane (respecting symmetry with respect to the real axis).

The linear state feedback control law, using the incremental input definition, is given by,

$$u_{av} = \overline{u}_{av} - k^T x_\delta = U - k^T (x - X) \tag{4.3}$$

In the context of switched power electronics models, such as dc-to-dc power converters, the derived average feedback control law (4.3) cannot be directly implemented on the nonlinear converter due to the nature itself of the actual control input u. The feedback controller is fed to the switched power converter via a $\Sigma - \Delta$ modulator as shown in the scheme of Figure 4.1

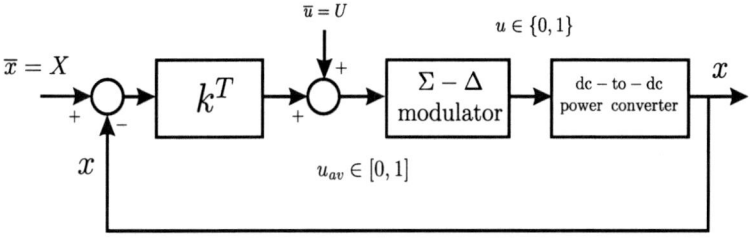

Fig. 4.1. Linear average state feedback via $\Sigma - \Delta$ modulation.

In dc-to-dc power converters, linearization based feedback controllers usually achieve regulation to the desired equilibrium point *even from initial conditions which are significantly far away from such an equilibrium*. This unique feature of power converters is perhaps at the root of the common practise, among power electronics engineers, of adopting the use of linear feedback controller design as a preferred control scheme rather than resorting to nonlinear feedback control techniques.

In particular, the linearization based feedback controllers perform a stabilization to the desired equilibrium point even when the motions of the controlled nonlinear system start at the origin of the state space, i.e., they are also suitable for the *start up* phase of the converters operation.

We will also use the time-invariant tangent linearization models, and the associated linear control laws, to achieve *trajectory tracking* in rest-to-rest, or equilibrium-to-equilibrium maneuvers. For this last special requirement, we invariably resort to the *flatness* property in order to specify the required nominal state and input trajectories associated with a particular trajectory tracking problem. The concept of flatness is explained further ahead in this chapter.

The linearized model of a nonlinear average model of the converter also leads to input-output linearized models of the system usually expressed in traditional transfer function form.

We show that, except for the Buck converter average model, such average input-output models invariably exhibit a non-minimum phase property when the incremental output capacitor voltage of the system is considered as the measured output of the system. On the other hand, the model obtained with the input inductor current, regarded as an output, exhibits a minimum phase property. An indirect regulation of the output voltage can then be feasibly achieved. For such indirect regulation possibility, we propose Proportional-Derivative (PD) controllers, which can be readily synthesized on the basis of the available state vector components. These average feedback controllers are, thus, based on a stable zero cancellation via dynamic linear state feedback. Naturally, such PD controllers are not feasible for input-output models based on the incremental average output capacitor voltage as measured output variable.

4.2.2 Pole Placement Based on Observer Design

Although, in principle, all variables in a circuit are measurable to a certain extent, it is also true that excess of measurement devices increases cost in circuit design and may contribute to performance and reliability degradation. Another common feature of switched power electronic devices, in which switched inductor currents are needed for control law synthesis, is the often discontinuous and chattering nature of such current variables. For this reason, it is, sometimes, not advisable to relay on inductor current measurements.

The preceding considerations lead to consider feedback laws based only on output system measurements. A common and traditional approach to output feedback control design is to resort to *state observers*, whether *full* state observers or *reduced* state observers. We provide several design examples concentrating only on reduced order observers.

In this chapter, we study the use of linear state observers for the control of traditional dc-to-dc power converters of the Buck, Boost and Buck-Boost types. The nonlinear average converter models are linearized about their equilibrium points and an observer is proposed for the average linearized model. The observed average state and the available output are then used in the synthesis of the linear full state feedback control law which is to regulate the nonlinear switched converter.

Consider the linearized average model of a dc-to-dc power converter

$$\dot{x}_\delta = Ax_\delta + bu_{av,\delta}$$
$$y_\delta = c^T x_\delta \qquad (4.4)$$

where y_δ is the incremental output, defined as $y - \overline{y}$ with $y = c^T x$ being the scalar average output of the system and $\overline{y} = Y$ is the equilibrium value of the output corresponding to the state equilibrium value $\overline{x} = X$, i.e., $\overline{y} = c^T \overline{x}$.

The linearized average converter system is assumed to be *observable* from the measured output $y_\delta = c^T x_\delta$. This means that the following property is valid

$$\text{rank } \mathcal{O} = \text{rank} \begin{bmatrix} c^T \\ c^T A \\ \vdots \\ c^T A^{n-1} \end{bmatrix} = n$$

The matrix $\mathcal{O}$ is known as the *observability* matrix of the system (4.4).

The Luenberger observer for such a system is represented by a dynamical system, built in the following form,

$$\dot{\widehat{x}}_\delta = A\widehat{x}_\delta + bu_{av,\delta} + l(y_\delta - \widehat{y}_\delta)$$
$$\widehat{y}_\delta = c^T \widehat{x}_\delta$$

where l is a (column) vector of constant gains. The estimation error $e_\delta = x_\delta - \widehat{x}_\delta$ evolves according to the dynamics

$$\dot{e}_\delta = (A - lc^T)e_\delta$$

The observability property of the average linearized system 4.4 guarantees that the eigenvalues of the matrix $(A - lc^T)$ can be arbitrarily placed in the complex plane (modulo symmetry with respect to the real axis) by appropriate choice of the entries in the vector l. In other words, the state estimation error is said to be stabilizable to zero by means of *output injection*. This guarantees global asymptotic stability of the origin of coordinates in the incremental estimation error space describing e_δ.

One of the fundamental results of linear systems theory is the *separation principle*. This principle states that the observer problem (i.e., that of selecting the gains in l that guarantee asymptotic estimation of the state variables x_δ) and the controller design problem (i.e., that of selecting the gains of the vector k^T to obtain an asymptotically stable system when the estimated states are used in the feedback law) can be independently solved, as if the estimated states were, in fact, the actual states.

This amounts to demonstrate that the following composite system is asymptotically stable to zero.

$$\dot{x}_\delta = Ax_\delta + bu_{av,\delta}$$
$$u_{av,\delta} = -k^T \widehat{x}_\delta$$
$$\dot{\widehat{x}}_\delta = A\widehat{x}_\delta + bu_{av,\delta} + lc^T(x_\delta - \widehat{x}_\delta)$$

provided the system $\dot{x}_\delta = (A - bk^T)x_\delta$ has the origin as an asymptotically stable equilibrium point.

The composite system is rewritten in matrix form as

$$\begin{bmatrix} \dot{x}_\delta \\ \dot{\hat{x}}_\delta \end{bmatrix} = \begin{bmatrix} A & -bk^T \\ lc^T & A - lc^T - bk^T \end{bmatrix} \begin{bmatrix} x_\delta \\ \hat{x}_\delta \end{bmatrix}$$

Transforming, via an invertible transformation, this system

$$\begin{bmatrix} x_\delta \\ e_\delta \end{bmatrix} = \begin{bmatrix} I & 0 \\ I & -I \end{bmatrix} \begin{bmatrix} x_\delta \\ \hat{x}_\delta \end{bmatrix}, \quad \begin{bmatrix} x_\delta \\ \hat{x}_\delta \end{bmatrix} = \begin{bmatrix} I & 0 \\ I & -I \end{bmatrix} \begin{bmatrix} x_\delta \\ e_\delta \end{bmatrix}$$

we obtain

$$\begin{bmatrix} \dot{x}_\delta \\ \dot{e}_\delta \end{bmatrix} = \begin{bmatrix} (A - bk^T) & bk^T \\ 0 & (A - lc^T) \end{bmatrix} \begin{bmatrix} x_\delta \\ e_\delta \end{bmatrix}$$

The stability of the overall closed loop system depends, in an decoupled manner, on the linear feedback stabilization of x_δ as if it were perfectly known and the stabilization of the state estimation error by means of output injection.

The average incremental output feedback control scheme, based on average incremental state estimation, for a dc-to-dc power converter, is depicted in Figure 4.2

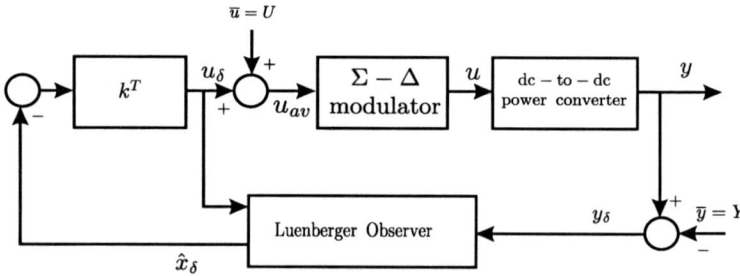

Fig. 4.2. Linear average output feedback for dc-to-dc power converters via $\Sigma - \Delta$ modulation.

4.2.3 Reduced Order Observers

Consider the linear system, expressed in composite form

$$\begin{bmatrix} \dot{x}_{1,\delta} \\ \dot{x}_{2,\delta} \end{bmatrix} = \begin{bmatrix} A_{11} & A_{12} \\ A_{21} & A_{22} \end{bmatrix} \begin{bmatrix} x_{1,\delta} \\ x_{2,\delta} \end{bmatrix} + \begin{bmatrix} b_1 \\ b_2 \end{bmatrix} u_\delta$$

$$y_\delta = [I \ 0] \begin{bmatrix} x_{1,\delta} \\ x_{2,\delta} \end{bmatrix} = x_{1,\delta}$$

where $x_{1,\delta} \in R^{n_1}$ and $x_{2,\delta} \in R^{n_2}$ with $n_1 + n_2 = n$. The elements in the matrices A and b are partitioned correspondingly to these dimensions. We assume that the pair of matrices $C = [I \ 0]$ and A are observable. We are interested in designing a *reduced order* observer, which only reconstructs, or estimates, the unmeasured state $x_{2,\delta}$ with an n_2 dimensional observer rather than a full order observer.

The $x_{2,\delta}$ system is then given by

$$\dot{x}_{2,\delta} = A_{21} y_\delta + A_{22} x_{2,\delta} + b_2 u_\delta$$

The first equation reads

$$\dot{y}_\delta = A_{11} y_\delta + A_{12} x_{2,\delta} + b_1 u_\delta$$

We may take the right hand side of the following expression:

$$A_{12} x_{2,\delta} = \dot{y}_\delta - A_{11} y_\delta - b_1 u_\delta \tag{4.5}$$

as a *virtual* measurement equation for $x_{2,\delta}$, in the $x_{2,\delta}$ subsystem, obtained through the artificial measurement map A_{12}.

Of course we have to avoid taking time derivatives on y_δ as suggested by this equation. If $z_\delta = A_{12} x_{2,\delta}$ is then considered as an extra measurement we propose the following observer for $x_{2,\delta}$,

$$\dot{\widehat{x}}_{2,\delta} = A_{21} y_\delta + A_{22} \widehat{x}_{2,\delta} + b_2 u_\delta + l_2(z_\delta - \widehat{z}_\delta)$$
$$z_\delta = A_{12} x_{2,\delta} = \dot{y}_\delta - A_{11} y_\delta - b_1 u_\delta$$
$$\widehat{z}_\delta = A_{12} \widehat{x}_{2,\delta}$$

Note that for such an observer the estimation error $e_{2,\delta} = x_{2,\delta} - \widehat{x}_{2,\delta}$ is given by subtracting, term by term, the following two equations:

$$\dot{x}_{2,\delta} = A_{21} y_\delta + A_{22} x_{2,\delta} + b_2 u_\delta$$
$$\dot{\widehat{x}}_{2,\delta} = A_{21} y_\delta + A_{22} \widehat{x}_{2,\delta} + b_2 u_\delta + l_2 A_{12}(x_{2,\delta} - \widehat{x}_{2,\delta})$$

we obtain

$$\dot{e}_{2,\delta} = (A_{22} - l_2 A_{12}) e_{2,\delta}$$

The desired asymptotic convergence property of $e_{2,\delta}$ to zero and, hence, the existence of an n_2-dimensional injection vector l_2, is established by determining the observability of the pair of matrices (A_{12}, A_{22}). This, incidentally, is a direct consequence of the assumed observability of the matrices $([I \ 0], A)$.

In order to obtain a feasible expression for the observer, which does not involve derivatives of the system output, y, we proceed as follows.

Substitution of the term $A_{12} x_{2,\delta}$, by its expression found in (4.5), into the obtained reduced order observer equation results in the following set of equalities:

$$\dot{\widehat{x}}_{2,\delta} = A_{21}y_\delta + A_{22}\widehat{x}_{2,\delta} + b_2 u_\delta + l_2(A_{12}x_{2,\delta} - A_{12}\widehat{x}_{2,\delta})$$
$$= A_{21}y_\delta + A_{22}\widehat{x}_{2,\delta} + b_2 u_\delta + l_2(\dot{y}_\delta - A_{11}y_\delta - b_1 u_\delta - A_{12}\widehat{x}_{2,\delta})$$
$$= (A_{21} - l_2 A_{11})y_\delta + (b_2 - l_2 b_1)u_\delta + (A_{22} - l_2 A_{12})\widehat{x}_{2,\delta} + l_2 \dot{y}_\delta$$

We define $\zeta_{2,\delta} = \widehat{x}_{2,\delta} - l_2 y_\delta$ and by adding and subtracting the quantity $-l_2 y_\delta$ to $\widehat{x}_{2,\delta}$, we obtain the reduced order observer for the unmeasured state $x_{2,\delta}$:

$$\dot{\zeta}_{2,\delta} = (A_{21} - l_2 A_{11})y_\delta + (b_2 - l_2 b_1)u_\delta$$
$$+ (A_{22} - l_2 A_{12})\zeta_{2,\delta} + (A_{22} - l_2 A_{12})l_2 y_\delta$$
$$\widehat{x}_{2,\delta} = \zeta_{2,\delta} + l_2 y_\delta$$

4.2.4 Flatness

In general, a nonlinear MIMO system is said to be *flat* if there exists a set of independent artificial outputs, of the same cardinality as the control input set, called the *flat outputs*, which *completely differentially parameterize* the system states, control inputs and natural outputs of the system. This means that *all* variables in the system can be expressed in terms of the flat outputs and a *finite number* of their time derivatives. As suspected, this property enormously facilitates the computation of nominal state and nominal input trajectories, once the flat output trajectories are established, in accordance with the desired control objectives of the underlying nonlinear system. The involvement of the concept of flatness is better explained in the need for obtaining *nominal* state and control input trajectories for the nonlinear average system model. These trajectories are thus easily computed when the system is known to be flat and the flat output has a clear physical meaning (this is usually the case). The method simply entitles to propose a trajectory for the *flat output* which is compatible with the desired rest-to-rest maneuver. The flat output trajectory uniquely determines the state and the control input trajectory for the nonlinear system.

In the context of linear dynamic systems, flatness, adopts its simpler relation to the controllability of the system. A linear system is flat if, and only if, it is controllable. In other words, flatness and controllability are equivalent in the context of linear systems.

For SISO systems, a system is flat if, and only if, we can find an artificial scalar output y which completely differentially parameterizes all the states and the input of the system.

Consider the linear system $\dot{x}_\delta = Ax_\delta + bu_\delta$ with full rank controllability matrix given by $\mathcal{C} = [b, Ab, \cdots A^{n-1}b]$, and define $z_\delta = \mathcal{C}^{-1}x_\delta$. We have that, in new coordinates z, the system is expressed as

$$\dot{z}_\delta = \mathcal{C}^{-1}A\mathcal{C}z_\delta + \mathcal{C}^{-1}bu_\delta$$

Let $\mathcal{F} = \mathcal{C}^{-1}A\mathcal{C}$ and $g = \mathcal{C}^{-1}b$. The form of these matrices is easily assessed from the relation

$$\mathcal{CF} = A\mathcal{C}, \quad \mathcal{C}g = b$$

In other words, $\mathcal{F}$ and g must be such that

$$[b, Ab, \cdots A^{n-1}b]\mathcal{F} = [Ab, A^2b, \cdots, A^nb]$$

and

$$[b, Ab, \cdots, A^{n-1}b]g = b$$

It follows, after taking into account Cayley-Hamilton's theorem:

$$A^n = -\alpha_1 A^{n-1} - \alpha_2 A^{n-2} - \cdots - \alpha_{n-1}A - \alpha_n I$$

that

$$\mathcal{F} = \begin{bmatrix} 0 & 0 & \cdots & 0 & -\alpha_n \\ 1 & 0 & \cdots & 0 & -\alpha_{n-1} \\ 0 & 1 & \cdots & 0 & -\alpha_{n-2} \\ \vdots & \vdots & \ddots & \vdots & \vdots \\ 0 & 0 & \cdots & 1 & -\alpha_1 \end{bmatrix}, \quad g = \begin{bmatrix} 1 \\ 0 \\ 0 \\ \vdots \\ 0 \end{bmatrix}$$

The transformed system is then given, explicitly, by the following set of differential equations

$$\dot{z}_{1,\delta} = -\alpha_n z_{n,\delta} + u_\delta$$
$$\dot{z}_{2,\delta} = z_{1,\delta} - \alpha_{n-1} z_{n,\delta}$$
$$\vdots$$
$$\dot{z}_{n-1,\delta} = z_{n-2,\delta} - \alpha_2 z_{n,\delta}$$
$$\dot{z}_{n,\delta} = z_{n-1,\delta} - \alpha_1 z_{n,\delta}$$

It is clear that the transformed state variable $z_{n,\delta}$ plays the role of the sought flat output variable, y_δ, capable of differentially parameterizing the rest of the state variables and the input.

Indeed, we obtain from the transformed equations, letting $y = z_n$,

$$z_{n-1,\delta} = \dot{y}_\delta + \alpha_1 y_\delta$$
$$z_{n-2,\delta} = \ddot{y}_\delta + \alpha_1 \dot{y}_\delta + \alpha_2 y_\delta$$
$$\vdots$$
$$z_{1,\delta} = y_\delta^{(n-1)} + \alpha_1 y_\delta^{(n-2)} + \cdots + \alpha_{n-1} y_\delta$$
$$u_\delta = y_\delta^{(n)} + \alpha_1 y_\delta^{(n-1)} + \cdots + \alpha_n y_\delta$$

The flat output is doubtlessly the transformed variable z_n, i.e.,

$$y_\delta = [0, 0, \cdots 0, 1] z_\delta$$

or, in terms of the original state variables

$$y_\delta = [0,\ 0,\ \cdots\ 0,\ 1]\mathcal{C}^{-1}x_\delta$$

Clearly any multiple factor of y_δ also completely differentially parameterizes the state variables and the input variable. The flat output is, thus, constituted by any multiple of the quantity obtained from the linear combination of the state variables formed with the last row of the inverse of the controllability matrix.

Note that the flat output y_δ is an *observable output* of the system.

Thanks to the differential parametrization provided by flatness, the feedback controller design problem is reduced to that corresponding to a pure chain of integrations.

Indeed, the state dependent input coordinate transformation

$$u_\delta = v_\delta + \alpha_1 y_\delta^{(n-1)} + \cdots + \alpha_n y_\delta$$

renders the system equivalent to the pure integration system

$$y_\delta^{(n)} = v_\delta \qquad (4.6)$$

with auxiliary input v_δ.

The successive time derivatives of the flat output are obtained as

$$\begin{bmatrix} y_\delta \\ \dot{y}_\delta \\ \vdots \\ y_\delta^{(n-1)} \end{bmatrix} = \begin{bmatrix} 0 & 0 & \cdots & 0 & 0 & 1 \\ 0 & 0 & \cdots & 0 & 1 & -\alpha_1 \\ 0 & 0 & \cdots & 1 & -\alpha_1 & -(\alpha_1+\alpha_2) \\ \vdots & \vdots & \cdots & \vdots & \vdots & \vdots \\ 1 & * & \cdots & * & * & * \end{bmatrix} \begin{bmatrix} z_{1,\delta} \\ z_{2,\delta} \\ \vdots \\ z_{n,\delta} \end{bmatrix}$$

The obtained time derivatives of the flat output y_δ can always be placed back in terms of the original state variables x_δ by simply using the (invertible) relation between z_δ and x_δ,

$$\begin{bmatrix} y_\delta \\ \dot{y}_\delta \\ \vdots \\ y_\delta^{(n-1)} \end{bmatrix} = \begin{bmatrix} 0 & 0 & \cdots & 0 & 0 & 1 \\ 0 & 0 & \cdots & 0 & 1 & -\alpha_1 \\ 0 & 0 & \cdots & 1 & -\alpha_1 & -(\alpha_1+\alpha_2) \\ \vdots & \vdots & \cdots & \vdots & \vdots & \vdots \\ 1 & * & \cdots & * & * & * \end{bmatrix} \mathcal{C}^{-1} x_\delta \qquad (4.7)$$

Naturally, in such simple system as those constituted by the average models of the dc-to-dc power converters such formula are seldom used since the corresponding relations may be worked out by hand.

The flat output usually has a nice physical interpretation. In terms of the linearized average models of most of the dc-to-dc power converters studied here, the flat outputs have the interpretation of an incremental linearized stored energy. In the Buck converter case, however, it just represents the output capacitor voltage.

Let the polynomial $p_d(s)$ in the complex variable s, with real coefficients $\{\gamma_1,\ldots,\gamma_{n-1}\}$, given by

$$p_d(s) = s^n + \gamma_1 s^{n-1} + \cdots + \gamma_n$$

be a *Hurwitz* polynomial. This means that $p_d(s)$ has all its roots strictly in the left portion of the complex plane.

A convenient full state linear feedback controller specifying the auxiliary output v_δ for the pure integration system 4.6 may be proposed to be,

$$v_\delta = -\gamma_1 y_\delta^{(n-1)} - \cdots - \gamma_n y_\delta$$

In other words, the feedback control law:

$$u_\delta = (\alpha_1 - \gamma_1) y_\delta^{(n-1)} + \cdots + (\alpha_n - \gamma_n) y_\delta \qquad (4.8)$$

renders a closed loop system whose characteristic polynomial is, precisely, $p_d(s)$.

Naturally, in view of the relation (4.7), the availability of the flat output y_δ, and its first $n-1$ time derivatives, is guaranteed as long as the state of the system x_δ is available for measurement. The feedback law (4.8) can then be synthesized as

$$u_{av} = \bar{u} + \begin{bmatrix} \theta_n & \theta_{n-1} & \cdots & \theta_1 \end{bmatrix} \begin{bmatrix} 0 & 0 & \cdots & 0 & 0 & 1 \\ 0 & 0 & \cdots & 0 & 1 & -\alpha_1 \\ 0 & 0 & \cdots & 1 & -\alpha_1 & -(\alpha_1+\alpha_2) \\ \vdots & \vdots & \cdots & \vdots & \vdots & \vdots \\ 1 & * & \cdots & * & * & * \end{bmatrix} C^{-1} x_\delta$$

with $\theta_j = \alpha_j - \gamma_j$, $j = 1, 2, \ldots n$.

If the states of the system are not readily available, but the flat output y_δ is, then thanks to the observability of the flat output, a Luenberger observer may be devised to asymptotically recover the full state of the system. The formulae developed above can be used except that x_δ is to be replaced by its asymptotic estimate. Similarly, if an observable output, other than the flat output, is the only available signal, the above formulae can still be used replacing the state x_δ by its asymptotic estimate.

4.2.5 Generalized Proportional Integral Controllers

If the design of a reduced asymptotic state observer needs to be avoided in the average feedback controller design, a possible alternative is represented by the use of *state reconstructors* based on iterated integration of inputs and outputs. The technique, also called Generalized Proportional Integral (GPI) control, is, fundamentally, a linear controller design technique. It bypasses the use of asymptotic observers and integrates the controller design into an input-output feedback approach.

A GPI controller may be shown to be equivalent to a *classical compensation network controller*. We shall explore the GPI control of average models of dc-to-dc power converters using the linearized model around a given equilibrium and, whenever tractable, we shall establish the relationship of these controllers with classical feedback compensation networks.

The GPI control technique is based on *integral reconstructors* of the state vector. Such reconstructors obtain the state variables as a finite linear combination of iterated integrals of inputs and outputs in compliance with the system model while regarding the unknown initial conditions, and other external perturbations of classical type, as being zero.

The obtained expressions for the state are then in error which may grow only in a time polynomial manner. Thanks to the superposition principle, the use of such a faulty estimator in any state based compensator design may be conveniently compensated (also in an unstable manner) using a compatible linear combination of iterated integrals of the output signal error, or of the input signal error, which guarantees overall closed loop asymptotic stability. It may be shown that the appropriate nesting of integrations in the controller expression, *à la* Hörmander, leads to an internally stable controller.

Consider the linear observable system $\dot{x}_\delta = Ax_\delta + bu_\delta$, $y_\delta = c^T x_\delta$. Neglecting the state initial conditions (formally setting $x_\delta(0) = 0$), we obtain

$$x_\delta = A \int_0^t x_\delta(\sigma) d\sigma + b \int_0^t u_\delta(\sigma) d\sigma$$

For simplicity we drop the integral limits and the integration variable in favor of the simpler notation

$$x_\delta = A(\int x_\delta) + b(\int u_\delta) \quad \text{1}$$

Iterating on the implicit expression for the state x_δ we find the following implicit expression for x in terms of $n-1$ integrals of x and the control input:

$$x_\delta = A^{n-1}(\int^{(n-1)} x_\delta) + \sum_1^{n-1} A^{j-1} b(\int^{(j)} u_\delta)$$

On the other hand, we may obtain the following string of output signal time derivatives

---

[1] We also let, $(\int^{(k)} \varphi_\delta) = \int_0^t \int_0^{\sigma_1} \cdots \int_0^{\sigma_{k-1}} \varphi_\delta(\sigma_k) d\sigma_k \cdots d\sigma_1$.

$$\begin{bmatrix} y_\delta \\ \dot{y}_\delta \\ \vdots \\ y_\delta^{(n-1)} \end{bmatrix} = \begin{bmatrix} c^T \\ c^T A \\ \vdots \\ c^T A^{n-1} \end{bmatrix} x + \begin{bmatrix} 0 & 0 & \vdots & 0 \\ cb & 0 & \vdots & 0 \\ cAb & cb & \cdots & 0 \\ \vdots & \vdots & \ddots & \vdots \\ cA^{n-2}b & cA^{n-3}b & \cdots & cb \end{bmatrix} \begin{bmatrix} u_\delta \\ \dot{u}_\delta \\ \vdots \\ u_\delta^{(n-2)} \end{bmatrix}$$

Integrating this expression a total of $n-1$ times, neglecting again initial conditions for the output derivatives and the input derivatives, we obtain:

$$\begin{bmatrix} (\int^{(n-1)} y_\delta) \\ (\int^{(n-2)} y_\delta) \\ \vdots \\ y_\delta \end{bmatrix} = \begin{bmatrix} c^T \\ c^T A \\ \vdots \\ c^T A^{n-1} \end{bmatrix} (\int^{(n-1)} x_\delta)$$

$$+ \begin{bmatrix} 0 & 0 & \vdots & 0 \\ cb & 0 & \vdots & 0 \\ cAb & cb & \cdots & 0 \\ \vdots & \vdots & \ddots & \vdots \\ cA^{n-2}b & cA^{n-3}b & \cdots & cb \end{bmatrix} \begin{bmatrix} (\int^{(n-1)} u_\delta) \\ (\int^{(n-2)} u_\delta) \\ \vdots \\ (\int u_\delta) \end{bmatrix}$$

which we rewrite, letting $\mathcal{O}$ denote the system observability matrix, as

$$\begin{bmatrix} (\int^{(n-1)} y_\delta) \\ (\int^{(n-2)} y_\delta) \\ \vdots \\ y_\delta \end{bmatrix} = \mathcal{O}(\int^{(n-1)} x_\delta) + M \begin{bmatrix} (\int^{(n-1)} u_\delta) \\ (\int^{(n-2)} u_\delta) \\ \vdots \\ (\int u_\delta) \end{bmatrix}$$

Eliminating the term with the iterated integral of x_δ between the two found expressions, we find

$$x_\delta = A^{n-1} \mathcal{O}^{-1} \left\{ \begin{bmatrix} (\int^{(n-1)} y_\delta) \\ (\int^{(n-2)} y_\delta) \\ \vdots \\ y_\delta \end{bmatrix} - M \begin{bmatrix} (\int^{(n-1)} u_\delta) \\ (\int^{(n-2)} u_\delta) \\ \vdots \\ (\int u_\delta) \end{bmatrix} \right\}$$

$$+ \sum_1^{n-1} A^{j-1} b (\int^{(j)} u_\delta)$$

Let $P(s^{-1})y$ and $Q(s^{-1})u$ denote suitable polynomial vectors of iterated integrations of the output and the input variables. We may adopt as a *structural estimate* of the state vector x, denoted by $\hat{x}$, the expression:

$$\widehat{x}_\delta = P(s^{-1})y_\delta + Q(s^{-1})u_\delta$$

The structurally estimated (or integrally reconstructed) state of the system in terms of finite linear combinations of iterated integrals of inputs and outputs exhibits a finite order time polynomial error with respect to the actual state vector value. Using the integral reconstruction of the state vector in place of the actual state in the linear state feedback control law produces an unstable closed loop system rendering the designed feedback control useless. One resorts then to the superposition principle and complements the estimated state feedback controller with a suitable additive signal comprising a linear combination of a sufficient number of iterated integrals of the incremental output errors, so that the destabilizing effect of the reconstruction error is cancelled in the closed loop system output differential equation description. This last statement of compensating unstable components is justified in the fact that a suitable integral action cancels the off-set effect caused by a constant estimation error (assumed to come from the integral reconstructor error). A suitably tuned double integration of the output stabilization, or tracking, error cancels an increasing, or decreasing, linear function of time appearing additively in the closed loop system as injected by the erroneous feedback. Three iterated integrals cancel a time parabolic term and so on. Naturally, the unstable cancellation requires some further algebraic manipulation in order to produce a stable controller. This is usually achieved through *iterated integration nesting* or by resorting to transfer function descriptions of the compensating subsystem comprising the reconstructed state feedback and the iterated integral compensation terms.

4.2.6 Passivity Based Control

The linearized models of the studied dc-to-dc power converters exhibit a clear "energy management" structure. The linearized system clearly exhibits the conservative part of the system, the dissipative part of the system and the energy acquisition part of the dynamics. It turns out that if internal resistances of inductors and switches are overlooked, all of the basic converter structures are under-damped, i.e., some degrees of freedom of the converter do not have any dissipative forces. Based on Lyapunov stability theory, we propose a desired time varying trajectory for the linearized dynamics state. This results in the need to inject damping into the desired system dynamics and to force the incremental energy (energy of the tracking error system) to be driven to zero by feedback.

The methodology results in an output dynamic feedback controller which induces a "shaped" closed loop energy and enhances the closed loop damping of the system. For this reason, the method is better known as the "Energy shaping + damping injection" (ESDI) methodology (see [48]).

It turns out that for the linearized models of the studied dc-to-dc power converters, the ESDI method produces simple to implement dynamic output feedback controllers.

4.2 Linear Feedback Control

Consider a linear time-invariant single input, controllable, system of the form: $\mathcal{P}\dot{x}_\delta = Ax_\delta + bu_\delta$, where $\mathcal{P}$ is a positive definite symmetric matrix. In most cases $\mathcal{P}$ is just the identity matrix or it can be removed by the invertible state coordinate transformation $z_\delta = \mathcal{P}^{-1}x_\delta$. We specifically assume that the (drift) vector Ax_δ of the linear system is such that the right hand side of the system exhibits the following particular structure:

$$\mathcal{P}\dot{x}_\delta = \mathcal{J}x_\delta - \mathcal{R}x_\delta + bu_\delta$$

where $\mathcal{J}$ is a skew-symmetric matrix, $\mathcal{R}$ is a symmetric, positive semi-definite, matrix, i.e.,

$$\mathcal{J}^T + \mathcal{J} = 0, \quad \mathcal{R}^T = \mathcal{R}$$

We say that the term $\mathcal{J}x_\delta$ represents the *conservative* forces and that the term $\mathcal{R}x_\delta$ represents the *dissipative forces* in the system. The term bu_δ is the energy acquisition term. A justification for advocating this terminology stems from the fact that if $x_\delta^T \mathcal{P} x_\delta$ represents a "state energy" then its time derivatives exhibits a zero term, $x_\delta^T \mathcal{J} x_\delta$, i.e., $\mathcal{J}x_\delta$ yields the *invariant part* of the energy, a negative semi-definite term $-x_\delta^T \mathcal{R} x_\delta$ and the term $x_\delta^T bu_\delta$ which is the one responsible for acquiring, or discarding, energy to control the system. Since the *passive* output of the system, denoted by y_δ, is defined as: $y_\delta = b^T x_\delta$, this last term is simply the product $y_\delta u_\delta$, also known as the *supply rate*.

Let $x_\delta^*(t)$ represent a state trajectory, which is deemed to be desirable. Controllability of the system guarantees the possibility of tracking such a state trajectory via a suitable feedback control law. Consider the following strictly positive tracking error energy function,

$$V(x_\delta - x_\delta^*(t)) = \frac{1}{2}(x_\delta - x_\delta^*(t))^T \mathcal{P}(x_\delta - x_\delta^*(t))$$

The time derivative of such an energy function is given by

$$\dot{V} = (x_\delta - x_\delta^*(t))^T (\dot{x}_\delta - \dot{x}_\delta^*(t))$$
$$= (x_\delta - x_\delta^*(t))^T (\mathcal{J}x_\delta - \mathcal{R}x_\delta + bu_\delta - \dot{x}_\delta^*(t))$$

Let us assume that the time derivative term, $\dot{x}_\delta^*(t)$, is taken to be

$$\dot{x}_\delta^*(t) = \mathcal{J}x_\delta^*(t) - \mathcal{R}x_\delta^*(t) + \mathcal{R}_I(x_\delta - x_\delta^*(t)) + bu_\delta \quad (4.9)$$

where $\mathcal{R}_I$ is a positive semi-definite matrix such that

$$\mathcal{R} + \mathcal{R}_I > 0$$

The *exogenous* system (4.9) represents a copy of the system with enhanced damping which is active only when the tracking error $e_\delta = x_\delta - x_\delta^*(t)$ is nonzero. It precisely coincides with the original system dynamics when the tracking error is null.

We obtain the following evaluation of the time derivative of V along the trajectories of the tracking error state $e_\delta = x_\delta - x_\delta^*(t)$:

$$\dot{V}(e) = -(x_\delta - x_\delta^*(t))^T (\mathcal{R} + \mathcal{R}_I)(x_\delta - x_\delta^*(t)) < 0$$

According to Lyapunov stability theory the error e_δ asymptotically converges to zero and, therefore, x_δ and x_δ^* converge to each other.

The dynamics (4.9) represents a *desired nominal dynamics*, or a *reference dynamics* that we would like the system to emulate. Thus, given a reference trajectory $x_\delta^*(t)$, and assuming that the state x_δ is measurable, we can compute the required control u_δ from one of the equations in the system (4.9) in which u_δ explicitly appears along with some components of the state x_δ. We can compute u_δ from the exogenous system equations:

$$u_\delta = \frac{b^T}{b^T b} [\mathcal{P}\ddot{x}_\delta^*(t) - \mathcal{J}x_\delta^*(t) + \mathcal{R}x_\delta^*(t) - \mathcal{R}_I(x_\delta - x_\delta^*(t))] \qquad (4.10)$$

which is simply obtained by pre-multiplying the exogenous system equations (4.9) by b^T and solving for u_δ.

Consider now the projection operator matrix

$$\mathcal{M} = \left(I - \frac{1}{b^T b} b b^T\right)$$

Note that $\mathcal{M}^2 = \mathcal{M}$, while $\mathcal{M}b = 0$. Thus $\mathcal{M}$ is a projection operator, along b, onto the orthogonal subspace to b. Thus, for any $z \in R^n$, $\mathcal{M}z \perp b$.

Using the projection operator, $\mathcal{M}$, on both sides of the reference trajectory dynamics (4.9), we obtain the following *redundant dynamic* system, representing the dynamic part of the required feedback controller.

$$\mathcal{M}\mathcal{P}\dot{x}_\delta^*(t) = \mathcal{M}\mathcal{J}x_\delta^*(t) - \mathcal{M}\mathcal{R}x_\delta^*(t) + \mathcal{M}\mathcal{R}_I(x_\delta - x_\delta^*(t)) \qquad (4.11)$$

Note that $b^T \mathcal{R}_I$ cannot be identically zero, for then the control u_δ in (4.10) does not depend on the system state x_δ and u_δ would be a *open loop* controller. Therefore $\mathcal{R}_I b \neq 0$ and b is not in the *null space* of the symmetric matrix $\mathcal{R}_I$. Since the null space and the range space of a symmetric matrix are *orthogonal*, it follows that b is in the range space of $\mathcal{R}_I$.

The effects of the complementary damping map $\mathcal{R}_I$ which are present in the $n-1$ dimensional subspace orthogonal to the range of b do not affect the feedback control actions since these pass through the input vector b before affecting the state evolution. Hence, it is superfluous to propose a matrix $\mathcal{R}_I$ whose range space does not coincide with the one-dimensional range space of the vector b. This yields, for any arbitrary positive constant γ, the need that the system satisfies following *dissipation matching condition*:

$$\mathcal{R} + \gamma b b^T > 0 \qquad (4.12)$$

which will be important in the next section.

4.2.7 A Hamiltonian Systems Viewpoint

Consider again the linear system, written out in the form

$$\dot{x} = \mathcal{J}x - \mathcal{R}x + bu$$

If the output of the system happens to be given by $y = b^T x$, it receives the name of *passive output*. This simple fact prompts us to view the previous version of the system as a *Hamiltonian system*.

Indeed, note that if we take $H = \frac{1}{2}x^T x$, and from the fact that the *column vector*: $\partial H/\partial x$ coincides with the state x, we may take H as a *storage function*. The system, with scalar output $y = b^T x$, has the natural representation as a *linear Hamiltonian system*:

$$\dot{x} = \mathcal{J}\frac{\partial H}{\partial x} - \mathcal{R}\frac{\partial H}{\partial x} + bu$$

$$y = b^T \frac{\partial H}{\partial x}$$

In fact, it is easy to show that any linear system in state space representation trivially enjoys such a Hamiltonian representation taking the stored energy to be $H(x) = \frac{1}{2}x^T x$. We have,

$$\dot{x} = A\frac{\partial H}{\partial x} + bu$$

The square matrix A can be decomposed into a symmetric part and a skew symmetric part as follows

$$A = \frac{1}{2}(A + A^T) + \frac{1}{2}(A - A^T)$$

Hence,

$$\dot{x} = \frac{1}{2}(A - A^T)\frac{\partial H}{\partial x} + \frac{1}{2}(A + A^T)\frac{\partial H}{\partial x} + bu$$

Let $\mathcal{J} = \frac{1}{2}(A - A^T)$ and define $\mathcal{R} = -\frac{1}{2}(A + A^T)$. The system is then represented as

$$\dot{x} = \mathcal{J}x - \mathcal{R}x + bu$$

The passive output is clearly $y = b^T x = b^T \frac{\partial H}{\partial x}$. Note that it will be assumed that $\mathcal{R} \geq 0$.

The total time derivative of the scalar storage function H, with due account of the fact that

$$\frac{\partial H}{\partial x^T}\mathcal{J}\frac{\partial H}{\partial x} = \left[\frac{\partial H}{\partial x}\right]^T \mathcal{J}\frac{\partial H}{\partial x} = 0,$$

and that $y = b^T \frac{\partial H}{\partial x}$, is given by:

$$\dot{H} = \frac{\partial H}{\partial x^T}\left[\mathcal{J}\frac{\partial H}{\partial x} - \mathcal{R}\frac{\partial H}{\partial x} + bu\right]$$

$$= -\frac{\partial H}{\partial x^T}\mathcal{R}\frac{\partial H}{\partial x} + yu$$

Integrating the previous expression, we obtain, by virtue of the positive semi-definite character of $\mathcal{R}$, the basic *passivity relation*

$$H(x(t)) - H(x(t_0)) \leq \int_{t_0}^{t} y(\sigma)u(\sigma)d\sigma$$

Suppose for a moment that the control objective is to stably drive the system state $x = \frac{\partial H}{\partial x}$ towards zero. Furthermore, note that due to the fact that H is a storage function, $H(0) = 0$. We assume also that H is positive definite (usually, it is only a positive semi-definite function in order to qualify as a storage function).

The passivity relation implies that along the trajectories of the system we have

$$\frac{d}{dt}H(x(t)) \leq yu = -\frac{\partial H}{\partial x^T}\mathcal{R}\frac{\partial H}{\partial x} + \frac{\partial H}{\partial x^T}bu$$

(note that y is scalar and, hence, $y^T = y$). We may clearly choose the control input as an *output feedback control* law of the form:

$$u = -\gamma b^T \frac{\partial H}{\partial x} = -\gamma y$$

where γ is a positive scalar quantity.

The total time derivative of the energy function $H > 0$ is given by

$$\dot{H} = -\frac{\partial H}{\partial x^T}\left[\mathcal{R} + \gamma bb^T\right]\frac{\partial H}{\partial x}$$

The time derivative of H is negative definite as long as the matrix $\mathcal{R}+\gamma bb^T$ is negative definite. This means that the proposed output feedback controller, $u = -\gamma y$, causes the origin to be a globally asymptotically stable equilibrium point. If, on the other hand, $\mathcal{R} + \gamma bb^T$ is only negative semi-definite. The origin is still an asymptotically stable equilibrium point provided the largest invariant set

$$\left\{x \mid \frac{\partial H}{\partial x^T}\left[\mathcal{R} + \gamma bb^T\right]\frac{\partial H}{\partial x} = x^T\left[\mathcal{R} + \gamma bb^T\right]x = 0\right\}$$

is represented only by $x = 0$.

The stability condition $\left[\mathcal{R} + \gamma bb^T\right] > 0$ is equivalent to the following matching condition:

$$\mathcal{N}(\mathcal{R}) \subset Im\,(b), \quad Im\,\mathcal{R} \oplus Im\,(b) = R^n$$

Note that since $\dim\{Im(b)\} = 1$, then for guaranteeing global asymptotic stability of the origin of coordinates, it is necessary that $\dim\{Im\,\mathcal{R}\}$, equals $n - 1$. Since this is seldom the case in converters of dimension greater than 2, then the La Salle's invariance principle, associated with the set of states satisfying the relation:

$$\left\{ x \mid \frac{\partial H}{\partial x^T} \left[\mathcal{R} + \gamma b b^T \right] \frac{\partial H}{\partial x} = 0 \right\} = \{0\}$$

must be verified.

The feedback control design approach, implied in the previous developments, can be easily extended to *stabilization* and *tracking* problems.

Indeed, if the desired equilibrium is given by $\overline{x}$, with corresponding equilibrium input $\overline{u}$, we can rewrite the linear system, after defining $e = x - \overline{x}$ and $e_u = u - \overline{u}$ and $e_y = y - \overline{y}$, as:

$$\dot{e} = \mathcal{J} \frac{\partial H(e)}{\partial e} - \mathcal{R} \frac{\partial H(e)}{\partial e} + b e_u$$

$$e_y = b^T \frac{\partial H(e)}{\partial e}$$

Note that with $H = 0.5 x^T x$, we have

$$x - \overline{x} = \frac{\partial H(x)}{\partial x} - \left. \frac{\partial H(x)}{\partial x} \right|_{\overline{x}} = \frac{\partial H(e)}{\partial e}$$

Clearly, the incremental control input $e_u = -\gamma e_y$, or, equivalently: $u = \overline{u} - \gamma e_y = \overline{u} - \gamma(y - \overline{y})$, yields, under the previous assumptions, the origin of the state error space, e, as an asymptotically stable equilibrium.

In some instances, the Hamiltonian model of the system adopts the following *modified* Hamiltonian form

$$\mathcal{D}\dot{x} = \mathcal{J} \frac{\partial H}{\partial x} - \mathcal{R} \frac{\partial H}{\partial x} + bu,$$

$$y = b^T \frac{\partial H}{\partial x}$$

with $\mathcal{D} = \mathcal{D}^T > 0$ and the usual structural characteristics: $\mathcal{J} + \mathcal{J}^T = 0$, $\mathcal{R} = \mathcal{R}^T \geq 0$.

Finally, in the multi-variable case, with $H(x) = \frac{1}{2} x^T x$

$$\dot{x} = \mathcal{J} \frac{\partial H}{\partial x} - \mathcal{R} \frac{\partial H}{\partial x} + Bu$$

with B a full rank $n \times m$ matrix, The passive outputs are readily given by

$$y = B^T \frac{\partial H}{\partial x}$$

4 Approximate Linearization Methods

The system is evidently passive since

$$\dot{H}(x) = -\frac{\partial H}{\partial x^T}\mathcal{R}\frac{\partial H}{\partial x} + y^T u \leq y^T u$$

The output feedback control law

$$u = -\Gamma y, \quad \Gamma = \Gamma^T > 0$$

yields,

$$\dot{H}(x) = -\frac{\partial H}{\partial x^T}\left[\mathcal{R} + B\Gamma B^T\right]\frac{\partial H}{\partial x}$$

The dissipation matching condition, guaranteeing exponential asymptotic stability of the origin, adopts the form:

$$\mathcal{R} + B\Gamma B^T > 0$$

4.3 The Buck Converter

4.3.1 Generalities about the Average Normalized Model

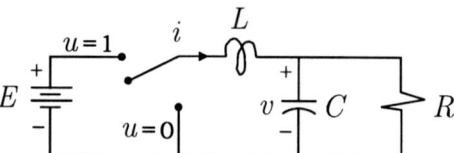

Fig. 4.3. The Buck converter.

Consider the average normalized model of the Buck converter circuit shown in Figure 4.3.

$$\dot{x}_1 = -x_2 + u_{av}$$
$$\dot{x}_2 = x_1 - \frac{x_2}{Q}$$

where x_1 is the normalized inductor current, x_2 is the normalized output voltage and u_{av} represents the average switch position function, necessarily restricted to continuously take values on the set $[0, 1]$.

The average system, which is linear, is clearly controllable and observable from each one of the states when taken as single system outputs.

Equilibrium point

The equilibrium of the average normalized state, parameterized in terms of the desired average equilibrium output voltage $\bar{x}_2 = V_d$ is given by

$$\bar{x}_1 = \frac{V_d}{Q}, \qquad \bar{x}_2 = V_d, \qquad \bar{u}_{av} = V_d$$

Hence, the desired equilibrium output voltage must satisfy the restriction:

$$0 < V_d < 1$$

Input-output model

Taking $y = x_1$ as the system output, the input-output relation is given by:

$$\ddot{y} + \frac{1}{Q}\dot{y} + y = \dot{u}_{av} + \frac{1}{Q}u_{av}$$

The system is clearly *minimum phase* from this output, since when $y = \bar{y} = V_d/Q$, the zero dynamics is given by

$$\dot{u}_{av} = -\frac{1}{Q}(u_{av} - Q\bar{y}) = -\frac{1}{Q}(u_{av} - V_d)$$

The system is stable with eigenvalues located at the points

$$s_{1,2} = -\frac{1}{2Q} \pm \sqrt{\frac{1}{4Q^2} - 1}$$

The two roots above are real and negative whenever $\frac{1}{2Q} > 1$. They are equal when $Q = 0.5$. Otherwise, the roots are complex, but they are still located in the stable portion of the complex plane.

Taking $y = x_2$ as the system output, the input-output relation is given by:

$$\ddot{y} + \frac{1}{Q}\dot{y} + y = u_{av}$$

The system is devoid of a zero dynamics, clearly indicating that $y = x_2$ is the flat output.

The system is also stable with eigenvalues located at the points

$$s_{1,2} = -\frac{1}{2Q} \pm \sqrt{\frac{1}{4Q^2} - 1}$$

The two roots are real and negative whenever $\frac{1}{2Q} > 1$, critical at $Q = 0.5$, and they are, otherwise, complex, but they are still stable.

4.3.2 Controller Design by Pole Placement

Recall the average normalized model of the Buck converter

$$\dot{x}_1 = -x_2 + u_{av}$$
$$\dot{x}_2 = x_1 - \frac{x_2}{Q}$$

We assume that both states are available for measurement. Suppose it is desired to regulate the state of the system to the equilibrium point $\bar{x}_1 = V_d/Q$, $\bar{x}_2 = V_d$.

Notice that the equilibrium state satisfies the relations:

$$0 = -\bar{x}_2 + \bar{u}_{av}$$
$$0 = \bar{x}_1 - \frac{\bar{x}_2}{Q}$$

Defining: $e_1 = x_1 - \bar{x}_1$, $e_2 = x_2 - \bar{x}_2$ and $e_{u_{av}} = u_{av} - \bar{u}_{av} = u_{av} - V_d$, we find, subtracting term by term both expressions, the state error dynamics

$$\dot{e}_1 = -e_2 + e_{u_{av}}$$
$$\dot{e}_2 = e_1 - \frac{e_2}{Q}$$

We devise the following state error feedback control law:

$$e_{u_{av}} = -k_1 e_1 - k_2 e_2$$

This choice yields the closed loop system:

$$\dot{e}_1 = -k_1 e_1 - (1 + k_2) e_2$$
$$\dot{e}_2 = e_1 - \frac{e_2}{Q}$$

which, in matrix form, reads

$$\dot{e} = \begin{bmatrix} -k_1 & -(1 + k_2) \\ 1 & -\frac{1}{Q} \end{bmatrix} e$$

The eigenvalues of the closed loop system are obtained from the solution of the characteristic equation:

$$s^2 + \left(\frac{1}{Q} + k_1\right)s + \frac{k_1}{Q} + k_2 + 1 = 0$$

By comparison with a desired characteristic equation of the form

$$s^2 + 2\zeta\omega_n s + \omega_n^2 = 0$$

we obtain the controller gains

$$k_1 = -\frac{1}{Q} + 2\zeta\omega_n, \qquad k_2 = -1 - \frac{1}{Q}\left(-\frac{1}{Q} + 2\zeta\omega_n\right) + \omega_n^2$$

The average state error feedback controller is then found to be:

$$u_{av} = \overline{u}_{av} + \left(\frac{1}{Q} - 2\zeta\omega_n\right)e_1 + \left(1 - \omega_n^2 - \frac{1}{Q^2} + \frac{2}{Q}\zeta\omega_n\right)e_2$$

with:

$$e_1 = x_1 - \frac{V_d}{Q}, \qquad e_2 = x_2 - V_d, \qquad \overline{u}_{av} = V_d$$

4.3.3 Proportional-Derivative Control via State Feedback

Recall the normalized average model of the Buck converter

$$\dot{x}_1 = -x_2 + u_{av}$$
$$\dot{x}_2 = x_1 - \frac{x_2}{Q}$$

Assume both states are available for measurement. Suppose it is desired to regulate the output voltage x_2 to a reference constant value $V_d \in (0,1)$. The input output relation, with $y = x_2$ was determined to be:

$$\ddot{y} + \frac{1}{Q}\dot{y} + y = u_{av}$$

A feedback controller that places the closed loop response in the form:

$$\ddot{y} + 2\zeta\omega_n\dot{y} + \omega_n^2(y - V_d) = 0$$

where ζ and ω_n represent desirable damping factor and natural frequency is given by the following proportional derivative controller with constant feedforward term:

$$u_{av} = (1 - \omega_n^2)y + \left(\frac{1}{Q} - 2\zeta\omega_n\right)\dot{y} + \omega_n^2 V_d$$

From the state space representation of the system we have that $y = x_2$ and $\dot{y} = x_1 - x_2/Q$. Placing the proportional derivative controller in terms of the states of the system yields:

$$u_{av} = \left(\frac{1}{Q} - 2\zeta\omega_n\right)x_1 + \left(1 - \omega_n^2 - \frac{1}{Q^2} + \frac{2}{Q}\zeta\omega_n\right)x_2 + \omega_n^2 V_d$$

Note that the derived controller completely coincides with the error state feedback controller previously designed.

Indeed,
$$u_{av} = \left(\frac{1}{Q} - 2\zeta\omega_n\right)x_1 + \left(1 - \omega_n^2 - \frac{1}{Q^2} + \frac{2}{Q}\zeta\omega_n\right)x_2 + \omega_n^2 V_d$$

subtracting the equilibrium state value to each state and adding the required terms so as not to alter the expression we have

$$u_{av} = \left(\frac{1}{Q} - 2\zeta\omega_n\right)\left(x_1 - \frac{V_d}{Q}\right) + \left(1 - \omega_n^2 - \frac{1}{Q^2} + \frac{2}{Q}\zeta\omega_n\right)(x_2 - V_d)$$
$$+ \omega_n^2 V_d + \left(\frac{1}{Q} - 2\zeta\omega_n\right)\frac{V_d}{Q} + \left(1 - \omega_n^2 - \frac{1}{Q^2} + \frac{2}{Q}\zeta\omega_n\right)V_d$$
$$= V_d + \left(\frac{1}{Q} - 2\zeta\omega_n\right)e_1 + \left(1 - \omega_n^2 - \frac{1}{Q^2} + \frac{2}{Q}\zeta\omega_n\right)e_2$$

The state stabilization problem is most efficiently accomplished by considerations on the input output relation associated to the variable x_2 alone.

Simulations

We used the following normalized parameter value $Q = 0.5$, the desired normalized voltage $V_r = 0.6$, and for the controller design we set:

$$\zeta = 0.81, \quad \omega_n = 0.9$$

Figure 4.4 shows the normalized average response of the Buck system to the proposed PD controller using full state feedback.

Figure 4.5 shows the Buck converter responses to a $\Sigma-\Delta$ modulator implementation of the average normalized state feedback control. Simulations were carried out with the Euler integration algorithm including a large integration step of 0.1 time units.

Figure 4.6 shows the responses of the same Buck converter to a $\Sigma - \Delta$ implementation of the average normalized state feedback control. Simulations were carried out with the Euler integration algorithm including a smaller integration step of 0.05 time units.

4.3.4 Trajectory Tracking

Consider now the problem of trajectory tracking for the output voltage. In particular, we are interested in driving the system from an initial equilibrium to a final equilibrium in a given amount of time. Let $V_r(\tau)$ be a time function defining the desired normalized voltage output reference trajectory starting at a constant value V_{r1} at time $\tau = \tau_1$ and ending at the value V_{r2} at time $\tau = \tau_2$ with $\tau_2 = \tau_1 + T$.

Recall the input output expression for the average normalized system:

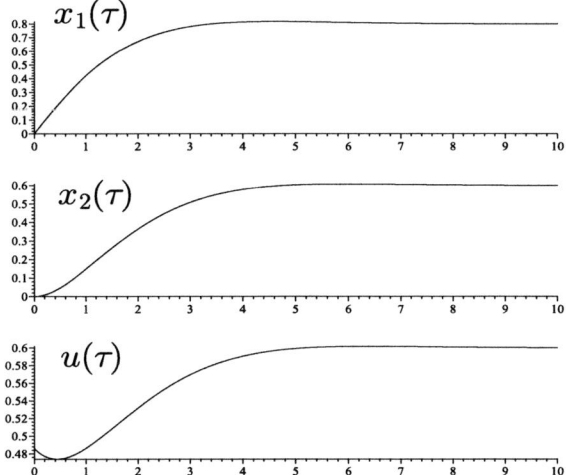

Fig. 4.4. Average PD-controlled performance of Buck converter.

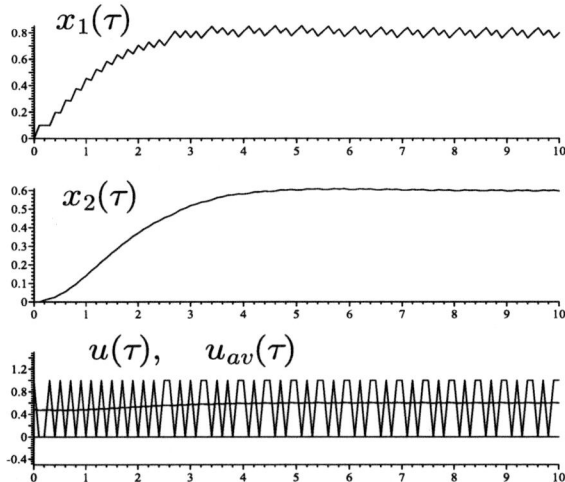

Fig. 4.5. Performance of PD-controlled Buck converter using a $\Sigma - \Delta$ modulator.

$$\ddot{y} + \frac{1}{Q}\dot{y} + y = u_{av}$$

A controller that imposes a desired characteristic polynomial for the tracking error dynamics $\ddot{e}_y + 2\zeta\omega_n\dot{e}_y + \omega_n^2 e_y = 0$, with $e_y = y - V_r(\tau)$, thus achieving the desired equilibrium transfer, is given by

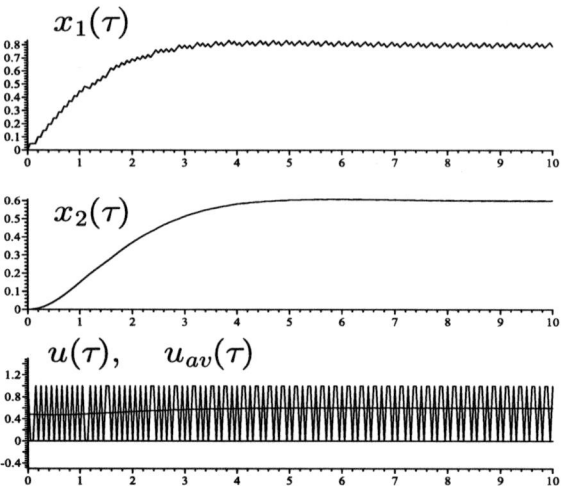

Fig. 4.6. PD-controlled responses of a Buck converter using a $\Sigma - \Delta$ modulator with small integration step.

$$u_{av} = (1 - \omega_n^2)y + \left(\frac{1}{Q} - 2\zeta\omega_n\right)\dot{y} + \omega_n^2 V_r(\tau) + 2\zeta\omega_n \dot{V}_r(\tau) + \ddot{V}_r(\tau)$$

This controller is seen to be equivalent to the following controller after adding and subtracting some feed-forward terms:

$$u_{av} = (1 - \omega_n^2)(y - V_r) + \left(\frac{1}{Q} - 2\zeta\omega_n\right)(\dot{y} - \dot{V}_r) + V_r(\tau) + \frac{1}{Q}\dot{V}_r(\tau) + \ddot{V}_r(\tau)$$

but, $\ddot{V}_r(\tau) + \frac{1}{Q}\dot{V}_r(\tau) + V_r(\tau)$ is just the nominal control input corresponding to the desired output trajectory. We call such a controller $u^*(\tau)$, or $u_r(\tau)$. We have:

$$u_{av} = (1 - \omega_n^2)(y - V_r(\tau)) + \left(\frac{1}{Q} - 2\zeta\omega_n\right)(\dot{y} - \dot{V}_r(\tau)) + u^*(\tau)$$

In terms of the tracking error we obtain:

$$u_{av} = (1 - \omega_n^2)e_y + \left(\frac{1}{Q} - 2\zeta\omega_n\right)\dot{e}_y + u^*(\tau)$$

Let $u_{av} - u^*(\tau) = e_{u_{av}}$, we get

$$e_{u_{av}} = (1 - \omega_n^2)e_y + \left(\frac{1}{Q} - 2\zeta\omega_n\right)\dot{e}_y$$

Notice that the control input

4.3 The Buck Converter

$$u_{av} = e_{u_{av}} + u^*(\tau) = e_{u_{av}} + \ddot{V}_r(\tau) + \frac{1}{Q}\dot{V}_r(\tau) + V_r(\tau)$$

yields the closed loop system

$$\ddot{e}_y + \frac{1}{Q}\dot{e}_y + e_y = e_{u_{av}}$$

The tracking error proportional derivative controller $e_{u_{av}}$ finally yields the desired tracking error dynamics

$$\ddot{e}_y + 2\zeta\omega_n\dot{e}_y + \omega_n^2 e_y = 0$$

Note that we would have obtained exactly the same result if we started with the system model along with the "reference model"

$$\dot{x}_1 = -x_2 + u_{av}$$
$$\dot{x}_2 = x_1 - \frac{x_2}{Q}$$

$$\dot{x}_1^*(\tau) = -x_2^*(\tau) + u^*(\tau)$$
$$\dot{x}_2^*(\tau) = x_1^*(\tau) - \frac{x_2^*(\tau)}{Q}$$

where evidently

$$x_1^*(\tau) = \dot{x}_2^*(\tau) + \frac{1}{Q}x_2^*(\tau)$$

The error system is just

$$\dot{e}_1 = -e_2 + e_{u_{av}}$$
$$\dot{e}_2 = e_1 - \frac{e_2}{Q}$$

where $e_1 = x_1 - x_1^*$ and $e_2 = x_2 - x_2^*(\tau) = y - V_r(\tau)$.

Letting $e_y = e_2$ we have the input output relationship

$$\ddot{e}_y + \frac{1}{Q}\dot{e}_y + e_y = e_{u_{av}}$$

The proportional derivative controller is the justified also from a state space viewpoint.

Simulations

We propose a controller for smoothly rising the steady state normalized average output voltage, used as a target in the previous simulation example, from the value: $V_r(\tau_1) = V_{r1} = 0.6$ towards the final desired value: $V_r(\tau_2) = V_{r2} = 0.8$, in, say, 5 normalized units of time.

150   4 Approximate Linearization Methods

We use the following reference trajectory:

$$V_r(\tau) = V_{r1} + (V_{r2} - V_{r1})\varphi(\tau, \tau_1, \tau_2)$$

where $\varphi(\tau, \tau_1, \tau_2)$ is a piecewise polynomial function interpolating between the values of 0 and 1, which is of the following form:

$$\varphi(\tau, \tau_1, \tau_2) = \begin{cases} 0 & \text{for } \tau \leq \tau_1 \\ \left(\frac{\tau-\tau_1}{\tau_2-\tau_1}\right)^5 \left[21 - 35\left(\frac{\tau-\tau_1}{\tau_2-\tau_1}\right) + 15\left(\frac{\tau-\tau_1}{\tau_2-\tau_1}\right)^2\right] & \text{for } \tau \in (\tau_1, \tau_2) \\ 1 & \text{for } \tau \geq \tau_2 \end{cases}$$

Figure 4.7 shows the trajectory tracking features of the average normalized Buck converter to a proportional derivative tracking controller.

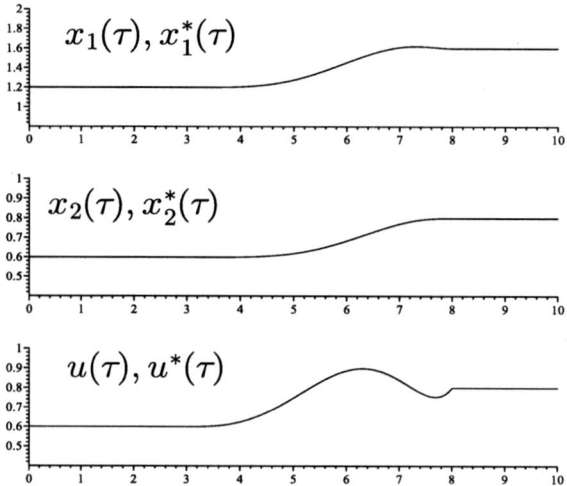

Fig. 4.7. Average PD controlled trajectory tracking for a Buck converter.

Figure 4.8 shows the trajectory tracking features of the switched normalized Buck converter model to a proportional derivative tracking controller implemented through a $\Sigma - \Delta$ modulator.

4.3.5 Fliess' Generalized Canonical Forms

A standard procedure in the control of linear systems consists in the so called transformation to *controller canonical form*. In our Buck example this amounts to a rather direct exercise in using invertible state coordinate transformations. For instance, in the average Buck converter system

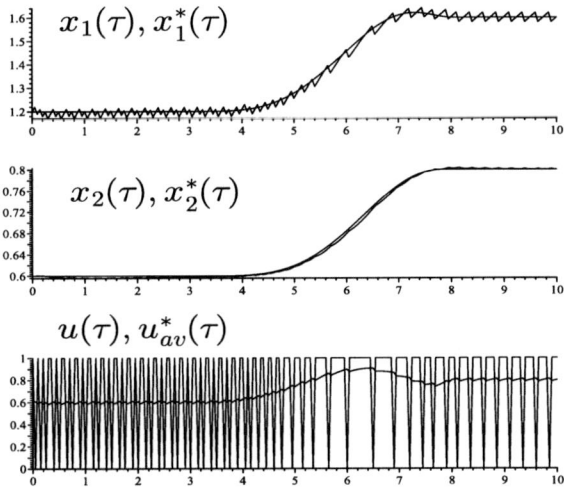

Fig. 4.8. PD controlled trajectory tracking for a Buck converter using a $\Sigma - \Delta$ modulator.

$$\dot{x}_1 = -x_2 + u_{av}$$
$$\dot{x}_2 = x_1 - \frac{1}{Q}x_2$$

The invertible state coordinate transformation:

$$\begin{bmatrix} z_1 \\ z_2 \end{bmatrix} = \begin{bmatrix} 0 & 1 \\ 1 & -\frac{1}{Q} \end{bmatrix} \begin{bmatrix} x_1 \\ x_2 \end{bmatrix}, \qquad \begin{bmatrix} x_1 \\ x_2 \end{bmatrix} = \begin{bmatrix} \frac{1}{Q} & 1 \\ 1 & 0 \end{bmatrix} \begin{bmatrix} z_1 \\ z_2 \end{bmatrix}$$

leads to the controlled system model

$$\dot{z} = \begin{bmatrix} 0 & 1 \\ -1 & -\frac{1}{Q} \end{bmatrix} z + \begin{bmatrix} 0 \\ 1 \end{bmatrix} u_{av}$$

Clearly, this canonical form is quite useful for control design purposes.

If we consider the average normalized system, along with an output function, then the *Fliess' generalized observability canonical form* may be computed. Since when the output is $y = x_2$, the resulting observability canonical form coincides with the controller canonical form (a fact intimately related to the *flatness* of the particular output x_2), we only revise the case when $y = x_1$.

Consider the following invertible input dependent state coordinate transformation of the average model:

$$z_1 = x_1, \qquad z_2 = -x_2 + u_{av}, \qquad x_1 = z_1, \qquad x_2 = -z_2 + u_{av}$$

We obtain

$$\dot{z}_1 = z_2$$
$$\dot{z}_2 = -z_1 - \frac{1}{Q}z_2 + \frac{1}{Q}u_{av} + \dot{u}_{av}$$
$$y = z_1$$

The nature of the zero dynamics of the output and the dynamic feedback controller design by stable zero cancellation follow quite readily from this canonical form.

4.3.6 State Feedback Control via Observer Design

Consider the normalized average model of the Buck converter circuit

$$\dot{x}_1 = -x_2 + u_{av}$$
$$\dot{x}_2 = x_1 - \frac{x_2}{Q}$$

where x_1 represents the normalized inductor current, x_2 stands for the normalized output voltage and u_{av} denotes the control input (switch position function).

The average system model was found to be controllable and observable from both states. In particular, the system is observable from the output capacitor voltage variable $y = x_2$.

An observer for the state variable x_1 could be tentatively proposed as follows:

$$\dot{\widehat{x}}_1 = -y + u + \lambda(x_1 - \widehat{x}_1)$$

with $\lambda > 0$ a design constant. Evidently, such an "observer" is not feasible, due to the fact that it requires the actual state x_1 for its realization, which is precisely the variable we would like to estimate. Nevertheless, the observer has a satisfying property: the estimation error $e_1 = x_1 - \widehat{x}_1$ exhibits the dynamics

$$\dot{e}_1 = -\lambda e_1$$

which implies exponential stability of the origin of the error space, as long as we choose $\lambda > 0$.

Let us replace the average current variable x_1 by its equivalent expression, obtained from the second equation of the Buck converter model:

$$x_1 = \dot{y} + \frac{y}{Q}$$

We obtain the observer

$$\dot{\widehat{x}}_1 = -y + u_{av} + \lambda \left(\dot{y} + \frac{y}{Q} - \widehat{x}_1 \right)$$

4.3 The Buck Converter

The additional complication with this observer is the presence of the time derivative of the output signal y. Something that can be achieved in practise nowadays, but which is not advisable in many many applications. For this reason, define the auxiliary variable $\zeta = \widehat{x}_1 - \lambda y$ and obtain, after some rearrangement, the stable dynamics of this auxiliary variable,

$$\dot{\zeta} = -\lambda \zeta - \left(1 - \frac{\lambda}{Q} + \lambda^2\right) y + u_{av}$$

The auxiliary variable ζ only requires for its synthesis, inputs, outputs, and being able to solve linear differential equations from arbitrary initial conditions. The synthesis of ζ as a dynamic system poses no problem whatsoever. From knowledge of ζ, the estimate of x_1 is readily obtained as

$$\widehat{x}_1 = \zeta + \lambda y$$

We propose the observer controller average feedback law:

$$u_{av} = V_d - k_1 \left(\widehat{x}_1 - \frac{V_d}{Q}\right) - k_2(x_2 - V_d)$$
$$\widehat{x}_1 = \zeta + \lambda y$$
$$\dot{\zeta} = -\lambda \zeta - \left(1 - \frac{\lambda}{Q} + \lambda^2\right) y + u_{av}$$

with k_1 and k_2 designed *as if* the state of the average system were fully available for measurement. In other words, we used the same gains, k_1 and k_2, found for the regulation of the converter through full state average feedback control and pole placement described earlier. As already mentioned, we chose $\lambda > 0$.

Simulations

Figure 4.9 shows the responses of the normalized Buck converter variables to the full state linear feedback controller synthesized with the help of a reduced order observer estimating the inductor current and implemented via a $\Sigma - \Delta$ modulator. The initial condition of the observer was, on purpose, chosen to be different to that of the initial normalized inductor current.

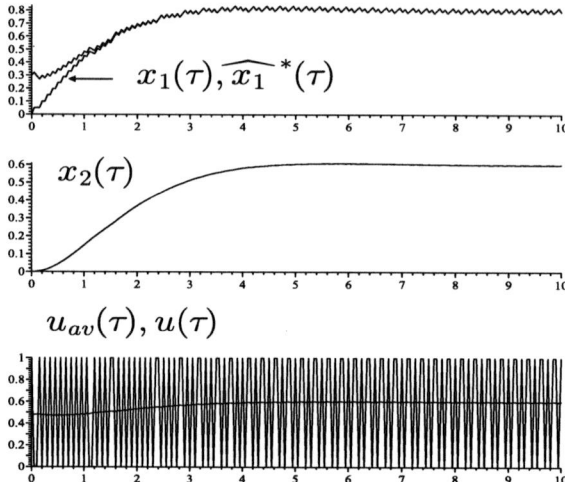

Fig. 4.9. Controlled responses of a Buck converter using a reduced order observer.

4.3.7 GPI Controller Design

Consider the average normalized model of the Buck converter with the output capacitor voltage x_2 taken as the measured output,

$$\dot{x}_1 = -x_2 + u_{av}$$
$$\dot{x}_2 = x_1 - \frac{x_2}{Q}$$
$$y = x_2$$

The average inductor current is given, modulo a constant additive initial condition term, by

$$\widehat{x}_1(t) = \int_0^t [u(\sigma) - y(\sigma)]\, d\sigma$$

A state feedback controller of the form,

$$u_{av} = V_d - k_1 x_1 - k_2 x_2$$

would be replaced by the GPI controller:

$$u_{av} = V_d - k_1 \widehat{x}_1 - k_2 y - k_3 \rho, \qquad \dot{\rho} = y - V_d$$

Note that the reconstructed state $\widehat{x}_1$ differs from the actual value of the average inductor current in a constant quantity determined by the initial condition $x_1(t_0) = x_{10}$. Indeed

$$\widehat{x}_1 = x_1 - x_{10}$$

This fact becomes useful at the closed loop stability analysis stage.
The closed loop system is then given by

$$\dot{x}_1 = -(y - V_d) - k_1(x_1 - x_{10}) - k_2 y - k_3 \rho$$
$$\dot{y} = x_1 - \frac{y}{Q}$$
$$\dot{\rho} = y - V_d$$

Eliminating x_1 we obtain

$$\ddot{y} = -(y - V_d) - k_1 \left(\dot{y} + \frac{1}{Q} y - x_{10} \right) - k_2 y - k_3 \rho - \frac{1}{Q} \dot{y}$$
$$\dot{\rho} = y - V_d$$

In other words,

$$\ddot{y} = -\left(\frac{1}{Q} + k_1 \right) \dot{y} - \left(k_2 + \frac{k_1}{Q} + 1 \right)(y - V_d) - k_3 \rho + \gamma$$
$$\dot{\rho} = y - V_d$$

with γ being a constant of the form: $\gamma = k_1 x_{10} - (k_2 + k_1/Q) V_d$

Finally, the closed loop system, in terms of the average output capacitor voltage variable is given by

$$y^{(3)} + \left(\frac{1}{Q} + k_1 \right) \ddot{y} + \left(k_2 + \frac{k_1}{Q} + 1 \right) \dot{y} + k_3(y - V_d) = 0$$

Evidently, the design gains k_1, k_2 and k_3 can be chosen to obtain a desired characteristic polynomial with roots located at preselected locations in the stable portion of the complex plane. For instance, letting the desired characteristic polynomial $p_d(s)$ be

$$p_d(s) = (s^2 + 2\xi \omega_n s + \omega_n^2)(s + p)$$
$$= s^3 + (p + 2\xi \omega_n) s^2 + (\omega_n^2 + 2\xi \omega_n p) s + \omega_n^2 p$$

We can immediately obtain the design gains by direct comparison of the closed loop characteristic polynomial with the desired characteristic polynomial

$$k_1 = p + 2\xi \omega_n - \frac{1}{Q}$$
$$k_2 = -\frac{1}{Q} \left(p + 2\xi \omega_n - \frac{1}{Q} \right) - 1 + \omega_n^2 + 2\xi \omega_n p$$
$$k_3 = \omega_n^2 p$$

The proposed GPI controller admits a reinterpretation in classical terms. Define $e_{u_{av}} = u_{av} - \bar{u}_{av} = u_{av} - V_d$ and let $e_y = y - V_d$. We may rewrite the proposed GPI controller as:

$$e_{u_{av}} = -k_1 \int_{t_0}^{t}(e_{u_{av}} - e_y)dt - k_2 e_y - k_3 \int_{t_0}^{t} e_y dt - k_2 V_d$$

In transfer function notation, the controller is obtained as

$$e_{u_{av}}(s) = -\left[\frac{k_2 s + (k_1 - k_3)}{s + k_3}\right] e_y(s)$$

The reinterpretation of the average GPI feedback control scheme as a classical compensating network is shown in Figure 4.10.

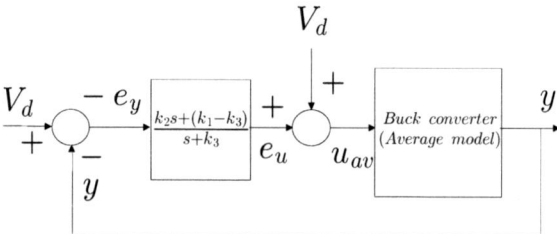

Fig. 4.10. The GPI controller as a classical compensation network.

The form of the classical controller corresponds to either a "lead" or a "lag" controller, depending, respectively, whether the zero of the controller transfer function is closer to the imaginary axis than the pole of the same transfer function.

Simulations

Figure 4.11 shows the performance of the GPI controller acting on a switched Buck converter.

The design parameters were set to be

$$V_d = 0.6, \quad \zeta = 0.81, \quad \omega_n = 1, \quad p = 1$$

For this choice the controller turned out to be a *lag* controller.

4.3.8 Passivity Based Control

Note that the normalized Buck converter model

$$\dot{x}_1 = -x_2 + u_{av}$$
$$\dot{x}_2 = x_1 - \frac{x_2}{Q}$$
$$y = x_2$$

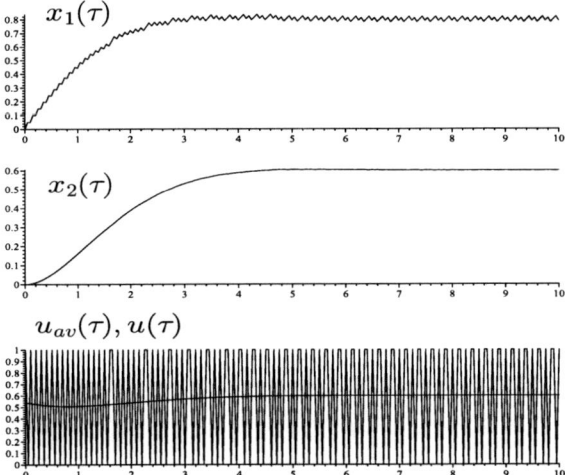

Fig. 4.11. Performance of GPI controlled Buck converter.

may be equivalently written in a manner that clearly depicts the energy structure of the system

$$\dot{x} = \begin{bmatrix} 0 & -1 \\ 1 & 0 \end{bmatrix} x + \begin{bmatrix} 0 & 0 \\ 0 & -\frac{1}{Q} \end{bmatrix} x + \begin{bmatrix} 1 \\ 0 \end{bmatrix} u_{av}$$

The first term clearly exhibits the conservative part of the dynamics, the second term is the dissipative part of the dynamics and the last one is the input port or energy acquisition term. We briefly denote these terms by $\dot{x} = \mathcal{J}x + \mathcal{R}x + bu_{av}$.

Consider the following energy function $V(x) = \frac{1}{2}(x - x_d(\tau))^T(x - x_d(\tau))$ where $x_d(\tau)$ represents a desired state trajectory.

The time derivative of the energy function, along the trajectories of the system is given by

$$\dot{V}(x) = (x - x_d)^T(\dot{x} - \dot{x}_d) = (x - x_d)^T(\mathcal{J}x + \mathcal{R}x + bu - \dot{x}_d)$$

Let $\dot{x}_d$ be given by

$$\dot{x}_d = \mathcal{J}x_d + \mathcal{R}x_d + \mathcal{R}_I(x - x_d) + bu_{av}$$

where $\mathcal{R}_I$ is a positive definite symmetric matrix, such that $\mathcal{R}_d = \mathcal{R} - \mathcal{R}_I$ is a symmetric, negative definite matrix. We have

$$\dot{V}(x) = (x - x_d)^T(\mathcal{J}(x - x_d) + \mathcal{R}_d(x - x_d))$$
$$= (x - x_d)^T \mathcal{R}_d(x - x_d) < 0$$

As a consequence the tracking error $(x - x_d)$ decreases asymptotically to zero.

The dynamics

$$\dot{x}_d = \mathcal{J} x_d + \mathcal{R} x_d + \mathcal{R}_I(x - x_d) + b u_{av}$$

yields both the control input and the dynamics of the required controller.

We set,

$$\mathcal{R}_I = \begin{bmatrix} R_I & 0 \\ 0 & 0 \end{bmatrix}, \quad R_I > 0$$

We obtain, then, the following dynamical controller, after replacing x_{2d} by the controller state variable ξ

$$u_{av} = \dot{x}_{1d} + \xi - R_I(x_1 - x_{1d})$$

$$\dot{\xi} = x_{1d} - \frac{1}{Q}\xi$$

In the previously derived controller, addressed as a *damping injection plus energy shaping controller*, or shortly, passivity based controller, the signal, $x_{1d}(t)$, represents a nominal desired trajectory for the average Buck converter circuit model.

Note that a nominal trajectory for x_1 of the form, $x_{1d}(\tau)$, may be completely specified in terms of a desired behavior for the average output capacitor voltage x_2, which we denote by $x_2^*(\tau)$. Indeed, x_1 is differentially parameterizable in terms of x_2 via the expression

$$x_1 = \dot{x}_2 + \frac{1}{Q}x_2$$

We thus obtain x_{1d} from the relation

$$x_{1d}(t) = \dot{x}_2^*(t) + \frac{1}{Q}x_2^*(\tau)$$

Simulations

We propose the derived passivity based controller for smoothly rising the steady state normalized average output voltage from the initial equilibrium value: $\bar{x}_2(\tau_1) = 0.6$ towards the final desired value: $\bar{x}_2(\tau_2) = 0.8$, in, say, $\tau_2 - \tau_1 = 5$ normalized units of time.

We use, as proposed earlier, the following reference trajectory:

$$x_2^*(\tau) = \bar{x}_2(\tau_1) + (\bar{x}_2(\tau_2) - \bar{x}_2(\tau_1))\varphi(\tau, \tau_1, \tau_2)$$

where $\varphi(\tau, \tau_1, \tau_2)$ is a piecewise polynomial function interpolating between the values of 0 and 1, which is of the following form:

4.3 The Buck Converter   159

$$\varphi(\tau,\tau_1,\tau_2) = \begin{cases} 0 & \text{for } \tau \le \tau_1 \\ \left(\frac{\tau-\tau_1}{\tau_2-\tau_1}\right)^5 \left[252 - 1050\left(\frac{\tau-\tau_1}{\tau_2-\tau_1}\right) + 1800\left(\frac{\tau-\tau_1}{\tau_2-\tau_1}\right)^2 \right. \\ \left. -1575\left(\frac{\tau-\tau_1}{\tau_2-\tau_1}\right)^3 + 700\left(\frac{\tau-\tau_1}{\tau_2-\tau_1}\right)^4 - 126\left(\frac{\tau-\tau_1}{\tau_2-\tau_1}\right)^5\right] \\ & \text{for } \tau \in (\tau_1,\tau_2) \\ 1 & \text{for } \tau \ge \tau_2 \end{cases}$$

For the simulations, we set: $\tau_1 = 3$ and $\tau_2 = 8$ with $R_I = 1$.

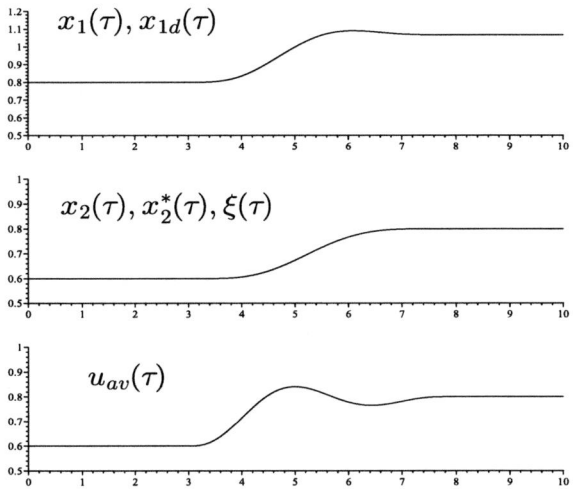

Fig. 4.12. Passivity based average controlled responses of Buck converter in a trajectory tracking task.

Figure 4.12 shows the trajectory tracking features of the average normalized Buck converter responding to a passivity based tracking controller achieving a rest-to-rest maneuver for the output capacitor voltage.

Figure 4.13 shows the trajectory tracking features of the switched normalized Buck converter, implemented with the help of a $\Sigma - \Delta$ modulator, responding to a passivity based tracking controller achieving a rest-to-rest maneuver of the output capacitor voltage.

4.3.9 The Hamiltonian Systems Viewpoint

Consider the positive definite storage function of the incremental variables $e_1 = x_1 - \overline{x}_1 = x_1 - \frac{V_d}{Q}$, $e_2 = x - V_d$, given by $H(e) = 0.5(e_1^2 + e_2^2)$.

$$\dot{e} = \begin{bmatrix} 0 & -1 \\ 1 & 0 \end{bmatrix} \frac{\partial H(e)}{\partial e} + \begin{bmatrix} 0 & 0 \\ 0 & -\frac{1}{Q} \end{bmatrix} \frac{\partial H(e)}{\partial e} + \begin{bmatrix} 1 \\ 0 \end{bmatrix} e_{u_{av}}$$

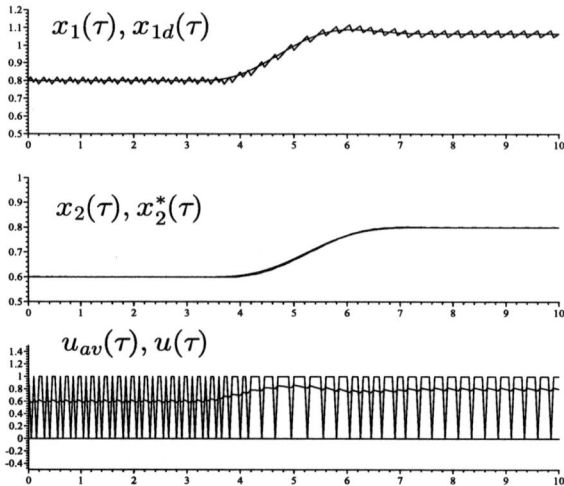

Fig. 4.13. Passivity based controlled responses of Buck converter in a tracking task implemented via a $\Sigma - \Delta$ modulator.

The passive output error variable e_y is given by

$$e_y = y - \overline{y} = e_1 = \begin{bmatrix} 1 & 0 \end{bmatrix} \frac{\partial H(e)}{\partial e}$$

The passivity based controller is simply obtained from $e_{u_{av}} = -\gamma e_y$, i.e.,

$$u_{av} = \overline{u}_{av} - \gamma(x_1 - \overline{x}_1) = V_d - \gamma \left(x_1 - \frac{V_d}{Q} \right)$$

The closed loop system is given by

$$\dot{x}_1 = -x_2 + V_d - \gamma \left(x_1 - \frac{V_d}{Q} \right)$$

$$\dot{x}_2 = x_1 - \frac{x_2}{Q}$$

$$y = x_1$$

This system is equivalent to the average Hamiltonian error system:

$$\dot{e} = \begin{bmatrix} 0 & -1 \\ 1 & 0 \end{bmatrix} \frac{\partial H(e)}{\partial e} + \begin{bmatrix} -\gamma & 0 \\ 0 & -\frac{1}{Q} \end{bmatrix} \frac{\partial H(e)}{\partial e}$$

The origin of the error space turns into a globally, exponentially asymptotically stable equilibrium point for the closed loop system.

$$u_{av} = \overline{u}_{av} - \gamma(x_1 - \overline{x}_1) = V_d - \gamma \left(x_1 - \frac{V_d}{Q} \right) \tag{4.13}$$

Note the striking similarity, and simplicity, of this controller with the one obtained by standard passivity considerations.

Simulations

For the simulations we have chosen $\gamma = 0.25$, $V_d = 0.6$ and $Q = 0.75$. Figure 4.14 shows the average system response to the linear passive output average feedback control, starting from zero initial conditions.

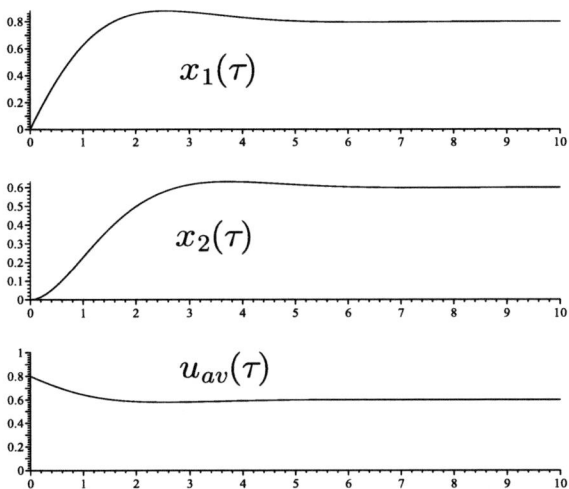

Fig. 4.14. Average responses of a Buck converter to a linear passivity based controller.

In order to test the versatility of this simple controller, we demanded a rest-to-rest trajectory tracking task. The corresponding controller turned out to be

$$u_{av}(\tau) = u^*_{av}(\tau) - \gamma[x_1 - x_1^*(\tau)] \qquad (4.14)$$

with $u^*_{av}(\tau)$ and $x_1^*(\tau)$ determined on the basis of the desired output voltage trajectory, $y^*(\tau) = x_2^*(\tau)$, by means of the differential parametrization:

$$x_1^*(\tau) = \dot{y}^*(\tau) + \frac{1}{Q}y^*(\tau)$$

$$u^*_{av}(\tau) = \ddot{y}^*(\tau) + \frac{1}{Q}\dot{y}^*(\tau) + y^*(\tau)$$

We set an equilibrium-to-equilibrium transfer from the initial value:

$$(\bar{x}_1(\tau_1),\ \bar{x}_2(\tau_1)) = (V_{d1}/Q, V_{d1}) = (0.8, 0.6)$$

to the final equilibrium value:

$$(\bar{x}_1(\tau_2),\ \bar{x}_2(\tau_2)) = (V_{d2}/Q, V_{d2}) = (0.8, 1.0667)$$

within a time interval $[\tau_1, \tau_2]$, defined by $\tau_1 = 3$, $\tau_2 = 8$, i.e., $\tau_2 - \tau_1 = 5$ normalized time units. Figure 4.15 shows the performance of the responses of the system to the average feedback controller actions, implemented through a $\Sigma - \Delta$ modulator.

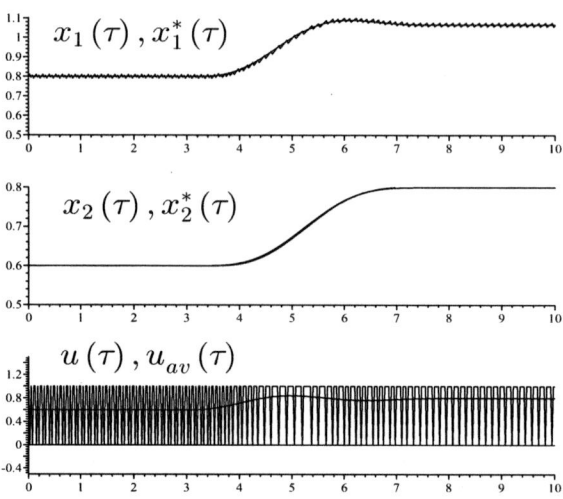

Fig. 4.15. Trajectory tracking performance of a Buck converter using linear passivity based controller and $\Sigma - \Delta$ modulation.

4.3.10 Implementation of the Linear Passivity Based Control for the Buck Converter

In this section, we implement and validate, using a $\Sigma - \Delta$-*modulator circuit*, the feedback controller obtained by standard passivity considerations (4.13) on the *Buck system* prototype.

Figure 4.16 shows a block diagram of the *Buck system* including a *control* block, and an *amplitude limiter circuit* block.

The *Buck system* and *driver* blocks, respectively, were already explained in detail in the Section 2.2. We will present the experimental implementation of a linear passive output average feedback control acting on the the Buck converter and implemented with the help of a $\Sigma - \Delta$-*modulator* circuit. We only describe here the synthesized *control circuit* and its associated *amplitude limiter circuit*.

Control block

In the *control circuit* block the average linear passivity based control strategy is realized. The electric current i and the voltage signals v are received from

Buck system

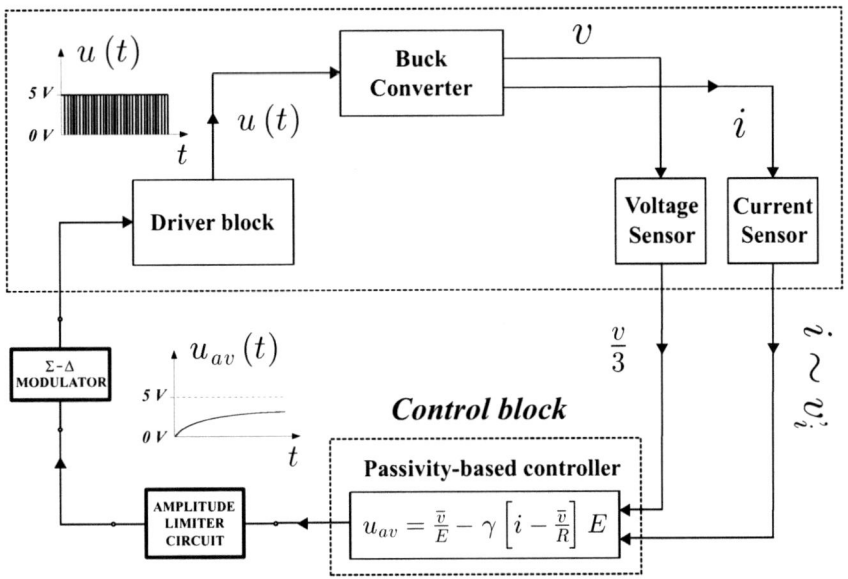

Fig. 4.16. Block diagram of the Buck power converter to a $\Sigma - \Delta$-modulator implementation of a linear passive output average feedback control.

the *Buck system* block (see Figure 2.5). Control strategy (4.13) is implemented using analog electronics lifting the normalization of the involved expressions. Using

$$x_1 = \frac{1}{E}\sqrt{\frac{L}{C}}i, \qquad x_2 = \frac{v}{E}, \qquad Q = R\sqrt{\frac{C}{L}}$$

we can rewrite (4.13) as:

$$u_{av} = \frac{\bar{v}}{E} - \gamma_{actual}\left[i - \frac{\bar{v}}{R}\right]E \qquad (4.15)$$

where $\gamma_{actual} > 0$, represents the de-normalized form of the normalized gain γ. The relation within γ_{actual} and γ normalized (denoted only by the symbol γ) after quite straightforward but tedious algebraic manipulations, is given by the following expression:

$$\gamma = \left[E^2\sqrt{\frac{C}{L}}\right]\gamma_{actual} \qquad (4.16)$$

this relation is also true for the Boost and Buck-Boost power converters.

Figure 4.17 shows the actual *control* block circuit. It also shows the transfer functions that realize the op-amps that achieve the actual implementation of the designed passivity based controller (4.15). The figure also depicts the inputs and the output signals of the *control* block.

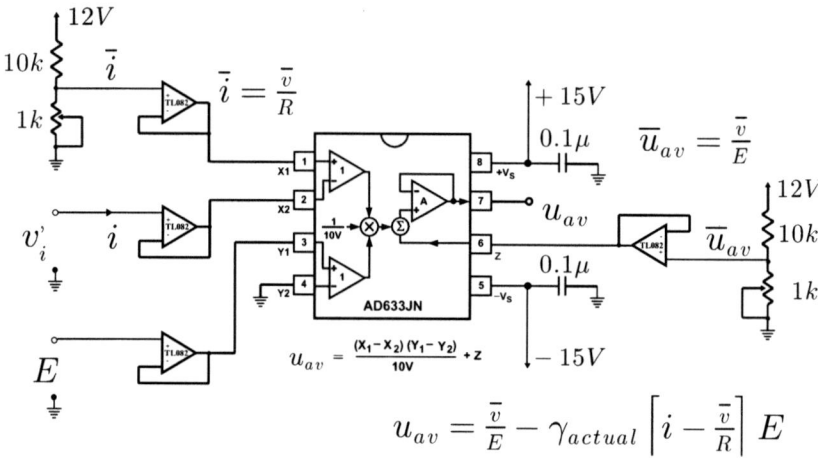

$$u_{av} = \frac{\bar{v}}{E} - \gamma_{actual} \left[\bar{i} - \frac{\bar{v}}{R} \right] E$$

Fig. 4.17. Control circuit structure implemented for the passivity based stabilizing controller.

The *control* block processes two signals coming from the Buck converter: the *inductor current signal* i, transformed into a proportional voltage signal v'_i using the LEM HAW 15-P current sensor (see Figure 2.7), and the *output voltage* v. On the other hand, the output of the *control* block is the denormalized input u_{av} given by (4.15).

Ideally, the average control output u_{av} is limited to take values into the interval $[0, 1]$. In practise this interval is actually a $[0,5]$ volt interval. The average control signal is transformed into an equivalent switching pulse signal, with amplitudes of 0 V and 5 V, thanks to the $\Sigma - \Delta$-*modulator* block. The output of this block commands the gate of the Mosfet IC NT2984.

Generally speaking, the limit bounds for u_{av} are exceed when the Buck converter operates in the transitory start-up state. Consequently, the $\Sigma - \Delta$-*modulator* block could suffer damages. Thus, we built a block that limits the input voltage for the $\Sigma - \Delta$-modulator block to the interval $[0\ V, 5\ V]$. For this task we using a *Amplitude limiter circuit*, which we explain next.

Amplitude limiter circuit

The *amplitude limiter circuit*, or *clipper*, is used to eliminate signal values which lie outside a certain voltage interval. In our case, we desire to limit the output voltage of the *control* block, u_{av}, to the interval $[0\ V, 1\ V]$ at all times.

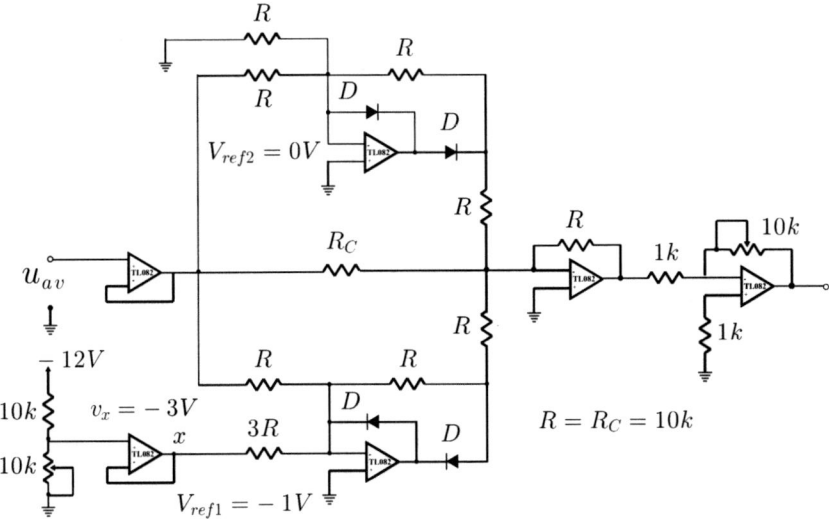

Fig. 4.18. Circuit diagram of an amplitude limiter circuit.

An *amplitude limiter circuit* consists of a *bipolar-output dead-zone circuit* and an adding resistor, R_C, (see [7]). Since the output of the first component is an *inverting precision limiter*, this output is connected to the input of an *inverting amplifier*, which has a closed loop gain given by:

$$A = \frac{v_o}{v_i} = -\frac{R_F}{R_i}$$

We set $A = -5$ to accomplish the re-scaling required at the input of the experimental $\Sigma - \Delta$-modulator block. As a consequence, the resulting output for the *amplitude limiter circuit* block is a *non-inverting precision limiter*, shown in Figure 4.18. Figure 4.19 shows the hardware implementation of the *amplitude limiter circuit* block. The actual *amplitude limiter circuit* block was designed with the help of six op-amps IC TL082.

Experimental results

Figure 4.20 depicts the experimental results portraying the closed loop response of the Buck converter system when the linear passivity based stabilizing controller is implemented. The controller and the system parameters were chosen to be:

$$L = 15.91 \text{ mH}, \quad C = 50 \text{ }\mu\text{F}, \quad R = 25 \text{ }\Omega, \quad E = 24 \text{ V}$$

with $\gamma_{actual} = 0.1$. We set an actual desired output voltage of $\overline{v} = 18$ V. This voltage determines a steady state current $\overline{i} = 0.72$ A, and $\overline{u}_{av} = 0.75$.

166   4 Approximate Linearization Methods

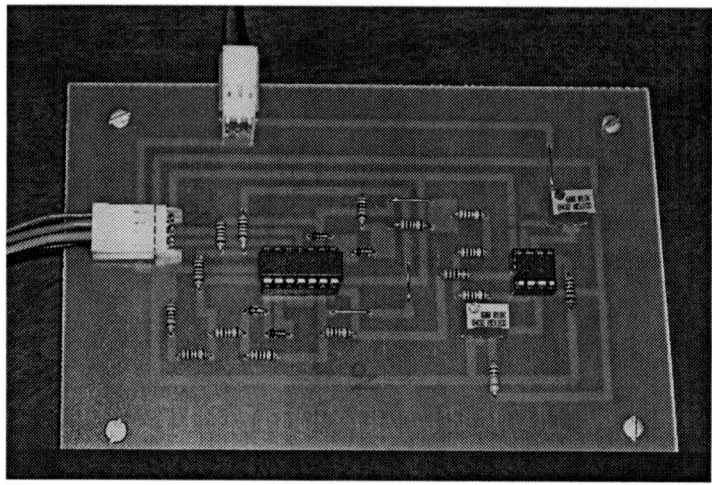

Fig. 4.19. Hardware implementation of the amplitude limiter circuit.

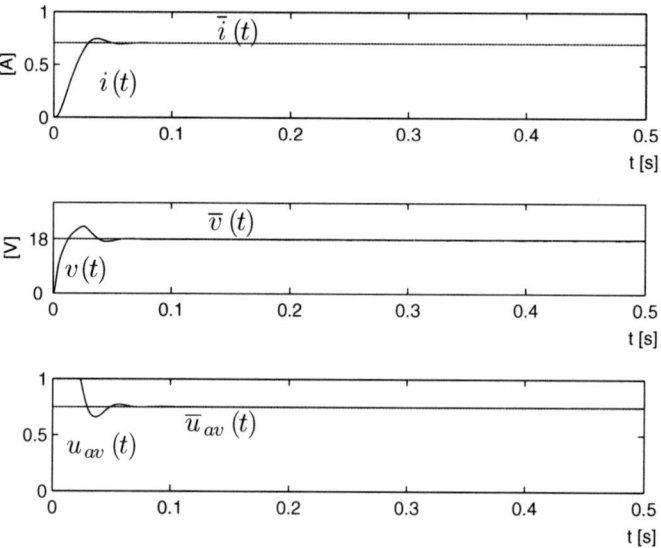

Fig. 4.20. Experimental closed loop response of the Buck power converter to a $\Sigma - \Delta$-modulator implementation of a passivity based stabilizing controller.

On the other hand the corresponding de-normalized controller (4.14) is given by

$$u_{av}(t) = u^*_{av}(t) - \gamma_{actual}\left[i - i^*(t)\right]E \qquad (4.17)$$

We implemented this controller using the PCI-6025E National Instruments card, in connection with the MATLAB®-Simulink® program. A nominal desired output voltage profile, exhibiting a rather smooth start for the dc-to-dc converter, was specified using an interpolating Bezier polynomial of tenth order, defined by:

$$F^*(t) = v^*(t) = \overline{v}(t_1) + [\overline{v}(t_2) - \overline{v}(t_1)]\varphi(t, t_1, t_2) \quad (4.18)$$

where $\varphi(t, t_1, t_2)$ is a piecewise polynomial function interpolating between the values of 0 and 1. This function is of the following form:

$$\varphi(t, t_1, t_2) = \begin{cases} 0 & \text{for } t \leq t_1 \\ \left(\frac{t-t_1}{t_2-t_1}\right)^5 \left[252 - 1050\left(\frac{t-t_1}{t_2-t_1}\right) + 1800\left(\frac{t-t_1}{t_2-t_1}\right)^2 \right. \\ \left. -1575\left(\frac{t-t_1}{t_2-t_1}\right)^3 + 700\left(\frac{t-t_1}{t_2-t_1}\right)^4 - 126\left(\frac{t-t_1}{t_2-t_1}\right)^5\right] \\ & \text{for } t \in (t_1, t_2) \\ 1 & \text{for } t \geq t_2 \end{cases} \quad (4.19)$$

We have used: $t_1 = 0.5$ s, $t_2 = 1.2$ s, i.e., $t_2 - t_1 = 0.7$ s, and $\gamma_{actual} = 0.18$.

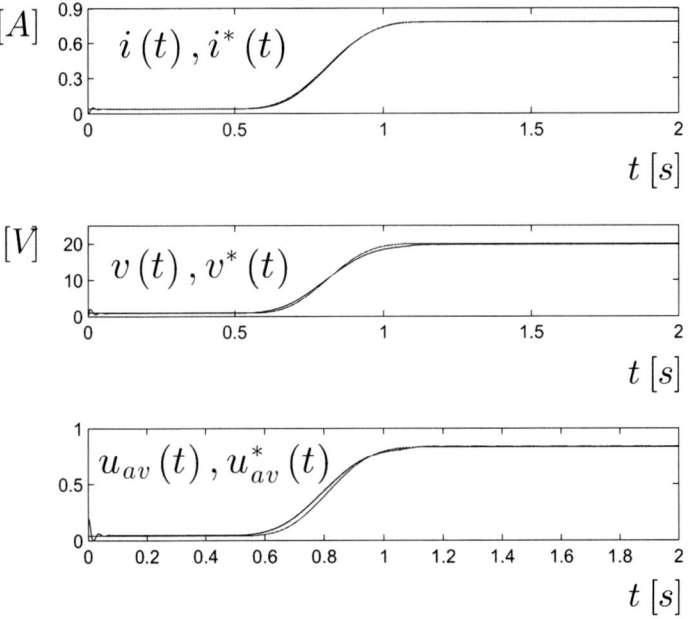

Fig. 4.21. Experimental input current, output voltage and average control input response for a nominal output voltage trajectory tracking task.

We set an equilibrium-to-equilibrium transfer from the initial value:

$$\left[\bar{i}(t_1), \bar{v}(t_1)\right] = \left[\frac{\bar{v}_{ini}}{R}, \bar{v}_{ini}\right] = [40 \text{ mA}, 1 \text{ V}]$$

to the final equilibrium value:

$$\left[\bar{i}(t_2), \bar{v}(t_2)\right] = \left[\frac{\bar{v}_{fin}}{R}, \bar{v}_{fin}\right] = [800 \text{ mA}, 20 \text{ V}]$$

Finally, the corresponding average control input signal generated by the linear feedback controller of the Buck converter varies between the initial and final values, respectively, $\bar{u}_{av}(t_1) = 41.667 \times 10^{-3}$ and $\bar{u}_{av}(t_2) = 0.83333$. The values for L, C, R and E were chosen to be exactly the same as in the previous experimental results about regulation.

Figure 4.21 depicts the experimental results which achieve the demanded rest to rest task.

4.4 The Boost Converter

4.4.1 Generalities about the Average Normalized Model

Consider the Boost dc-to-dc power converter shown in Figure 4.22

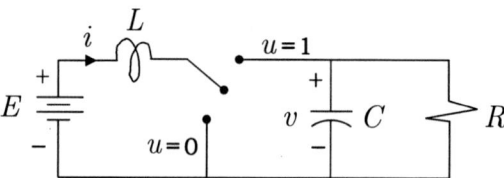

Fig. 4.22. The Boost converter.

The average normalized model of this system is given by

$$\dot{x}_1 = -u_{av} x_2 + 1$$
$$\dot{x}_2 = u_{av} x_1 - \frac{1}{Q} x_2$$

with x_1 being the normalized inductor current, x_2 represents the normalized output voltage and u_{av} is the control input (switch position function).

Equilibrium points

The average state equilibrium point is obtained from the preceding model, assuming that the control input variable continuously takes values in the interval $(0, 1)$. Parameterizing the equilibrium in terms of the desired equilibrium output voltage, $\bar{x}_2 = V_d > 0$, we have:

$$\bar{x}_1 = \frac{V_d^2}{Q}, \qquad \bar{u}_{av} = \frac{1}{V_d}$$

Since $Q > 0$ then $\bar{x}_1 > 0$ and then we also have:

$$1 < V_d < \infty$$

Approximate linearization

The linearization of the average state system around the average equilibrium point is given by

$$\dot{x}_{1\delta} = -\frac{1}{V_d} x_{2\delta} - V_d u_{av,\delta}$$

$$\dot{x}_{2\delta} = \frac{1}{V_d} x_{1\delta} - \frac{1}{Q} x_{2\delta} + \frac{V_d^2}{Q} u_{av,\delta}$$

where

$$x_{1\delta} = x_1 - \frac{V_d^2}{Q}, \qquad x_{2\delta} = x_2 - V_d, \qquad u_{av,\delta} = u_{av} - \frac{1}{V_d}$$

In matrix form, $\dot{x}_\delta = A x_\delta + b u_{av,\delta}$, we have,

$$\dot{x}_\delta = \begin{bmatrix} 0 & -\frac{1}{V_d} \\ \frac{1}{V_d} & -\frac{1}{Q} \end{bmatrix} x_\delta + \begin{bmatrix} -V_d \\ \frac{V_d^2}{Q} \end{bmatrix} u_{av,\delta}$$

The linearized system is found controllable,

$$\mathcal{C} = \begin{bmatrix} -V_d & -\frac{V_d}{Q} \\ \frac{V_d^2}{Q} & -\left(1 + \frac{V_d^2}{Q^2}\right) \end{bmatrix}, \qquad \det \mathcal{C} = V_d \left(1 + 2\frac{V_d^2}{Q^2}\right)$$

The system is also observable from any of the state variables. Indeed, when $y_\delta = x_{2\delta}$, we have,

$$\mathcal{O} = \begin{bmatrix} 0 & 1 \\ \frac{1}{V_d} & -\frac{1}{Q} \end{bmatrix}, \qquad \det \mathcal{O} = -\frac{1}{V_d}$$

In a similar form, when $y_\delta = x_{1\delta}$, we have:

$$\mathcal{O} = \begin{bmatrix} 1 & 0 \\ 0 & -\frac{1}{V_d} \end{bmatrix}, \qquad \det \mathcal{O} = -\frac{1}{V_d}$$

Input-output model

The average normalized input-output relation depends on the defined system output. Taking $y_\delta = x_{2\delta}$ as the system output, we obtain the following input-output model,

$$\ddot{y}_\delta + \frac{1}{Q}\dot{y}_\delta + \frac{1}{V_d^2}y_\delta = -u_{av,\delta} + \frac{V_d^2}{Q}\dot{u}_{av,\delta}$$

which clearly says that the system is stable with unstable *zero dynamics*

$$-u_{av,\delta} + \frac{V_d^2}{Q}\dot{u}_{av,\delta} = 0, \qquad \dot{u}_{av,\delta} = \left(\frac{Q}{V_d^2}\right)u_{av,\delta}$$

In other words, the transfer function of the linearized system is given by

$$\frac{y_\delta(s)}{u_{av,\delta}(s)} = G_\delta(s) = \frac{\left(\frac{V_d^2}{Q}\right)s - 1}{s^2 + \frac{1}{Q}s + \frac{1}{V_d^2}}$$

Then, clearly, the incremental average normalized output capacitor voltage $y_\delta = x_{2\delta}$ is a *non-minimum phase* output.

Taking the output of the system to be $y_\delta = x_{1\delta}$, we have, after some algebraic manipulations:

$$\ddot{y}_\delta + \frac{1}{Q}\dot{y}_\delta + \frac{1}{V_d^2}y_\delta = -\left(\frac{2V_d}{Q}u_{av,\delta} + V_d\dot{u}_{av,\delta}\right)$$

which clearly says that the system is stable with stable *zero dynamics*

$$\frac{2V_d}{Q}u_{av,\delta} + V_d\dot{u}_{av,\delta} = 0, \qquad \dot{u}_{av,\delta} = -\left(\frac{2}{Q}\right)u_{av,\delta}$$

The transfer function of the average linearized system is given by,

$$\frac{y_\delta(s)}{u_{av,\delta}(s)} = G_\delta(s) = -\frac{V_d\left(s + \frac{2}{Q}\right)}{s^2 + \frac{1}{Q}s + \frac{1}{V_d^2}}$$

The output $y_\delta = x_{1\delta}$ is then a *minimum phase* output.

Flatness

The system, being controllable, is also *differentially flat*. The flat output is any variable proportional to the following variable z:

$$z = [0\ 1]\mathcal{C}^{-1}x_\delta = \frac{-\frac{V_d^2}{Q}x_{1\delta} - V_d x_{2\delta}}{V_d\left(1 + \frac{2V_d^2}{Q^2}\right)}$$

i.e., we can take as a flat output

$$F_\delta = \frac{V_d^2}{Q} x_{1\delta} + V_d x_{2\delta} = \overline{x}_1 x_{1\delta} + \overline{x}_2 x_{2\delta}$$

i.e., the flat output is the linearized total normalized stored energy $0.5(x_1^2+x_2^2)$. The first order time derivative of F_δ is just

$$\dot{F}_\delta = x_{1\delta} - \frac{2V_d}{Q} x_{2\delta}$$

and the second order time derivative,

$$\ddot{F}_\delta = \left(\frac{2V_d}{Q^2} - \frac{1}{V_d}\right) x_{2\delta} - \frac{2}{Q} x_{1\delta} - V_d \left(1 + \frac{2V_d^2}{Q^2}\right) u_{av,\delta}$$

The differential parametrization of the state and input variables in terms of the incremental normalized flat output F_δ is given by

$$x_{1\delta} = \frac{1}{(1+\frac{2V_d^2}{Q^2})} \left[\dot{F}_\delta + \left(\frac{2}{Q}\right) F_\delta\right]$$

$$x_{2\delta} = \frac{1}{V_d\left(1+\frac{2V_d^2}{Q^2}\right)} \left[F_\delta - \left(\frac{V_d^2}{Q}\right) \dot{F}_\delta\right]$$

$$u_{av,\delta} = -\frac{1}{V_d\left(1+\frac{2V_d^2}{Q^2}\right)} \left[\ddot{F}_\delta + \frac{1}{Q}\dot{F}_\delta + \frac{1}{V_d^2}F_\delta\right]$$

Thus, under the input coordinate transformation:

$$u_{av,\delta} = -\frac{1}{V_d\left(1+\frac{2V_d^2}{Q^2}\right)} \left[v_\delta + \frac{1}{Q}\dot{F}_\delta + \frac{1}{V_d^2}F_\delta\right]$$

the normalized Boost system is seen to be equivalent to the second order pure integration system:

$$\ddot{F}_\delta = v_\delta$$

The previously given differential parametrization allows one to reconfirm several of the already established properties of the tangent linearization system for the Boost converter.

For instance, letting $x_{1\delta} = 0$, we find the *stable* zero dynamics

$$\dot{F}_\delta = -\left(\frac{2}{Q}\right) F_\delta$$

For $x_{2\delta} = 0$, we find that the zero dynamics is *unstable* and given by,

$$\dot{F}_\delta = \left(\frac{Q}{V_d^2}\right) F_\delta$$

Similarly, letting $u_{av,\delta} = 0$, we find that the free incremental response is asymptotically stable:

$$\ddot{F}_\delta + \frac{1}{Q}\dot{F}_\delta + \frac{1}{V_d^2}F_\delta = 0$$

The linearized system eigenvalues are given by the roots of the characteristic equation:

$$s^2 + \left(\frac{1}{Q}\right)s + \frac{1}{V_d^2} = 0$$

Indeed,

$$s_{1,2} = -\frac{1}{2Q} \pm \sqrt{\frac{1}{4Q^2} - \frac{1}{V_d^2}}$$

The roots are both real and strictly negative as long as

$$\frac{1}{V_d} < \frac{1}{2Q}$$

otherwise they are still located in the left half of the complex plane, but no longer real.

4.4.2 Control via State Feedback

Consider the linearized state space model of the average Boost converter around the state equilibrium point, $\bar{x}_1 = V_d^2/Q$, $\bar{x}_2 = V_d$

$$\dot{x}_{1\delta} = -\frac{1}{V_d}x_{2\delta} - V_d u_{av,\delta}$$

$$\dot{x}_{2\delta} = \frac{1}{V_d}x_{1\delta} - \frac{1}{Q}x_{2\delta} + \frac{V_d^2}{Q}u_{av,\delta}$$

The system was found to be controllable, hence stabilizable by means of linear state feedback to the origin of the incremental state space.

The feedback controller:

$$u_{av,\delta} = -k_1 x_{1\delta} - k_2 x_{2\delta}$$

yields the average closed loop system in matrix form

$$\dot{x}_\delta = \begin{bmatrix} k_1 V_d & -\left(\frac{1}{V_d} - k_2 V_d\right) \\ \left(\frac{1}{V_d} - k_1 \frac{V_d^2}{Q}\right) & -\left(\frac{1}{Q} + k_2 \frac{V_d^2}{Q}\right) \end{bmatrix} x_\delta$$

The characteristic polynomial of the average closed loop system is found to be,

$$p(s) = s^2 + \left(\frac{1}{Q} + k_2\frac{V_d^2}{Q} + k_1 V_d\right)s + \frac{1}{V_d^2} - k_2 - 2k_1\frac{V_d}{Q}$$

By equating the characteristic polynomial to a desired one, given in traditional form: $p_d(s) = s^2 + 2\zeta\omega_n s + \omega_n^2$, we obtain

$$k_1 = \frac{1}{V_d\left(1 - \frac{2V_d^2}{Q^2}\right)}\left[\frac{2}{Q} + 2\zeta\omega_n + \frac{V_d^2}{Q}\omega_n^2\right]$$

$$k_2 = -\frac{1}{V_d\left(1 - \frac{2V_d^2}{Q^2}\right)}\left[\frac{2V_d}{Q}\left(-\frac{1}{Q} + 2\zeta\omega_n\right) + V_d\left(\omega_n^2 - \frac{1}{V_d^2}\right)\right]$$

Simulations

It is desired to control the original nonlinear Boost converter towards the state equilibrium point

$$\bar{x}_1 = 3, \qquad \bar{x}_2 = V_d = 1.5$$

The corresponding equilibrium value for the control input is $\bar{u}_{av} = \frac{1}{V_d} = 0.6666$.

In this instance, we have taken the quality parameter $Q = 0.75$, and the linearized feedback controller design parameters as:

$$\zeta = 0.81, \qquad \omega_n = 1.2$$

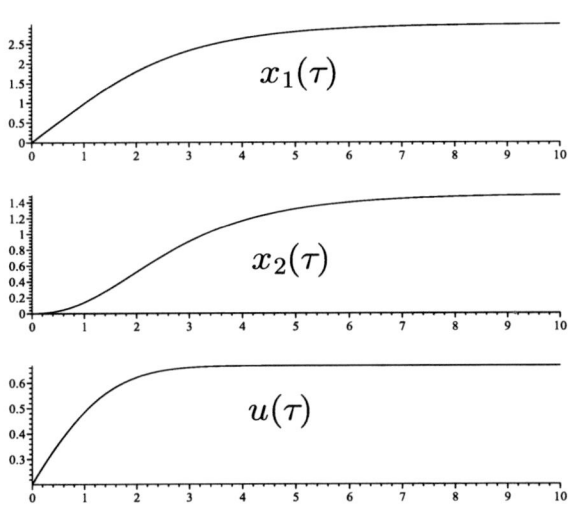

Fig. 4.23. Average responses of a linear feedback controlled Boost converter.

Figure 4.23 shows the performance of the average nonlinear Boost circuit to the action of the linear stabilizing feedback controller.

Figure 4.24 shows the performance of the switched nonlinear Boost circuit to the action of the linear stabilizing feedback controller implemented through a $\Sigma - \Delta$ modulator.

174   4 Approximate Linearization Methods

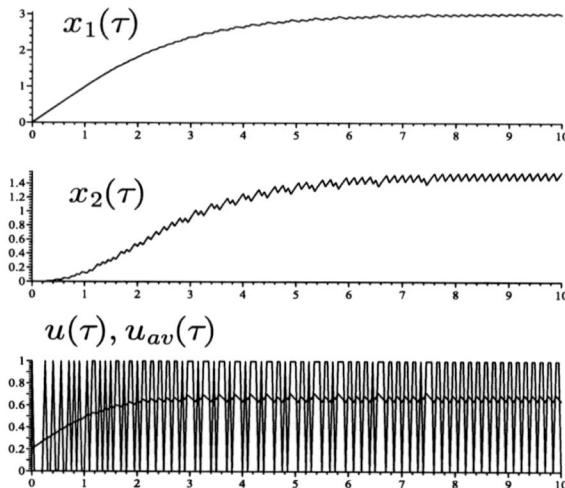

Fig. 4.24. Responses of a linear feedback controlled Boost converter using a $\Sigma - \Delta$ modulation implementation.

4.4.3 Proportional-Derivative State Feedback Control

Consider the state space model of the average normalized Boost converter linearized around the state equilibrium point $\bar{x}_1 = V_d^2/Q$, $\bar{x}_2 = V_d$

$$\dot{x}_{1\delta} = -\frac{1}{V_d} x_{2\delta} - V_d u_{av,\delta}$$

$$\dot{x}_{2\delta} = \frac{1}{V_d} x_{1\delta} - \frac{1}{Q} x_{2\delta} + \frac{V_d^2}{Q} u_{av,\delta}$$

The corresponding input-output incremental model when the output y_δ is taken to be $x_{1\delta}$ is given by,

$$\ddot{y}_\delta + \frac{1}{Q}\dot{y}_\delta + \frac{1}{V_d^2} y_\delta = -V_d \left(\frac{2}{Q} u_{av,\delta} + \dot{u}_{av,\delta} \right)$$

Notice that the following dynamic PD controller stabilizes the incremental system state to the origin of coordinates,

$$\dot{u}_{av,\delta} + \frac{2}{Q} u_{av,\delta} = -\frac{1}{V_d} \left[\left(\frac{1}{Q} - 2\zeta\omega_n \right) \dot{y}_\delta + \left(\frac{1}{V_d^2} - \omega_n^2 \right) y_\delta \right]$$

The closed loop dynamics is readily found to be

$$\ddot{y}_\delta + 2\zeta\omega_n \dot{y}_\delta + \omega_n^2 y_\delta = 0$$

The dynamic PD controller is transformed into a dynamic state feedback controller by replacing $y_\delta = x_{1\delta}$ and $\dot{y}_\delta = \left(-\frac{1}{V_d} x_{2\delta} - V_d u_{av,\delta} \right)$. We obtain

$$\dot{u}_{av,\delta} = -\left(\frac{1}{Q} + 2\zeta\omega_n\right)u_{av,\delta}$$
$$-\frac{1}{V_d}\left[\left(\frac{1}{V_d^2} - \omega_n^2\right)x_{1\delta} - \frac{1}{V_d}\left(\frac{1}{Q} - 2\zeta\omega_n\right)x_{2\delta}\right]$$

This average linear controller is to be used in the control of the actual switched nonlinear Boost circuit system. We use the average dynamic feedback controller.

$$u_{av} = \overline{u}_{av} + u_{av,\delta}$$
$$\dot{u}_{av,\delta} = -\left(\frac{1}{Q} + 2\zeta\omega_n\right)u_{av,\delta}$$
$$-\frac{1}{V_d}\left[\left(\frac{1}{V_d^2} - \omega_n^2\right)\left(x_1 - \frac{V_d^2}{Q}\right) - \frac{1}{V_d}\left(\frac{1}{Q} - 2\zeta\omega_n\right)(x_2 - V_d)\right]$$

with $\overline{u}_{av} = \frac{1}{V_d}$ and

$$u = 0.5(1 + \text{sign}(e)), \quad \frac{de}{dt} = u_{av} - u$$

Simulations

It is desired to control the original nonlinear Boost converter towards the state equilibrium point
$$\overline{x}_1 = 3, \quad \overline{x}_2 = V_d = 1.5$$

The corresponding equilibrium value for the control input is $\overline{u}_{av} = \frac{1}{V_d} = 0.6666$.

In this instance, we have taken the quality parameter $Q = 0.75$, and the linearized feedback controller design parameters as:

$$\zeta = 0.81, \quad \omega_n = 1.5$$

Figure 4.25 depicts the performance of the dynamic state feedback controller to an initial equilibrium perturbation in the average normalized Boost converter system.

Figure 4.26 depicts the performance of the switched dynamic state feedback controller for a Boost converter system.

The proposed dynamical proportional derivative feedback controller for the average Boost converter circuit is also capable of achieving the desired state equilibrium value from zero initial conditions.

The simulations depicting the average controlled responses of the Boost system, started from zero initial conditions, are shown in Figure 4.27.

Figure 4.28 depicts the performance of the dynamic state feedback controller in the switched normalized Boost converter system starting from zero initial conditions.

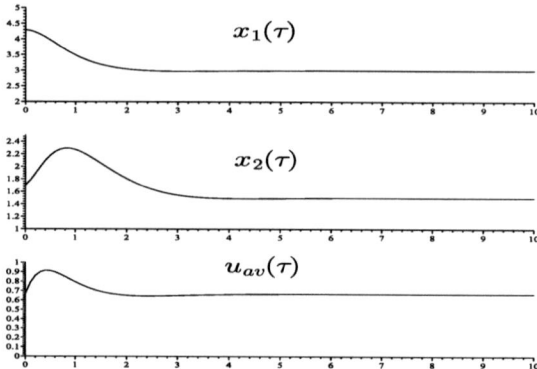

Fig. 4.25. Performance of average dynamic state feedback controller for a Boost converter.

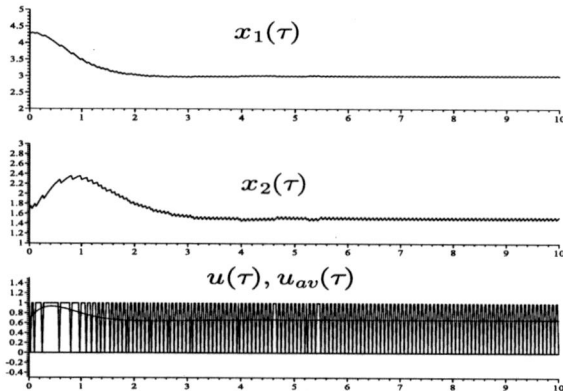

Fig. 4.26. Performance of dynamic state feedback controller for a Boost converter using a $\Sigma - \Delta$ modulator.

4.4.4 Trajectory Tracking

We now examine the possibilities of relaying on the linearized tangent average model of the Boost converter to perform trajectory tracking maneuvers implying excursions of the state trajectory which significantly take the state away from the equilibrium point.[2]

---

[2] We clarify that the considered linearized tangent model is computed around a specific equilibrium point. A second possibility is given by the tangent linearization around the given desired state and corresponding input trajectories, yielding a time-varying linear incremental system model. The treatment of this challenging case falls somewhat outside the scope of this book. The interested reader is referred to [72]

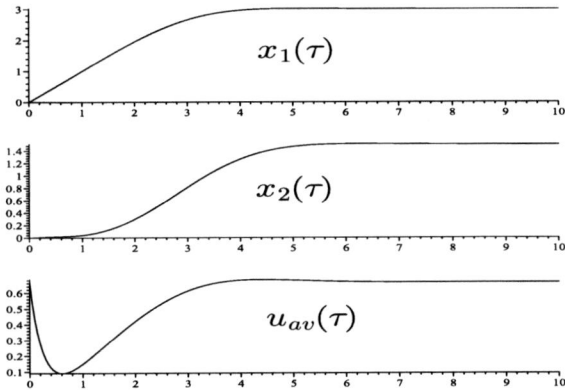

Fig. 4.27. Start-up performance of dynamic state feedback controlled average Boost converter.

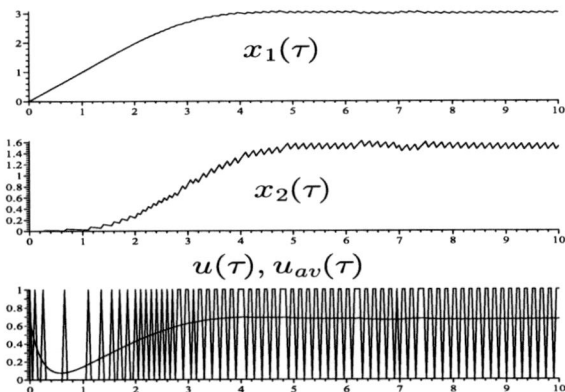

Fig. 4.28. Start-up performance of dynamic state feedback controlled Boost converter.

We propose to use the already studied linear average state feedback controller

$$u_{av,\delta} = -k_1 x_{1\delta} - k_2 x_{2\delta}$$

with

$$k_1 = \frac{1}{V_d \left(1 - \frac{2V_d^2}{Q^2}\right)} \left[-\frac{2}{Q} + 2\zeta\omega_n + \frac{V_d^2}{Q}\omega_n^2\right]$$

$$k_2 = -\frac{1}{V_d \left(1 - \frac{2V_d^2}{Q^2}\right)} \left[\frac{2V_d}{Q}\left(-\frac{1}{Q} + 2\zeta\omega_n\right) + V_d\left(\omega_n^2 - \frac{1}{V_d^2}\right)\right]$$

Instead of considering the incremental state variables $x_{1\delta}$, $x_{2\delta}$ and the incremental input variable $u_{av,\delta}$ as representing, respectively, a stabilization error, and a control input error, we consider them to represent a normalized trajectory tracking error:

$$x_{1\delta} = x_1 - x_1^*(\tau), \qquad x_{2\delta} = x_2 - x_2^*(\tau), \qquad u_{av,\delta} = u_{av} - u^*(\tau)$$

The problem reduces then to specify a suitable normalized state trajectory $x^*(\tau)$ and its corresponding nominal control input trajectory $u^*(\tau)$ for the average normalized nonlinear Boost circuit.

$$\dot{x}_1 = -u_{av}x_2 + 1$$
$$\dot{x}_2 = u_{av}x_1 - \frac{x_2}{Q}$$

The state trajectory planning is most efficiently carried out in terms of the average total stored energy of the system

$$F = \frac{1}{2}\left[x_1^2 + x_2^2\right]$$

The time derivatives of F, along the system trajectories, are given by

$$\dot{F} = x_1 - \frac{x_2^2}{Q}$$

$$\ddot{F} = \left(1 + \frac{2}{Q^2}x_2^2\right) - u_{av}x_2\left(1 + \frac{2}{Q}x_1\right)$$

From these relations we obtain the following parametrization of the state variables

$$x_1 = -\frac{Q}{2} + \sqrt{\frac{Q^2}{4} + Q\dot{F} + 2F}$$

$$x_2 = \sqrt{Q\left(-\dot{F} - \frac{Q}{2} + \sqrt{\frac{Q^2}{4} + Q\dot{F} + 2F}\right)}$$

$$u_{av} = \frac{\left(1 + \frac{2}{Q^2}x_2^2\right) - \ddot{F}}{x_2\left(1 + \frac{2}{Q}x_1\right)}$$

Thus, if a trajectory is prescribed for the average total stored energy which corresponds to a transfer between initial and final normalized state equilibria, then the state trajectories and the nominal input trajectories are completely determined. These trajectories would be useful in the specification of the proposed full state feedback trajectory tracking controller.

Suppose we want to transfer the system from the initial equilibrium point $\overline{x}_1(\tau_1) = \frac{V_{d1}^2}{Q}$, $\overline{x}_2(\tau_1) = V_{d1}$ towards the final equilibrium $\overline{x}_1(\tau_2) = \frac{V_{d2}^2}{Q}$, $\overline{x}_2(\tau_2) = V_{d2}$ during an interval of time $[\tau_1, \tau_2]$. The corresponding average total stored energy values are given by

4.4 The Boost Converter 179

$$\overline{F}(\tau_1) = \frac{1}{2}\left[\frac{V_{d1}^4}{Q^2} + V_{d1}^2\right], \quad \overline{F}(\tau_2) = \frac{1}{2}\left[\frac{V_{d2}^4}{Q^2} + V_{d2}^2\right]$$

One proceeds to prescribe a nominal normalized average trajectory $F^*(\tau)$ for the average total stored energy F using a smooth interpolating polynomial function on the time interval $[\tau_1, \tau_2]$. Once $F^*(\tau)$ is prescribed, the nominal average control input and the nominal state trajectories are easily computed.

$$x_1^*(\tau) = -\frac{Q}{2} + \sqrt{\frac{Q^2}{4} + Q\dot{F}^*(\tau) + 2F^*(\tau)}$$

$$x_2^*(\tau) = \sqrt{Q\left(-\dot{F}^*(\tau) - \frac{Q}{2} + \sqrt{\frac{Q^2}{4} + Q\dot{F}^*(\tau) + 2F^*(\tau)}\right)}$$

$$u_{av}^*(\tau) = \frac{1 + \frac{2}{Q^2}[x_2^*(\tau)]^2 - \ddot{F}^*(\tau)}{x_2^*(\tau)\left[1 + \frac{2}{Q}x_1^*(\tau)\right]}$$

It is desired to transfer the normalized state equilibrium point from the initial value $\overline{x}_1 = 3$, $\overline{x}_2 = V_{d1} = 1.5$ towards the final desired value $\overline{x}_1 = 5.333$, $\overline{x}_2 = V_{d2} = 2.0$ in 10 normalized units of time. The state feedback controller previously designed for stabilization is seen to also accomplish the desired equilibrium-to-equilibrium transfer as the following figure depicts.

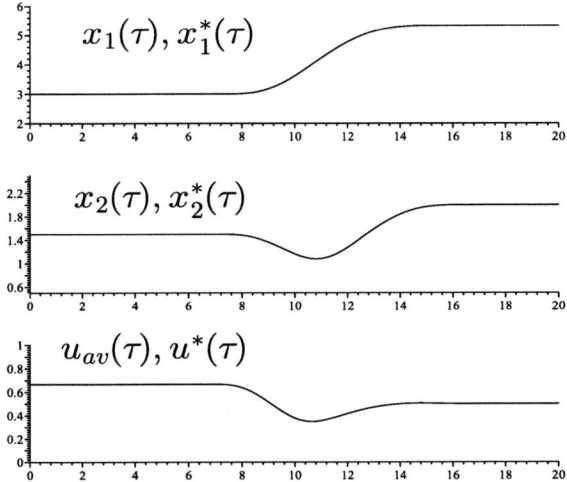

Fig. 4.29. Equilibrium to equilibrium transfer using a linear state feedback controller.

The total stored energy was planned to undergo a rest to rest transfer from the initial value of 5.625 towards the final value of 16.222 in 10 normalized time units. We used an interpolation polynomial of the form:

180   4 Approximate Linearization Methods

$$F^*(\tau) = F(\tau_1) + [F(\tau_2) - F(\tau_1)]\,\varphi(\tau, \tau_1, \tau_2)$$

with

$$\varphi(\tau, \tau_1, \tau_2) = \left(\frac{\tau - \tau_1}{\tau_2 - \tau_1}\right)^5 \left[r_1 - r_2\left(\frac{\tau - \tau_1}{\tau_2 - \tau_1}\right) + \cdots - r_6 \left(\frac{\tau - \tau_1}{\tau_2 - \tau_1}\right)^5\right]$$

$$r_1 = 252, \qquad r_2 = 1050, \qquad r_3 = 1800$$
$$r_4 = 1575, \qquad r_5 = 700, \qquad r_6 = 126$$

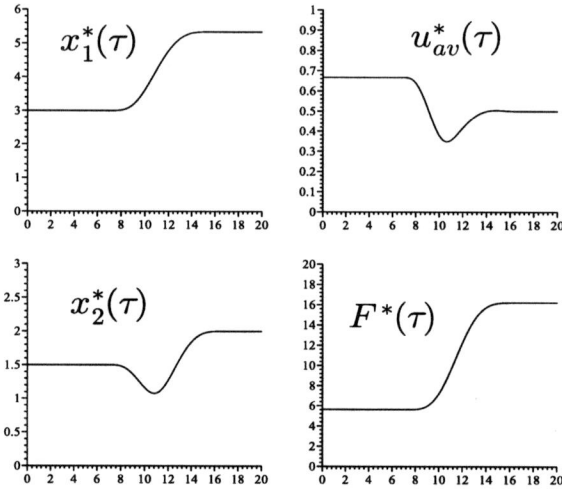

Fig. 4.30. Average rest-to-rest trajectory tracking for the Boost converter.

Figure 4.30 shows the normalized average state and control input trajectories of the nonlinear Boost circuit computed on the basis of an off-line planned total average stored energy reference trajectory $F^*(\tau)$. Such a trajectory is computed in correspondence with the transfer from an initial average equilibrium output voltage towards a final desired average equilibrium output voltage.

Note that the response of the output variable $x_2^*(\tau)$ exhibits the typical non-minimum phase behavior.

Simulations

Figure 4.31 depicts the response of the nonlinear normalized switched Boost converter circuit to the control action of a full state incremental feedback controller computed on the basis of the linearized tangent average system complemented with the nominal control input.

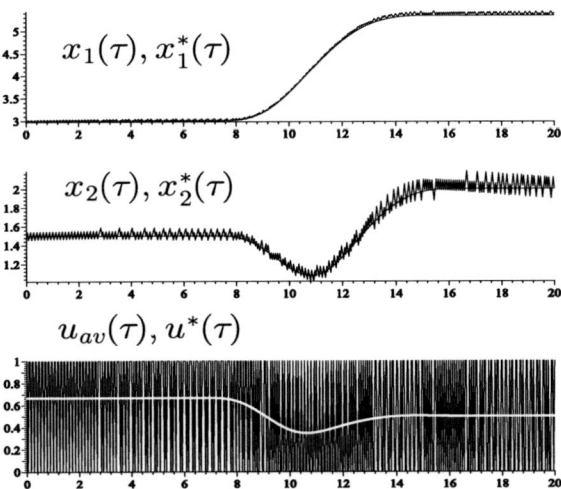

Fig. 4.31. Trajectory tracking performance of Boost converter to a linear state feedback controller.

4.4.5 Fliess' Generalized Canonical Form

Consider the linearized normalized average Boost converter system:

$$\dot{x}_{1\delta} = -\frac{1}{V_d}x_{2\delta} - V_d u_{av,\delta}$$
$$\dot{x}_{2\delta} = \frac{1}{V_d}x_{1\delta} - \frac{1}{Q}x_{2\delta} + \frac{V_d^2}{Q}u_{av,\delta}$$
$$y_\delta = x_{1\delta}$$

The input dependent state coordinate transformation

$$z_{1\delta} = x_{1\delta}, \qquad z_{2\delta} = -\frac{1}{V_d}x_{2\delta} - V_d u_{av,\delta}$$
$$x_{1\delta} = z_{1\delta}, \qquad x_{2\delta} = -V_d z_{2\delta} - V_d^2 u_{av,\delta}$$

leads to the generalized observability canonical form, given by

$$\dot{z}_{1\delta} = z_{2\delta}$$
$$\dot{z}_{2\delta} = -\frac{1}{V_d^2}z_{1\delta} - \frac{1}{Q}z_{2\delta} - \frac{V_d}{Q}u_{av,\delta} - V_d \dot{u}_{av,\delta}$$

Letting $y_\delta = z_{1\delta}$, the incremental input-output model for the average normalized Boost system is readily obtained as

$$\ddot{y}_\delta + \frac{1}{Q}\dot{y}_\delta + \frac{1}{V_d^2}y_\delta = -\frac{V_d}{Q}u_{av,\delta} - V_d \dot{u}_{av,\delta} \qquad (4.20)$$

Zero dynamics and average dynamic feedback controller design by stable zero cancellation follows readily from this model.

4.4.6 State Feedback Control via Observer Design

Consider the tangent linearization of the average normalized Boost converter system around an average normalized equilibrium point,

$$\dot{x}_{1\delta} = -\frac{1}{V_d}x_{2\delta} - V_d u_{av,\delta}$$

$$\dot{x}_{2\delta} = \frac{1}{V_d}x_{1\delta} - \frac{1}{Q}x_{2\delta} + \frac{V_d^2}{Q}u_{av,\delta}$$

where

$$x_{1\delta} = x_1 - \frac{V_d^2}{Q}, \qquad x_{2\delta} = x_2 - V_d, \qquad u_{av,\delta} = u_{av} - \frac{1}{V_d}$$

Considering $y_\delta = x_{2\delta}$, as a system output, a reduced order observer for $x_{1\delta}$ is readily obtained as,

$$\dot{\hat{x}}_{1\delta} = -\frac{1}{V_d}y_\delta - V_d u_{av,\delta} + \lambda(x_{1\delta} - \hat{x}_{1\delta})$$

with

$$x_{1\delta} = V_d\left[\dot{y}_\delta + \frac{1}{Q}y_\delta - \frac{V_d^2}{Q}u_{av,\delta}\right]$$

Define $\zeta = \hat{x}_{1\delta} - \lambda V_d y_\delta$. We have the following stable dynamics for ζ:

$$\dot{\zeta} = -\lambda\zeta - \left(\frac{1}{V_d} - \frac{\lambda V_d}{Q} + \lambda^2 V_d\right)y_\delta - \left(V_d + \frac{\lambda V_d^3}{Q}\right)u_{av,\delta}$$

and the estimate of the average normalized incremental inductor current $x_{1\delta}$ is given by

$$\hat{x}_{1\delta} = \zeta + \lambda V_d y_\delta$$

Simulations

Figure 4.32 shows the response of the normalized Boost converter to the full state linear feedback controller synthesized with the help of a reduced order observer estimating the incremental inductor current.

The initial condition of the observer dynamics was, on purpose, chosen to be different from zero to obtain an initial average normalized incremental inductor current which was also different from zero.

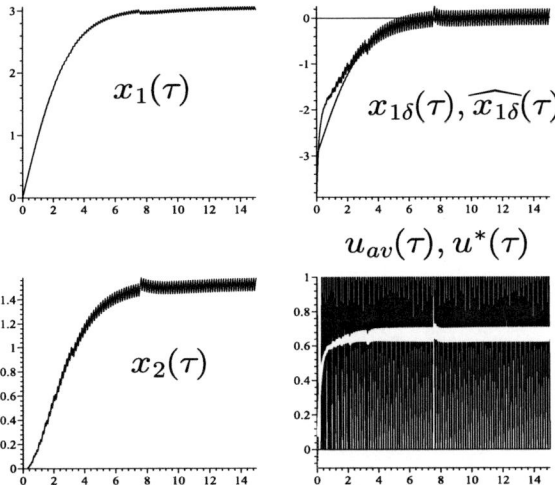

Fig. 4.32. Average responses of Boost converter to linear state controller achieved via a reduced order observer.

4.4.7 GPI Controller Design

We consider the linearized model of the Boost converter around a desired average equilibrium point given, as already stated, by

$$\overline{x}_1 = \frac{V_d^2}{Q}, \qquad \overline{x}_2 = V_d, \qquad \overline{u}_{av} = \frac{1}{V_d}$$

$$\dot{x}_{1\delta} = -\frac{1}{V_d} x_{2\delta} - V_d u_{av,\delta}$$
$$\dot{x}_{2\delta} = \frac{1}{V_d} x_{1\delta} - \frac{1}{Q} x_{2\delta} + \frac{V_d^2}{Q} u_{av,\delta}$$
$$y_{2\delta} = x_{2\delta}$$

A structural estimate of the unmeasured incremental average inductor current $x_{1\delta}$ is simply given by

$$\widehat{x}_{1\delta} = -\int_{t_0}^{\tau} \left[\frac{1}{V_d} y_\delta(\sigma) + V_d u_{av,\delta}(\sigma) \right] d\sigma$$

The structural estimate of $x_{1\delta}$ differs from its actual average value in a constant quantity

$$x_{1\delta} = \widehat{x}_{1\delta} + x_{1\delta}(\tau_0) = \widehat{x}_{1\delta} + x_{10\delta}$$

The incremental controller $u_{av,\delta} = -k_1 x_{1\delta} - k_2 x_{2\delta}$ is now replaced by the controller based on the structural estimate plus some output error integral compensation term.

$$u_{av,\delta} = -k_1 \widehat{x}_{1\delta} - k_2 y_\delta - k_3 \rho_\delta$$
$$\dot{\rho}_\delta = y_\delta$$

We use, however, in the forthcoming analysis, the relation: $\widehat{x}_{1\delta} = x_{1\delta} - x_{10\delta}$. The closed loop system is hence given by

$$\dot{x}_{1\delta} = V_d k_1 x_{1\delta} + \left(k_2 V_d - \frac{1}{V_d}\right) y_\delta + k_3 V_d \rho_\delta - V_d k_1 x_{10\delta}$$

$$\dot{y}_\delta = \left(\frac{1}{V_d} - k_1 \frac{V_d^2}{Q}\right) x_{1\delta} - \left(\frac{1}{Q} + k_2 \frac{V_d^2}{Q}\right) y_\delta - \frac{V_d^2}{Q} k_3 \rho_\delta + k_1 \frac{V_d^2}{Q} x_{10\delta}$$

$$\dot{\rho}_\delta = y_\delta$$

Clearly, the equilibrium point of the closed loop linear system is given by

$$\overline{y}_\delta = 0, \qquad \overline{x}_{1\delta} = 0, \qquad \overline{\rho}_\delta = \frac{k_1}{k_3} x_{10\delta}$$

The closed loop system is of the form $\dot{x}_\delta = Ax_\delta + v_\delta$, with $v_\delta = constant$. The eigenvalues of the A matrix are obtained from the relation:

$$\det \begin{bmatrix} s - V_d k_1 & -\left(k_2 V_d - \frac{1}{V_d}\right) & -k_3 V_d \\ -\left(\frac{1}{V_d} - k_1 \frac{V_d^2}{Q}\right) & s + \left(\frac{1}{Q} + k_2 \frac{V_d^2}{Q}\right) & k_3 \frac{V_d^2}{Q} \\ 0 & -1 & s \end{bmatrix} = 0$$

The characteristic polynomial is then given by

$$p(s) = s^3 - \left(V_d k_1 - \frac{V_d^2}{Q} k_2 - \frac{1}{Q}\right)s^2 - \left(2\frac{V_d}{Q}k_1 + k_2 - \frac{V_d^2}{Q}k_3 - \frac{1}{V_d^2}\right)s - k_3$$

Equating the coefficients of the closed loop characteristic polynomial to those of a desired characteristic polynomial $p_d(s)$ of the form:

$$p_d(s) = s^3 + (2\zeta\omega_n + p)s^2 + (2\zeta\omega_n p + \omega_n^2)s + \omega_n^2 p$$

we obtain the following system of linear equations for the k gains

$$\begin{bmatrix} -V_d & \frac{V_d^2}{Q} & 0 \\ -2\frac{V_d}{Q} & -1 & \frac{V_d^2}{Q} \\ 0 & 0 & -1 \end{bmatrix} \begin{bmatrix} k_1 \\ k_2 \\ k_3 \end{bmatrix} = \begin{bmatrix} -\frac{1}{Q} + 2\zeta\omega_n + p \\ -\frac{1}{V_d^2} + 2\zeta\omega_n p + \omega_n^2 \\ \omega_n^2 p \end{bmatrix}$$

i.e.,

$$\begin{bmatrix} k_1 \\ k_2 \\ k_3 \end{bmatrix} = \begin{bmatrix} -Q^2 \frac{1/Q + 2\zeta\omega_n + p}{V_d(Q^2 + 2V_d^2)} - V_d Q \frac{-\frac{1}{V_d^2} + 2\zeta\omega_n p + \omega_n^2}{Q^2 + 2V_d^2} - \frac{V_d^3 \omega_n^2 p}{Q^2 + 2V_d^2} \\ 2Q \frac{1/Q + 2\zeta\omega_n + p}{Q^2 + 2V_d^2} - Q^2 \frac{-\frac{1}{V_d^2} + 2\zeta\omega_n p + \omega_n^2}{Q^2 + 2V_d^2} - \frac{V_d^2 Q \omega_n^2 p}{Q^2 + 2V_d^2} \\ -\omega_n^2 p \end{bmatrix}$$

The proposed GPI controller is rewritten as a classical output feedback compensator given by:

$$u_{av,\delta}(s) = -\left[\frac{k_2 s - \left(\frac{1}{V_d}k_1 - k_3\right)}{s - V_d k_1}\right] y_\delta(s)$$

4.4.8 Passivity Based Control

The normalized average linearized Boost converter system is written in the "energy revealing" form: $\dot{x}_\delta = \mathcal{J} x_\delta + \mathcal{R} x_\delta + b u_{av,\delta}$. Indeed

$$\dot{x}_\delta = \begin{bmatrix} 0 & -\frac{1}{V_d} \\ \frac{1}{V_d} & 0 \end{bmatrix} x_\delta + \begin{bmatrix} 0 & 0 \\ 0 & -\frac{1}{Q} \end{bmatrix} x_\delta + \begin{bmatrix} -V_d \\ \frac{V_d^2}{Q} \end{bmatrix} u_{av,\delta}$$

Using the procedure developed in the previous passivity based controller design example, we obtain the following damped copy of the desired system behavior (we choose $\mathcal{R}_I > 0$),

$$\dot{x}_\delta^* = \begin{bmatrix} 0 & -\frac{1}{V_d} \\ \frac{1}{V_d} & 0 \end{bmatrix} x_\delta^* + \begin{bmatrix} 0 & 0 \\ 0 & -\frac{1}{Q} \end{bmatrix} x_\delta^* + \begin{bmatrix} \mathcal{R}_I & 0 \\ 0 & 0 \end{bmatrix}(x_\delta - x_\delta^*) + \begin{bmatrix} -V_d \\ \frac{V_d^2}{Q} \end{bmatrix} u_{av,\delta}$$

The tracking error $e_\delta = x_\delta - x_\delta^*$ evolves according to

$$\dot{e}_\delta = \begin{bmatrix} 0 & -\frac{1}{V_d} \\ \frac{1}{V_d} & 0 \end{bmatrix} e_\delta - \begin{bmatrix} \mathcal{R}_I & 0 \\ 0 & \frac{1}{Q} \end{bmatrix} e_\delta$$

The stability of the error dynamics is assessed by the evolution of the Lyapunov function candidate $V(e_\delta) = 0.5(e_{1\delta}^2 + e_{2\delta}^2)$. We obtain, along the trajectories of the incremental average tracking error variable $e(t)$

$$\dot{V}(e_\delta) = -\mathcal{R}_I e_{1\delta}^2 - \frac{1}{Q}e_{2\delta}^2 \leq -2\left[\min\left\{\mathcal{R}, \frac{1}{Q}\right\}\right] V(e_\delta)$$

The incremental tracking error exponentially converges to zero.

The passivity based dynamic average incremental feedback controller is then given by:

$$u_{av,\delta} = -\frac{1}{V_d}\left[\dot{x}_{1\delta}^*(\tau) + \frac{1}{V_d}\zeta_\delta - \mathcal{R}_I(x_{1\delta} - x_{1\delta}^*(\tau))\right]$$

$$\dot{\zeta}_\delta = -\frac{1}{Q}\zeta_\delta + \frac{1}{V_d}x_{1\delta}^*(\tau) + \frac{V_d^2}{Q}u_{av,\delta}$$

Simulations

We set as our desired average trajectory for the incremental variable $x_{1\delta}^*(\tau)$ the value of zero. This means that we are interested in using the controller for stabilization purposes.

We have set the controller design parameter $\mathcal{R}_I$ to be 1. The desired average output voltage $V_d = 1.6$ and $Q = 0.75$.

The average control input is then prescribed to be

$$u_{av} = \overline{u}_{av} + u_{av,\delta}$$

with $u_{av,\delta}$ as given by the passivity based average controller. Note that for initial conditions which imply a large value of the incremental average control input value, $u_{av,\delta}$, actual average control input saturations are possible (i.e., $u_{av} \notin [0, 1]$).

Figure 4.33 depicts the response of the average Boost converter model to the incremental passivity based average feedback controller. Controller saturation is dully accounted for.

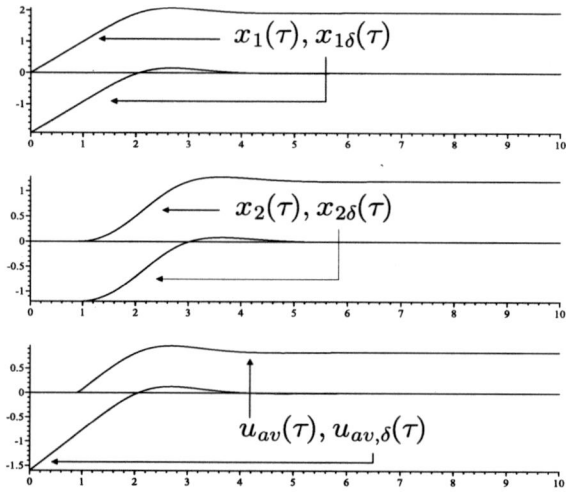

Fig. 4.33. Average responses of Boost converter to the linear passivity based controller.

Figure 4.34 depicts the response of the switched Boost converter model to the incremental passivity based average feedback controller implemented through a $\Sigma - \Delta$ modulator. Controller saturation is clearly exhibited in the switched controller behavior from the start-up operation from zero initial conditions. Nevertheless, the feedback option manages to stabilize the system around the desired equilibrium.

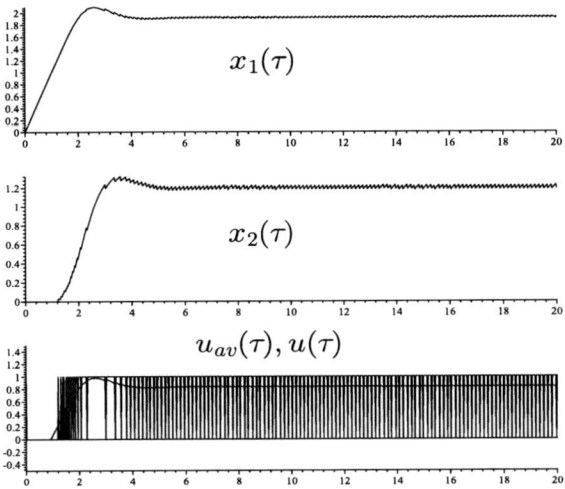

Fig. 4.34. Responses of switched Boost converter to the linearization based passivity based controller.

4.4.9 The Hamiltonian Systems Viewpoint

Take as an average energy function of the linearized system the quantity

$$H(x_\delta) = \frac{1}{2}\left[x_{1\delta}^2 + x_{2\delta}^2\right]$$

The passive output is clearly

$$y_\delta = -V_d x_{1\delta} + \frac{V_d^2}{Q} x_{2\delta}$$

Note that the dissipation matching condition is satisfied. For this, write it in the form:

$$\mathcal{R} + \gamma bb^T = \begin{bmatrix} \gamma V_d^2 & -\gamma \frac{V_d^3}{Q} \\ -\gamma \frac{V_d^3}{Q} & \left(\frac{1}{Q} + \gamma \frac{V_d^4}{Q^2}\right) \end{bmatrix} > 0$$

A linear stabilizing controller is then given by

$$u_{av,\delta} = -\gamma y_\delta = \gamma V_d x_{1\delta} - \gamma \frac{V_d^2}{Q} x_{2\delta}$$

Simulation

We set $\gamma = 0.25$ and as a desired average normalized equilibrium point to be, $\overline{x}_2 = V_d = 1.2$ and $\overline{x}_1 = V_d^2/Q = 1.92$ with $Q = 0.75$.

188   4 Approximate Linearization Methods

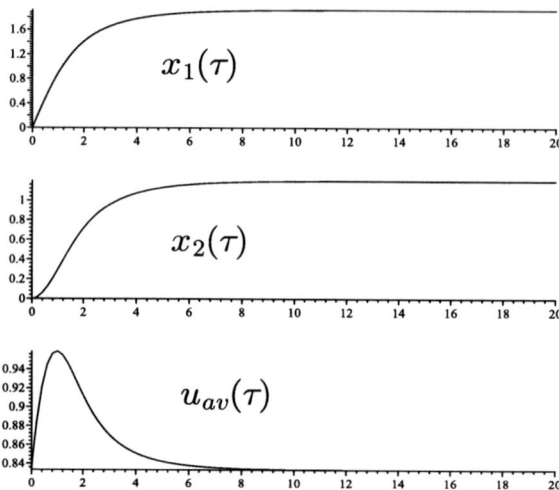

Fig. 4.35. Average responses of the Boost converter to linear static passivity based controller.

Figure 4.35 depicts the average controlled response of the nonlinear Boost converter to the proposed incremental passivity based output feedback control.

Figure 4.36 depicts the switched controlled response of the nonlinear Boost converter to the proposed incremental passivity based output feedback control implemented through a $\Sigma - \Delta$ modulator.

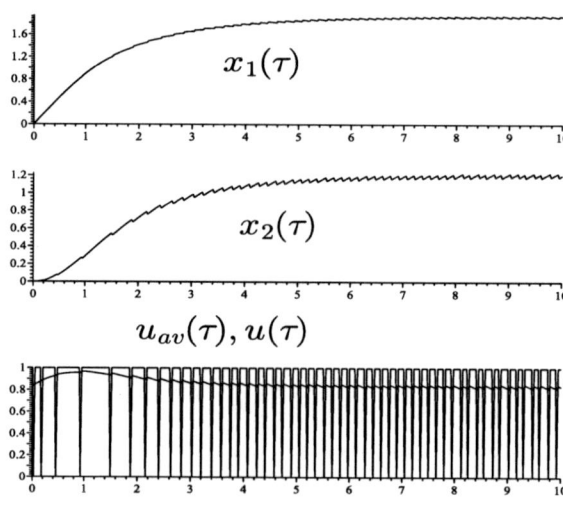

Fig. 4.36. Responses of the switched Boost converter to linear static passivity based controller.

4.5 The Buck-Boost Converter

4.5.1 Generalities about the Model

Consider the Buck-Boost dc-to-dc power converter, shown in Figure 4.37.

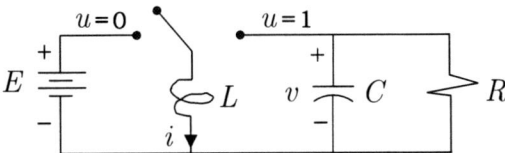

Fig. 4.37. The Buck-Boost converter.

The normalized average model of this system is given by

$$\dot{x}_1 = u_{av}x_2 + (1 - u_{av})$$
$$\dot{x}_2 = -u_{av}x_1 - \frac{1}{Q}x_2$$

where x_1 is the normalized average inductor current, x_2 is normalized average output voltage and u_{av} is the average control input.

The average state equilibrium point is obtained from the preceding model, assuming that the average control input variable is constant. Parameterizing the equilibrium in terms of the desired equilibrium output voltage, $\bar{x}_2 = V_d < 0$, we have:

$$\bar{x}_1 = -\frac{(1-V_d)V_d}{Q}, \quad \bar{u}_{av} = \frac{1}{1-V_d}$$

Since $Q > 0$ and $V_d < 0$, then $\bar{x}_1 > 0$ and then since $\bar{u}_{av} \in (0, 1)$, we also have:

$$0 > V_d > -\infty$$

Approximate linearization

The tangent linearization of the average normalize Buck-Boost system model, around the average equilibrium point is given by

$$\dot{x}_{1\delta} = \frac{1}{1-V_d}x_{2\delta} - (1-V_d)u_{av,\delta}$$
$$\dot{x}_{2\delta} = -\frac{1}{1-V_d}x_{1\delta} - \frac{1}{Q}x_{2\delta} + \left(\frac{(1-V_d)V_d}{Q}\right)u_{av,\delta}$$

where

4 Approximate Linearization Methods

$$x_{1\delta} = x_1 + \frac{(1-V_d)V_d}{Q}, \quad x_{2\delta} = x_2 - V_d, \quad u_{av,\delta} = u_{av} - \frac{1}{1-V_d}$$

In matrix form: $\dot{x}_\delta = Ax_\delta + bu_{av,\delta}$ we have,

$$\dot{x}_\delta = \begin{bmatrix} 0 & \frac{1}{1-V_d} \\ -\frac{1}{1-V_d} & -\frac{1}{Q} \end{bmatrix} x_\delta + \begin{bmatrix} -(1-V_d) \\ \frac{(1-V_d)V_d}{Q} \end{bmatrix} u_{av,\delta}$$

We summarize the fundamental properties of the average normalized linearized system:

The system is controllable

$$C = \begin{bmatrix} -(1-V_d) & \frac{V_d}{Q} \\ \frac{(1-V_d)V_d}{Q} & 1 - \frac{(1-V_d)V_d}{Q^2} \end{bmatrix}, \quad \det C = -(1-V_d)\left[1 - \frac{(1-2V_d)V_d}{Q^2}\right]$$

The system is observable from any state variable. Indeed, when the output of the system is set to be: $y_\delta = x_{2\delta}$, we have,

$$\mathcal{O} = \begin{bmatrix} 0 & 1 \\ \frac{1}{1-V_d} & -\frac{1}{Q} \end{bmatrix}, \quad \det \mathcal{O} = \frac{1}{1-V_d}$$

In a similar form, when the output is regarded to be, $y_\delta = x_{1\delta}$, we have:

$$\mathcal{O} = \begin{bmatrix} 1 & 0 \\ 0 & \frac{1}{1-V_d} \end{bmatrix}, \quad \det \mathcal{O} = \frac{1}{1-V_d}$$

Input-output models

Consider the linearized output, $y_\delta = x_{2\delta}$,

$$\ddot{y}_\delta + \frac{1}{Q}\dot{y}_\delta + \frac{1}{(1-V_d)^2}y_\delta = u_{av,\delta} + \frac{(1-V_d)V_d}{Q}\dot{u}_{av,\delta}$$

which clearly says that the linearized system is stable with unstable *zero dynamics*

$$u_{av,\delta} + \frac{(1-V_d)V_d}{Q}\dot{u}_{av,\delta} = 0, \quad \dot{u}_{av,\delta} = -\left(\frac{Q}{(1-V_d)V_d}\right)u_{av,\delta}$$

In other words, the transfer function of the linearized system is given by:

$$\frac{y_\delta(s)}{u_{av,\delta}(s)} = G_\delta(s) = -\frac{\left(\frac{(1-V_d)V_d}{Q}\right)s + 1}{s^2 + \frac{1}{Q}s + \frac{1}{(1-V_d)^2}}$$

Then $y_\delta = x_{2\delta}$ is a *non-minimum phase* output.

4.5 The Buck-Boost Converter

Consider now the output, $y_\delta = x_{1\delta}$. We have, after some algebraic manipulations:

$$\ddot{y}_\delta + \frac{1}{Q}\dot{y}_\delta + \frac{1}{(1-V_d)^2}y_\delta = -\left(\frac{1-2V_d}{Q}\right)u_{av,\delta} - (1-V_d)\dot{u}_{av,\delta}$$

which clearly says that the linearized system is stable with stable *zero dynamics*

$$\frac{1-2V_d}{Q}u_{av,\delta} + (1-V_d)\dot{u}_{av,\delta} = 0, \qquad \dot{u}_{av,\delta} = -\left(\frac{1-2V_d}{Q(1-V_d)}\right)u_{av,\delta}$$

The transfer function of the linearized system is then given by

$$\frac{y_\delta(s)}{u_{av,\delta}(s)} = G_\delta(s) = -\frac{(1-V_d)s + \frac{1-2V_d}{Q}}{s^2 + \frac{1}{Q}s + \frac{1}{(1-V_d)^2}}$$

The output $y_\delta = x_{1\delta}$ is then a *minimum phase* output.

Flatness

The average normalized system, being controllable, is also *differentially flat*. The flat output may be chosen to be

$$F_\delta = -\left[\frac{(1-V_d)V_d}{Q}\right]x_{1\delta} - (1-V_d)x_{2\delta}$$

i.e., the flat output is the linearization of the following quantity: $0.5[x_1^2 + (1-x_2)^2]$, which is related to the normalized total stored energy.

The first order time derivative of F_δ is just

$$\dot{F}_\delta = x_{1\delta} + \left(\frac{1-2V_d}{Q}\right)x_{2\delta}$$

and the second order time derivative,

$$\ddot{F}_\delta = -\frac{1-2V_d}{Q(1-V_d)}x_{1\delta} + \left[\frac{1}{1-V_d} + \frac{1-2V_d}{Q^2}\right]x_{2\delta} - (1-V_d)\left[1 - \frac{V_d}{Q}\right]u_{av,\delta}$$

The differential parametrization of the average incremental state and input variables in terms of the incremental normalized flat output F_δ is given by,

$$x_{1\delta} = \frac{1}{(1-V_d)\left[1 - \frac{V_d(1-2V_d)}{Q^2}\right]}\left[\frac{1-2V_d}{Q}F_\delta + (1-V_d)\dot{F}_\delta\right]$$

$$x_{2\delta} = -\frac{1}{(1-V_d)\left[1 - \frac{V_d(1-2V_d)}{Q^2}\right]}\left[F_\delta + \left(\frac{(1-V_d)V_d}{Q}\right)\dot{F}_\delta\right]$$

$$u_{av,\delta} = -\frac{1}{V_d\left(1 + \frac{2V_d^2}{Q^2}\right)}\left[\ddot{F}_\delta + \frac{1}{Q}\dot{F}_\delta + \frac{1}{V_d^2}F_\delta\right]$$

Thus, under the input coordinate transformation:

$$u_{av,\delta} = -\frac{1}{V_d\left(1 + \frac{2V_d^2}{Q^2}\right)}\left[v_\delta + \frac{1}{Q}\dot{F}_\delta + \frac{1}{V_d^2}F_\delta\right]$$

the normalized average model of the linearized Buck-Boost converter system is seen to be equivalent to the second order pure integration system:

$$\ddot{F}_\delta = v_\delta$$

The previously given differential parametrization allows one to reconfirm several of the already established properties of the tangent linearization system for the Buck-Boost converter. For instance, letting $x_{1\delta} = 0$, we find the corresponding *stable* zero dynamics

$$\dot{F}_\delta = -\left[\frac{1-2V_d}{Q(1-V_d)}\right]F_\delta$$

For $x_{2\delta} = 0$, we find that the zero dynamics is *unstable*

$$\dot{F}_\delta = -\left[\frac{Q}{(1-V_d)V_d}\right]F_\delta, \quad (V_d < 0)$$

Similarly, letting $u_{av,\delta} = 0$, we find that the free incremental response is asymptotically stable:

$$\ddot{F}_\delta + \frac{1}{Q}\dot{F}_\delta + \frac{1}{V_d^2}F_\delta = 0$$

The average linearized system eigenvalues are given by the two solutions of the characteristic equation:

$$s^2 + \left(\frac{1}{Q}\right)s + \frac{1}{(1-V_d)^2} = 0$$

Indeed,

$$s_{1,2} = -\frac{1}{2Q} \pm \sqrt{\frac{1}{4Q^2} - \frac{1}{(1-V_d)^2}}$$

The roots of the characteristic polynomial are both real and strictly negative as long as

$$\frac{1}{1-V_d} < \frac{1}{2Q}$$

otherwise, they are still located in the left half of the complex plane but no longer real.

4.5.2 State Feedback Controller Design

We consider now the tangent linearization model of the average normalized Buck-Boost converter system:

$$\dot{x}_1 = u_{av}x_2 + (1 - u_{av})$$
$$\dot{x}_2 = -u_{av}x_1 - \frac{x_2}{Q}$$

around the equilibrium point:

$$\bar{x}_1 = -\frac{(1-V_d)V_d}{Q}, \quad \bar{x}_2 = V_d < 0, \quad \bar{u}_{av} = \frac{1}{1-V_d}$$

The linearization of the average model is given by,

$$\dot{x}_{1\delta} = \frac{1}{1-V_d}x_{2\delta} - (1-V_d)u_{av,\delta}$$
$$\dot{x}_{2\delta} = -\frac{1}{1-V_d}x_{1\delta} - \frac{1}{Q}x_{2\delta} + \left(\frac{(1-V_d)V_d}{Q}\right)u_{av,\delta}$$

where,

$$x_{1\delta} = x_1 + \frac{(1-V_d)V_d}{Q}, \quad x_{2\delta} = x_2 - V_d, \quad u_{av,\delta} = u_{av} - \frac{1}{1-V_d}.$$

We seek an average linear state feedback controller of the form:

$$u_{av,\delta} = -k_1 x_{1\delta} - k_2 x_{2\delta}$$

which drives the average stabilization error state x_δ to zero in an exponentially stable fashion. We design such a controller with the help of the average tangent linearization system and will use, for the average nonlinear system, the control input

$$u_{av} = \frac{1}{1-V_d} - k_1\left(x_1 + \frac{(1-V_d)V_d}{Q}\right) - k_2(x_2 - V_d)$$

The closed loop tangent system is given by

$$\dot{x}_\delta = \begin{bmatrix} k_1(1-V_d) & \frac{1}{1-V_d} + k_2(1-V_d) \\ -\left(\frac{1}{1-V_d} + k_1\frac{V_d(1-V_d)}{Q}\right) & -\left(\frac{1}{Q} + k_2\frac{V_d(1-V_d)}{Q}\right) \end{bmatrix} x_\delta$$

whose characteristic polynomial is given by

$$p(s) = s^2 + \left[\frac{1}{Q} - k_1(1-V_d) + k_2\left(\frac{V_d(1-V_d)}{Q}\right)\right]s$$
$$+ \left[\frac{1}{(1-V_d)^2} + k_2 - k_1\left(\frac{1-2V_d}{Q}\right)\right]$$

Equating this polynomial to a desired closed loop polynomial of the form: $p_d(s) = s^2 + 2\zeta\omega_n s + \omega_n^2$, we obtain the feedback gains for the linear controller

$$k_1 = r\left[-\frac{1}{Q} + 2\zeta\omega_n - \frac{V_d(1-V_d)}{Q}\left(\omega_n^2 - \frac{1}{(1-V_d)^2}\right)\right]$$

$$k_2 = r\left[\frac{1-2V_d}{Q}\left(-\frac{1}{Q} + 2\zeta\omega_n\right) - (1-V_d)\left(\omega_n^2 - \frac{1}{(1-V_d)^2}\right)\right]$$

where

$$r = -\frac{1}{(1-V_d)(1 - \frac{V_d(1-2V_d)}{Q})}$$

Simulations

We performed simulations to assess the effectiveness of the proposed full state feedback controller, computed on the basis of the tangent linearized system, to accomplish a stabilization around a normalized equilibrium value for initial conditions set at the origin of coordinates.

We used the following parameters and design values

$$Q = 0.75, \qquad V_d = -1.5, \qquad \zeta = 0.81, \qquad \omega_n = 0.65$$

It turns out that for linearized closed loop natural frequencies, ω_n, which demand faster responses the average control input initially takes negative values. This would cause a temporary saturation to zero of the corresponding switched controller. For this reason a slower response is proposed.

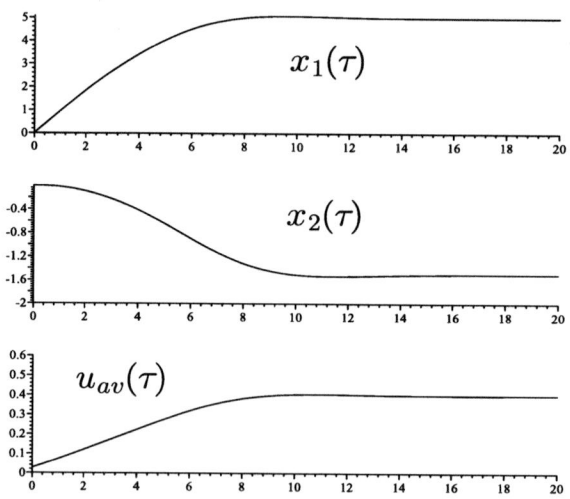

Fig. 4.38. Response of average Buck-Boost converter to linear state feedback controller.

Figure 4.38 depicts the response of the nonlinear average Buck-Boost converter circuit to the control action of a full state incremental feedback controller computed on the basis of the linearized tangent average system complemented with the nominal equilibrium control input.

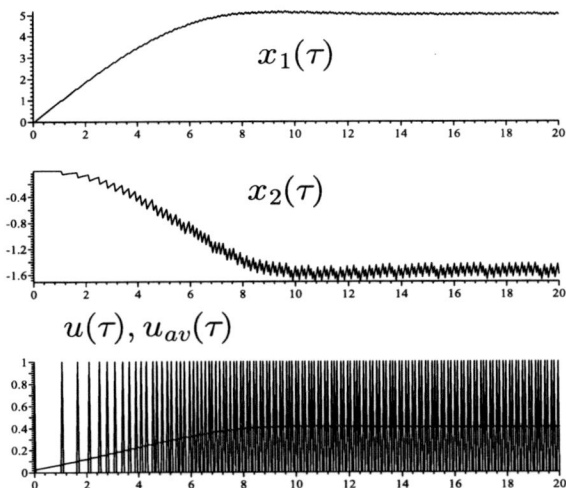

Fig. 4.39. Response of switched Buck-Boost converter to linear state feedback controller.

Figure 4.39 depicts the response of the nonlinear switched Buck-Boost converter circuit to the control action of a full state incremental feedback controller computed on the basis of the linearized tangent average system complemented with the nominal equilibrium control input. The switched control is implemented with the help of a $\Sigma - \Delta$ modulator.

A faster demanded response for the linearized dynamics is obtained with a higher value of ω_n. We let this design parameter be $\omega_n = 0.85$. Figure 4.40 depicts the saturation of the switched delta modulation controller which temporarily looses the sliding mode behavior of the $\Sigma - \Delta$ modulator. Nevertheless, after the sliding mode is recovered in the $\Sigma - \Delta$ modulator, the system converges towards the specified equilibrium state.

4.5.3 Dynamic Proportional-Derivative State Feedback Control

Recall the input-output average linearized model of the normalized Buck-Boost converter when the output y_δ is taken to be $x_{1\delta}$:

$$\ddot{y}_\delta + \frac{1}{Q}\dot{y}_\delta + \frac{1}{(1-V_d)^2}y_\delta = -\left(\frac{1-2V_d}{Q}\right)u_{av,\delta} - (1-V_d)\dot{u}_{av,\delta}$$

Clearly, a dynamic proportional derivative feedback controller of the form,

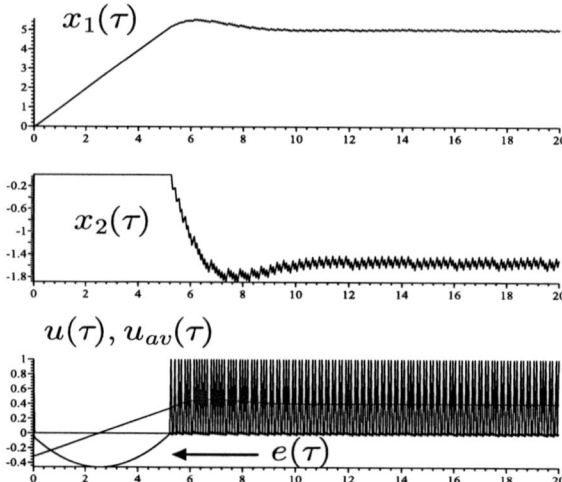

Fig. 4.40. Response of switched Buck-Boost converter to linear state feedback controller with input saturation.

$$\dot{u}_{av,\delta} = -\left(\frac{1-2V_d}{Q(1-V_d)}\right) u_{av,\delta}$$
$$- \frac{1}{1-V_d}\left[\left(\frac{1}{Q} - 2\zeta\omega_n\right)\dot{y}_\delta + \left(\frac{1}{(1-V_d)^2} - \omega_n^2\right) y_\delta\right]$$

yields a closed loop system of the desired classical form,

$$\ddot{y}_\delta + 2\zeta\omega_n \dot{y}_\delta + \omega_n^2 y_\delta = 0$$

In terms of the incremental state variables, the average dynamical PD controller is rewritten as:

$$\dot{u}_{av,\delta} = -\left(\frac{1-2V_d}{Q(1-V_d)} + 2\zeta\omega_n - \frac{1}{Q}\right) u_{av,\delta}$$
$$- \frac{1}{1-V_d}\left[\frac{1}{1-V_d}\left(\frac{1}{Q} - 2\zeta\omega_n\right) x_{2\delta} + \left(\frac{1}{(1-V_d)^2} - \omega_n^2\right) x_{1\delta}\right]$$

The control to be applied to the average nonlinear Buck-Boost circuit is of the form:

$$u_{av} = \frac{1}{1-V_d} + u_{av,\delta}$$

while the actual switched control is synthesized with the help of a $\Sigma - \Delta$ modulator.

Simulations

We set the normalized value $V_d = -1.5$ as a desired capacitor voltage value.
The controller gains were set to be, as usual

$$\zeta = 0.81, \quad \omega_n = 0.9, \quad Q = 0.75$$

For the controller design we used the tangent linearized model about the equilibrium point

$$\overline{x}_2 = V_d = -1.5, \quad \overline{x}_1 = -\frac{V_d(1-V_d)}{Q} = 5$$

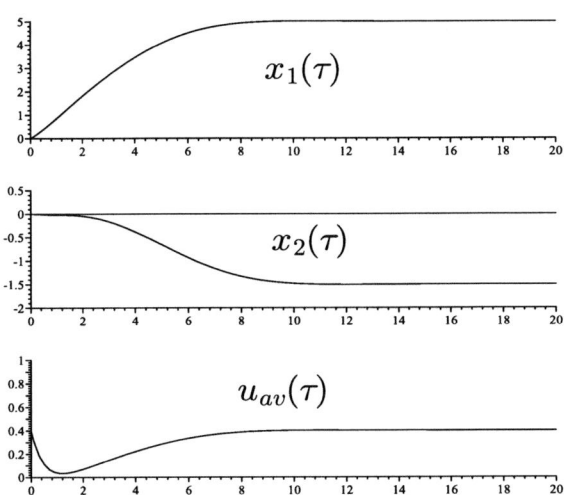

Fig. 4.41. Average responses of Buck-Boost converter to dynamic PD control.

Figure 4.41 depicts the response of the nonlinear average normalized Buck-Boost converter circuit to the proposed dynamical PD control action. The controller is synthesized as a full state incremental feedback controller on the basis of the linearized tangent input output model of the average system, complemented with the nominal control input.

Figure 4.42 depicts the response of the nonlinear switched Buck-Boost converter circuit to the control action of a full state incremental feedback controller of the PD type computed on the basis of the linearized tangent input output model of the average system complemented with the nominal control input.

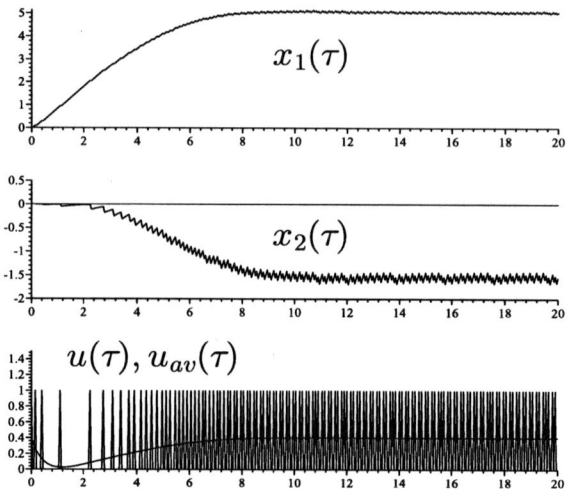

Fig. 4.42. Responses of switched Buck-Boost converter to dynamic PD control.

4.5.4 Trajectory Tracking

As it was done in the Boost converter, we now examine the possibilities of re-laying on the linearized tangent average model to perform trajectory tracking maneuvers which imply excursions of the state trajectory which significantly take the state away from the equilibrium point.

We propose to use the already considered state feedback controller

$$u_{av,\delta} = -k_1 x_{1\delta} - k_2 x_{2\delta}$$

with

$$k_1 = r\left[-\frac{1}{Q} + 2\zeta\omega_n - \frac{V_d(1-V_d)}{Q}\left(\omega_n^2 - \frac{1}{(1-V_d)^2}\right)\right]$$

$$k_2 = r\left[\frac{1-2V_d}{Q}\left(-\frac{1}{Q} + 2\zeta\omega_n\right) - (1-V_d)\left(\omega_n^2 - \frac{1}{(1-V_d)^2}\right)\right]$$

where

$$r = -\frac{1}{(1-V_d)\left(1 - \frac{V_d(1-2V_d)}{Q}\right)}$$

The incremental normalized state variables $x_{1\delta}$, $x_{2\delta}$ and the incremental input variable $u_{av,\delta}$ are now regarded as state and control input trajectory tracking errors:

$$x_{1\delta} = x_1 - x_1^*(\tau), \qquad x_{2\delta} = x_2 - x_2^*(\tau), \qquad u_{av,\delta} = u_{av} - u^*(\tau)$$

The problem reduces then to specify a suitable state trajectory $x^*(\tau)$ and its corresponding nominal control input trajectory $u^*(t)$ for the nonlinear normalized average Buck-Boost circuit model.

$$\dot{x}_1 = u_{av}x_2 + (1 - u_{av})$$
$$\dot{x}_2 = -u_{av}x_1 - \frac{x_2}{Q}$$

The state trajectory planning may be carried out in terms of the following quantity, related to the average total stored energy of the system

$$F = \frac{1}{2}\left[x_1^2 + (x_2 - 1)^2\right]$$

The time derivatives of F, along the system trajectories, are given by

$$\dot{F} = x_1 + \frac{x_2(1 - x_2)}{Q}$$
$$\ddot{F} = -u_{av}\left[1 + \frac{1}{Q}(1 - 2x_2)x_1 - x_2\right] - \frac{(1 - 2x_2)x_2}{Q^2}$$

From the previous expressions, we should be able to obtain a parametrization of the state and input variables

$$x_1 = \varphi_1(F, \dot{F}), \qquad x_2 = \varphi_2(F, \dot{F}), \qquad u_{av} = \vartheta(F, \dot{F}, \ddot{F}) \qquad (4.21)$$

However, these functions are not easy to find, as a full fourth degree algebraic equation must be solved. This prompts the concept of *implicit differential parameterizations*, a topic beyond the scope of this book.

4.5.5 Fliess' Generalized Canonical Forms

The normalized average linearized model of the Buck-Boost converter with output function $y_\delta = x_{1\delta}$ is given by:

$$\dot{x}_{1\delta} = \frac{1}{1 - V_d}x_{2\delta} - (1 - V_d)u_{av,\delta}$$
$$\dot{x}_{2\delta} = -\frac{1}{1 - V_d}x_{1\delta} - \frac{1}{Q}x_{2\delta} + \left(\frac{(1 - V_d)V_d}{Q}\right)u_{av,\delta}$$
$$y_\delta = x_{1\delta}$$

The, invertible, input dependent state coordinate transformation

$$\begin{bmatrix} z_{1\delta} \\ z_{2\delta} \end{bmatrix} = \begin{bmatrix} 1 & 0 \\ 0 & \frac{1}{1-V_d} \end{bmatrix}\begin{bmatrix} x_{1\delta} \\ x_{2\delta} \end{bmatrix} + \begin{bmatrix} 0 \\ -(1 - V_d) \end{bmatrix}u_{av,\delta}$$
$$\begin{bmatrix} x_{1\delta} \\ x_{2\delta} \end{bmatrix} = \begin{bmatrix} 1 & 0 \\ 0 & (1 - V_d) \end{bmatrix}\begin{bmatrix} z_{1\delta} \\ z_{2\delta} \end{bmatrix} + \begin{bmatrix} 0 \\ (1 - V_d)^2 \end{bmatrix}u_{av,\delta}$$

Leads to the Fliess' generalized observability canonical form for the average linearized Buck-Boost converter

$$\dot{z}_{1\delta} = z_{2\delta}$$
$$\dot{z}_{2\delta} = -\frac{1}{(1-V_d)^2}z_{1\delta} - \frac{1}{Q}z_{2\delta} - \frac{(1-V_d)^2}{Q}u_{av,\delta} - (1-V_d)\dot{u}_{av,\delta}$$
$$y_\delta = z_{1\delta}$$

In a similar fashion, for the average output $y_\delta = x_{2\delta}$, we obtain,

$$\dot{z}_{1\delta} = z_{2\delta}$$
$$\dot{z}_{2\delta} = -\frac{1}{(1-V_d)^2}z_{1\delta} - \frac{1}{Q}z_{2\delta} - u_{av,\delta} - \frac{V_d(1-V_d)}{Q}\dot{u}_{av,\delta}$$
$$y_\delta = z_{2\delta}$$

by using the following invertible, input dependent, state coordinate transformation

$$\begin{bmatrix} z_{1\delta} \\ z_{2\delta} \end{bmatrix} = \begin{bmatrix} 0 & 1 \\ -\frac{1}{1-V_d} & -\frac{1}{Q} \end{bmatrix} \begin{bmatrix} x_{1\delta} \\ x_{2\delta} \end{bmatrix} + \begin{bmatrix} 0 \\ \frac{V_d(1-V_d)}{Q} \end{bmatrix} u_{av,\delta}$$

The properties of the linearized system zero dynamics and the task of designing dynamic linear feedback controllers follows quite directly from this type of average normalized model.

4.5.6 Control via Observer Design

Consider the tangent linearization model of the average normalized Buck-Boost converter dynamics around an equilibrium point given by:

$$\overline{x}_1 = -\frac{(1-V_d)V_d}{Q}, \quad \overline{u}_{av} = \frac{1}{1-V_d}$$

This linearized average model is given by,

$$\dot{x}_{1\delta} = \frac{1}{1-V_d}x_{2\delta} - (1-V_d)u_{av,\delta}$$
$$\dot{x}_{2\delta} = -\frac{1}{1-V_d}x_{1\delta} - \frac{1}{Q}x_{2\delta} + \left(\frac{(1-V_d)V_d}{Q}\right)u_{av,\delta}$$

where

$$x_{1\delta} = x_1 + \frac{(1-V_d)V_d}{Q}, \quad x_{2\delta} = x_2 - V_d, \quad u_{av,\delta} = u_{av} - \frac{1}{1-V_d}$$

Recall that the system was found to be observable from the (non-minimum phase) output, $y_\delta = x_2 - V_d$.

Consider then a Luenberger reduced order observer for the unmeasured incremental average normalized inductor current:

$$\dot{\widehat{x}}_{1\delta} = \frac{1}{1-V_d}y_\delta - (1-V_d)u_{av,\delta} + \lambda(x_{1\delta} - \widehat{x}_{1\delta})$$

with $x_{1\delta}$ obtained from the second equation of the linearized average converter model,
$$x_{1\delta} = (1 - V_d)\left[-\dot{y}_\delta - \frac{1}{Q}y_\delta + \frac{(1-V_d)V_d}{Q}u_{av,\delta}\right]$$

Define: $\zeta = x_{1\delta} + \lambda(1 - V_d)y_\delta$, and obtain the estimate of the incremental inductor current, $\widehat{x_{1\delta}}$, as
$$\widehat{x_{1\delta}} = \zeta - \lambda(1 - V_d)y_\delta$$

with ζ being the solution of the stable system:
$$\dot{\zeta} = -\lambda\zeta + \left[\frac{1}{1-V_d} - \lambda\frac{1-V_d}{Q} + \lambda^2(1-V_d)\right]y_\delta - \left[(1-V_d) - \lambda\frac{(1-V_d)^2 V_d}{Q}\right]u_{av,\delta}$$

Simulations

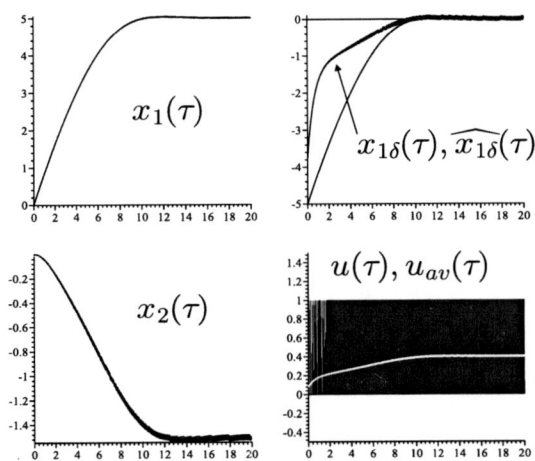

Fig. 4.43. Response of the switched Buck-Boost converter to a linear feedback controlled using a reduced order observer.

Figure 4.43 depicts the response of the nonlinear switched Buck-Boost converter circuit to the control action of a full state incremental feedback controller synthesized on the basis of a reduced order observer for the average inductor current of the linearized tangent system model. The switched input is obtained from a $\Sigma - \Delta$ modulator.

4.5.7 GPI Controller Design

Consider the linearized model of the average normalized Buck-Boost converter, around a desired average equilibrium point given by,

$$\overline{x}_1 = -\frac{(1-V_d)V_d}{Q}, \qquad \overline{x}_2 = V_d, \qquad \overline{u}_{av} = \frac{1}{1-V_d}$$

$$\dot{x}_{1\delta} = \frac{1}{1-V_d}x_{2\delta} - (1-V_d)u_{av,\delta}$$
$$\dot{x}_{2\delta} = -\frac{1}{1-V_d}x_{1\delta} - \frac{1}{Q}x_{2\delta} + \frac{(1-V_d)V_d}{Q}u_{av,\delta}$$
$$y_{2\delta} = x_{2\delta}$$

A structural estimate of the unmeasured incremental average inductor current $x_{1\delta}$ is simply given by

$$\widehat{x}_{1\delta} = \int_{\tau_0}^{\tau}\left[\frac{1}{1-V_d}y_\delta(\sigma) - (1-V_d)u_{av,\delta}(\sigma)\right]d\sigma$$

The structural estimate of $x_{1\delta}$ differs from its actual average value in a constant quantity

$$x_{1\delta} = \widehat{x}_{1\delta} + x_{1\delta}(\tau_0) = \widehat{x}_{1\delta} + x_{10\delta}$$

The incremental state feedback controller $u_{av,\delta} = -k_1 x_{1\delta} - k_2 x_{2\delta}$ is now replaced by the controller based on the structural estimate plus some output error integral compensation term.

$$u_{av,\delta} = -k_1\widehat{x}_{1\delta} - k_2 y_\delta - k_3\rho_\delta$$
$$\dot{\rho}_\delta = y_\delta$$

We use, however, in the forthcoming analysis, the relation: $\widehat{x}_{1\delta} = x_{1\delta} - x_{10\delta}$. The closed loop system is hence given by

$$\dot{x}_{1\delta} = k_1(1-V_d)x_{1\delta} + \left(k_2(1-V_d) + \frac{1}{1-V_d}\right)y_\delta + k_3(1-V_d)\rho_\delta - k_1(1-V_d)x_{10\delta}$$
$$\dot{y}_\delta = -\left(\frac{1}{1-V_d} + k_1\frac{(1-V_d)V_d}{Q}\right)x_{1\delta} - \left(\frac{1}{Q} + k_2\frac{(1-V_d)V_d}{Q}\right)y_\delta$$
$$\qquad - k_3\frac{(1-V_d)V_d}{Q}\rho_\delta + k_1\frac{(1-V_d)V_d}{Q}x_{10\delta}$$
$$\dot{\rho}_\delta = y_\delta$$

Clearly, the equilibrium point of the closed loop linear system is given by:

$$\overline{y}_\delta = 0, \qquad \overline{x}_{1\delta} = 0, \qquad \overline{\rho}_\delta = \frac{k_1}{k_3}x_{10\delta}$$

4.5 The Buck-Boost Converter

The closed loop system is of the form $\dot{x}_\delta = Ax_\delta + v_\delta$, with $v_\delta = constant$. The eigenvalues of the A matrix are obtained from the relation:

$$\det \begin{bmatrix} s - k_1(1-V_d) & -\left(k_2(1-V_d) + \frac{1}{1-V_d}\right) & -k_3(1-V_d) \\ \left(\frac{1}{1-V_d} + k_1 \frac{(1-V_d)V_d}{Q}\right) & s + \left(\frac{1}{Q} + k_2 \frac{(1-V_d)V_d}{Q}\right) & k_3 \frac{(1-V_d)V_d}{Q} \\ 0 & -1 & s \end{bmatrix} = 0$$

The characteristic polynomial is then given by

$$p(s) = s^3 + \left(\frac{1}{Q} - k_1(1-V_d) + k_2 \frac{(1-V_d)V_d}{Q}\right)s^2$$
$$+ \left(\frac{1}{(1-V_d)^2} - k_1 \frac{(1-2V_d)}{Q} + k_2 + k_3 \frac{(1-V_d)V_d}{Q}\right)s + k_3$$

Equating the coefficients of the closed loop characteristic polynomial to those of a desired characteristic polynomial $p_d(s)$ of the form:

$$p_d(s) = s^3 + (2\zeta\omega_n + p)s^2 + (2\zeta\omega_n p + \omega_n^2)s + \omega_n^2 p$$

we obtain the following system of linear equations for the k gains,

$$\begin{bmatrix} -(1-V_d) & \frac{(1-V_d)V_d}{Q} & 0 \\ -\frac{(1-2V_d)}{Q} & 1 & \frac{(1-V_d)V_d}{Q} \\ 0 & 0 & 1 \end{bmatrix} \begin{bmatrix} k_1 \\ k_2 \\ k_3 \end{bmatrix} = \begin{bmatrix} a \\ b \\ c \end{bmatrix}$$

i.e.

$$\begin{bmatrix} k_1 \\ k_2 \\ k_3 \end{bmatrix} = \begin{bmatrix} -Q^2 \frac{a}{(Q^2 - V_d + 2V_d^2)(1-V_d)} + V_d Q \frac{b}{(Q^2 - V_d + 2V_d^2)} - \frac{(1-V_d)V_d^2 c}{(Q^2 - V_d + 2V_d^2)} \\ Q^2 \frac{b}{(Q^2 - V_d + 2V_d^2)} - Q \frac{(1-2V_d)a}{(Q^2 - V_d + 2V_d^2)(1-V_d)} - Q \frac{(1-V_d)V_d c}{(Q^2 - V_d + 2V_d^2)} \\ c \end{bmatrix}$$

where

$$a = -\frac{1}{Q} + 2\zeta\omega_n + p, \qquad b = -\frac{1}{(1-V_d)^2} + 2\zeta\omega_n p + \omega_n^2, \qquad c = \omega_n^2 p$$

The proposed GPI controller is rewritten as a classical output feedback compensator given by:

$$u_{av,\delta}(s) = - \left[\frac{k_2 s + \left(\frac{1}{1-V_d}k_1 + k_3\right)}{s - k_1(1-V_d)} \right] y_\delta(s)$$

4.5.8 Passivity Based Control

The average linearized Buck-Boost converter system is written in the "energy revealing" form: $\dot{x}_\delta = \mathcal{J}x_\delta + \mathcal{R}x_\delta + bu_{av,\delta}$. Indeed

$$\dot{x}_\delta = \begin{bmatrix} 0 & \frac{1}{1-V_d} \\ -\frac{1}{1-V_d} & 0 \end{bmatrix} x_\delta + \begin{bmatrix} 0 & 0 \\ 0 & -\frac{1}{Q} \end{bmatrix} x_\delta + \begin{bmatrix} -(1-V_d) \\ \frac{(1-V_d)V_d}{Q} \end{bmatrix} u_{av,\delta}$$

Using the procedure developed in the previous passivity based controllers design examples, we obtain the following damped copy of the desired system behavior (we choose $\mathcal{R}_I > 0$),

$$\dot{x}_\delta^* = \begin{bmatrix} 0 & \frac{1}{1-V_d} \\ -\frac{1}{1-V_d} & 0 \end{bmatrix} x_\delta^* + \begin{bmatrix} 0 & 0 \\ 0 & -\frac{1}{Q} \end{bmatrix} x_\delta^* + \begin{bmatrix} \mathcal{R}_I & 0 \\ 0 & 0 \end{bmatrix} (x_\delta - x_\delta^*) + \begin{bmatrix} -(1-V_d) \\ \frac{(1-V_d)V_d}{Q} \end{bmatrix} u_{av,\delta}$$

The tracking error $e_\delta = x_\delta - x_\delta^*$ evolves according to

$$\dot{e}_\delta = \begin{bmatrix} 0 & \frac{1}{1-V_d} \\ -\frac{1}{1-V_d} & 0 \end{bmatrix} e_\delta - \begin{bmatrix} \mathcal{R}_I & 0 \\ 0 & \frac{1}{Q} \end{bmatrix} e_\delta$$

The stability of the error dynamics is assessed by the evolution of the Lyapunov function candidate $V(e_\delta) = 0.5(e_{1\delta}^2 + e_{2\delta}^2)$. We obtain, along the trajectories of the incremental average tracking error variable $e(\tau)$

$$\dot{V}(e_\delta) = -\mathcal{R}_I e_{1\delta}^2 - \frac{1}{Q}e_{2\delta}^2 \leq -2\left[\min\left\{\mathcal{R}, \frac{1}{Q}\right\}\right] V(e_\delta)$$

The incremental tracking error exponentially converges to zero.

The passivity based dynamic average incremental feedback controller is then given by:

$$u_{av,\delta} = -\frac{1}{1-V_d}\left[\dot{x}_{1\delta}^*(\tau) - \frac{1}{1-V_d}\zeta_\delta - \mathcal{R}_I(x_{1\delta} - x_{1\delta}^*(\tau))\right]$$

$$\dot{\zeta}_\delta = -\frac{1}{Q}\zeta_\delta - \frac{1}{1-V_d}x_{1\delta}^*(\tau) + \frac{(1-V_d)V_d}{Q}u_{av,\delta}$$

Simulations

We set as our desired average trajectory for the incremental variable $x_{1\delta}^*(\tau)$ the value of zero. This means that we are interested in using the controller for stabilization purposes.

We have set the controller design parameter $\mathcal{R}_I$ to be 1. The desired average output voltage $V_d = -1.5$ and $Q = 0.75$.

The average control input is then prescribed to be

$$u_{av} = \overline{u}_{av} + u_{av,\delta}$$

with $u_{av,\delta}$ as given by the passivity based average controller. Note that for initial conditions which imply a large value of the incremental average control input value, $u_{av,\delta}$, actual average control input saturations are possible (i.e., $u_{av} \notin [0,1]$).

Figure 4.44 depicts the response of the average Buck-Boost converter model to the incremental passivity based average feedback controller. Controller saturation is dully accounted for.

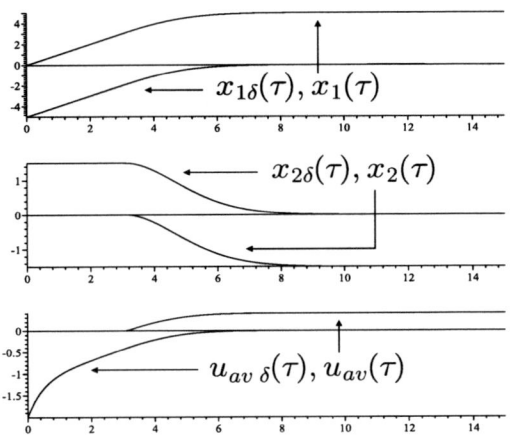

Fig. 4.44. Average responses of the Buck-Boost converter to passivity based dynamic feedback control.

Figure 4.45 depicts the response of the switched Buck-Boost converter model, to the incremental passivity based average feedback controller implemented through a $\Sigma - \Delta$ modulator. Controller saturation is clearly exhibited in the switched controller behavior.

4.5.9 The Hamiltonian Systems Viewpoint

The average linearized Buck-Boost converter system is written in the Hamiltonian form:

$$\dot{x}_\delta = \mathcal{J}\frac{\partial H(x_\delta)}{\partial x_\delta} + \mathcal{R}\frac{\partial H(x_\delta)}{\partial x_\delta} + bu_{av,\delta}$$

$$y_\delta = b^T\frac{\partial H(x_\delta)}{\partial x_\delta}$$

with $H(x_\delta) = \frac{1}{2}\left[x_{1\delta}^2 + x_{2\delta}^2\right]$.

Indeed,

$$\dot{x}_\delta = \begin{bmatrix} 0 & \frac{1}{1-V_d} \\ -\frac{1}{1-V_d} & 0 \end{bmatrix}\frac{\partial H(x_\delta)}{\partial x_\delta} + \begin{bmatrix} 0 & 0 \\ 0 & -\frac{1}{Q} \end{bmatrix}\frac{\partial H(x_\delta)}{\partial x_\delta} + \begin{bmatrix} -(1-V_d) \\ \frac{(1-V_d)V_d}{Q} \end{bmatrix}u_{av,\delta}$$

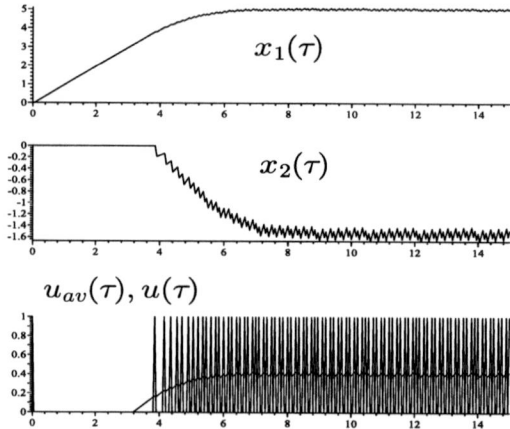

Fig. 4.45. Responses of the switched Buck-Boost converter to passivity based dynamic feedback control.

The passive output is given by

$$y_\delta = b^T \frac{\partial H(x_\delta)}{\partial x_\delta} = -(1-V_d)x_{1\delta} + \frac{(1-V_d)V_d}{Q}x_{2\delta}$$

The dissipation matching condition adopts the form:

$$\mathcal{R} + \gamma bb^T = \begin{bmatrix} \gamma(1-V_d)^2 & -\gamma\frac{(1-V_d)^2 V_d}{Q} \\ -\gamma\frac{(1-V_d)^2 V_d}{Q} & \frac{1}{Q} + \gamma\frac{(1-V_d)^2 V_d^2}{Q^2} \end{bmatrix} > 0$$

The average incremental passive output stabilizing feedback controller is readily obtained as:

$$u_{av,\delta} = -\gamma y_\delta = \gamma(1-V_d)x_{1\delta} - \gamma\frac{(1-V_d)V_d}{Q}x_{2\delta} \qquad (4.22)$$

Using the definition of the incremental average control:

$$u_{av,\delta} = u_{av} - \overline{u}_{av}$$

whit

$$\overline{u}_{av} = \frac{1}{1-V_d}$$

and

$$\overline{x}_1 = -\frac{(1-V_d)V_d}{Q}, \qquad \overline{x}_2 = V_d$$

we can write (4.22) as:

$$u_{av} = \frac{1}{1-V_d} + \gamma\left[(1-V_d)\left(x_1 + \frac{(1-V_d)V_d}{Q}\right) - \frac{(1-V_d)V_d}{Q}(x_2 - V_d)\right]$$

Reducing terms in the previous expression, we obtained:

$$u_{av} = \frac{1}{1-V_d} + \gamma \left[x_1 + \frac{V_d}{Q}(1-x_2) \right] (1 - V_d) \qquad (4.23)$$

The incremental average closed loop system is given by

$$\dot{x}_\delta = \begin{bmatrix} 0 & \frac{1}{1-V_d} \\ -\frac{1}{1-V_d} & 0 \end{bmatrix} \frac{\partial H(x_\delta)}{\partial x_\delta} - \begin{bmatrix} \gamma(1-V_d)^2 & -\gamma\frac{(1-V_d)^2 V_d}{Q} \\ -\gamma\frac{(1-V_d)^2 V_d}{Q} & \frac{1}{Q} + \gamma\frac{(1-V_d)^2 V_d^2}{Q^2} \end{bmatrix} \frac{\partial H(x_\delta)}{\partial x_\delta}$$

Simulations

We set: $V_d = -1.5$, $Q = 0.75$, $\gamma = 0.25$. The average passive output feedback controller manages to stabilize the nonlinear switched converter from the zero initial conditions, in spite of initial controller saturation.

Figure 4.46 depicts the response of the nonlinear system to the proposed average static passivity based controller implemented through a $\Sigma - \Delta$ modulator.

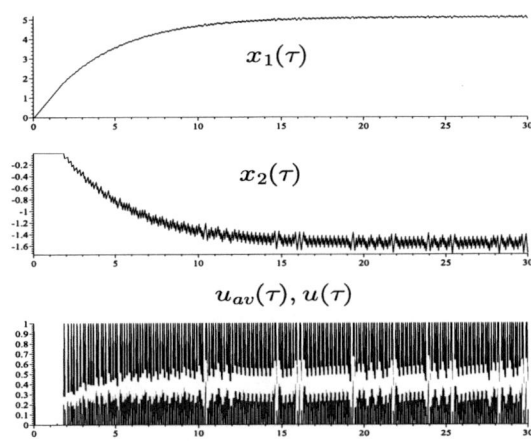

Fig. 4.46. Responses of switched Buck-Boost converter to static passivity based control based on the linearized model.

4.5.10 Experimental Passivity based Control of the Buck-Boost Converter

Here we propose to implement in the Buck-Boost experimental prototype, the developed normalized static passivity based controller (4.23) computed on the basis of the linearized average converter model and using the $\Sigma-\Delta$-modulator implementation scheme.

In Figure 4.47 we show the corresponding block diagram. It illustrates all the components of the system that we built with its respective *control* block already inserted. We remark that there are four important blocks, similar to previous implementations, in this case we have the following blocks: *Buck-Boost system*, $\Sigma - \Delta$-*modulator, amplitude limiter circuit* and the *control*.

Here we only will explain the hardware implementation of the *control* block for the Buck-Boost converter when it is controlled through a linearization based static passivity based feedback control, implemented through a $\Sigma - \Delta$-modulator block, which was presented at the end of the Chapter 3.

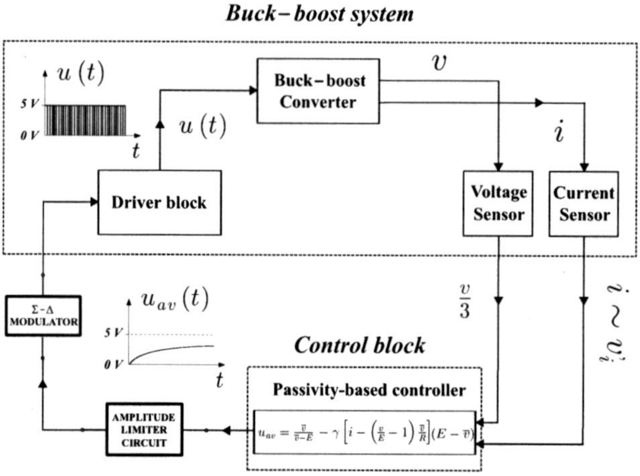

Fig. 4.47. Block diagram of the Buck-Boost power converter with a $\Sigma - \Delta$-modulator implementation of a linear static passivity based average feedback control.

Control block

The control strategy (4.23) is implemented using analog electronics while noticing that

$$x_1 = \frac{1}{E}\sqrt{\frac{L}{C}}i, \qquad x_2 = \frac{v}{E}, \qquad Q = R\sqrt{\frac{C}{L}}$$

We rewrite (4.23) in non-normalized form as:

$$u_{av} = \frac{E}{E - \overline{v}} + \gamma_{actual}\left[i - \left(\frac{v}{E} - 1\right)\frac{\overline{v}}{R}\right](E - \overline{v}) \qquad (4.24)$$

where $\gamma_{actual} > 0$, represents the de-normalized form of the gain γ.

Figure 4.48 shows the actual *control* block. It also shows the transfer functions that realize the op-amps for achieving the actual implementation of the de-normalized linear static passivity based controller.

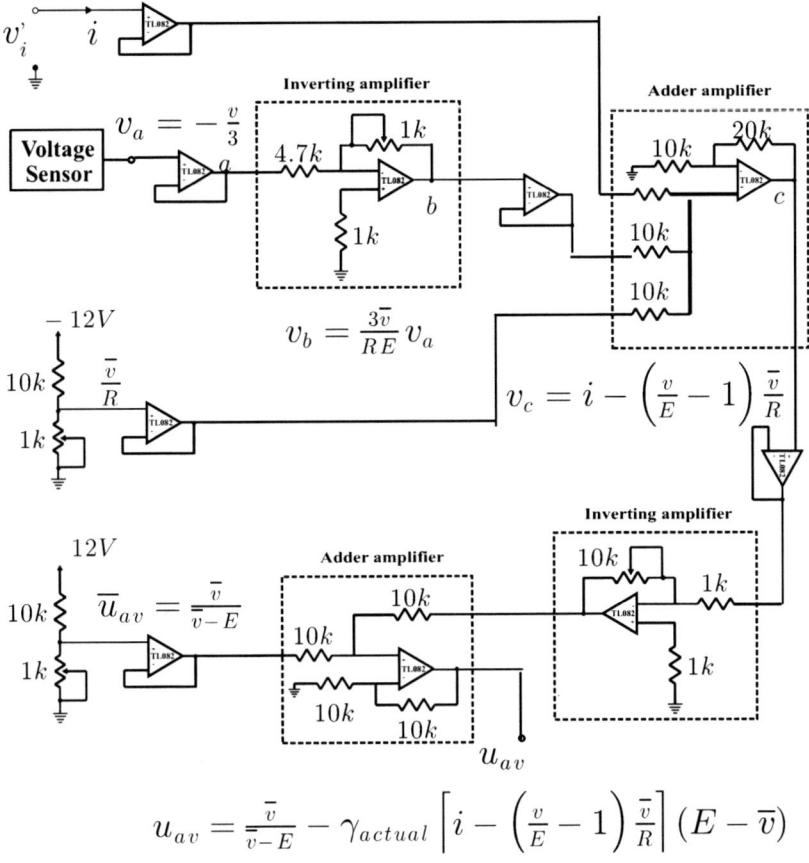

Fig. 4.48. Control circuit structure implemented for the static passivity based stabilizing controller based on the linearized model of the Buck-Boost converter.

The *control* block receives two signals: the *inductor current signal* i, which is transformed into a proportional voltage signal v'_i using the LEM HAW 15-P current sensor (see Figure 2.7), and the *output voltage* v, both from the *Buck-Boost system*. On the other hand, the output of the *control* block is the de-normalized *average control* u_{av} (4.24).

An *amplitude limiter circuit* block is placed between the *control* block output and the $\Sigma - \Delta$-*modulator* block input. The resultant conditioned average control input signal, generated by the *amplitude limiter circuit* block, is transformed at a switched pulse signal with values of 0 V and 5 V into the $\Sigma - \Delta$-*modulator* block. The output of the modulator feeds the Mosfet IC NT2984 acting as switch.

210   4 Approximate Linearization Methods

Experimental results

The corresponding experimental results obtained on the developed experimental test bench are shown in Figure 4.49. The figure depicts the closed loop response of the Buck-Boost system for the implemented passivity based stabilizing controller. The values of the components for this system were set to be:

$$L = 15.91 \text{ mH}, \quad C = 470 \text{ } \mu\text{F}, \quad R = 52 \text{ } \Omega, \quad E = 12 \text{ V}$$

with $\gamma_{actual} = 0.1$. We set an actual desired output voltage of $\bar{v} = -24$ V. This voltage determines a steady state current $\bar{i} = 1.385$ A, and $\bar{u}_{av} = 0.666$.

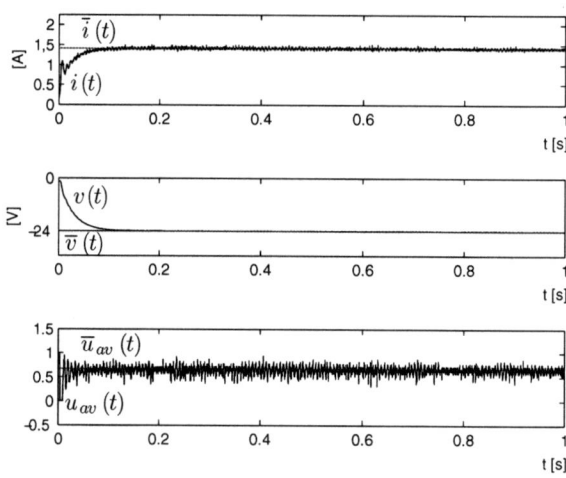

Fig. 4.49. Experimental closed loop response of the Buck-Boost power converter to a $\Sigma - \Delta$-modulator implementation of a static passivity based stabilizing controller computed on the basis of the linearized converter model.

4.6 The Cúk Converter

4.6.1 Generalities about the Model

The dc-to-dc power converter known as the Cúk converter is shown in Figure 4.50.

Consider the average normalized model of the Cúk converter

$$\dot{x}_1 = -(1 - u_{av})x_2 + 1$$
$$\dot{x}_2 = (1 - u_{av})x_1 + u_{av}x_3$$
$$\alpha_1 \dot{x}_3 = -u_{av}x_2 - x_4$$
$$\alpha_2 \dot{x}_4 = x_3 - \frac{1}{Q}x_4$$

4.6 The Cúk Converter

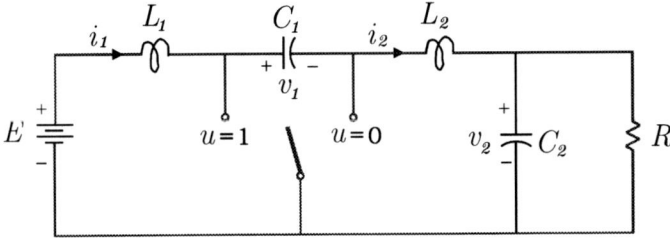

Fig. 4.50. The Cúk converter.

The average normalized tangent linearization model of this converter, around the equilibrium point:

$$\bar{x}_1 = \frac{V_d^2}{Q}, \quad \bar{x}_2 = 1 - V_d, \quad \bar{x}_3 = \frac{V_d}{Q}, \quad \bar{x}_4 = V_d, \quad \bar{u}_{av} = -\frac{V_d}{1 - V_d}$$

is given by,

$$\dot{x}_{1\delta} = -\frac{1}{1 - V_d}x_{2\delta} + (1 - V_d)u_{av,\delta}$$

$$\dot{x}_{2\delta} = \frac{1}{1 - V_d}x_{1\delta} - \frac{V_d}{1 - V_d}x_{3\delta} + \frac{V_d(1 - V_d)}{Q}u_{av,\delta}$$

$$\alpha_1 \dot{x}_{3\delta} = \frac{V_d}{1 - V_d}x_{2\delta} - x_{4\delta} - (1 - V_d)u_{av,\delta}$$

$$\alpha_2 \dot{x}_{4\delta} = x_{3\delta} - \frac{1}{Q}x_{4\delta}$$

where

$$x_{1\delta} = x_1 - \bar{x}_1, \quad x_{2\delta} = x_2 - \bar{x}_2, \quad x_{3\delta} = x_3 - \bar{x}_3, \quad x_{4\delta} = x_4 - \bar{x}_4$$

and

$$u_{av,\delta} = u_{av} - \bar{u}_{av}$$

The linearized average normalized system is found to be controllable and hence flat.

A rather fast test for the observability of the incremental average output $y_\delta = x_{4\delta}$ is as follows:

Assuming we know $u_{av,\delta}$, y_δ and its time derivatives. We see that all state variables will be known or computable. Indeed, from the last equation we would find: $x_{3\delta}$ as $\alpha_2 \dot{y}_\delta + (1/Q)y_\delta$. Knowing now $x_{3\delta}$, the state variable $x_{2\delta}$ can be computed from the third equation and, finally, $x_{1\delta}$ is computed from the second equation. We conclude that all state variables may be computed in terms of the incremental input, the incremental output, and a finite number of its time derivatives. The average system is, hence, observable from the incremental output capacitor voltage variable $x_{4\delta}$.

Similarly, the system is observable from the incremental average input inductor current, $x_{1\delta}$.

To investigate the nature of the *zero dynamics* corresponding to these two observable outputs we proceed as follows:

Let $y_\delta = x_{4\delta} = 0$, then, from the last equation of the linearized model, we get that, $x_{3\delta} = 0$. Hence, from the third equation we find that the average incremental control input must be given by,

$$u_{av,\delta} = \frac{V_d}{(1-V_d)^2} x_{2\delta}$$

Substituting this expression for the incremental control input in the first two linearized equations leads to the closed loop system:

$$\dot{x}_{1\delta} = -x_{2\delta}$$
$$\dot{x}_{2\delta} = \frac{1}{1-V_d} x_{1\delta} + \frac{V_d^2}{Q(1-V_d)} x_{2\delta}$$

which is unstable with characteristic polynomial given by

$$p(s) = s^2 - \frac{V_d^2}{Q(1-V_d)} s + \frac{1}{1-V_d}$$

Let $y_\delta = x_{1\delta} = 0$, then, from the first equation of the linearized model, we get that, the incremental average control input must be

$$u_{av,\delta} = \left[\frac{1}{(1-V_d)^2}\right] x_{2\delta}$$

Substituting this expression for the incremental control input in the rest of the linearized equations, and using the fact that $x_{1\delta} = 0$, leads to the closed loop system:

$$\dot{x}_{2\delta} = -\frac{V_d}{1-V_d} x_{3\delta} + \frac{V_d}{(1-V_d)Q} x_{2\delta}$$
$$\alpha_1 \dot{x}_{3\delta} = -x_{2\delta} - x_{4\delta}$$
$$\alpha_2 \dot{x}_{4\delta} = x_{3\delta} - \frac{1}{Q} x_{4\delta}$$

which is stable with characteristic polynomial given by

$$p(s) = s^3 + \frac{1-(1+\alpha_2)V_d}{\alpha_2(1-V_d)Q} s^2 - \frac{(\alpha_1 + \alpha_2 Q^2)V_d - (1-V_d)Q^2}{\alpha_1 \alpha_2 (1-V_d)Q^2} s - \frac{2V_d}{\alpha_1 \alpha_2 (1-V_d)Q}$$

The average linearized Cúk converter system exhibits the incremental output voltage as a *non-minimum* phase output and the incremental input current as a *minimum phase* output.

This fact, which is common to many classical dc-to-dc power converters, prompts *indirect* feedback control of the converter by regulating the incremental input inductor current to the desired equilibrium and letting the asymptotically stable zero dynamics take care of the internal and actual output voltage behavior.

4.6.2 The Hamiltonian System Approach

We write the average linearized Cúk converter model in the modified Hamiltonian form

$$\mathcal{P}\dot{x}_\delta = \mathcal{J}\frac{\partial H}{\partial x_\delta} - \mathcal{R}\frac{\partial H}{\partial x_\delta} + bu_{av,\delta}$$

with $H(x_\delta) = \frac{1}{2}x_\delta^T x_\delta$ and $\mathcal{P} = \text{diag}(1,1,\alpha_1,\alpha_2)$.

We have:

$$\begin{bmatrix} 1 & 0 & 0 & 0 \\ 0 & 1 & 0 & 0 \\ 0 & 0 & \alpha_1 & 0 \\ 0 & 0 & 0 & \alpha_2 \end{bmatrix} \dot{x}_\delta = \begin{bmatrix} 0 & -\frac{1}{1-V_d} & 0 & 0 \\ \frac{1}{1-V_d} & 0 & -\frac{V_d}{1-V_d} & 0 \\ 0 & \frac{V_d}{1-V_d} & 0 & -1 \\ 0 & 0 & 1 & 0 \end{bmatrix} \frac{\partial H}{\partial x_\delta}$$

$$- \begin{bmatrix} 0 & 0 & 0 & 0 \\ 0 & 0 & 0 & 0 \\ 0 & 0 & 0 & 0 \\ 0 & 0 & 0 & \frac{1}{Q} \end{bmatrix} \frac{\partial H}{\partial x_\delta} + \begin{bmatrix} (1-V_d) \\ \frac{V_d(1-V_d)}{Q} \\ -(1-V_d) \\ 0 \end{bmatrix} u_{av,\delta}$$

The dissipation matching condition is not strictly satisfied and takes the form

$$\mathcal{R} + \gamma bb^T = \begin{bmatrix} \gamma(1-V_d)^2 & \gamma\frac{V_d(1-V_d)^2}{Q} & -\gamma(1-V_d)^2 & 0 \\ \gamma\frac{V_d(1-V_d)^2}{Q} & \gamma\frac{V_d^2(1-V_d)^2}{Q^2} & -\gamma\frac{V_d(1-V_d)^2}{Q} & 0 \\ -\gamma(1-V_d)^2 & -\gamma\frac{V_d(1-V_d)^2}{Q} & \gamma(1-V_d)^2 & 0 \\ 0 & 0 & 0 & \frac{1}{Q} \end{bmatrix} \geq 0$$

The passive output is given by,

$$y_\delta = (1-V_d)x_{1\delta} + \frac{V_d(1-V_d)}{Q}x_{2\delta} - (1-V_d)x_{3\delta}$$

The set of vectors which are in the null space of the matrix, $\mathcal{R} + \gamma bb^T$, are of the form: $z = [x_{1\delta}\ x_{2\delta}\ x_{3\delta}\ 0]$ such that $\xi_\delta = x_{1\delta} + \frac{V_d}{Q}x_{2\delta} - x_{3\delta} = 0$, i.e., they lay in a subspace of R^4 and corresponds to $y_\delta = (1-V_d)\xi_\delta = 0$. This means the nonlinear system is controlled by the equilibrium input: $\overline{u}_{av} = -V_d/(1-V_d)$, i.e., the incremental system is controlled by $u_{av,\delta} = 0$. The only trajectory of the incremental system with $x_{4\delta} = 0$ and $u_{av,\delta} = 0$ corresponds to the origin.

This is compatible with the fact that in order for the closed loop incremental average system to have the origin as an asymptotically stable equilibrium, the trajectories of the system should have no other equilibrium than the origin itself. The origin of the average output feedback controlled system is, according to LaSalle's theorem, an asymptotically stable equilibrium.

The output feedback control law (with design parameter $\gamma > 0$)

$$u_{av,\delta} = -\gamma \left[(1-V_d)x_{1\delta} + \frac{V_d(1-V_d)}{Q}x_{2\delta} - (1-V_d)x_{3\delta} \right]$$

The average control to be implemented is synthesized as

$$u_{av} = -\frac{V_d}{1-V_d} - \gamma \left[(1-V_d)\left(x_1 - \frac{V_d^2}{Q}\right) \right.$$
$$\left. + \frac{V_d(1-V_d)}{Q}(x_2 - (1-V_d)) - (1-V_d)\left(x_3 - \frac{V_d}{Q}\right) \right] \quad (4.25)$$

Simulations

We consider a simple non-normalized average Cúk converter model with the following parameter values

$$L_1 = 30 \text{ mH}, \quad C_1 = 150 \text{ }\mu\text{F}, \quad L_2 = 30 \text{ mH}, \quad C_2 = 50 \text{ }\mu\text{F},$$
$$R = 10 \text{ }\Omega, \quad E = 100 \text{ V}$$

and the design parameter to $\gamma = 1$.

It is assumed that it is desired to drive the average output voltage to $\bar{v}_2 = -200$ V, with corresponding steady state values of the currents and internal capacitor voltage given by:

$$\bar{i}_1 = 40 \text{ A}, \quad \bar{v}_1 = 300 \text{ V}, \quad \bar{i}_2 = -20 \text{ A}$$

and

$$\bar{u}_{av} = 0.666$$

Figure 4.51 shows the average response of the average Cúk converter model to the average static passivity based feedback controller computed on the basis of the average normalized system linearization.

4.7 The Zeta Converter

4.7.1 Generalities about the Model

Consider the dc-to-dc power converter known as the Zeta converter, shown in Figure 4.52. The average normalized model of this system is given by

$$\frac{dx_1}{d\tau} = -(1-u_{av})x_2 + u_{av}$$
$$\frac{dx_2}{d\tau} = (1-u_{av})x_1 - u_{av}x_3$$
$$\alpha_1 \frac{dx_3}{d\tau} = u_{av}x_2 - x_4 + u_{av}$$
$$\alpha_2 \frac{dx_4}{d\tau} = x_3 - \frac{1}{Q}x_4$$

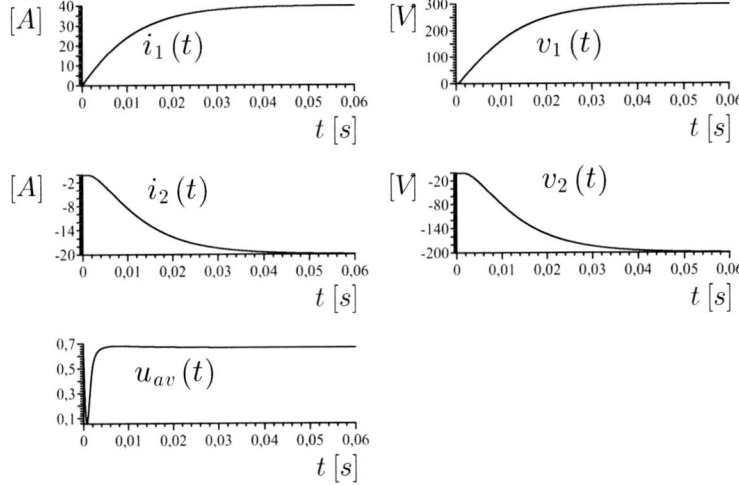

Fig. 4.51. Average responses of Cúk converter model to static passivity based control based on approximate linearization.

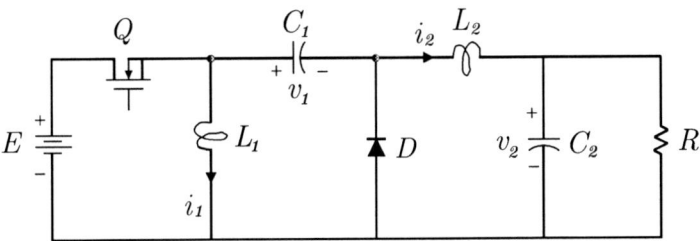

Fig. 4.52. The Zeta converter.

The average normalized state equilibrium point is obtained from the preceding model, assuming that the control input variable continuously takes values in the interval $(0,1)$. Parameterizing the equilibrium in terms of the desired equilibrium output voltage, $\bar{x}_4 = V_d$, we have:

$$\bar{x}_1 = \frac{V_d^2}{Q}, \quad \bar{x}_2 = V_d, \quad \bar{x}_3 = \frac{V_d}{Q}, \quad \bar{x}_4 = V_d, \quad \bar{u}_{av} = \frac{V_d}{1+V_d}$$

Hence due to the average control input hard limitations $(0 < u_{av} < 1)$ we have that the equilibrium output voltage must satisfy the positive output voltage restriction:

$$V_d > 0$$

The tangent linearization of the average state system around the average equilibrium point is given by

$$\dot{x}_{1\delta} = -\frac{1}{1+V_d}x_{2\delta} + (1+V_d)u_{av,\delta}$$

$$\dot{x}_{2\delta} = \frac{1}{1+V_d}x_{1\delta} - \frac{V_d}{1+V_d}x_{3\delta} - \frac{V_d}{Q}(1+V_d)u_{av,\delta}$$

$$\alpha_1\dot{x}_{3\delta} = \frac{V_d}{1+V_d}x_{2\delta} - x_{4\delta} + (1+V_d)u_{av,\delta}$$

$$\alpha_2\dot{x}_{4\delta} = x_{3\delta} - \frac{1}{Q}x_{4\delta}$$

where

$$x_{1\delta} = x_1 - \frac{V_d^2}{Q}, \quad x_{2\delta} = x_2 - V_d, \quad x_{3\delta} = x_3 - \frac{V_d}{Q}, \quad x_{4\delta} = x_4 - V_d$$

and

$$u_{av,\delta} = u_{av} - \frac{V_d}{1+V_d}$$

In matrix form: $\dot{x}_\delta = Ax_\delta + bu_{av,\delta}$ we have,

$$\dot{x}_\delta = \begin{bmatrix} 0 & -\frac{1}{1+V_d} & 0 & 0 \\ \frac{1}{1+V_d} & 0 & -\frac{V_d}{1+V_d} & 0 \\ 0 & \frac{V_d}{\alpha_1(1+V_d)} & 0 & -\frac{1}{\alpha_1} \\ 0 & 0 & \frac{1}{\alpha_2} & -\frac{1}{\alpha_2 Q} \end{bmatrix} x_\delta + \begin{bmatrix} (1+V_d) \\ -\frac{V_d}{Q}(1+V_d) \\ \frac{1}{\alpha_1}(1+V_d) \\ 0 \end{bmatrix} u_{av,\delta}$$

The system is found to be controllable and hence flat, since the controllability matrix, $\mathcal{C} = \begin{bmatrix} B & Ab & A^2b & A^3b \end{bmatrix}$, has full range, i.e., rank $[\mathcal{C}] = 4$.

A rather fast test for observability of the incremental average output $y_\delta = x_{4\delta}$ is as follows:

Assuming we know $u_{av,\delta}$, y_δ and its time derivatives, we see that all state variables will be known or computable. Indeed, from the last equation we would find: $x_{3\delta}$ as $\alpha_2\dot{y}_\delta + \frac{1}{Q}y_\delta$. Knowing now $x_{3\delta}$, the state variable $x_{2\delta}$ can be computed from the third equation and $x_{1\delta}$ from the second equation. We conclude that all state variables are computable in terms of the incremental input, the incremental output, and a finite number of its time derivatives. The average system is, hence, observable from the incremental output capacitor voltage variable $x_{4\delta}$.

Similarly, the system is observable from the incremental average input inductor current, $x_{1\delta}$.

To investigate the nature of the *zero dynamics* corresponding to these two observable outputs we proceed as follows:

Let $y_\delta = x_{4\delta} = 0$, then, from the last equation of the linearized model, we get that, $x_{3\delta} = 0$. Hence, from the third equation we find that the control input must be

$$u_{av,\delta} = -\frac{V_d}{(1+V_d)^2}x_{2\delta}$$

4.7 The Zeta Converter

Substituting this expression for the incremental control input in the first two linearized equations leads to the closed loop system:

$$\dot{x}_{1\delta} = -x_{2\delta}$$
$$\dot{x}_{2\delta} = \frac{1}{1+V_d}x_{1\delta} + \frac{1}{Q}\frac{V_d^2}{(1+V_d)}x_{2\delta}$$

which is unstable with characteristic polynomial given by

$$p(s) = s^2 - \frac{1}{Q}\frac{V_d^2}{1+V_d}s + \frac{1}{1+V_d}$$

Let $y_\delta = x_{1\delta} = 0$, then, from the first equation of the linearized model, we get that, the control input must be

$$u_{av,\delta} = \frac{1}{(1+V_d)^2}x_{2\delta}$$

Substituting this expression for the incremental control input in the rest of the linearized equations, and using the fact that $x_{1\delta} = 0$, leads to the closed loop system:

$$\dot{x}_{2\delta} = -\frac{V_d}{Q}\frac{1}{(1+V_d)}x_{2\delta} - \frac{V_d}{1+V_d}x_{3\delta}$$
$$\alpha_1\dot{x}_{3\delta} = x_{2\delta} - x_{4\delta}$$
$$\alpha_2\dot{x}_{4\delta} = x_{3\delta} - \frac{1}{Q}x_{4\delta}$$

which is stable with characteristic polynomial given by

$$p(s) = s^3 + \eta s^2 + \mu s + \kappa$$

where

$$\eta = \frac{1}{Q}\left(\frac{1}{\alpha_2} + \frac{V_d}{1+V_d}\right)$$
$$\mu = \frac{\alpha_1 V_d + (1+(1+\alpha_2)V_d)Q^2}{\alpha_1\alpha_2(1+V_d)Q^2}$$
$$\kappa = \frac{2V_d}{\alpha_1\alpha_2(1+V_d)Q}$$

since the Routh-Hurwitz array for this polynomial is given by

$$\begin{array}{c|cc} s^3 & 1 & \mu \\ s^2 & \eta & \kappa \\ s & \mu - \frac{\kappa}{\eta} & \\ s^0 & \kappa & \end{array}$$

and all the coefficients in first column all have the same sign, i.e.,

$$\eta > 0$$

$$\mu - \frac{\kappa}{\eta} = \frac{1}{\alpha_2} \left[\frac{V_d}{(V_d + 1) Q^2} + \frac{1 + (2 + (1 + \alpha_2^2) V_d) V_d}{\alpha_1 (1 + V_d) (1 + (1 + \alpha_2) V_d)} \right] > 0$$

$$\kappa > 0$$

Therefore, all the roots of $p(s)$ have negative real parts. Thus, the average linearized Zeta converter system exhibits the incremental output voltage as a *non-minimum* phase output and the incremental input current as a *minimum phase* output.

This fact, which is common to many classical dc-to-dc power converters, prompts *indirect* feedback control of the converter by regulating the incremental input inductor current to the desired equilibrium and letting the asymptotically stable zero dynamics take care of the internal and actual output voltage behavior.

4.7.2 The Hamiltonian System Approach

We write the average linearized Zeta converter model in the modified Hamiltonian form

$$\mathcal{P}\dot{x}_\delta = \mathcal{J}\frac{\partial H}{\partial x_\delta} - \mathcal{R}\frac{\partial H}{\partial x_\delta} + b u_{av,\delta}$$

with $H(x_\delta) = \frac{1}{2} x_\delta^T x_\delta$ and $\mathcal{P} = \mathrm{diag}(1, 1, \alpha_1, \alpha_2)$.
We have:

$$\begin{bmatrix} 1 & 0 & 0 & 0 \\ 0 & 1 & 0 & 0 \\ 0 & 0 & \alpha_1 & 0 \\ 0 & 0 & 0 & \alpha_2 \end{bmatrix} \dot{x}_\delta = \begin{bmatrix} 0 & -\frac{1}{1+V_d} & 0 & 0 \\ \frac{1}{1+V_d} & 0 & -\frac{V_d}{1+V_d} & 0 \\ 0 & \frac{V_d}{(1+V_d)} & 0 & -1 \\ 0 & 0 & 1 & 0 \end{bmatrix} \frac{\partial H}{\partial x_\delta}$$

$$- \begin{bmatrix} 0 & 0 & 0 & 0 \\ 0 & 0 & 0 & 0 \\ 0 & 0 & 0 & 0 \\ 0 & 0 & 0 & \frac{1}{Q} \end{bmatrix} \frac{\partial H}{\partial x_\delta} + \begin{bmatrix} (1 + V_d) \\ -\frac{V_d}{Q}(1 + V_d) \\ (1 + V_d) \\ 0 \end{bmatrix} u_{av,\delta}$$

The dissipation matching condition is not strictly satisfied and takes the form

$$\mathcal{R} + \gamma bb^T = (1 + V_d)^2 \begin{bmatrix} \gamma & -\gamma\frac{V_d}{Q} & \gamma & 0 \\ -\gamma\frac{V_d}{Q} & \gamma\frac{V_d^2}{Q^2} & -\gamma\frac{V_d}{Q} & 0 \\ \gamma & -\gamma\frac{V_d}{Q} & \gamma & 0 \\ 0 & 0 & 0 & \frac{1}{(1+V_d)^2 Q} \end{bmatrix} \geq 0$$

The passive output is given by,

$$y_\delta = (1+V_d)\,x_{1\delta} - \frac{V_d}{Q}(1+V_d)\,x_{2\delta} + (1+V_d)\,x_{3\delta}$$

The set of vectors which are in the null space of the matrix, $\mathcal{R}+\gamma bb^T$, are of the form: $z = \begin{bmatrix} x_1 & x_{2\delta} & x_{3\delta} & 0 \end{bmatrix}$ such that $\xi_\delta = x_{1\delta} - \frac{V_d}{Q}x_{2\delta} + x_{3\delta}$, i.e., they lay in a subspace of R^4 and corresponds to $y_\delta = (1+V_d)\,\xi_\delta = 0$. This means the nonlinear system is controlled by the equilibrium input: $\bar{u}_{av} = \frac{V_d}{1+V_d}$, i.e., the incremental system is controlled by $u_{av,\delta} = 0$. The only trajectory of the incremental system with $x_{4\delta} = 0$ and $u_{av,\delta} = 0$ corresponds to the origin.

This is compatible with the fact that in order for the closed loop incremental average system to have the origin as an asymptotically stable equilibrium, the trajectories of the system should have no other equilibrium than the origin itself. The origin of the average output feedback controlled system is, hence, an asymptotically stable equilibrium.

The output feedback control law (with design parameter $\gamma > 0$)

$$u_{av,\delta} = -\gamma b^T \frac{\partial H(x_\delta)}{\partial x_\delta} = -\gamma \left[(1+V_d)\,x_{1\delta} - \frac{V_d(1+V_d)}{Q}x_{2\delta} + (1+V_d)\,x_{3\delta}\right]$$

The average control to be implemented is synthesized as

$$u_{av} = \frac{V_d}{1+V_d} - \gamma \left[(1+V_d)\left(x_1 - \frac{V_d^2}{Q}\right)\right.$$
$$\left. - \frac{V_d(1+V_d)}{Q}(x_2 - V_d) + (1+V_d)\left(x_3 - \frac{V_d}{Q}\right)\right]$$

Simulations

Figure 4.53 shows the average response of the Zeta converter model to the incremental average passive output feedback controller for a typical average Zeta converter circuit model ($L_1 = 600\ \mu H$, $C_1 = 10\ \mu F$, $L_2 = 600\ \mu H$, $C_2 = 10\ \mu F$, $R = 40\ \Omega$ and $E = 100$ V). We set a desired steady state output voltage of $\bar{v}_2 = 200$ V and the corresponding steady state variables yields: $\bar{i}_1 = 10$ A, $\bar{v}_1 = 200$ V, $\bar{i}_2 = 5$ A and $\bar{u}_{av} = 0.666$. This parameters values yields $Q = 5.164$, $\alpha_1 = \alpha_2 = 1$ and the time normalization factor was found to be $\sqrt{L_1 C_1} = 7.746 \times 10^{-5}$ s. The design parameter we make equal to $\gamma = 1$.

4.8 The Quadratic Buck Converter

4.8.1 Generalities about the Model

The dc-to-dc power converter known as the quadratic Buck converter has the following average normalized model:

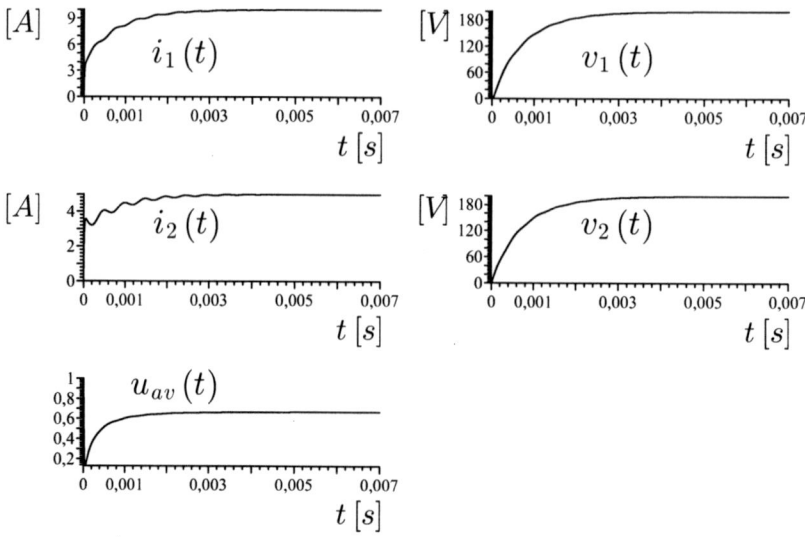

Fig. 4.53. Average response of the average Zeta converter model to the incremental average passive output feedback controller.

$$\frac{dx_1}{d\tau} = -x_2 + u_{av}$$

$$\frac{dx_2}{d\tau} = x_1 - u_{av}x_3$$

$$\alpha_1 \frac{dx_3}{d\tau} = u_{av}x_2 - x_4$$

$$\alpha_2 \frac{dx_4}{d\tau} = x_3 - \frac{1}{Q}x_4$$

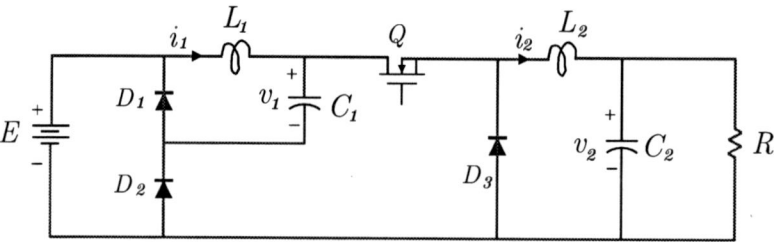

Fig. 4.54. Switch regulated dc-to-dc quadratic Buck power converter.

The average equilibrium point of the system is, in terms of the desired output equilibrium voltage $\bar{x}_4 = V_d$, given by:

4.8 The Quadratic Buck Converter

$$\bar{x}_1 = \frac{(V_d)^{\frac{3}{2}}}{Q}, \quad \bar{x}_2 = \sqrt{V_d}, \quad \bar{x}_3 = \frac{V_d}{Q}, \quad \bar{x}_4 = V_d$$

$$\bar{u}_{av} = \sqrt{V_d}$$

Hence, due to the average control input hard limitations ($0 < u_{av} < 1$) we have that the equilibrium output voltage must satisfy the positive output voltage restriction:

$$0 < V_d < 1$$

Tangent linearization of the average model, around the equilibrium point, leads to the following set of equations,

$$\dot{x}_{1\delta} = -x_{2\delta} + u_{av,\delta}$$

$$\dot{x}_{2\delta} = x_{1\delta} - \sqrt{V_d}x_{3\delta} - \frac{V_d}{Q}u_{av,\delta}$$

$$\alpha_1\dot{x}_{3\delta} = \sqrt{V_d}x_{2\delta} - x_{4\delta} + \sqrt{V_d}u_{av,\delta}$$

$$\alpha_2\dot{x}_{4\delta} = x_{3\delta} - \frac{1}{Q}x_{4\delta}$$

where

$$x_{1\delta} = x_1 - \frac{(V_d)^{\frac{3}{2}}}{Q}, \quad x_{2\delta} = x_2 - \sqrt{V_d}, \quad x_{3\delta} = x_3 - \frac{V_d}{Q}, \quad x_{4\delta} = x_4 - V_d$$

$$u_{av,\delta} = u_{av} - \sqrt{V_d}$$

In matrix form: $\dot{x}_\delta = Ax_\delta + bu_{av,\delta}$ we have,

$$\dot{x}_\delta = \begin{bmatrix} 0 & -1 & 0 & 0 \\ 1 & 0 & -\sqrt{V_d} & 0 \\ 0 & \frac{\sqrt{V_d}}{\alpha_1} & 0 & -\frac{1}{\alpha_1} \\ 0 & 0 & \frac{1}{\alpha_2} & -\frac{1}{\alpha_2 Q} \end{bmatrix} x_\delta + \begin{bmatrix} 1 \\ -\frac{V_d}{Q} \\ \frac{\sqrt{V_d}}{\alpha_1} \\ 0 \end{bmatrix} u_{av,\delta}$$

The system is controllable and, hence flat

$$\mathcal{C} = \begin{bmatrix} b & Ab & A^2b & A^3b \end{bmatrix}$$

$$= \begin{bmatrix} 1 & \frac{V_d}{Q} & -\frac{\beta_1}{\alpha_1} & -\frac{V_d\beta_2}{\alpha_1 Q} \\ -\frac{V_d}{Q} & \frac{\beta_1}{\alpha_1} & \frac{V_d\beta_2}{\alpha_1 Q} & \frac{V_d(1+\alpha_2 V_d)-\alpha_1^2\alpha_2}{\alpha_1^2\alpha_2} \\ \frac{\sqrt{V_d}}{\alpha_1} & -\frac{V_d^{\frac{3}{2}}}{\alpha_1 Q} & -\frac{\sqrt{V_d}(1-\beta_4)}{\alpha_1^2\alpha_2} & \frac{\sqrt{V_d}[1+\alpha_2 V_d(1+\alpha_2\beta_2)]}{\alpha_1^2\alpha_2^2 Q} \\ 0 & \frac{\sqrt{V_d}}{\alpha_1\alpha_2} & -\frac{\sqrt{V_d}(1+\alpha_2 V_d)}{\alpha_1\alpha_2^2 Q} & \frac{\sqrt{V_d}[\beta_3-\alpha_2 Q^2(1-\beta_4)]}{\alpha_1^2\alpha_2^3 Q^2} \end{bmatrix}$$

The controllability matrix has full range, i.e., rank $[\mathcal{C}] = 4$, where:

$$\beta_1 = \alpha_1 - V_d, \quad \beta_2 = \alpha_1 + V_d, \quad \beta_3 = \alpha_1(1+\alpha_2 V_d), \quad \beta_4 = \alpha_2(\alpha_1 - V_d)$$

A rather fast test for observability of the incremental average output $y_\delta = x_{4\delta}$ is as follows:

Assuming we know $u_{av,\delta}$, y_δ and its time derivatives, we see that all state variables will be known or computable. Indeed, from the last equation we would find: $x_{3\delta}$ as $\alpha_2 \dot{y}_\delta + \frac{1}{Q} y_\delta$. Knowing now $x_{3\delta}$, the state variable $x_{2\delta}$ can be computed from the third equation and $x_{1\delta}$ from the second equation. We conclude that all state variables are computable in terms of the incremental input, the incremental output, and a finite number of its time derivatives. The average system is, hence, observable from the incremental output capacitor voltage variable $x_{4\delta}$.

Similarly, the system is observable from the incremental average input inductor current, $x_{1\delta}$.

To investigate the nature of the *zero dynamics* corresponding to these two observable outputs we proceed as follows:

Let $y_\delta = x_{4\delta} = 0$, then, from the last equation of the linearized model, we get that, $x_{3\delta} = 0$. Hence, from the third equation we find that the control input must be

$$u_{av,\delta} = -x_{2\delta}$$

Substituting this expression for the incremental control input in the first two linearized equations leads to the closed loop system:

$$\dot{x}_{1\delta} = -2x_{2\delta}$$
$$\dot{x}_{2\delta} = x_{1\delta} + \frac{V_d}{Q} x_{2\delta}$$

which is unstable with characteristic polynomial given by

$$p(s) = s^2 - \frac{V_d}{Q} s + 2$$

Let $y_\delta = x_{1\delta} = 0$, then, from the first equation of the linearized model, we get that, the control input must be

$$u_{av,\delta} = x_{2\delta}$$

Substituting this expression for the incremental control input in the rest of the linearized equations, and using the fact that $x_{1\delta} = 0$, leads to the closed loop system:

$$\dot{x}_{2\delta} = -\frac{V_d}{Q} x_{2\delta} - \sqrt{V_d} x_{3\delta}$$
$$\alpha_1 \dot{x}_{3\delta} = 2\sqrt{V_d} x_{2\delta} - x_{4\delta}$$
$$\alpha_2 \dot{x}_{4\delta} = x_{3\delta} - \frac{1}{Q} x_{4\delta}$$

which is stable with characteristic polynomial given by

$$p(s) = s^3 + \frac{(1+\alpha_2 V_d)}{\alpha_2 Q}s^2 + \frac{\alpha_1 V_d + Q^2(1+2\alpha_2 V_d)}{\alpha_1 \alpha_2 Q^2}s + \frac{3V_d}{\alpha_1 \alpha_2 Q}$$

The average linearized quadratic Buck converter system exhibits the incremental output voltage as a *non-minimum* phase output and the incremental input current as a *minimum phase* output.

This fact, which is common to many classical dc-to-dc power converters, prompts *indirect* feedback control of the converter by regulating the incremental input inductor current to the desired equilibrium and letting the asymptotically stable zero dynamics take care of the internal and actual output voltage behavior.

4.8.2 State Feedback Controller Design

We consider now the tangent linearization model of the average normalized quadratic Buck converter system:

$$\frac{dx_1}{d\tau} = -x_2 + u_{av}$$

$$\frac{dx_2}{d\tau} = x_1 - u_{av} x_3$$

$$\alpha_1 \frac{dx_3}{d\tau} = u_{av} x_2 - x_4$$

$$\alpha_2 \frac{dx_4}{d\tau} = x_3 - \frac{1}{Q}x_4$$

around the equilibrium point parameterizing in terms of the desired constant input control, i.e., $\overline{u}_{av} = U$:

$$\overline{x}_1 = \frac{U^3}{Q}, \quad \overline{x}_2 = U, \quad \overline{x}_3 = \frac{U^2}{Q}, \quad \overline{x}_4 = U^2$$

given by

$$\dot{x}_\delta = \begin{bmatrix} 0 & -1 & 0 & 0 \\ 1 & 0 & -U & 0 \\ 0 & \frac{1}{\alpha_1}U & 0 & -\frac{1}{\alpha_1} \\ 0 & 0 & \frac{1}{\alpha_2} & -\frac{1}{\alpha_2 Q} \end{bmatrix} x_\delta + \begin{bmatrix} 1 \\ -\frac{1}{Q}U^2 \\ \frac{1}{\alpha_1}U \\ 0 \end{bmatrix} u_{av,\delta}$$

where

$$x_{1\delta} = x_1 - \frac{U^3}{Q}, \quad x_{2\delta} = x_2 - U, \quad x_{3\delta} = x_3 - \frac{U^2}{Q}, \quad x_{4\delta} = x_4 - U^2$$

$$u_{av,\delta} = u_{av} - U$$

In this case the controllability matrix again has full rank, hence the tangent linearization of the system is controllable, and is given by:

$$\mathcal{C} = \begin{bmatrix} b & Ab & A^2b & A^3b \end{bmatrix}$$

$$= \begin{bmatrix} 1 & \frac{U^2}{Q} & \frac{U^2}{\alpha_1}-1 & -\frac{(\alpha_1+U^2)U^2}{\alpha_1 Q} \\ -\frac{U^2}{Q} & \frac{\alpha_1-U^2}{\alpha_1} & \frac{(\alpha_1+U^2)U^2}{\alpha_1 Q} & \frac{(1+\alpha_2 U^2)U^2-\alpha_1^2\alpha_2}{\alpha_1^2\alpha_2} \\ \frac{U}{\alpha_1} & -\frac{U^3}{\alpha_1 Q} & \frac{[1-\alpha_2(\alpha_1-U^2)]U}{\alpha_1^2\alpha_2} & \frac{[\alpha_2(\alpha_1\alpha_2+\alpha_2 U^2+1)U^2+1]U}{\alpha_1^2\alpha_2^2 Q} \\ 0 & \frac{U}{\alpha_1\alpha_2} & -\frac{(1+\alpha_2 U^2)U}{\alpha_1\alpha_2^2 Q} & \frac{[\alpha_1\alpha_2^2 Q^2+(\alpha_1-\alpha_2 Q^2)(1+\alpha_2 U^2)]U}{\alpha_1^2\alpha_2^3 Q^2} \end{bmatrix}$$

We seek for a linear state feedback controller of the form

$$u_{av,\delta} = -\mathbf{K}\mathbf{x}_\delta = -K_1 x_{1\delta} - K_2 x_{2\delta} - K_3 x_{3\delta} - K_4 x_{4\delta}$$

which drives the stabilization error state x_δ to zero in an exponentially stable fashion.

We can obtain the design gains, $\mathbf{K}$, by direct application from the formula proposed by Ackermann (see [33]), given by:

$$\mathbf{K} = \begin{bmatrix} 0 & 0 & 0 & 1 \end{bmatrix} \mathcal{C}^{-1} \alpha_c(A) \qquad (4.26)$$

where the inverse of the controllability matrix for this system is:

$$\mathcal{C}^{-1} = \frac{1}{\det \mathcal{C}} \text{ adjugate } \mathcal{C}$$

$$= \frac{\alpha_1^4 \alpha_2^3 Q^4}{\beta_2 \alpha_1^2 + \beta_1 \alpha_1 + \beta_0} \begin{bmatrix} * & * & * & * \\ * & * & * & * \\ * & * & * & * \\ \gamma_{41} & \gamma_{42} & \gamma_{43} & \gamma_{44} \end{bmatrix} \qquad (4.27)$$

with the coefficients β and γ defined for the following relationships:

$$\beta_2 = 2\left(U^4 + Q^2\right)\left(\alpha_2 U^2 + 2\alpha_2^2 Q^2 + 1\right) U^2$$
$$\beta_1 = \left[3\alpha_2 U^8 + \left(4\alpha_2^2 Q^2 + 3\right) U^6 - 7\alpha_2 Q^2 U^4 \right.$$
$$\left. - \left(8\alpha_2^2 Q^2 + 3\right) Q^2 U^2 - 4\alpha_2 Q^4\right] U^2$$
$$\beta_0 = \left[3\alpha_2 U^6 + \left(4\alpha_2^2 Q^2 + 6\right) U^4 + 4\alpha_2 Q^2 U^2 + Q^2\right] Q^2 U^2$$

$$\gamma_{41} = \frac{\alpha_1\alpha_2 U^6 + (\alpha_1+\alpha_2 Q^2) U^4 + Q^2(2-3\alpha_1\alpha_2) U^2 - \alpha_1 Q^2}{\alpha_1^3 \alpha_2^2 Q^3} U^2$$

$$\gamma_{42} = \frac{2\alpha_1\alpha_2 U^4 + 2(\alpha_1+\alpha_2 Q^2) U^2 + (1-2\alpha_1\alpha_2) Q^2}{\alpha_1^3 \alpha_2^2 Q^2} U^2$$

$$\gamma_{43} = \frac{\alpha_1\alpha_2 U^6 + (\alpha_1+\alpha_2 Q^2) U^4 + Q^2(\alpha_1\alpha_2-1) U^2 + \alpha_1 Q^2}{\alpha_1^2 \alpha_2^2 Q^3} U$$

$$\gamma_{44} = \frac{2\alpha_1\alpha_2 U^6 + [2\alpha_2(Q^2+\alpha_1^2)-\alpha_1] U^4 + (1-4\alpha_1\alpha_2) Q^2 U^2 - \alpha_1(1-2\alpha_1\alpha_2) Q^2}{\alpha_1^3 \alpha_2 Q^2} U$$

4.8 The Quadratic Buck Converter

The parameters " $*$ " in general are different of zero, but they are not matter in the calculation of the gains **K** when the Ackermann's formula is used since the vector $\begin{bmatrix} 0 & 0 & 0 & 1 \end{bmatrix}$ only extracts the last row of the adjugate of the controllability matrix.

We propound that the tangent linearization of the average state system has its poles, in closed loop, in the roots of the desired characteristic polynomial:

$$P_d(s) = \left(s^2 + 2\xi w_n s + w_n^2\right)^2$$
$$= s^4 + 4\xi w_n s^3 + 2w_n^2\left(1 + 2\xi^2\right)s^2 + 4\xi w_n^3 s + w_n^4$$

hence $\alpha_c(A)$ is reduces to the following:

$$\alpha_c(A) = A^4 + 4\xi w_n A^3 + 2w_n^2\left(1 + 2\xi^2\right)A^2 + 4\xi w_n^3 A + w_n^4 I$$

$$= \begin{bmatrix} a_{11} & a_{12} & a_{13} & a_{14} \\ a_{21} & a_{22} & a_{23} & a_{24} \\ a_{31} & a_{32} & a_{33} & a_{34} \\ a_{41} & a_{42} & a_{43} & a_{44} \end{bmatrix} \quad (4.28)$$

where:

$$a_{11} = w_n^4 - 2\left(1 + 2\xi^2\right)w_n^2 + \frac{U^2}{\alpha_1} + 1$$

$$a_{21} = 4\left[w_n^2 - \frac{1}{\alpha_1}\left(\alpha_1 + U^2\right)\right]\xi w_n$$

$$a_{31} = \frac{1}{\alpha_1}\left[2\left(1 + 2\xi^2\right)w_n^2 - \frac{1 + \alpha_2 U^2}{\alpha_1 \alpha_2} - 1\right]U$$

$$a_{41} = \frac{1}{\alpha_1 \alpha_2}\left[4\xi w_n - \frac{1}{\alpha_2 Q}\right]U$$

$$a_{12} = 4\frac{\alpha_1\left(1 - w_n^2\right) + U^2}{\alpha_1}\xi w_n$$

$$a_{22} = w_n^4 - 2\frac{\left(1 + 2\xi^2\right)\left(\alpha_1 + U^2\right)}{\alpha_1}w_n^2 + \frac{U^2}{\alpha_1^2 \alpha_2} + \frac{\left(\alpha_1 + U^2\right)^2}{\alpha_1^2}$$

$$a_{32} = \frac{4\xi U}{\alpha_1}w_n^3 - 4\frac{\left[1 + \left(\alpha_1 + U^2\right)\alpha_2\right]\xi U}{\alpha_1^2 \alpha_2}w_n + \frac{U}{\alpha_1^2 \alpha_2^2 Q}$$

$$a_{42} = \frac{2U}{\alpha_1 \alpha_2}\left(1 + 2\xi^2\right)w_n^2 - \frac{4\xi U}{\alpha_1 \alpha_2^2 Q}w_n + \frac{[\alpha_1 - Q^2 \alpha_2\left(1 + \alpha_2\left(\alpha_1 + U^2\right)\right)]U}{\alpha_1^2 \alpha_2^3 Q^2}$$

$$a_{13} = 2\left(1+2\xi^2\right)U\omega_n^2 - \frac{\left[1+\alpha_2\left(\alpha_1+U^2\right)\right]U}{\alpha_1\alpha_2}$$

$$a_{23} = -4U\xi\omega_n^3 + 4\frac{\left[1+\alpha_2\left(\alpha_1+U^2\right)\right]\xi U}{\alpha_1\alpha_2}\omega_n - \frac{U}{\alpha_1\alpha_2^2 Q}$$

$$a_{33} = \omega_n^4 - 2\frac{\left(1+2\xi^2\right)\left(1+\alpha_2 U^2\right)}{\alpha_1\alpha_2}\omega_n^2 + \frac{4}{Q}\frac{\xi}{\alpha_1\alpha_2^2}\omega_n + \frac{\frac{\left(1+\alpha_2 U^2\right)^2}{\alpha_1\alpha_2^2}+U^2-\frac{1}{\alpha_2^3 Q^2}}{\alpha_1}$$

$$a_{43} = 4\frac{\xi}{\alpha_2}\omega_n^3 - 2\frac{1+2\xi^2}{\alpha_2^2 Q}\omega_n^2 - 4\frac{\left(Q^2\alpha_2-\alpha_1+Q^2 U^2\alpha_2^2\right)\xi}{\alpha_1\alpha_2^3 Q^2}\omega_n + \frac{1+\alpha_2 U^2-\frac{\alpha_1-Q^2\alpha_2}{\alpha_2 Q^2}}{\alpha_1\alpha_2^3 Q}$$

$$a_{14} = -4\frac{\xi U}{\alpha_1}\omega_n + \frac{U}{\alpha_1\alpha_2 Q}$$

$$a_{24} = 2\frac{\left(1+2\xi^2\right)U}{\alpha_1}\omega_n^2 - 4\frac{\xi U}{\alpha_1\alpha_2 Q}\omega_n + \frac{U}{\alpha_1^2}\left[\frac{\alpha_1-\alpha_2 Q^2}{\alpha_2^2 Q^2} - \left(\alpha_1+U^2\right)\right]$$

$$a_{34} = -4\frac{\xi}{\alpha_1}\omega_n^3 + 2\frac{1+2\xi^2}{\alpha_1\alpha_2 Q}\omega_n^2 + 4\frac{\left[\alpha_2\left(1+\alpha_2 U^2\right)Q^2-\alpha_1\right]\xi}{\alpha_1^2\alpha_2^2 Q^2}\omega_n + \frac{\alpha_1-\alpha_2\left(2+\alpha_2 U^2\right)Q^2}{\alpha_1^2\alpha_2^3 Q^3}$$

$$a_{44} = \omega_n^4 - 4\frac{\xi}{\alpha_2 Q}\omega_n^3 + 2\frac{\left(\alpha_1-\alpha_2 Q^2\right)\left(1+2\xi^2\right)}{\alpha_1\alpha_2^2 Q^2}\omega_n^2 + 4\frac{\left(2\alpha_2 Q^2-\alpha_1\right)\xi}{\alpha_1\alpha_2^3 Q^3}\omega_n$$

$$+\frac{\alpha_1\left(\alpha_1-3\alpha_2 Q^2\right)+\alpha_2^2\left(1+\alpha_2 U^2\right)Q^4}{\alpha_1^2\alpha_2^4 Q^4}$$

Finally, substituting (4.27) and (4.28) in the Ackermann's formula (4.26), **K** is given by:

$$\mathbf{K} = \frac{\alpha_1^4\alpha_2^3 Q^4}{\beta_2\alpha_1^2+\beta_1\alpha_1+\beta_0}\left[\delta_{11}\ \delta_{12}\ \delta_{13}\ \delta_{14}\right]$$

where:

$$\delta_{11} = a_{11}\gamma_{41} + a_{21}\gamma_{42} + a_{31}\gamma_{43} + a_{41}\gamma_{44}$$
$$\delta_{12} = a_{12}\gamma_{41} + a_{22}\gamma_{42} + a_{32}\gamma_{43} + a_{42}\gamma_{44}$$
$$\delta_{13} = a_{13}\gamma_{41} + a_{23}\gamma_{42} + a_{33}\gamma_{43} + a_{43}\gamma_{44}$$
$$\delta_{14} = a_{14}\gamma_{41} + a_{24}\gamma_{42} + a_{34}\gamma_{43} + a_{44}\gamma_{44}$$

Hence, the feedback linear control law that stabilizes the incremental model around the origin is given then for:

$$u_{av,\delta} = -\frac{\alpha_1^4\alpha_2^3 Q^4}{\beta_2\alpha_1^2+\beta_1\alpha_1+\beta_0}\left[\delta_{11}x_{1\delta}+\delta_{12}x_{2\delta}+\delta_{13}x_{3\delta}+\delta_{14}x_{4\delta}\right]$$

and the linear controller that stabilizes the original nonlinear system to the equilibrium point $(\bar{x}_1, \bar{x}_2, \bar{x}_3, \bar{x}_4)$, it is obtained substituting the incremental variables in the linear controller for their values in function of the original variables. Thus the average control to be implemented is synthesized as:

$$u_{av} = U - \frac{\alpha_1^4 \alpha_2^3 Q^4}{\beta_2 \alpha_1^2 + \beta_1 \alpha_1 + \beta_0} \times$$
$$[\delta_{11}(x_1 - \overline{x}_1) + \delta_{12}(x_2 - \overline{x}_2) + \delta_{13}(x_3 - \overline{x}_3) + \delta_{14}(x_4 - \overline{x}_4)]$$

4.8.3 The Hamiltonian System Approach

We write the average linearized quadratic Buck converter model in the modified Hamiltonian form

$$\mathcal{P}\dot{x}_\delta = \mathcal{J}\frac{\partial H}{\partial x_\delta} - \mathcal{R}\frac{\partial H}{\partial x_\delta} + bu_{av,\delta}$$

with $H(x_\delta) = \frac{1}{2}x_\delta^T x_\delta$ and $\mathcal{P} = \text{diag}(1, 1, \alpha_1, \alpha_2)$.
We have:

$$\begin{bmatrix} 1 & 0 & 0 & 0 \\ 0 & 1 & 0 & 0 \\ 0 & 0 & \alpha_1 & 0 \\ 0 & 0 & 0 & \alpha_2 \end{bmatrix} \dot{x}_\delta = \begin{bmatrix} 0 & -1 & 0 & 0 \\ 1 & 0 & -\sqrt{V_d} & 0 \\ 0 & \sqrt{V_d} & 0 & -1 \\ 0 & 0 & 1 & 0 \end{bmatrix} \frac{\partial H(x_\delta)}{\partial x_\delta}$$

$$- \begin{bmatrix} 0 & 0 & 0 & 0 \\ 0 & 0 & 0 & 0 \\ 0 & 0 & 0 & 0 \\ 0 & 0 & 0 & \frac{1}{Q} \end{bmatrix} \frac{\partial H(x_\delta)}{\partial x_\delta} + \begin{bmatrix} 1 \\ -\frac{V_d}{Q} \\ \sqrt{V_d} \\ 0 \end{bmatrix} u_{av,\delta}$$

The dissipation matching condition is not strictly satisfied and takes the form

$$\mathcal{R} + \gamma bb^T = \begin{bmatrix} \gamma & -\gamma\frac{V_d}{Q} & \gamma\sqrt{V_d} & 0 \\ -\gamma\frac{V_d}{Q} & \gamma\left(\frac{V_d}{Q}\right)^2 & -\gamma\frac{V_d^{\frac{3}{2}}}{Q} & 0 \\ \gamma\sqrt{V_d} & -\gamma\frac{V_d^{\frac{3}{2}}}{Q} & \gamma V_d & 0 \\ 0 & 0 & 0 & \frac{1}{Q} \end{bmatrix} \geq 0$$

The passive output is given by,

$$y = b^T \frac{\partial H(x_\delta)}{\partial x_\delta} = x_{1\delta} - \frac{V_d}{Q}x_{2\delta} + \sqrt{V_d}x_{3\delta}$$

The set of vectors which are in the null space of the matrix, $\mathcal{R} + \gamma bb^T$, are of the form: $z = \begin{bmatrix} x_{1\delta} & x_{2\delta} & x_{3\delta} & 0 \end{bmatrix}$ such that $\xi_\delta = x_{1\delta} - \frac{V_d}{Q}x_{2\delta} + \sqrt{V_d}x_{3\delta}$, i.e., they lay in a subspace of R^4 and corresponds to $y_\delta = \xi_\delta = 0$. This means the nonlinear system is controlled by the equilibrium input: $\overline{u}_{av} = \sqrt{V_d}$, i.e., the incremental system is controlled by $u_{av,\delta} = 0$. The only trajectory of the incremental system with $x_{4\delta} = 0$ and $u_{av,\delta} = 0$ corresponds to the origin.

This is compatible with the fact that in order for the closed loop incremental average system to have the origin as an asymptotically stable equilibrium, the trajectories of the system should have no other equilibrium than the origin itself. The origin of the average output feedback controlled system is, hence, an asymptotically stable equilibrium.

The output feedback control law (with design parameter $\gamma > 0$) is

$$u_{av,\delta} = -\gamma b^T \frac{\partial H(x_\delta)}{\partial x_\delta} = -\gamma \left[x_{1\delta} - \frac{V_d}{Q} x_{2\delta} + \sqrt{V_d} x_{3\delta} \right]$$

The average control to be implemented is synthesized as

$$u_{av} = \sqrt{V_d} - \gamma \left[\left(x_1 - \frac{(V_d)^{\frac{3}{2}}}{Q} \right) - \frac{V_d}{Q} \left(x_2 - \sqrt{V_d} \right) + \sqrt{V_d} \left(x_3 - \frac{V_d}{Q} \right) \right]$$

Simulations

We consider a quadratic Buck converter with the following parameters:

$$L_1 = 1.5 \text{ H}, \quad C_1 = 10 \text{ } \mu\text{F}, \quad L_2 = 600 \text{ } \mu\text{H}, \quad C_2 = 10 \text{ } \mu\text{F},$$

$$R = 40 \text{ } \Omega, \quad E = 100 \text{ V}$$

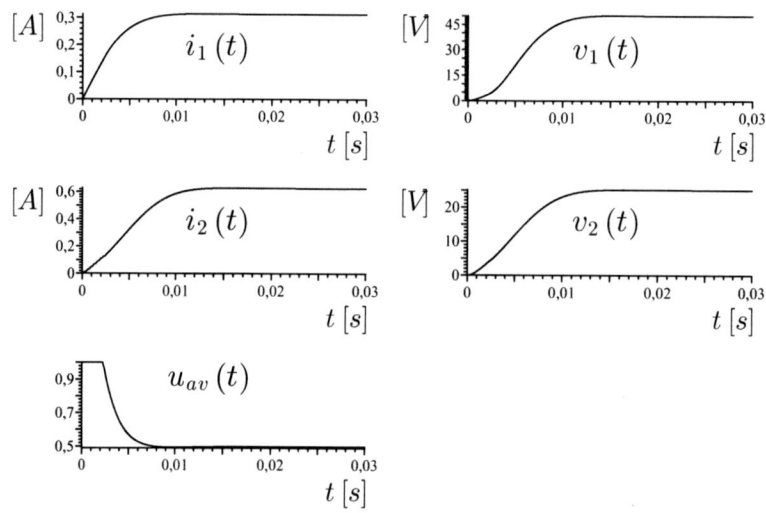

Fig. 4.55. Response of the average quadratic Buck converter model to the incremental passivity based average feedback controller.

It is desired to regulate the output capacitor voltage to the equilibrium value: $\bar{v}_2 = 25$ V.

The equilibrium values of the currents and the internal capacitor voltage are given by

$$\bar{i}_1 = 0.3125 \text{ A}, \qquad \bar{v}_1 = 50 \text{ V}, \qquad \bar{i}_2 = 0.625 \text{ A}$$

and $\bar{u}_{av} = 0.5$.

Figure 4.55 depicts the average response of the average quadratic Buck converter model to the incremental average passive output feedback controller when the design parameter we make equal to $\gamma = 1$.

4.9 The Boost-Boost Converter

4.9.1 Generalities about the Model

Consider the average normalized model of a multi-variable Boost-Boost converter, with the following simplification: $\alpha_1 = \alpha_2 = 1$, we have

$$\dot{x}_1 = -u_{1av}x_2 + 1$$
$$\dot{x}_2 = u_{1av}x_1 - \frac{1}{Q_1}x_2 - x_3$$
$$\dot{x}_3 = x_2 - u_{2av}x_4$$
$$\dot{x}_4 = u_{2av}x_3 - \frac{1}{Q_L}x_4$$

The equilibrium point of the system, for a desired set of output average equilibrium voltages $\bar{x}_2 = V_{2d}$ and $\bar{x}_4 = V_{4d}$, is given by,

$$\bar{x}_1 = \frac{V_{2d}^2}{Q_1} + \frac{V_{4d}^2}{Q_L}, \qquad \bar{x}_2 = V_{2d}, \qquad \bar{x}_3 = \frac{V_{4d}^2}{Q_L V_{2d}}, \qquad \bar{x}_4 = V_{4d}$$

$$\bar{u}_{1,av} = \frac{1}{V_{2d}}, \qquad \bar{u}_{2,av} = \frac{V_{2d}}{V_{4d}}$$

The linearized system around such an equilibrium point is given by,

$$\dot{x}_{1\delta} = -\frac{1}{V_{2d}}x_{2\delta} - V_{2d}u_{1av,\delta}$$
$$\dot{x}_{2\delta} = \frac{1}{V_{2d}}x_{1\delta} - \frac{x_{2\delta}}{Q_1} - x_{3\delta} + \left(\frac{V_{2d}^2}{Q_1} + \frac{V_{4d}^2}{Q_L}\right)u_{1av,\delta}$$
$$\dot{x}_{3\delta} = x_{2\delta} - \frac{V_{2d}}{V_{4d}}x_{4\delta} - V_{4d}u_{2av,\delta}$$
$$\dot{x}_{4\delta} = \frac{V_{2d}}{V_{4d}}x_{3\delta} - \frac{x_{4\delta}}{Q_L} + \frac{V_{4d}^2}{V_{2d}Q_L}u_{2av,\delta}$$

The system is controllable, and the variables $x_{1\delta}$ and $x_{3\delta}$ are observable minimum-phase outputs, while $x_{2\delta}$ and $x_{4\delta}$ are non-minimum phase outputs.

4 Approximate Linearization Methods

A decoupled average incremental feedback control policy proposes feedback control actions which, for each input utilize only variables pertaining to its particular subsystem. In this case, one would like to propose a controller of the form

$$u_{1av,\delta} = -k_1 x_{1\delta} - k_2 x_{2\delta}$$
$$u_{2av,\delta} = -k_3 x_{3\delta} - k_4 x_{4\delta}$$

to obtain the closed loop linearized average dynamics

$$\dot{x}_\delta = \begin{bmatrix} k_1 V_{2d} & k_2 V_{2d} - \frac{1}{V_{2d}} & 0 & 0 \\ \frac{1}{V_{2d}} - k_1(\frac{V_{2d}^2}{Q_1} + \frac{V_{4d}^2}{Q_L}) & -\frac{1}{Q_1} - k_2(\frac{V_{2d}^2}{Q_1} + \frac{V_{4d}^2}{Q_L}) & -1 & 0 \\ 0 & 1 & k_3 V_{4d} & -\frac{V_{2d}}{V_{4d}} + k_4 V_{4d} \\ 0 & 0 & \frac{V_{2d}}{V_{4d}} - k_3 \frac{V_{4d}^2}{V_{2d} Q_L} & -\frac{1}{Q_L} - k_4 \frac{V_{4d}^2}{V_{2d} Q_L} \end{bmatrix} x_\delta$$

The closed loop system fourth order characteristic polynomial exhibits quite a complex expression from which nonlinear equations would have to be solved for the required feedback gains k_1, k_2, k_3 and k_4. We adopt a decoupled design strategy by placing the poles of each block through the corresponding gains and locate the poles of each subsystem deep into the stable region of the complex plane.

The control gains are obtained by forcing the block diagonal matrices of the closed loop system to have their eigenvalues at desired locations: Thus, we have chosen k_1 and k_3 so that the eigenvalues of the sub-matrix

$$\begin{bmatrix} k_1 V_{2d} & k_2 V_{2d} - \frac{1}{V_{2d}} \\ \frac{1}{V_{2d}} - k_1(\frac{V_{2d}^2}{Q_1} + \frac{V_{4d}^2}{Q_L}) & -\frac{1}{Q_1} - k_2(\frac{V_{2d}^2}{Q_1} + \frac{V_{4d}^2}{Q_L}) \end{bmatrix}$$

were located in the stable region of the complex plane.

Similarly, we have chosen the gains k_3 and k_4 so that the eigenvalues of the sub-matrix:

$$\begin{bmatrix} k_3 V_{4d} & -\frac{V_{2d}}{V_{4d}} + k_4 V_{4d} \\ \frac{V_{2d}}{V_{4d}} - k_3 \frac{V_{4d}^2}{V_{2d} Q_L} & -\frac{1}{Q_L} - k_4 \frac{V_{4d}^2}{V_{2d} Q_L} \end{bmatrix}$$

were located at desired locations in the stable region of the complex plane.

Such gains are obtained by equating the characteristic polynomial of each sub-matrix to the desired polynomials

$$p_{id}(s) = s^2 + 2\zeta_i \omega_{in} s + \omega_{in}^2, \quad i = 1, 2$$

We obtain the following expressions for the gains:

$$k_1 = \frac{-(-\frac{1}{Q_1} + 2\zeta_1 \omega_{n1}) - (\frac{V_{4d}^2}{Q_L} + \frac{V_{2d}^2}{Q_1})(-\frac{1}{V_{2d}^2} + \omega_{n1}^2)}{V_{2d} + (\frac{V_{4d}^2}{Q_L} + \frac{V_{2d}^2}{Q_1})(\frac{V_{4d}^2}{V_{2d} Q_L} + 2\frac{V_{2d}}{Q_1})}$$

$$k_2 = \frac{(\frac{V_{4d}^2}{V_{2d}Q_L} + 2\frac{V_{2d}^2}{Q_1})(-\frac{1}{Q_1} + 2\zeta_1\omega_{n1}) - V_{2d}(-\frac{1}{V_{2d}^2} + \omega_{n1}^2)}{V_{2d} + (\frac{V_{4d}^2}{Q_L} + \frac{V_{2d}^2}{Q_1})(\frac{V_{4d}^2}{V_{2d}Q_L} + 2\frac{V_{2d}}{Q_1})}$$

$$k_3 = \frac{-V_{2d}(-\frac{1}{Q_L} + 2\zeta_2\omega_{n2}) - (\frac{V_{4d}^2}{V_{2d}Q_L})(-\frac{V_{2d}^2}{V_{4d}^2} + \omega_{n2}^2)}{V_{2d}V_{4d} + 2\frac{V_{4d}^3}{V_{2d}Q_L^2}}$$

$$k_4 = \frac{2\frac{V_{4d}}{Q_L}(-\frac{1}{Q_L} + 2\zeta_2\omega_{n2}) - V_{4d}(-\frac{V_{2d}^2}{V_{4d}^2} + \omega_{n2}^2)}{V_{2d}V_{4d} + 2\frac{V_{4d}^3}{V_{2d}Q_L^2}}$$

Simulations

It is desired to bring the normalized capacitor voltages V_{2d} and V_{4d} to the values $V_{2d} = 1.5$ and $V_{4d} = 2$ in a Boost-Boost converter with $Q_1 = 0.5$, $Q_2 = 0.75$. We used the proposed decoupled feedback control laws for closed loop pole placement with the following parameters:

$$\zeta_1 = \zeta_2 = 0.81, \qquad \omega_{n1} = \omega_{n2} = 1$$

Figure 4.56 depicts the average response of the Boost-Boost converter to the block decoupled linear feedback control law stabilizing the system state variables to the desired equilibrium from the origin of coordinates.

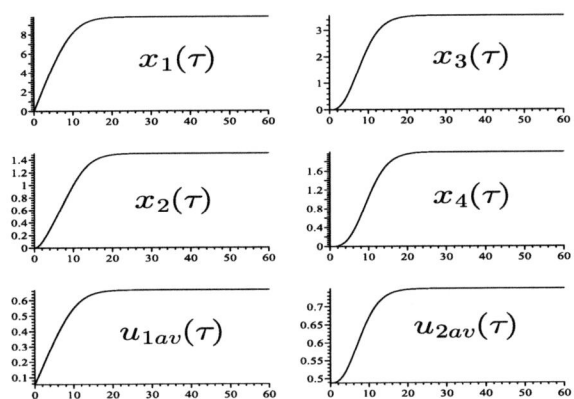

Fig. 4.56. Average performance of Boost-Boost converter to stabilizing block decoupled linear feedback.

Figure 4.57 depicts the response of the switched Boost-Boost converter to the block decoupled linear average feedback control law forcing the state

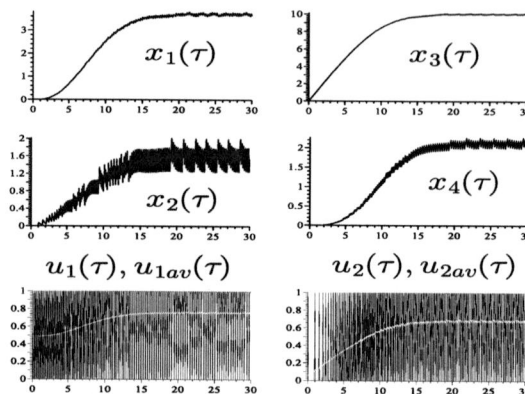

Fig. 4.57. Average performance of Boost-Boost converter to stabilizing block decoupled linear feedback.

variables to the desired equilibrium from the origin of coordinates. A $\Sigma - \Delta$ modulator was used for the implementation.

Figure 4.58 depicts the response of the switched Boost-Boost converter to the saturated block decoupled linear average feedback control law forcing the state variables to the desired equilibrium from the origin of coordinates. ($\zeta_1 = \zeta_2 = 0.81$, $\omega_{n1} = \omega_{n2} = 1.2$)

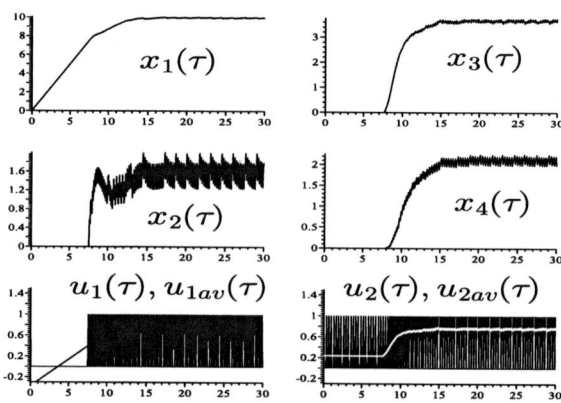

Fig. 4.58. Performance of switched Boost-Boost converter to stabilizing block decoupled linear feedback from zero start-up conditions.

4.9.2 The Hamiltonian System Approach

The normalized, linearized, average system is written in Hamiltonian form

$$\dot{x} = \begin{bmatrix} 0 & -\frac{1}{V_{2d}} & 0 & 0 \\ \frac{1}{V_{2d}} & 0 & -1 & 0 \\ 0 & 1 & 0 & -\frac{V_{2d}}{V_{4d}} \\ 0 & 0 & \frac{V_{2d}}{V_{4d}} & 0 \end{bmatrix} \frac{\partial H}{\partial x} - \begin{bmatrix} 0 & 0 & 0 & 0 \\ 0 & \frac{1}{Q_1} & 0 & 0 \\ 0 & 0 & 0 & 0 \\ 0 & 0 & 0 & \frac{1}{Q_L} \end{bmatrix} \frac{\partial H}{\partial x}$$

$$+ \begin{bmatrix} -V_{2d} & 0 \\ \left(\frac{V_{2d}^2}{Q_1} + \frac{V_{4d}^2}{Q_L}\right) & 0 \\ 0 & -V_{4d} \\ 0 & \frac{V_{4d}^2}{V_{2d}Q_L} \end{bmatrix} u_{av,\delta}$$

The average model clearly depicts the two constitutive blocks with an input decoupled structure and a decoupled dissipative map structure. Note the simple state interaction represented by the off-diagonal blocks in the conservative map of the system.

The nearly decoupled structure of the system motivates the search for a decoupled output feedback structure.

Note that by choosing: the Γ matrix in a diagonal form: $\Gamma = \text{diag}[\gamma_1, \gamma_2]$, with $\gamma_1, \gamma_2 > 0$, the dissipation matching condition takes the following natural block-decoupled form:

$$\mathcal{R} + B\Gamma B^T =$$
$$\begin{bmatrix} \gamma_1 V_{2d}^2 & -\gamma_1 V_{2d}\left(\frac{V_{2d}^2}{Q_1} + \frac{V_{4d}^2}{Q_L}\right) & 0 & 0 \\ -\gamma_1 V_{2d}\left(\frac{V_{2d}^2}{Q_1} + \frac{V_{4d}^2}{Q_L}\right) & \frac{1}{Q_1} + \gamma_1\left(\frac{V_{2d}^2}{Q_1} + \frac{V_{4d}^2}{Q_L}\right) & 0 & 0 \\ 0 & 0 & \gamma_2 V_{4d}^2 & -\gamma_2 \frac{V_{4d}^3}{Q_L V_{2d}} \\ 0 & 0 & -\gamma_2 \frac{V_{4d}^3}{Q_L V_{2d}} & \frac{1}{Q_L} + \gamma_2 \frac{V_{4d}^4}{Q_L^2 V_{2d}^2} \end{bmatrix} > 0$$

The passive outputs are given by

$$y_\delta = B^T \frac{\partial H(x_\delta)}{\partial x_\delta}$$

which, in explicit form yields:

$$y_{1\delta} = -V_{2d}x_{1\delta} + \left(\frac{V_{2d}^2}{Q_1} + \frac{V_{4d}^2}{Q_L}\right)x_{2\delta}, \quad y_{1\delta} = -V_{4d}x_{3\delta} + \frac{V_{4d}^2}{QV_{2d}}x_{4\delta}$$

Each passive output involves state variables which are ascribed to his own converter block. The average passive output feedback control policy can be proposed to be decoupled, as follows

$$u_{1av,\delta} = \gamma_1 V_{2d}x_{1\delta} - \gamma_1 \left(\frac{V_{2d}^2}{Q_1} + \frac{V_{4d}^2}{Q_L}\right)x_{2\delta}, \quad u_{2av,\delta} = \gamma_2 V_{4d}x_{3\delta} - \gamma_2 \frac{V_{4d}^2}{Q_L V_{2d}}x_{4\delta}$$

234   4 Approximate Linearization Methods

Simulations

We prescribed a normalized equilibrium point corresponding to the following parameter values and desired average normalized capacitor voltages:

$$Q_1 = 0.5, \qquad Q_L = 0.75, \qquad V_{2d} = 1.5, \qquad V_{4d} = 2.0$$

The feedback gains were chosen to be: $\gamma_1 = 0.05$, $\gamma_2 = 0.2$

Figure 4.59 depicts the response of the average nonlinear model of the Boost-Boost circuit to the decoupled average static exact stabilization error dynamics passive output feedback controller.

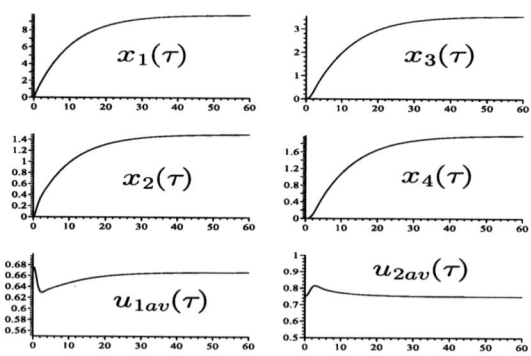

Fig. 4.59. Boost-Boost converter response to static passivity based controller.

Figure 4.60 depicts the response of the switched nonlinear Boost-Boost circuit model to the decoupled average passive output feedback controller implemented through a $\Sigma - \Delta$ modulator.

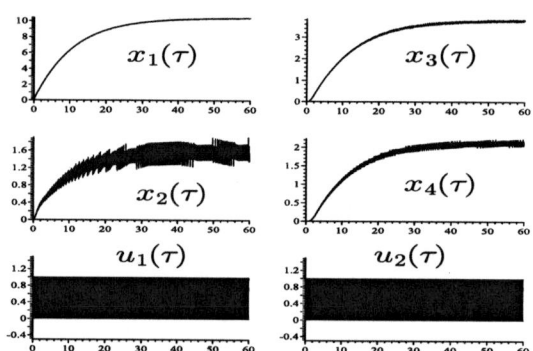

Fig. 4.60. Boost-Boost converter switched response to static passivity based controller.

5
Nonlinear Methods in the Control of Power Electronics Devices

5.1 Introduction

In this chapter, we explore several nonlinear feedback controller design techniques for DC-to-DC power converters. The variety of nonlinear control techniques which is available nowadays is vast. Therefore, we concentrate only in a few methods where controllers may be relatively simpler. We concentrate on the following possibilities for nonlinear feedback control design: Feedback state linearization, input output linearization, flatness, passivity based control, dynamic feedback control by input-output linearization and exact tracking, or stabilization, error passive output feedback, which we also address as, static linear passivity based control. In fact, within this last energy-based formulation, it becomes readily apparent that an interesting class of nonlinear systems, which includes SISO and MIMO DC-to-DC power converters, can be properly semi-globally stabilized by means of linear, time-invariant, feedback of the exact stabilization error dynamics. Moreover, trajectory tracking can also be achieved by linear feedback but now using time-varying gains. This useful result, intuitively used by many practitioners in the Power Electronics field, seems to be completely unnoticed in the existing control literature. A key point in establishing these results comes from the possibilities of exactly expressing the stabilization and tracking error dynamics in a stability invariant, isomorphic, manner to the tangent linearization model of all studied DC-to-DC power converters.

We also devote some attention, in this chapter, to the problem of state estimation in DC-to-DC power converters. Full order observers, and reduced order observers, are developed for the most common DC-to-DC power converters. The underlying feature of these two types of observers is the dependence of the estimation error dynamics on the control input. Contrary to the linear systems case, the presence of the input signal in the estimation error dynamics is not disturbing at all since, systematically, such external influence is invariably constrained to the conservative portion of the system energy managing structure, thus playing no active role in the asymptotic convergence properties of the estimated states to the actual states. In reduced order observers, a similar invariance of the estimation error dynamics with respect to the explicit dependance on control inputs can also be established, but this time in terms of appropriate control dependent time scalings of the error dynamics.

5.2 Control of DC-to-DC Power Converters via Feedback Linearization

We start out by revisiting the elementary aspects of geometric control theory, as applied to SISO nonlinear systems of the form $\dot{x} = f(x) + g(x)u_{av}$ with output $y = h(x)$ where f and g are smooth vector fields and h is a smooth scalar output function. This smooth system has the interpretation of an *average model* of a certain DC-to-DC power converter regulated by a single switch. All our considerations are local in nature, i.e., they are valid on an open neighborhood of an arbitrary representative point of the system state x. A complete account of the geometric theory of nonlinear systems is available from the excellent book by Isidori [31]. The reader may also benefit from the clear exposition in Khalil [37].

5.2.1 Isidori's Canonical Form

Let the scalar output function $y = h(x)$ be a smooth function. Assume that the system is *relative degree* equals to r, which is a strictly positive integer, not greater than the order n of the system, i.e., $1 \leq r \leq n$. This relative degree assumption means that the r-th time derivative of the output function y is the first higher order time derivative of y that explicitly exhibits, in its expression, the control input function u_{av}. In other words:

$$\frac{\partial y^{(j)}}{\partial u_{av}} = 0, \quad j = 1, \cdots, r-1, \quad \frac{\partial y^{(r)}}{\partial u_{av}} \neq 0$$

This means, in particular, that

$$\begin{bmatrix} y \\ \dot{y} \\ \vdots \\ y^{(r-1)} \end{bmatrix} = \begin{bmatrix} h(x) \\ L_f h(x) \\ \vdots \\ L_f^{r-1} h(x) \end{bmatrix}$$

The absence of any influence of the average control input u_{av} on the first $r-1$ time derivatives of y is valid thanks to the following fact

$$L_g L_f^j h(x) = 0, \quad j = 0, 1, 2, \cdots, r-2$$

Clearly then, the explicit appearance of u_{av} in the r-th time derivative of y is due to the condition

$$LgL_f^{r-1}h(x) \neq 0$$

We will propose a full state coordinate transformation which will explicitly exhibit the integration input-output structure of the system. For this, we shall adopt the first r coordinate functions of such transformation, to be of the form:

$$z = \Phi(x) = \begin{bmatrix} h(x) \\ L_f h(x) \\ \vdots \\ L_f^{(r-1)} h(x) \end{bmatrix}$$

Note that these first r coordinates are necessarily *independent* of each other. In fact, there is no linear combination of such r coordinates, using non-zero constant coefficients, that can be made identically zero. Indeed, let $\gamma_1, \cdots, \gamma_r$ be non-zero constant parameters, and assume, contrary to what we want to prove, that

$$q(x) = \sum_{j=1}^{r} \gamma_j L_f^{j-1} h(x) = 0$$

Taking the directional derivative of $q(x)$ with respect to the vector field g, we obtain: $\gamma_r L_g L_f^{r-1} h = 0$, which implies, by virtue of the relative degree r assumption, $L_g L_f^{r-1} h \neq 0$, that, necessarily, $\gamma_r = 0$. It follows that $q(x)$ is then given by,

$$q(x) = \sum_{j=1}^{r-1} \gamma_j L_f^{j-1} h(x)$$

Taking now the directional derivative with respect to f and then with respect to g of $q(x)$, i.e., taking the iterated directional derivative $L_g L_f$ of $q(x)$, we obtain now that $\gamma_{r-1} L_g L_f^{r-1} h(x) = 0$ and hence, necessarily $\gamma_{r-1} = 0$ since $L_g L_f^{r-1} h(x) \neq 0$. In this manner, we soon conclude that all the γ_js are necessarily zero, which is a contradiction.

The rest of the variables in the transformation to be defined, in number of $n - r$, may be arbitrarily chosen, as long as they are independent among themselves and independent also of the first r variables. These additional variables do not yield any specially interesting structure to the transformed equations and, therefore, we generically lump them into the $n - r$ dimensional vector which will still satisfy a nonlinear set of differential equations (linearly) involving the control input u_{av} and the first r variables z. We then define

$$\begin{bmatrix} \eta_1 \\ \eta_2 \\ \vdots \\ \eta_{n-r} \end{bmatrix} = \begin{bmatrix} \psi_1(x) \\ \psi_2(x) \\ \vdots \\ \psi_{n-r}(x) \end{bmatrix}$$

The complete state coordinate transformation

$$\begin{bmatrix} z \\ \eta \end{bmatrix} = \begin{bmatrix} \Phi(x) \\ \Psi(x) \end{bmatrix} = \Theta(x)$$

yields a set of differential equations which are in general nonlinear. Clearly, the first r new coordinates $z = (z_1, \cdots, z_r)$ satisfy the suggestive set of equations which may be exactly turned into a set of linear equations after a state dependent input coordinate transformation. This first r equations are:

$$\dot{z}_1 = z_2$$
$$\dot{z}_2 = z_3$$
$$\vdots$$
$$\dot{z}_r = L_f^r(h \circ \Theta^{-1})(z, \eta) + L_g L_f^{r-1}(h \circ \Theta^{-1})(z, \eta) u_{av} \quad (5.1)$$

The rest of the transformed equations are given by an expression of the form

$$\dot{\eta} = A(z, \eta) + B(z, \eta) u_{av}$$

One can, of course, arrange the nature of the transformation $\Psi(x)$, so as to eliminate the influence of the control input u_{av} in the last $n - r$ transformed equations, but this is not really necessary to understand the fundamental structure of the system, which is represented by the first r transformed equations.

The transformed system reads as follows:

$$\dot{z}_1 = z_2$$
$$\dot{z}_2 = z_3$$
$$\vdots$$
$$\dot{z}_r = L_f^r(h \circ \Theta^{-1})(z, \eta) + L_g L_f^{r-1}(h \circ \Theta^{-1})(z, \eta) u_{av}$$
$$\dot{\eta} = A(z, \eta) + B(z, \eta) u_{av}$$
$$y = z_1$$

which is here addressed as *Isidori's canonical form* for nonlinear SISO systems.

5.2.2 Input-Output Feedback Linearization

The invertible, state-dependent, input coordinate transformation:

$$L_f^r(h \circ \Theta^{-1})(z, \eta) + L_g L_f^{r-1}(h \circ \Theta^{-1})(z, \eta) u_{av} = v_{av} \quad (5.2)$$

leading to,

$$u_{av} = \frac{v_{av} - L_f^r(h \circ \Theta^{-1})(z, \eta)}{L_g L_f^{r-1}(h \circ \Theta^{-1})(z, \eta)} \quad (5.3)$$

yields a specially simple system in the first r coordinates,

$$\dot{z}_1 = z_2$$
$$\dot{z}_2 = z_3$$
$$\vdots$$
$$\dot{z}_r = v_{av}$$
$$\dot{\eta} = \Gamma(z, \eta, v_{av})$$
$$y = z_1$$

where $\Gamma(z, \eta, v_{av})$ is obtained from the use of the input transformation (5.3), in the last $n - r$ equations. i.e.,

$$\Gamma(z, \eta, v_{av}) = A(z, \eta) + B(z, \eta) \begin{bmatrix} v_{av} - L_f^r(h \circ \Theta^{-1})(z, \eta) \\ L_g L_f^{r-1}(h \circ \Theta^{-1})(z, \eta) \end{bmatrix}$$

Let $\bar{y}$ be a desired equilibrium point of the system output. Corresponding to this constant equilibrium value, one has the equilibrium value of the transformed variable z, which we denote by $\bar{z}$. Given the nature of the components of z the variables: $z_2, \cdots, z_r$ clearly have as equilibrium value, the value of zero. We denote by $\bar{z}$ the vector $\bar{z} = (\bar{y}, 0, \cdots, 0)$.

It is relatively simple to device a feedback control law for v_{av}, depending on a finite number of time derivatives of $y = z_1$, which stabilizes the output of the system, $y = z_1$, to the desired constant value, $\bar{y}$. Such a controller could be devised, in transformed coordinates, as

$$v_{av}(z) = -\alpha_0(z_1 - \bar{y}) - \alpha_1 z_2 - \cdots - \alpha_{r-1} z_r$$

where the α_js are design gains, properly chosen. Clearly, this controller requires the generation of several time derivatives of y, represented here by the variables $z_2, z_3, \cdots, z_r$. But, we should not forget that these required signals are also functions of the states of the system through the transformation function $(z, \eta) = \Theta(x)$. If all state variables are available for measurement, such a control law can be synthesized nonlinearly in terms of x. We choose the set of design constant parameters: $\alpha_0, \cdots, \alpha_{r-1}$ in an appropriate fashion, so as to make the linear closed loop subsystem, expressed in z coordinates, exhibit an asymptotically stable behavior. This simply entitles to have the input-output linear closed loop system

$$y^{(r)} + \alpha_{r-1} y^{(r-1)} + \cdots + \alpha_1 \dot{y} + \alpha_0 (y - \bar{y}) = 0$$

have a characteristic polynomial with all roots in the left half of the complex plane.

Extreme caution should be placed, however, on the use of this controller design method since the silent presence of the closed loop differential equations for the state variables η,

$$\dot{\eta} = \Gamma(z, \eta, v_{av}(z))$$

may not exhibit a convenient closed loop behavior. We say that the system output, y, is a *non-minimum phase output variable* whenever the induced dynamics (also called the *zero dynamics*)

$$\dot{\eta} = \Gamma(\bar{z}, \eta, v_{av}(z))$$

is unstable. Otherwise, we say that the output is a *minimum phase output*. Only in minimum phase cases we may actually attempt an exact input-output linearization controller implementation for the regulation of the output towards a desired equilibrium value, or for the tracking of a desired trajectory. This limitation will become evident in the control of some average models of DC-to-DC power converters.

Figure 5.1 depicts the exact input-output linearization feedback control scheme for a stabilization task around a constant output equilibrium value of the DC-to-DC power converter implemented through a $\Sigma - \Delta$ modulator.

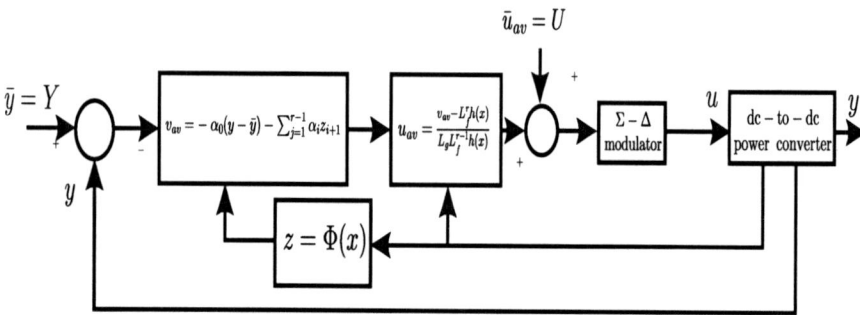

Fig. 5.1. Stabilization scheme in an exact average input-output linearization scheme for a DC-to-DC power converter.

We address the previous feedback control design technique as exact *input-output feedback linearization*. It is one of the most explored nonlinear feedback control design technique in the area of average control of DC-to-DC power converters. The more ambitious alternative of exact "input to state" feedback linearization results in more involved controllers but, certainly, with some definite advantages. We explore this new alternative in the next paragraphs.

5.2.3 State Feedback Linearization

Exact state feedback linearization may be achieved when the output of the system $y = h(x)$ is relative degree equals to n. i.e., when there is no zero dynamics associated with the output y. Since, seldom the system output, y, satisfies this requirement, one proceeds to search for such linearizing, or flat, output function $h(x)$. In the SISO case it becomes quite *systematic* to carry out this task. The rule is that if the matrix

5.2 Feedback Linearization

$$C(x) = [g, ad_f g, ad_f^2 g, \cdots, ad_f^{n-1} g]$$

which we call, somewhat abusively, "the controllability matrix" of the system, is full rank, and the set of vector fields $\{g, ad_f g, ad_f^2 g, \cdots, ad_f^{n-2} g\}$ is involutive then the row gradient of the flat output h, denoted in local coordinates by $dh = \partial h/\partial x^T$, is given by the last row of the inverse of the controllability matrix $C(x)$ multiplied by an arbitrary nonzero scalar factor $\gamma(x)$. i.e.,

$$dh = \frac{\partial h}{\partial x^T} = \gamma(x)[0 \; 0 \; \cdots \; 1]C^{-1}(x)$$

To prove this result, we proceed as follows: We realize that if the seek an output function $y = h(x)$ which is to be relative degree n, then the following set of relations must be satisfied

$$L_g L_f^j h(x) = 0, \quad j = 0, 1, \cdots, n-2, \quad L_g L_f^{n-1} h(x) = \gamma(x) \neq 0$$

The first $n-1$ relations lead to a rather involved set of higher order partial differential equations which are not easy to solve. Using the well known relation,

$$L_f L_g h(x) - L_g L_f h(x) = L_{[f,g]} h(x)$$

and the notation $[f, g] = ad_f g$, we readily obtain an equivalence between the condition $L_g L_f h(x) = 0$ (representing a second order linear partial differential equation for the unknown function h) and the condition $L_{ad_f g} h(x) = 0$ (representing only a first order linear partial differential equation for h). Using now the following relation $L_{ad_f g} L_f h - L_f L_{ad_f g} h = L_{[f, ad_f g]} h = L_{ad_f^2 g} h$, the condition: $L_g L_f^2 h(x) = 0$ is found to be equivalent to $L_{ad_f^2 g} h = 0$. We find that the set of equalities: $L_g L_f^j h(x) = 0, \; j = 0, 1, \cdots, n-2$ is equivalent to the set of equalities $L_{ad_f^j g} h, j = 0, 1, 2, \cdots, n-2$ and that the equality $L_g L_f^{n-1} h(x) = \gamma(x)$ is equivalent to $L_{ad_f^{n-1} g} h = \gamma(x)$. As a result, we obtain the following set of linear partial differential equations for h

$$\frac{\partial h}{\partial x^T} [g, ad_f g, \cdots, ad_f^{n-2} g, ad_f^{n-1} g] = [0 \; \cdots \; 0 \; \gamma(x)]$$

The *integrability* of the vector fields: $\{g, ad_f g, \cdots, ad_f^{n-2} g\}$, i.e., the existence of a smooth scalar function $h(x)$ such that all these vector fields conform a tangent plane, of local fixed dimension $n-1$, to the level sets $h(x) = constant$ is guaranteed, thanks to Frobenius's theorem, if and only if the set of vector fields $\{g, ad_f g, \cdots, ad_f^{n-2} g\}$ conform an *involutive set*. In other words, there exists a smooth scalar function $h(x)$ that "integrates" all the vector fields such that its row gradient, $dh = \frac{\partial h}{\partial x^T}$, annihilates the given set of vector fields. The result follows.

The state coordinate transformation, $z = \Phi(x)$, defined by

$$z = \Phi(x) = \begin{bmatrix} h(x) \\ L_f h(x) \\ \vdots \\ L_f^{(n-1)} h(x) \end{bmatrix}$$

yields the transformed system in the following particular form:

$$\dot{z}_1 = z_2$$
$$\dot{z}_2 = z_3$$
$$\vdots$$
$$\dot{z}_n = L_f^n(h \circ \Phi^{-1})(z) + L_g L_f^{r-1}(h \circ \Phi^{-1})(z) u_{av} \quad (5.4)$$

The invertible, state dependent, input coordinate transformation given by

$$L_f^n(h \circ \Phi^{-1})(z) + L_g L_f^{r-1}(h \circ \Phi^{-1})(z) u_{av} = v_{av}$$

takes the system into the *Brunovsky's canonical form*

$$\dot{z}_1 = z_2$$
$$\dot{z}_2 = z_3$$
$$\vdots$$
$$\dot{z}_n = v_{av} \quad (5.5)$$

The transformed system (5.5) is equivalent to the pure integration system

$$y^{(n)} = v_{av}$$

Given a trajectory tracking task, for the original state x of the system, denoted by $x^*(t)$, this trajectory can be immediately translated into a corresponding trajectory tracking objective for the linearizing, or flat, output y, that we will denote by $y^*(t)$. This is easy to establish by reading the first entry of the nominal transformed coordinates $z^* = \Phi(x^*)$. More frequently, a flat output trajectory tracking task is demanded and specified in the form of of $y^*(t)$. The time derivatives of this reference trajectory are readily established to conform the nominal value $z^*(t)$ of the transformed state variable z. The corresponding state reference trajectory is immediately obtained using the inverse transformation $x^* = \Phi^{-1}(z)$

A trajectory tracking feedback controller for the Brunovsky system (5.5), is easily specified as follows:

$$v_{av} = z_n^*(t) - \sum_{j=1}^{n} \alpha_{j-1}(z_j - z_j^*(t)) \quad (5.6)$$

where the constant coefficients α are set so that the corresponding closed loop linear system for the stabilization error $e = y - y^*(t) = z_1 - z_1^*(t)$, exhibits a characteristic polynomial of the form

$$p_d(s) = s^n + \alpha_{n-1} s^{n-1} + \cdots + \alpha_0$$

with all roots in the left half of the complex plane.

5.2.4 The Boost Converter

Consider the average normalized model of the Boost converter

$$\dot{x}_1 = -u_{av} x_2 + 1$$
$$\dot{x}_2 = u_{av} x_1 - \frac{x_2}{Q}$$

Suppose it is desired to regulate the average value of the output capacitor voltage x_2 towards the average equilibrium value $\overline{y} = V_d$. The corresponding average equilibrium values for x_1 and u are given, respectively, by $\overline{x}_1 = V_d^2/Q$ and $\overline{u}_{av} = 1/V_d$.

Input-Output Feedback Linearization

Direct Method

Take as the output function of the system, the average normalized output capacitor voltage $y = x_2$.

The input-output feedback linearization is achieved by forcing the equation for x_2 to represent a linear dynamics with $\overline{y} = V_d$ representing an asymptotically exponentially stable equilibrium point. We set then the average control input to

$$u_{av} = \frac{y/Q - \lambda(y - \overline{y})}{x_1}$$

with the restriction, $0 \leq u_{av} \leq 1$, being enforced in closing the loop and λ being a strictly positive scalar constant.

The value of the average feedback function u_{av}, corresponding to the steady state of the average output variable y is given by $u_{av}(\overline{y}) = \overline{y}/(Qx_1)$. The corresponding zero dynamics of the controlled output function y is then obtained as

$$\dot{x}_1 = -\left(\frac{\overline{y}^2}{Q}\right)\frac{1}{x_1} + 1 = -\left(\frac{V_d^2}{Q}\right)\frac{1}{x_1} + 1$$

The average zero dynamics of the output capacitor voltage variable is therefore unstable. To show this, we simply resort to approximate linearization and find that around the equilibrium point $\overline{x}_1 = V_d^2/Q$ the tangent linearization of the dynamics for x_1 satisfies

$$\dot{x}_{1\delta} = \left(\frac{Q}{V_d^2}\right) x_{1\delta}$$

which is clearly unstable due to the strict positivity of the parameter Q. The feedback controller locally linearizing the input-output dynamics is therefore not feasible when the output function is represented by the average normalized output capacitor voltage since this is a *non-minimum phase* variable.

Indirect Method

Consider now the case when the output is given by the average input inductor current $y = x_1$. Suppose it is desired to regulate the average normalized current to the desired equilibrium value $\overline{x}_1 = \frac{V_d^2}{Q}$. We have that the linearizing controller is given by

$$u_{av} = \frac{1 + \lambda(x_1 - V_d^2/Q)}{x_2}$$

and the control input corresponding to the steady state value of the output $y = x_1$ is simply given by $u_{av}(\overline{y}) = 1/x_2$. The corresponding zero dynamics is readily found to be

$$\dot{x}_2 = \frac{V_d^2}{Qx_2} - \frac{x_2}{Q} = -\frac{1}{Qx_2}(x_2^2 - V_d^2) = -\frac{1}{Qx_2}(x_2 - V_d)(x_2 + V_d)$$

Consider the Lyapunov function candidate which is positive definite in the state space of the zero dynamics $V(x_2) = \frac{1}{2}(x_2 - V_d)^2 > 0$. The time derivative of this function along the solutions of the differential equation describing x_2 is given by

$$\dot{V}(x_2) = -\frac{(x_2 + V_d)}{Qx_2}(x_2 - V_d)^2$$

which is negative definite for all positive values of the state x_2. The equilibrium point $\overline{x}_2 = V_d$ is asymptotically stable. The control in this case is indirectly regulating the output voltage towards the equilibrium value $\overline{x}_2 = V_d$.

State Feedback Linearization

We carry out the search for the linearizing, or flat, output $y = h(x)$. In this case, we have the following definition of the vector fields characterizing the average system

$$f(x) = \begin{bmatrix} 1 \\ -\frac{x_2}{Q} \end{bmatrix}, \quad g(x) = \begin{bmatrix} -x_2 \\ x_1 \end{bmatrix}$$

The vector g is by itself trivially involutive since $[g, g] = 0$. One of the conditions for feedback linearization is clearly satisfied. The vector columns of the "controllability matrix" and its determinant are computed to be,

5.2 Feedback Linearization

$$C(x) = [g, ad_f g] = \begin{bmatrix} -x_2 & \frac{x_2}{Q} \\ x_1 & 1+\frac{x_1}{Q} \end{bmatrix}, \quad \det C = -x_2(1+2\frac{x_1}{Q})$$

The system satisfies the linear independence property as long as the evolution of the variables are away from the condition $\det C = 0$ represented here by $x_2 = 0$ and $x_1 = -Q/2$. Neither condition corresponds with the control objective which are to regulate the output voltage to a positive equilibrium value with a corresponding positive equilibrium value for the average inductor current. We therefore assume that $\det C \neq 0$ along the controlled trajectories leading towards stabilization. The gradient of the linearizing output is computed to be

$$\frac{\partial h}{\partial x^T} = \frac{\gamma(x)}{\det C}[0\ 1]\begin{bmatrix} 1+\frac{x_1}{Q} & -\frac{x_2}{Q} \\ -x_1 & -x_2 \end{bmatrix} = -\frac{\gamma(x)}{\det C}[x_1\ x_2]$$

We can freely set $\gamma(x)$ to be $\gamma(x) = -\det C$ since the system is assumed to be controllable throughout its trajectories towards stabilization around the desired average equilibrium point.

The linearizing output thus satisfies the pair of linear partial differential equations

$$\frac{\partial h}{\partial x_1} = x_1, \quad \frac{\partial h}{\partial x_2} = x_2$$

and a possible expression for the required linearizing output y is given by

$$y = \frac{1}{2}[x_1^2 + x_2^2]$$

The flat output, or linearizing output, is represented by the average total stored energy in the circuit. We have

$$\dot{y} = x_1 - \frac{x_2^2}{Q}$$

$$\ddot{y} = \left(1 + \frac{2}{Q^2}x_2^2\right) - u_{av}x_2\left(1 + \frac{2}{Q}x_1\right)$$

Note that the expression multiplying the average control input u_{av} is that of $\det C$. The state dependent, locally invertible, input coordinate transformation with auxiliary input variable v_{av}, defined by

$$\left(1 + \frac{2}{Q^2}x_2^2\right) - u_{av}x_2\left(1 + \frac{2}{Q}x_1\right) = v_{av}, \quad u_{av} = \frac{\left(1 + \frac{2}{Q^2}x_2^2\right) - v_{av}}{x_2\left(1 + \frac{2}{Q}x_1\right)}$$

yields a linear second order system for the average total stored energy,

$$\ddot{y} = v_{av}$$

An average controller stabilizing the energy towards a desired equilibrium value $\bar{y}$, is given by

$$v_{av} = -2\zeta\omega_n \dot{y} - \omega_n^2(y - \bar{y}) = -2\zeta\omega_n\left(x_1 - \frac{x_2^2}{Q}\right) - \omega_n^2\left(\frac{1}{2}\left[x_1^2 + x_2^2\right] - \bar{y}\right)$$

where ζ and ω_n are design parameters establishing the performance quality characteristics of the closed loop response of the second order linear system governing the evolution of the average total stored energy. We set $\zeta, \omega_n > 0$ to obtain an asymptotically exponentially stable equilibrium point for the closed loop system.

The full average state, locally linearizing, feedback controller is then obtained as

$$u_{av} = \frac{\left(1 + \frac{2}{Q^2}x_2^2\right) + \left[2\zeta\omega_n\left(x_1 - \frac{x_2^2}{Q}\right) + \omega_n^2\left(\frac{1}{2}\left[x_1^2 + x_2^2\right] - \bar{y}\right)\right]}{x_2\left(1 + \frac{2}{Q}x_1\right)}$$

The desired equilibrium value for the total stored energy is most conveniently parameterized in terms of the desired average output equilibrium voltage value V_d, as

$$\bar{y} = \frac{1}{2}\left[\frac{V_d^4}{Q^2} + V_d^2\right]$$

5.2.5 The Buck-Boost Converter

Consider now the average normalized model of the Buck-Boost converter

$$\dot{x}_1 = u_{av}x_2 + (1 - u_{av})$$
$$\dot{x}_2 = -u_{av}x_1 - \frac{x_2}{Q}$$

It is desired to regulate the average value of the output capacitor voltage x_2 towards the average equilibrium value $\bar{y} = V_d < 0$. The corresponding average equilibrium values for x_1 and u are given, respectively, by $\bar{x}_1 = -V_d(1-V_d)/Q$ and $\bar{u}_{av} = 1/(1 - V_d)$.

Input-Output Feedback Linearization

Direct Method

Take as the output function of the system, the average normalized output capacitor voltage $y = x_2$.

The input-output feedback linearization is achieved by inducing the variable x_2 to be described by a linear dynamics, with $\bar{y} = V_d < 0$, representing an asymptotically exponentially stable equilibrium point. We set then the average control input to

$$u_{av} = \frac{\lambda(y - V_d) - y/Q}{x_1}$$

with $\lambda > 0$ and the restriction, $0 \leq u_{av} \leq 1$, being enforced in the closed loop system.

The value of the average feedback function $u_{av}(x)$, corresponding to the steady state of the average output variable y is given by $u_{av}(\overline{y}) = -\overline{y}/(Qx_1) = -V_d/(Qx_1)$. The corresponding zero dynamics of the controlled output function y is then obtained as

$$\dot{x}_1 = 1 + \frac{V_d(1 - V_d)}{Qx_1}$$

The average zero dynamics of the output capacitor voltage variable is found to be unstable. To demonstrate this fact, we simply resort to approximate linearization and find that around the equilibrium point: $\overline{x}_1 = -V_d(1 - V_d)/Q$, the tangent linearization of the dynamics for x_1 satisfies,

$$\dot{x}_{1\delta} = -\frac{Q}{V_d(1 - V_d)} x_{1\delta}$$

which is clearly unstable due to the strict positivity of the parameter Q and the negativity of the average normalized desired voltage V_d. The feedback controller locally linearizing the input output dynamics is therefore not feasible when the output function is represented by the average normalized output capacitor voltage since this is a *non-minimum phase* variable.

Indirect Method

Consider now the case when the output is given by the average input inductor current $y = x_1$. Suppose it is desired to regulate the average normalized current to the desired (positive) equilibrium value $\overline{x}_1 = -\frac{V_d(1-V_d)}{Q}$. We have that the average feedback linearizing controller is given by

$$u_{av} = \frac{1 + \lambda(x_1 + V_d(1 - V_d)/Q)}{1 - x_2}$$

and the control input corresponding to the steady state value of the output $y = x_1$ is simply given by $u_{av}(\overline{y}) = 1/(1 - x_2)$. The corresponding zero dynamics is readily found to be

$$\dot{x}_2 = \frac{V_d(1 - V_d)}{Q(1 - x_2)} - \frac{x_2}{Q} = -\frac{x_2 - V_d}{Q(1 - x_2)}(1 - x_2 - V_d)$$

Consider the Lyapunov function candidate which is positive definite in the state space of the zero dynamics $V(x_2) = \frac{1}{2}(x_2 - V_d)^2 > 0$. The time derivative of this function along the solutions of the differential equation describing x_2 is given by

$$\dot{V}(x_2) = -\frac{(x_2 - V_d)^2}{Q(1 - x_2)}(1 - x_2 - V_d)$$

which is negative definite for all negative values of the state x_2 in its operating region. The equilibrium point $\bar{x}_2 = V_d < 0$ is therefore asymptotically stable.

State Feedback Linearization

We now look for the linearizing, or flat, output $y = h(x)$. In this particular case we have,

$$f(x) = \begin{bmatrix} 1 \\ -\frac{x_2}{Q} \end{bmatrix}, \qquad g(x) = \begin{bmatrix} -(1 - x_2) \\ -x_1 \end{bmatrix}$$

The vector g clearly satisfies $[g, g] = 0$ indicating the involutivity of the set constituted by the vector field g alone. The first condition for feedback linearization is then clearly satisfied. The column vectors of the "controllability matrix" and its determinant are computed to be

$$\mathcal{C} = [g, ad_f g] = \begin{bmatrix} -(1 - x_2) & -\frac{x_2}{Q} \\ -x_1 & -1 - \frac{x_1}{Q} \end{bmatrix}, \quad \det \mathcal{C} = (1 - x_2)\left(1 + \frac{x_1}{Q}\right) - \frac{x_1 x_2}{Q}$$

The system satisfies the linear independence property as long as the evolution of the variables are away from the condition $\det \mathcal{C} = 0$. We assume that $\det \mathcal{C} \neq 0$ along the controlled trajectories leading towards stabilization. The gradient of the linearizing output is computed to be

$$\frac{\partial h}{\partial x^T} = \frac{\gamma(x)}{\det \mathcal{C}}[0 \; 1]\begin{bmatrix} -\left(1 + \frac{x_1}{Q}\right) & \frac{x_2}{Q} \\ x_1 & -(1 - x_2) \end{bmatrix} = \frac{\gamma(x)}{\det \mathcal{C}}[x_1 \; -(1 - x_2)]$$

We can freely set $\gamma(x)$ to be $\gamma(x) = \det \mathcal{C}$, given that the system is assumed to be controllable throughout its trajectories towards stabilization around the desired average equilibrium point.

The linearizing output thus satisfies the pair of linear partial differential equations

$$\frac{\partial h}{\partial x_1} = x_1, \qquad \frac{\partial h}{\partial x_2} = -(1 - x_2)$$

and a possible expression for the required linearizing output y is given by

$$y = \frac{1}{2}\left[x_1^2 + (1 - x_2)^2\right]$$

We have,

$$\dot{y} = x_1 - \frac{(1-x_2)x_2}{Q}$$

$$\ddot{y} = 1 + \frac{x_2(1-x_2)}{Q^2} + \frac{x_2^2}{Q} - u_{av}\left[(1-x_2)\left(1-\frac{x_1}{Q}\right) + \frac{x_1 x_2}{Q}\right]$$

Note that the expression multiplying the average input u_{av} is, not coincidentally, that of $\det \mathcal{C}$. The state dependent, locally invertible, input coordinate transformation with auxiliary input variable v_{av}, defined by

$$1 + \frac{x_2(1-x_2)}{Q^2} + \frac{x_2^2}{Q} - u_{av}\left[(1-x_2)\left(1-\frac{x_1}{Q}\right) + \frac{x_1 x_2}{Q}\right] = v_{av},$$

$$u_{av} = \frac{1 + \frac{x_2(1-x_2)}{Q^2} + \frac{x_2^2}{Q} - v_{av}}{(1-x_2)\left(1-\frac{x_1}{Q}\right) + \frac{x_1 x_2}{Q}}$$

yields a linear second order system for the average total stored energy,

$$\ddot{y} = v_{av}$$

An average auxiliary state feedback controller stabilizing the average total stored energy towards a desired equilibrium value $\bar{y}$, is given by

$$v_{av} = -2\zeta\omega_n \dot{y} - \omega_n^2(y - \bar{y})$$
$$= -2\zeta\omega_n\left(x_1 - \frac{(1-x_2)x_2}{Q}\right) - \omega_n^2\left(\frac{1}{2}\left[x_1^2 + (1-x_2)^2\right] - \bar{y}\right)$$

We set $\zeta, \omega_n > 0$ to obtain an asymptotically exponentially stable equilibrium point for the closed loop system.

The full state feedback locally linearizing controller is then obtained as

$$u_{av} = \frac{1 + \frac{x_2(1-x_2)}{Q^2} + \frac{x_2^2}{Q} + \left\{2\zeta\omega_n\left(x_1 - \frac{(1-x_2)x_2}{Q}\right) + \omega_n^2\left(\frac{1}{2}\left[x_1^2 + (1-x_2)^2\right] - \bar{y}\right)\right\}}{(1-x_2)\left(1-\frac{x_1}{Q}\right) + \frac{x_1 x_2}{Q}}$$

The desired equilibrium value for the total stored energy is most conveniently parameterized in terms of the desired average output equilibrium voltage value, V_d, as

$$\bar{y} = \frac{1}{2}(1-V_d)^2\left[\frac{V_d^2}{Q^2} + 1\right]$$

5.2.6 The Cúk Converter

Consider the average normalized model of the Cúk converter

$$\dot{x}_1 = -(1-u_{av})x_2 + 1$$
$$\dot{x}_2 = (1-u_{av})x_1 + u_{av}x_3$$
$$\alpha_1 \dot{x}_3 = -u_{av}x_2 - x_4$$
$$\alpha_2 \dot{x}_4 = x_3 - \frac{1}{Q}x_4$$

For this system is desired to regulate the average value of the output capacitor voltage x_4 towards the average equilibrium value $\bar{y} = V_d < 0$. The corresponding average equilibrium values for x_1, x_2, x_3 and u_{av} are given, respectively, by:

$$\bar{x}_1 = \frac{V_d^2}{Q}, \quad \bar{x}_2 = 1 - V_d, \quad \bar{x}_3 = \frac{V_d}{Q}, \quad \bar{u}_{av} = \frac{V_d}{V_d - 1}$$

Input-Output Feedback Linearization

Direct Method

Take as the output function of the system, the average normalized output capacitor voltage $y = x_4$.

The input-output feedback linearization is achieved by forcing the equation for x_4 to represent a linear dynamics with $\bar{y} = V_d < 0$ representing an asymptotically exponentially stable equilibrium point. We set then the average control input to

$$u_{av} = \frac{-\frac{1}{\alpha_1 \alpha_2} y - \frac{1}{\alpha_2 Q} \dot{y} - v_{av}}{\frac{1}{\alpha_1 \alpha_2} x_2}$$

where

$$v_{av} = -\lambda_0 (y - \bar{y}) - \lambda_1 \dot{y}$$

with the restriction, $0 \leq u_{av} \leq 1$, being enforced in closing the loop and λ_0 and λ_1 being strictly positives scalars constants.

The value of the average feedback function u_{av}, corresponding to the steady state of the average output variable y is given by $u_{av}(\bar{y}) = -\bar{y}/x_2$. It follows that the ideal behavior of the x_3 variable corresponds itself to a constant value, i.e., $x_3 = \bar{x}_3 = V_d/Q$. The corresponding zero dynamics of the controlled output function y is then obtained as

$$\dot{x}_1 = 1 - (x_2 + V_d)$$
$$\dot{x}_2 = \left(1 + \frac{V_d}{x_2}\right) x_1 - \frac{V_d^2}{Q} \frac{1}{x_2}$$

which has by equilibrium point to $(\bar{x}_1, \bar{x}_2) = \left(V_d^2/Q, 1 - V_d\right)$. The average zero dynamics of the output capacitor voltage variable is therefore unstable. To show this we simply resort to approximate linearization and find that around the equilibrium point $(\bar{x}_1, \bar{x}_2)$ the tangent linearization of the dynamics for x_1 and x_2 satisfies

$$\begin{bmatrix} \dot{x}_{1\delta} \\ \dot{x}_{2\delta} \end{bmatrix} = \begin{bmatrix} 0 & -1 \\ \frac{1}{(1-V_d)} & \frac{V_d^2}{Q(1-V_d)} \end{bmatrix} \begin{bmatrix} x_{1\delta} \\ x_{2\delta} \end{bmatrix}$$

whose characteristic polynomial is just obtained as:

$$p(s) = s^2 - \frac{1}{Q}\frac{V_d^2}{(1-V_d)}s + \frac{1}{(1-V_d)}$$

which is clearly unstable because it has at least one unstable root in the complex plane. The feedback controller locally linearizing the input output dynamics is therefore not feasible when the output function is represented by the average normalized output capacitor voltage since this is a *non-minimum phase* variable.

Indirect Method

Consider now the case when the output is given by the average input inductor current $y = x_1$. Suppose it is desired to regulate the average normalized current to the desired equilibrium value $\bar{x}_1 = \frac{V_d^2}{Q}$. We have that the linearizing controller is given by

$$u_{av} = \frac{x_2 - 1 - \lambda(x_1 - V_d^2/Q)}{x_2}$$

and the control input corresponding to the steady state value of the output $y = x_1$ is simply given by $u_{av}(\bar{y}) = 1 - 1/x_2$, which under non-saturated operating conditions satisfies

$$0 < u_{av}(\bar{y}) < 1 \tag{5.7}$$

and, hence $x_2 \in (1, \infty)$.

The corresponding zero dynamics is readily found to be

$$\dot{x}_2 = \frac{V_d^2}{Q}\frac{1}{x_2} + \left(1 - \frac{1}{x_2}\right)x_3$$
$$\alpha_1 \dot{x}_3 = 1 - x_2 - x_4$$
$$\alpha_2 \dot{x}_4 = x_3 - \frac{1}{Q}x_4 \tag{5.8}$$

which has by equilibrium point to

$$\bar{x}_2 = 1 - V_d, \qquad \bar{x}_3 = \frac{1}{Q}V_d, \qquad \bar{x}_4 = V_d$$

It is important to say that the zero dynamics, given by Equation 5.8, is the same that we obtained in (3.11), then consider the Lyapunov function candidate

$$V(x_2, x_3, x_4) = \frac{1}{2}\left[(x_2 - \bar{x}_2)^2 + \alpha_1(x_3 - \bar{x}_3)^2 + \alpha_2(x_4 - \bar{x}_4)^2\right] + \gamma$$
$$+ \int_0^\tau \frac{[x_2(\sigma) - \bar{x}_2][x_3(\sigma) - \bar{x}_3]}{x_2(\sigma)}d\sigma$$

with γ being a strictly positive constant parameter, which is assumed to be sufficiently large so that V is strictly positive, and $x_2 \in (1, \infty)$ by (5.7). The time derivative of V, along the solution of the system of differential equation yields the following expression:

$$\dot{V}(x_2, x_3, x_4) = -\frac{1}{Q}(x_4 - \overline{x}_4)^2 + \overline{x}_3 \frac{(x_2 - \overline{x}_2)^2}{x_2} \leq 0$$

and by LaSalle's theorem, it is then clear that the average normalized input inductor current x_1, taken as a system output, is a *locally minimum phase output*. We, thus, attempt an *indirect regulation* of the converter average normalized output voltage, x_4, towards the desired value $\overline{x}_4 = V_d$. This is accomplished by primarily regulating the inductor current x_1 towards its corresponding average equilibrium value, $\overline{x}_1 = \frac{V_d^2}{Q}$.

State Feedback Linearization

Starting out from the normalized average model of the Cúk converter, we show that such a system cannot be written in Brunovsky's canonical form, i.e., there is no artificial output function of the states of the form $y = h(x)$, for which the system is exactly linearizable.

The bilinear set of equations describing the average normalized Cúk converter is of the general form:

$$\dot{x} = f(x) + g(x) u_{av}$$

where the *drift* vector field, $f(x)$, and the *control* vector field $g(x)$, are given by:

$$f(x) = \begin{bmatrix} 1 - x_2 \\ x_1 \\ -\frac{1}{\alpha_1} x_4 \\ \frac{1}{\alpha_2}\left(x_3 - \frac{x_4}{Q}\right) \end{bmatrix}, \quad g(x) = \begin{bmatrix} x_2 \\ -x_1 + x_3 \\ -\frac{1}{\alpha_1} x_2 \\ 0 \end{bmatrix}$$

In this case, the conditions for the existence of a flat output are the following:

1. The set of vector fields $\{g, ad_f g, ad_f^2 g, ad_f^3 g\}$ is linearly independent
2. The reduced set of vector fields $\{g, ad_f g, ad_f^2 g\}$ is *involutive*.

Computing the involved vector fields $ad_f g$, $ad_f^2 g$ and $ad_f^3 g$, we obtain:

5.2 Feedback Linearization

$$ad_f g = [f, g] = \frac{\partial g}{\partial x} f - \frac{\partial f}{\partial x} g = \begin{bmatrix} x_3 \\ -1 - \frac{1}{\alpha_1} x_4 \\ -\frac{1}{\alpha_1} x_1 \\ \frac{1}{\alpha_1 \alpha_2} x_2 \end{bmatrix}$$

$$ad_f^2 g = [f, ad_f g] = \frac{\partial ad_f g}{\partial x} f - \frac{\partial f}{\partial x} ad_f g = \begin{bmatrix} -1 - \frac{2}{\alpha_1} x_4 \\ -\left(1 + \frac{1}{\alpha_1 \alpha_2}\right) x_3 + \frac{1}{\alpha_1 \alpha_2 Q} x_4 \\ -\frac{1}{\alpha_1} \left[1 - \left(1 + \frac{1}{\alpha_1 \alpha_2}\right) x_2 \right] \\ \frac{1}{\alpha_1 \alpha_2} \left[2 x_1 + \frac{1}{\alpha_2 Q} x_2 \right] \end{bmatrix}$$

$$ad_f^3 g = [f, ad_f^2 g] = \begin{bmatrix} \frac{-(\alpha_1 \alpha_2 + 3) Q x_3 + 3 x_4}{\alpha_1 \alpha_2 Q} \\ \frac{\alpha_1^2 \alpha_2^2 Q^2 + \alpha_1 Q x_3 + [\alpha_2 (3 \alpha_1 \alpha_2 + 1) Q^2 - \alpha_1] x_4}{\alpha_1^2 \alpha_2^2 Q^2} \\ \frac{\alpha_2 (\alpha_1 \alpha_2 + 3) Q x_1 + x_2}{\alpha_1^2 \alpha_2^2 Q} \\ \frac{3 \alpha_1 \alpha_2 (\alpha_2 Q + x_1) Q - [\alpha_2 (3 \alpha_1 \alpha_2 + 1) Q^2 - \alpha_1] x_2}{\alpha_1^2 \alpha_2^3 Q^2} \end{bmatrix}$$

It is quite straightforward to show that the set of vector fields:

$$\{g, ad_f g, ad_f^2 g, ad_f^3 g\}$$

is linearly independent. Indeed, the "controllability matrix" $\mathcal{C}(x)$, for the Cúk, converter, defined by:

$$\mathcal{C}(x) = \begin{bmatrix} g, ad_f g, ad_f^2 g, ad_f^3 g \end{bmatrix}$$

is full rank, i.e., $rank\ [\mathcal{C}(x)] = 4$.

For exact feedback linearization, the reduced set of vector fields:

$$\{g, ad_f g, ad_f^2 g\} \tag{5.9}$$

must be involutive. We form then the following set of matrices,

$$\{g, ad_f g, ad_f^2 g, [g, ad_f g], [g, ad_f^2 g], [ad_f g, ad_f^2 g]\}$$

It is found that the rank of each one of these matrices is equal to 3. However, the vector field, $[g, ad_f g]$, which is given by:

$$[g, ad_f g] = \frac{\partial ad_f g}{\partial x} g - \frac{\partial g}{\partial x} ad_f g = \begin{bmatrix} 1 - \frac{1}{\alpha_1} (x_2 - x_4) \\ \frac{1}{\alpha_1} x_1 + x_3 \\ -\frac{1}{\alpha_1} \left(1 + x_2 + \frac{1}{\alpha_1} x_4 \right) \\ -\frac{1}{\alpha_1 \alpha_2} (x_1 - x_3) \end{bmatrix}$$

yields the following matrix to be of rank 4

$$\{g, ad_f g, ad_f^2 g, [g, ad_f g]\}$$

The set of vector fields (5.9) is not *involutive*. We conclude that there not exists output function of the form $h(x)$, such that the Cúk converter model is flat from this output.

5.2.7 The Sepic Converter

Consider now the average normalized model of the Sepic converter

$$\dot{x}_1 = -(1 - u_{av})(x_2 + x_4) + 1$$
$$\dot{x}_2 = (1 - u_{av})x_1 - u_{av}x_3$$
$$\alpha_1\dot{x}_3 = u_{av}x_2 - (1 - u_{av})x_4$$
$$\alpha_2\dot{x}_4 = (1 - u_{av})(x_1 + x_3) - \frac{1}{Q}x_4$$

It is desired to regulate the average value of the output capacitor voltage x_4 towards the average equilibrium value $\bar{y} = V_d$. The corresponding average equilibrium values for x_1, x_2, x_3 and u_{av} are given, respectively, by

$$\bar{x}_1 = \frac{V_d^2}{Q}, \quad \bar{x}_2 = 1, \quad \bar{x}_3 = \frac{V_d}{Q}, \quad \bar{u}_{av} = \frac{V_d}{V_d + 1}$$

Input-Output Feedback Linearization

Direct Method

Take as the output function of the system, the average normalized output capacitor voltage $y = x_4$.

The input-output feedback linearization is achieved by inducing the variable x_2 to be described by a linear dynamics with $\bar{y} = V_d$ representing an asymptotically exponentially stable equilibrium point. We set then the average control input to

$$u_{av} = \frac{\alpha_2\lambda(y - \bar{y}) - y/Q + x_1 + x_3}{x_1 + x_3}$$

with $\lambda > 0$ and the restriction, $0 \leq u_{av} \leq 1$, being enforced in the closed loop system.

The value of the average feedback function u_{av}, corresponding to the steady state of the average output variable y is given by $u_{av}(\bar{y}) = 1 - \frac{V_d}{Q(x_1+x_3)}$. The corresponding zero dynamics of the controlled output function y is then obtained as

$$\dot{x}_1 = 1 - \frac{V_d}{Q}\left(\frac{V_d + x_2}{x_1 + x_3}\right)$$
$$\dot{x}_2 = -x_3 + \frac{V_d}{Q}$$
$$\alpha_1\dot{x}_3 = x_2 - \frac{V_d}{Q}\left(\frac{V_d + x_2}{x_1 + x_3}\right)$$

5.2 Feedback Linearization

The equilibrium point of the zero dynamics is clearly given by

$$\overline{x}_1 = \frac{V_d^2}{Q}, \quad \overline{x}_2 = 1, \quad \overline{x}_3 = \frac{V_d}{Q}$$

The average zero dynamics of the output capacitor voltage variable is found to be unstable. To demonstrate this fact, we simply resort to approximate linearization and find that around the equilibrium point $(\overline{x}_1, \overline{x}_2, \overline{x}_3)$ the tangent linearization of the dynamics for (x_1, x_2, x_3) satisfies

$$\begin{bmatrix} \dot{x}_{1\delta} \\ \dot{x}_{2\delta} \\ \dot{x}_{3\delta} \end{bmatrix} = \begin{bmatrix} \frac{Q}{V_d(V_d+1)} & -\frac{1}{(1+V_d)} & \frac{Q}{V_d(V_d+1)} \\ 0 & 0 & -1 \\ \frac{Q}{\alpha_1 V_d(1+V_d)} & \frac{V_d}{\alpha_1(1+V_d)} & \frac{Q}{\alpha_1 V_d(1+V_d)} \end{bmatrix} \begin{bmatrix} x_{1\delta} \\ x_{2\delta} \\ x_{3\delta} \end{bmatrix}$$

whose associated characteristic polynomial is given by

$$p(s) = s^3 - \frac{(1+\alpha_1)Q}{\alpha_1 V_d (1+V_d)} s^2 + \frac{V_d}{\alpha_1 (1+V_d)} s - \frac{Q}{\alpha_1 V_d (1+V_d)}$$

which is unstable due to the strict positivity of the parameters Q, α_1 and the positivity of V_d. The feedback controller locally linearizing the input output dynamics is therefore not feasible when the output function is represented by the average normalized output capacitor voltage since this is a *non-minimum phase* variable.

Indirect Method

Consider now the case when the output is given by the average input inductor current $y = x_1$. Suppose it is desired to regulate the average normalized current to the desired equilibrium value $\overline{x}_1 = \frac{V_d^2}{Q}$. We have that the linearizing controller is given by

$$u_{av} = \frac{x_2 + x_4 - 1 - \lambda(y - \overline{y})}{x_2 + x_4}$$

and the control input corresponding to the steady state value of the output $y = x_1$ is simply given by $u_{av} = 1 - \frac{1}{x_2+x_4}$. The corresponding zero dynamics is found to be

$$\dot{x}_2 = \frac{V_d^2/Q + x_3}{x_2 + x_4} - x_3$$
$$\alpha_1 \dot{x}_3 = x_2 - 1$$
$$\alpha_2 \dot{x}_4 = \frac{V_d^2/Q + x_3}{x_2 + x_4} - \frac{1}{Q} x_4$$

The equilibrium point of the zero dynamics is clearly given by

$$\bar{x}_2 = 1, \quad \bar{x}_3 = \frac{V_d}{Q}, \quad \bar{x}_4 = V_d$$

In this case the average zero dynamics of the output inductor current variable is found to be stable. To demonstrate this fact, we simply resort to approximate linearization and find that around the equilibrium point $(\bar{x}_2, \bar{x}_3, \bar{x}_4)$ the tangent linearization of the dynamics for (x_2, x_3, x_4) satisfies

$$\begin{pmatrix} \dot{x}_{1\delta} \\ \dot{x}_{2\delta} \\ \dot{x}_{3\delta} \end{pmatrix} = \begin{pmatrix} -\frac{V_d}{Q(1+V_d)} & -\frac{V_d}{(1+V_d)} & -\frac{V_d}{Q(1+V_d)} \\ \frac{1}{\alpha_1} & 0 & 0 \\ -\frac{V_d}{\alpha_2 Q(1+V_d)} & \frac{1}{\alpha_2(1+V_d)} & -\frac{(1+2V_d)}{\alpha_2 Q(1+V_d)} \end{pmatrix} \begin{pmatrix} x_{1\delta} \\ x_{2\delta} \\ x_{3\delta} \end{pmatrix}$$

which has the following associated characteristic polynomial

$$p(s) = s^3 + \underbrace{\frac{1 + (2 + \alpha_2) V_d}{\alpha_2 (1 + V_d) Q}}_{=: \eta} s^2 + \underbrace{\frac{(\alpha_1 + \alpha_2 Q^2) V_d}{\alpha_1 \alpha_2 (1 + V_d) Q^2}}_{=: \mu} s + \underbrace{\frac{2 V_d}{\alpha_1 \alpha_2 (1 + V_d) Q}}_{=: \kappa}$$

And applying the Routh-Hurwitz stability criteria for $p(s)$, the corresponding Routh-Hurwitz array is given by

$$\begin{array}{c|cc} s^3 & 1 & \mu \\ s^2 & \eta & \kappa \\ s & \frac{\eta \mu - \kappa}{\eta} & \\ s^0 & \kappa & \end{array}$$

and since there are no change of signs in the first column of the Routh-Hurwitz array, i.e.,

$$\frac{1 + (2 + \alpha_2) V_d}{\alpha_2 (1 + V_d) Q} > 0$$

$$\frac{\eta \mu - \kappa}{\eta} > 0$$

$$\frac{2 V_d}{\alpha_1 \alpha_2 (1 + V_d) Q} > 0$$

then the equilibrium point $(\bar{x}_2, \bar{x}_3, \bar{x}_4)$ is therefore locally asymptotically stable.

State Feedback Linearization

Consider the average normalized model of the Sepic converter which is, as in the previous case, of the form:

$$\dot{x} = f(x) + g(x) u_{av}$$

with $f(x)$ and $g(x)$ given by:

$$f(x) = \begin{bmatrix} 1 - (x_2 + x_4) \\ x_1 \\ -\frac{1}{\alpha_1} x_4 \\ \frac{1}{\alpha_2}\left(x_1 + x_3 - \frac{x_4}{Q}\right) \end{bmatrix}, \quad g(x) = \begin{bmatrix} x_2 + x_4 \\ -(x_1 + x_3) \\ \frac{1}{\alpha_1}(x_2 + x_4) \\ \frac{1}{\alpha_2}(x_1 + x_3) \end{bmatrix}$$

We next show that the normalized average model does not admit a state and input coordinate transformation which exactly linearizes the system.

We compute the following vector fields:

$$ad_f g = [f, g] = \frac{\partial g}{\partial x} f - \frac{\partial f}{\partial x} g = \begin{bmatrix} -x_3 - \frac{1}{\alpha_2 Q} x_4 \\ -1 + \frac{1}{\alpha_1} x_4 \\ \frac{1}{\alpha_1}\left(x_1 - \frac{1}{\alpha_2 Q} x_4\right) \\ -\frac{1}{\alpha_2}\left(1 + \frac{1}{\alpha_2 Q} x_1 + \frac{1}{\alpha_1} x_2 + \frac{1}{\alpha_2 Q} x_3\right) \end{bmatrix}$$

$$ad_f^2 g = [f, ad_f g] = \frac{\partial ad_f g}{\partial x} f - \frac{\partial f}{\partial x} ad_f g = \begin{bmatrix} \gamma_{11} \\ \gamma_{21} \\ \gamma_{31} \\ \gamma_{41} \end{bmatrix}$$

where

$$\gamma_{11} = -1 - \frac{1}{\alpha_2} - \frac{2}{\alpha_2^2 Q} x_1 - \frac{1}{\alpha_1 \alpha_2} x_2 - \frac{2}{\alpha_2^2 Q} x_3 + \left(\frac{2}{\alpha_1} + \frac{1}{\alpha_2^2 Q^2}\right) x_4$$

$$\gamma_{21} = \frac{1}{\alpha_1 \alpha_2} x_1 + \left(\frac{1}{\alpha_1 \alpha_2} + 1\right) x_3 + \frac{1}{\alpha_2 Q}\left(1 - \frac{1}{\alpha_1}\right) x_4$$

$$\gamma_{31} = \frac{1}{\alpha_1}\left[\left(1 - \frac{1}{\alpha_2}\right) - \frac{2}{\alpha_2^2 Q} x_1 - \left(1 + \frac{1}{\alpha_1 \alpha_2}\right) x_2 - \frac{2}{\alpha_2^2 Q} x_3 - \left(1 - \frac{1}{\alpha_2^2 Q^2}\right) x_4\right]$$

$$\gamma_{41} = \frac{1}{\alpha_2} \begin{bmatrix} -\frac{2}{\alpha_2 Q} - \left(\frac{1}{\alpha_2^2 Q^2} + \frac{2}{\alpha_1}\right) x_1 + \frac{1}{\alpha_2 Q}\left(1 - \frac{1}{\alpha_1}\right) x_2 \\ -\left(\frac{1}{\alpha_2^2 Q^2} - 1\right) x_3 + \frac{2}{\alpha_2 Q}\left(\frac{1}{\alpha_1} + 1\right) x_4 \end{bmatrix}$$

We do not write down the vector $ad_f^3 g$, due to its involved component expressions. The computation of such a vector field is carried out, nevertheless, accordingly to:

$$ad_f^3 g = [f, ad_f^2 g] = \frac{\partial ad_f^2 g}{\partial x} f - \frac{\partial f}{\partial x} ad_f^2 g$$

It is quite straightforward to demonstrate, making use of well known symbolic manipulation computer packages, that the controllability matrix

$$\mathcal{C}(x) = [g, ad_f g, ad_f^2 g, ad_f^3 g]$$

is full rank 4.

On the other hand, we can also verify, with some further work, that the set of vectors

$$\{g, ad_f g, ad_f^2 g\}$$

is not *involutive*. As a consequence, the Sepic converter average normalized dynamics is not exactly linearizable by means of a nonlinear state coordinate transformation and a state dependent input coordinate transformation.

5.2.8 The Zeta Converter

Consider the average normalized model of the Zeta converter

$$\dot{x}_1 = -(1 - u_{av})x_2 + u_{av}$$
$$\dot{x}_2 = (1 - u_{av})x_1 - u_{av}x_3$$
$$\alpha_1\dot{x}_3 = u_{av}x_2 - x_4 + u_{av}$$
$$\alpha_2\dot{x}_4 = x_3 - \frac{1}{Q}x_4$$

It is desired to regulate the average value of the output capacitor voltage x_4 towards the average equilibrium value $\bar{y} = V_d$. The corresponding average equilibrium values for x_1, x_2, x_3 and u_{av} are given, respectively, by:

$$\bar{x}_1 = \frac{V_d^2}{Q}, \quad \bar{x}_2 = V_d, \quad \bar{x}_3 = \frac{V_d}{Q}, \quad \bar{u}_{av} = \frac{V_d}{V_d + 1}$$

Input-Output Feedback Linearization

Direct Method

Take as the output function of the system, the average normalized output capacitor voltage $y = x_4$.

The input-output feedback linearization is achieved by forcing the equation for x_4 to represent a linear dynamics with $\bar{y} = V_d$ representing an asymptotically exponentially stable equilibrium point. We set then the average control input to

$$u_{av} = \frac{\alpha_1\alpha_2 v_{av} + \frac{\alpha_1}{Q}\dot{y} + y}{1 + x_2}$$

where

$$v_{av} = -\lambda_0(y - \bar{y}) - \lambda_1 \dot{y}$$

with the restriction, $0 \leq u_{av} \leq 1$, being enforced in closing the loop and λ_0 and λ_1 being strictly positives scalars constants.

The value of the average feedback function u_{av}, corresponding to the steady state of the average output variable y is given by $\bar{u}_{av}(\bar{y}) = \bar{y}/(1 + x_2)$. It follows that the ideal behavior of the x_3 variable corresponds itself to a constant value, i.e., $x_3 = \bar{x}_3 = V_d/Q$. The corresponding zero dynamics of the controlled output function y is then obtained as

$$\dot{x}_1 = -x_2 + V_d$$
$$\dot{x}_2 = x_1 - \left(\frac{V_d/Q + x_1}{1 + x_2}\right)V_d$$

which has by equilibrium point to $(\bar{x}_1, \bar{x}_2) = (V_d^2/Q, V_d)$. The average zero dynamics of the output capacitor voltage variable is therefore unstable. To show this we simply resort to approximate linearization and find that around the equilibrium point $(\overline{x}_1, \overline{x}_2)$ the tangent linearization of the dynamics for x_1 and x_2 satisfies

$$\begin{bmatrix} \dot{x}_{1\delta} \\ \dot{x}_{2\delta} \end{bmatrix} = \begin{bmatrix} 0 & -1 \\ \frac{1}{1+V_d} & \frac{V_d^2}{Q(1+V_d)} \end{bmatrix} \begin{bmatrix} x_{1\delta} \\ x_{2\delta} \end{bmatrix}$$

whose characteristic polynomial is just obtained as:

$$p(s) = s^2 - \frac{1}{Q}\frac{V_d^2}{1+V_d}s + \frac{1}{1+V_d}$$

which is clearly unstable because it has at least one unstable root in the complex plane. The feedback controller locally linearizing the input output dynamics is therefore not feasible when the output function is represented by the average normalized output capacitor voltage since this is a *non-minimum phase* variable.

Indirect Method

Consider now the case when the output is given by the average input inductor current $y = x_1$. Suppose it is desired to regulate the average normalized current to the desired equilibrium value $\overline{x}_1 = V_d^2/Q$. We have that the linearizing controller is given by

$$u_{av} = \frac{x_2 - \lambda(x_1 - V_d^2/Q)}{1 + x_2}$$

and the control input corresponding to the steady state value of the output $y = x_1$ is simply given by $u_{av}(\overline{y}) = x_2/(1+x_2)$, which under non-saturated operating conditions satisfies

$$0 < u_{av}(\overline{y}) < 1 \tag{5.10}$$

thus $x_2 > 0$.

The corresponding zero dynamics is found to be

$$\dot{x}_2 = \frac{V_d^2/Q - x_2 x_3}{1 + x_2}$$
$$\alpha_1 \dot{x}_3 = x_2 - x_4$$
$$\alpha_2 \dot{x}_4 = x_3 - \frac{1}{Q} x_4 \tag{5.11}$$

which has by equilibrium point to

$$\overline{x}_2 = V_d, \quad \overline{x}_3 = \frac{V_d}{Q}, \quad \overline{x}_4 = V_d$$

It is important to say that the zero dynamics given by Equation 5.11 is the same that we obtained in (3.13), then consider the candidate Lyapunov function

$$V(x_2, x_3, x_4) = \frac{1}{2}\left[(x_2 - \overline{x}_2)^2 + \alpha_1(x_3 - \overline{x}_3)^2 + \alpha_2(x_4 - \overline{x}_4)^2\right] + \gamma$$
$$- \int_0^\tau \frac{[x_2(\sigma) - \overline{x}_2][x_3(\sigma) - \overline{x}_3]}{[1 + x_2(\sigma)]} d\sigma$$

with γ being a strictly positive constant parameter, which is assumed to be sufficiently large so that V is strictly positive, with $x_2 > 0$ by (5.10). The time derivative of V, along the solution of the system of differential equation yields the following expression:

$$\dot{V}(x_2, x_3, x_4) = -\frac{1}{Q}(x_4 - \overline{x}_4)^2 - \overline{x}_3\frac{(x_2 - \overline{x}_2)^2}{(1 + x_2)} \leq 0$$

and by LaSalle's theorem, it is then clear that the average normalized input inductor current x_1, taken as a system output, is a *locally minimum phase output*. We, thus, attempt an *indirect regulation* of the converter average normalized output voltage, x_4, towards the desired value $\overline{x}_4 = V_d$. This is accomplished by primarily regulating the inductor current x_1 towards its corresponding average equilibrium value, $\overline{x}_1 = V_d^2/Q$.

State Feedback Linearization

In this case we have

$$f(x) = \begin{bmatrix} -x_2 \\ x_1 \\ -\frac{1}{\alpha_1}x_4 \\ \frac{1}{\alpha_2}\left(x_3 - \frac{x_4}{Q}\right) \end{bmatrix}, \quad g(x) = \begin{bmatrix} 1 + x_2 \\ -(x_1 + x_3) \\ \frac{1}{\alpha_1}(1 + x_2) \\ 0 \end{bmatrix}$$

The controllability matrix

$$\mathcal{C}(x) = [g, ad_f g, ad_f^2 g, ad_f^3 g]$$

is found to be full rank 4.

The set of vector fields

$$\{g, ad_f g, ad_f^2 g\} = \left\{ \begin{bmatrix} 1 + x_2 \\ -x_1 - x_3 \\ \frac{(1+x_2)}{\alpha_1} \\ 0 \end{bmatrix}; \begin{bmatrix} -x_3 \\ \frac{(x_4 - \alpha_1)}{\alpha_1} \\ \frac{1}{\alpha_1}x_1 \\ -\frac{(1+x_2)}{\alpha_1\alpha_2} \end{bmatrix}; \begin{bmatrix} -1 + \frac{2}{\alpha_1}x_4 \\ \frac{(1+\alpha_1\alpha_2)Qx_3 - x_4}{\alpha_1\alpha_2 Q} \\ -\frac{1+(1+\alpha_1\alpha_2)x_2}{\alpha_1^2\alpha_2} \\ -\frac{(1+2\alpha_2 Qx_1 + x_2)}{\alpha_1\alpha_2^2 Q} \end{bmatrix} \right\}$$

is linearly independent, of course, but it is found to be non-involutive. This means that the system is not exactly feedback linearizable.

5.2.9 The Quadratic Buck Converter

It is left as an exercise, for the interested reader, that the average normalized quadratic Buck converter model, given by:

$$\dot{x}_1 = -x_2 + u_{av}$$
$$\dot{x}_2 = x_1 - u_{av}x_3$$
$$\alpha_1\dot{x}_3 = u_{av}x_2 - x_4$$
$$\alpha_2\dot{x}_4 = x_3 - \frac{x_4}{Q}$$

is not feedback linearizable by means of state feedback linearization, nor via an input-output feedback linearization at direct form.

5.3 Passivity Based Control of DC-to-DC Power Converters

The feedback controller design presented in this section follows the line of the *energy shaping plus damping injection* Passivity Based Control, as advocated in the book by Ortega et al. [48].

We postulate a rather general average model of DC-to-DC power converters regulated by a single switch. These are described, in general, by a system of the form:

$$\mathcal{A}\dot{x} = \mathcal{J}(u_{av})x - \mathcal{R}x + bu_{av} + \mathcal{E}$$

where $\mathcal{A}$ is a diagonal positive definite matrix, $\mathcal{J}(u_{av})$ is a skew symmetric matrix for all u_{av} and it is, moreover, an affine function of u of the form $\mathcal{J}_0 + \mathcal{J}_1 u$. This term represents the *conservative forces* of the system. The matrix $\mathcal{R}$ is a positive semi-definite symmetric matrix representing the *dissipation terms* of the circuit model. b is, generally speaking, a constant vector and it may contain some components which are dependent on the external constant sources. The term $\mathcal{E}$ also represents external constant voltage sources. The vector $x \in R^n$ is assumed to be available for measurement.

The control task is to track a given reference state trajectory, $x^*(t)$ which is to be determined on the basis of knowledge of the system structure and a specified output trajectory tracking, or stabilization, task.

One starts with a Lyapunov function candidate in the tracking error $e = (x - x^*(t))$, of the form

$$V(e) = \frac{1}{2}e^T \mathcal{A}e = \frac{1}{2}(x - x*(t))\mathcal{A}(x - x^*(t))$$

The time derivative of such a function along the trajectories of the system is given by

$$\dot{V}(e) = (x - x*(t))^T \mathcal{A}(\dot{x} - \dot{x}^*(t))$$
$$= (x - x*(t))^T ([\mathcal{J}(u_{av}) - \mathcal{R}]x + bu_{av} + \mathcal{E} - \mathcal{A}\dot{x}^*(t))$$

Setting

$$\mathcal{A}\dot{x}^*(t) = \mathcal{J}(u_{av})x^*(t) - \mathcal{R}x^*(t) + bu_{av} + \mathcal{E} + \mathcal{R}_I(x - x^*(t))$$

with $\mathcal{R}_I$ being a symmetric positive definite, or positive semi-definite, matrix satisfying the condition: $\mathcal{R} + \mathcal{R}_I > 0$. The above choice of the reference trajectory dynamics yields, in view of the fact that $e^T \mathcal{J}(u)e = 0$ for all u, the following evaluation of the time derivative of the Lyapunov function $V(e)$,

$$\dot{V}(e) = e^T(\mathcal{J}(u_{av})e - \mathcal{R}e - \mathcal{R}_I e) = -e^T(\mathcal{R} + \mathcal{R}_I)e < 0$$

The tracking error e has the origin as an asymptotically stable equilibrium point. The stability of such desired equilibrium point for the tracking error may be determined to be even of exponential nature. Indeed, let κ_A and $\kappa_{\mathcal{R}+\mathcal{R}_I}$ be, respectively, the smallest eigenvalues of the positive definite symmetric matrices A and $\mathcal{R} + \mathcal{R}_I$. Let $\kappa = min\{\kappa_A, \kappa_{\mathcal{R}+\mathcal{R}_I}\}$, we then have

$$\dot{V}(e) = -e^T(\mathcal{R} + \mathcal{R}_I)e \leq -\kappa\, V(e)$$

The tracking error e has then the origin as an asymptotically exponentially stable equilibrium point.

Note that the symmetric matrix $\mathcal{R}_I$ complements the stability features of the damping matrix $\mathcal{R}$ originally in the system. The condition $\mathcal{R} + \mathcal{R}_I > 0$ must be regarded as a sort of *dissipation matching condition* since the structure of the matrix b is responsible for achieving such a damping in a feedback manner, then the range space of $\mathcal{R}_I$ and the range space of b are not, necessarily, independent.

The system

$$\mathcal{A}\dot{x}^*(t) = \mathcal{J}(u_{av})x^*(t) - \mathcal{R}x^*(t) + bu_{av} + \mathcal{E} + \mathcal{R}_I(x - x^*(t))$$

plays the role of an exogenous controlled system which mimics the energy structure of the system and adds some extra damping term of the form $\mathcal{R}_I(x - x^*(t))$. This damping "injection" complements the dissipation of the original system in the tracking error dynamics. The exogenous system is a controlled system which plays the role of a reference model system with a fundamentally enhanced dissipation structure. Defining a desired reference trajectory for a relative degree one minimum phase output (state) variable in the reference model, the control input can be immediately computed in closed loop form while the rest of the reference states play the role of the dynamics of the reference variables, other than the chosen minimum phase output, constituting the feedback controller.

5.3.1 The Boost Converter

Consider the Boost converter shown in Figure 5.2

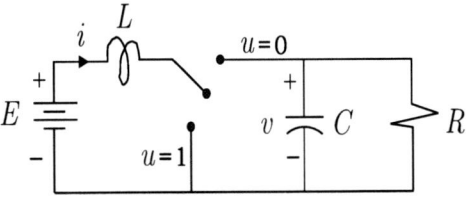

Fig. 5.2. The Boost converter.

The average normalized model of this converter is given by

$$\dot{x}_1 = -(1 - u_{av})x_2 + 1$$
$$\dot{x}_2 = (1 - u_{av})x_1 - \frac{1}{Q}x_2$$

Given a constant average control input $\bar{u}_{av} = U$ the corresponding equilibrium values for the average input current, $\bar{x}_1$, and average output voltage, $\bar{x}_2$, are given by:

$$\bar{x}_1 = \frac{1}{Q}\frac{1}{(1-U)^2}, \quad \bar{x}_2 = \frac{1}{(1-U)}$$

Using these expressions, we can write:

$$\bar{x}_1 = \frac{1}{Q}\bar{x}_2^2$$

Since the average normalized voltage variable x_2 is a *non-minimum phase output*, it is highly advisable to indirectly control the capacitor average voltage through the corresponding inductor current x_1, which is, in fact, a *minimum phase output*.

The Boost Converter as a Passive System

Considered the following *storage* function

$$H(x) = \frac{1}{2}\left(x_1^2 + x_2^2\right)$$

The time derivative of $H(x)$ along the regulated evolution of the system is given by

$$\dot{H}(x) = x_1\dot{x}_1 + x_2\dot{x}_2 = -\frac{1}{Q}x_2^2 + x_1 \leq x_1$$

It results in a passive relation involving the normalized external "input" E, which here corresponds with the additive scalar constant "1" in the normalized model, and the average normalized input current variable x_1. Direct integration yields the following passivity inequality,

$$H\left[x\left(t\right)\right] - H\left[x\left(0\right)\right] \leq \int_0^t x_1\left(\sigma\right) d\sigma$$

Passivity Based Control for the Boost Converter

The average normalized Boost circuit model may be written in the following matrix form:

$$\begin{bmatrix} \dot{x}_1 \\ \dot{x}_2 \end{bmatrix} = \begin{bmatrix} 0 & -(1-u_{av}) \\ (1-u_{av}) & 0 \end{bmatrix} \begin{bmatrix} x_1 \\ x_2 \end{bmatrix} + \begin{bmatrix} 0 & 0 \\ 0 & -\frac{1}{Q} \end{bmatrix} \begin{bmatrix} x_1 \\ x_2 \end{bmatrix} + \begin{bmatrix} 1 \\ 0 \end{bmatrix}$$

The exogenous system corresponding to the desired state variables x_d is thus given by,

$$\begin{bmatrix} \dot{x}_{1d} \\ \dot{x}_{2d} \end{bmatrix} = \begin{bmatrix} 0 & -(1-u_{av}) \\ (1-u_{av}) & 0 \end{bmatrix} \begin{bmatrix} x_{1d} \\ x_{2d} \end{bmatrix} + \begin{bmatrix} 0 & 0 \\ 0 & -\frac{1}{Q} \end{bmatrix} \begin{bmatrix} x_{1d} \\ x_{2d} \end{bmatrix} + \begin{bmatrix} 1 \\ 0 \end{bmatrix}$$
$$- \begin{bmatrix} -R_1 & 0 \\ 0 & 0 \end{bmatrix} \begin{bmatrix} x_1 - x_{1d} \\ x_2 - x_{2d} \end{bmatrix}$$

with $R_1 > 0$, being a positive scalar, and $x_d = \begin{bmatrix} x_{1d} & x_{2d} \end{bmatrix}^T$.

An indirect stabilization of the output voltage may be accomplished by setting: $x_{1d} = \overline{x}_1 = constant$. We obtain, from the first equation of the exogenous normalized model above, the following dynamical controller expression for the average input u_{av}:

$$u_{av} = -\frac{1}{\xi_2}\left[1 + R_1\left(x_1 - \overline{x}_1\right)\right] + 1$$

$$\dot{\xi}_2 = \left(1 - u_{av}\right)\overline{x}_1 - \frac{1}{Q}\xi_2$$

where the introduced variable ξ_2, has been used in replacement of the variable x_{2d} in the exogenous model. The differential equation for ξ_2 represents the dynamic part of the feedback controller.

We propose to use the average designed passivity based controller in a switched implementation using a $\Sigma - \Delta$-modulator.

$$u = \frac{1}{2}\left[1 + \text{sign } z\right], \quad \dot{z} = u_{av} - u$$

$$u_{av} = -\frac{1}{\xi_2}\left[1 + R_1\left(x_1 - \overline{x}_1\right)\right] + 1$$

$$\dot{\xi}_2 = \left(1 - u_{av}\right)\overline{x}_1 - \frac{1}{Q}\xi_2$$

Simulations

A typical Boost converter circuit is characterized by the following parameters:

$$L = 20 \text{ mH}, \quad C = 20 \text{ }\mu\text{F}, \quad R = 30 \text{ }\Omega, \quad E = 15 \text{ V}$$

These given set of parameter values yield the following normalized parameter values,

$$Q = 0.9487, \quad \sqrt{LC} = 6.3246 \times 10^{-4} \text{ s}$$

It is desired to reach an actual steady state equilibrium voltage of value $\bar{v} = 30$ V. The corresponding actual steady state current and control input, respectively, are given by $\bar{i} = 2$ A and $\bar{u}_{av} = 0.5$.

Figure 5.3 depicts the average responses of the Boost converter model to the passivity based control synthesized via the energy shaping and damping injection method.

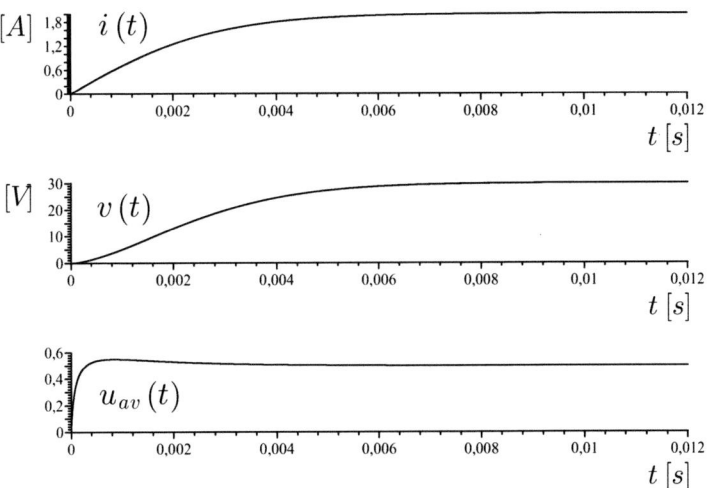

Fig. 5.3. Average response of Boost converter to passivity based control.

Figure 5.4 shows the closed loop response of the switched Boost power converter to a $\Sigma-\Delta$-modulator implementation of a passivity based stabilizing controller. The controller and the system parameters were chosen to be exactly the same as in the previous simulation.

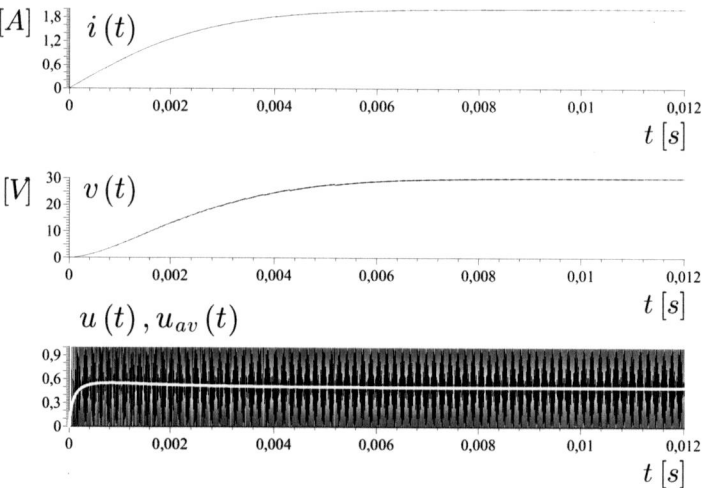

Fig. 5.4. Response of switched Boost converter to passivity based control implemented through a $\Sigma - \Delta$ modulator.

5.3.2 The Buck-Boost Converter

Consider the average normalized model of the Buck-Boost converter circuit

$$\dot{x}_1 = (1 - u_{av})x_2 + u_{av}$$
$$\dot{x}_2 = -(1 - u_{av})x_1 - \frac{1}{Q}x_2$$

As in the case of the Boost converter, x_2, is a *non-minimum phase output* for the average system. On the other hand, x_1 is a *minimum phase output* for the average system.

Given a constant average control input $u_{av} = U$ the corresponding equilibrium values for the average input current, $\overline{x}_1$, and average output voltage, $\overline{x}_2$, are:

$$\overline{x}_1 = \frac{1}{Q}\frac{U}{(1-U)^2}, \qquad \overline{x}_2 = -\frac{U}{(1-U)}$$

using these expressions, we can write:

$$\overline{x}_1 = (\overline{x}_2 - 1)\frac{\overline{x}_2}{Q}$$

Thus, if we desire to regulate x_2 towards an equilibrium value $\overline{x}_2$, then, such a regulation can be indirectly accomplished by regulating x_1 towards its corresponding equilibrium value.

The Buck-Boost Converter as a Passive System

Considered the following average storage function,

$$H(x) = \frac{1}{2}\left(x_1^2 + x_2^2\right)$$

The time derivative of $H(x)$ along the regulated evolution of the Buck-Boost system trajectories is given by

$$\dot{H}(x) = x_1\dot{x}_1 + x_2\dot{x}_2 = -\frac{1}{Q}x_2^2 + u_{av}x_1 \leq u_{av}x_1$$

The system is then a passive system and, in this case, the passivity relation directly involves the average control input u_{av} and the average current variable x_1. The integration of the time derivative of H, produces the following passivity inequality:

$$H[x(t)] - H[x(0)] \leq \int_0^t x_1(\sigma)\,u_{av}(\sigma)\,d\sigma$$

As in the previous case, we proceed to design a dynamic output feedback controller using the energy shaping plus damping injection controller design method.

Letting, $x_{1d} = \bar{x}_1 = constant$, we obtain the following average dynamical output feedback controller expression

$$u_{av} = \frac{1}{\xi_2 - 1}\left[R_1(x_1 - \bar{x}_1) + \xi_2\right]$$

$$\dot{\xi}_2 = -(1 - u_{av})\bar{x}_1 - \frac{1}{Q}\xi_2$$

where ξ_2, has replaced the variable x_{2d} in the exogenous system description.

Simulations

A typical set of parameters for the Buck-Boost converter circuit are:

$$L = 20\text{ mH}, \quad C = 20\ \mu\text{F}, \quad R = 30\ \Omega, \quad E = 15\text{ V}$$

These parameter values yield:

$$Q = 0.9487, \quad \sqrt{LC} = 6.3246 \times 10^{-4}\text{ s}$$

We set a desired steady state voltage of $\bar{v} = -22.5$ V, with the corresponding steady state current:

$$\bar{i} = 1.875\text{ A}$$

while the average equilibrium input turns out to be given by $\bar{u}_{av} = 0.6$.

268   5 Nonlinear Methods

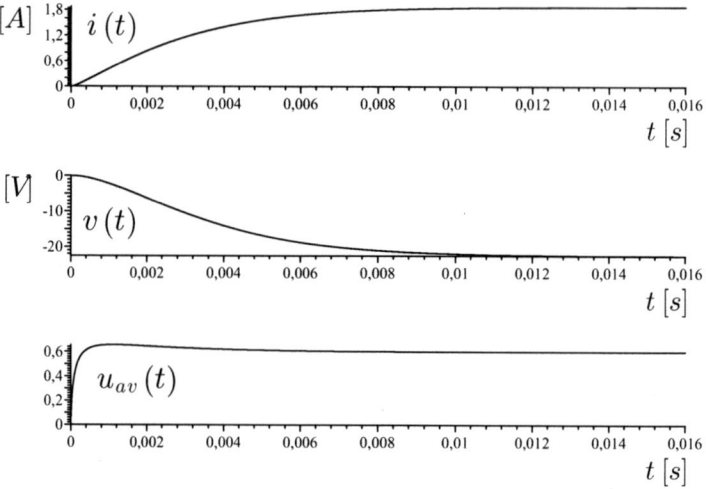

Fig. 5.5. Average responses of Buck-Boost converter to passivity based control.

Figure 5.5 shows the non-normalized system response of the Buck-Boost power converter, to the average passivity based controller designed on the basis of energy shaping plus damping injection.

Figure 5.6 shows the closed loop response of a Buck-Boost power converter to a $\Sigma - \Delta$-modulator implementation of the passivity based stabilizing controller.

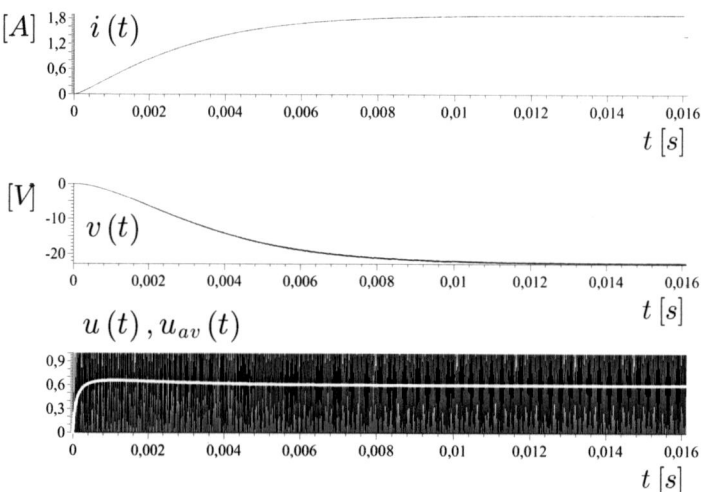

Fig. 5.6. Responses of switched Buck-Boost converter to passivity based control implemented through a $\Sigma - \Delta$ modulator.

5.3.3 The Cúk Converter

The average normalized model of the Cúk DC-to-DC power converter is given by,

$$\dot{x}_1 = -(1 - u_{av})x_2 + 1$$
$$\dot{x}_2 = (1 - u_{av})x_1 + u_{av}x_3$$
$$\alpha_1 \dot{x}_3 = -u_{av}x_2 - x_4$$
$$\alpha_2 \dot{x}_4 = x_3 - \frac{1}{Q}x_4$$

Given a constant average control input $u_{av} = U$, the corresponding equilibrium values for the average input currents, $\overline{x}_1$ and $\overline{x}_3$, and the average output voltages, $\overline{x}_2$ and $\overline{x}_4$, are given by,

$$\overline{x}_1 = \frac{1}{Q}\frac{U^2}{(1-U)^2}, \quad \overline{x}_2 = \frac{1}{1-U}, \quad \overline{x}_3 = -\frac{1}{Q}\frac{U}{(1-U)}, \quad \overline{x}_4 = -\frac{U}{1-U}$$

Using these expressions, we can write:

$$\overline{x}_1 = \frac{\overline{x}_4^2}{Q}, \quad \overline{x}_2 = 1 - \overline{x}_4, \quad \overline{x}_3 = \frac{\overline{x}_4}{Q}$$

In the Cúk converter circuit dynamics, the output voltage x_4 is found to be a *non-minimum phase output*, so it is preferable to indirectly control the average capacitor output voltage x_4 through the average inductor current x_1, which is a *minimum phase output*.

Considered the following average storage function,

$$H(x) = \frac{1}{2}\left(x_1^2 + x_2^2 + \alpha_1 x_3^2 + \alpha_2 x_4^2\right)$$

The time derivative of $H(x)$ along the controlled motions of the system is given by,

$$\dot{H}(x) = x_1\dot{x}_1 + x_2\dot{x}_2 + \alpha_1 x_3\dot{x}_3 + \alpha_4 x_4\dot{x}_4 = -\frac{1}{Q}x_4^2 + x_1 \leq x_1$$

then the average Cúk system is a passive system between the normalized input, E, here represented by the constant "1" and the input current variable x_1. The integration of the time derivative of H, produces the traditional passivity inequality,

$$H[x(t)] - H[x(0)] \leq \int_0^t x_1(\sigma)\, d\sigma$$

The auxiliary, *exogenous*, system, of the average Cúk converter model is then given

$$\dot{x}_d = \begin{pmatrix} 0 & -1+u_{av} & 0 & 0 \\ 1-u_{av} & 0 & \frac{u_{av}}{\alpha_1} & 0 \\ 0 & -\frac{u_{av}}{\alpha_1} & 0 & -\frac{1}{\alpha_1\alpha_2} \\ 0 & 0 & \frac{1}{\alpha_1\alpha_2} & 0 \end{pmatrix} \frac{\partial H(x_d)}{\partial x_d} + \begin{pmatrix} 0 & 0 & 0 & 0 \\ 0 & 0 & 0 & 0 \\ 0 & 0 & 0 & 0 \\ 0 & 0 & 0 & -\frac{1}{\alpha_2^2 Q} \end{pmatrix} \frac{\partial H(x_d)}{\partial x_d}$$

$$+ \begin{pmatrix} 1 \\ 0 \\ 0 \\ 0 \end{pmatrix} - \begin{pmatrix} -R_1 & 0 & 0 & 0 \\ 0 & 0 & 0 & 0 \\ 0 & 0 & 0 & 0 \\ 0 & 0 & 0 & 0 \end{pmatrix} \begin{pmatrix} x_1 - x_{1d} \\ x_2 - x_{2d} \\ \alpha_1(x_3 - x_{3d}) \\ \alpha_2(x_4 - x_{4d}) \end{pmatrix}$$

with $R_1 > 0$, being a positive scalar, and:

$$x_d = [x_{1d}, x_{2d}, x_{3d}, x_{4d}]^T, \qquad \partial H(x_d)/\partial x_d = [x_{1d}, x_{2d}, \alpha_1 x_{3d}, \alpha_2 x_{4d}]^T$$

Letting, $x_{1d} = \overline{x}_1 = constant$, we obtain the following dynamical average feedback controller expression for the average input u_{av}:

$$u_{av} = -\frac{1}{\xi_2}[1 + R_1(x_1 - \overline{x}_1)] + 1$$

$$\dot{\xi}_2 = (1 - u_{av})\overline{x}_1 + \xi_3 u_{av}$$

$$\dot{\xi}_3 = -\frac{1}{\alpha_1}\xi_2 u_{av} - \frac{1}{\alpha_1}\xi_4$$

$$\dot{\xi}_4 = \frac{1}{\alpha_2}\xi_3 - \frac{1}{\alpha_2 Q}\xi_4$$

where the variables ξ_2, ξ_3 and ξ_4, acting as the dynamical controller states, have replaced the auxiliary state variables x_{2d}, x_{3d} and x_{4d}, respectively.

Simulations

We take a Cúk converter with the following parameter values

$$L_1 = 30 \text{ mH}, \qquad C_1 = 150 \text{ }\mu\text{F}, \qquad L_2 = 30 \text{ mH}, \qquad C_2 = 50 \text{ }\mu\text{F},$$

$$R = 10 \text{ }\Omega, \qquad E = 100 \text{ V}$$

We set as desired steady state voltage $\overline{v}_2 = -200$ V, with corresponding steady state values of the currents and internal capacitor voltage given by:

$$\overline{i}_1 = 40 \text{ A}, \qquad \overline{v}_1 = 300 \text{ V}, \qquad \overline{i}_2 = -20 \text{ A}$$

and

$$\overline{u}_{av} = 0.666$$

Figure 5.7 depicts the average Cúk converter system response to the actions of a passivity based controller using the energy shaping plus damping injection method.

5.3 Passivity Based Control 271

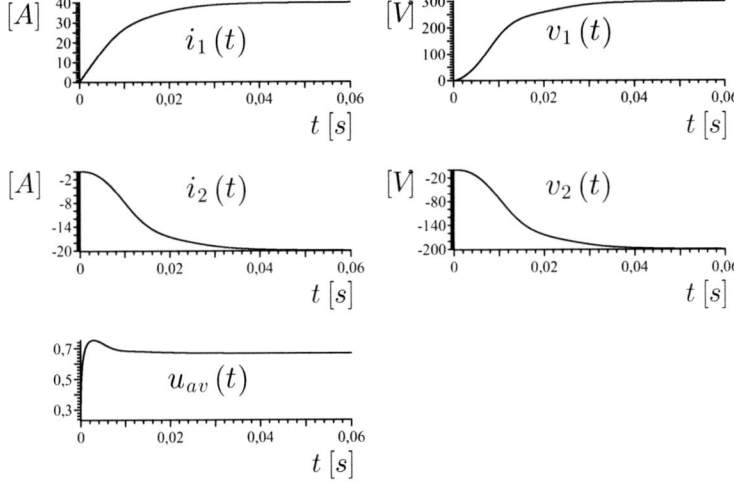

Fig. 5.7. Average responses of Cúk converter to a passivity based controller.

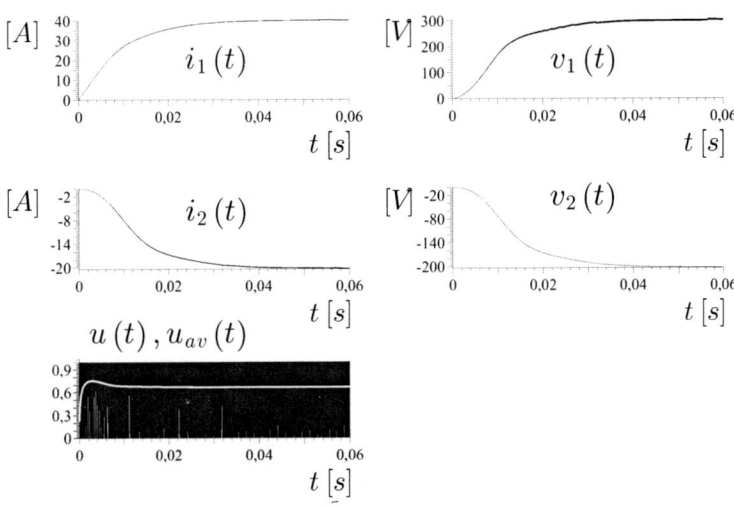

Fig. 5.8. Responses of switched Cúk converter to a passivity based controller implemented with a $\Sigma - \Delta$ modulator.

Figure 5.8 shows the closed loop response of a Cúk power converter to a passivity based stabilizing controller implemented with the help of a $\Sigma - \Delta$-modulator.

5.3.4 The Sepic Converter

The average normalized model of the Sepic converter circuit is given by

$$\dot{x}_1 = -(1 - u_{av})(x_2 + x_4) + 1$$
$$\dot{x}_2 = (1 - u_{av})x_1 - u_{av}x_3$$
$$\alpha_1 \dot{x}_3 = u_{av}x_2 - (1 - u_{av})x_4$$
$$\alpha_2 \dot{x}_4 = (1 - u_{av})(x_1 + x_3) - \frac{1}{Q}x_4$$

Given a constant average value of the output capacitor voltage x_4 towards the average equilibrium value V_d, the corresponding equilibrium values for the average input currents, $\bar{x}_1$ and $\bar{x}_3$, and the average output voltage $\bar{x}_2$, are:

$$\bar{x}_1 = \frac{V_d^2}{Q}, \quad \bar{x}_2 = 1, \quad \bar{x}_3 = \frac{V_d}{Q}, \quad \bar{u}_{av} = \frac{V_d}{V_d + 1}$$

In the average normalized model of the Sepic converter, the output voltage x_4 is a *non-minimum phase output* while x_1 is a *minimum phase output*. We proceed to indirectly control the capacitor output voltage x_4 through the inductor current x_1.

Consider the following storage function

$$H(x) = \frac{1}{2}\left(x_1^2 + x_2^2 + \alpha_1 x_3^2 + \alpha_2 x_4^2\right)$$

The Sepic converter is shown to satisfy a passivity inequality by considering the time derivative of the average stored energy $H(x)$ along the controlled motions of the system. Indeed,

$$\dot{H}(x) = x_1 \dot{x}_1 + x_2 \dot{x}_2 + \alpha_1 x_3 \dot{x}_3 + \alpha_4 x_4 \dot{x}_4 = -\frac{1}{Q}x_4^2 + x_1 \leq x_1$$

The integration of this expression produces the following passivity inequality,

$$H[x(t)] - H[x(0)] \leq \int_0^t x_1(\sigma)\, d\sigma$$

The explicit exogenous system for the Sepic converter, corresponding to the desired state variables x_d, is thus given by:

$$\dot{x}_d = \begin{pmatrix} 0 & -1 + u_{av} & 0 & -\frac{1-u_{av}}{\alpha_2} \\ 1 - u_{av} & 0 & -\frac{u_{av}}{\alpha_1} & 0 \\ 0 & \frac{u_{av}}{\alpha_1} & 0 & -\frac{1}{\alpha_1}\frac{1-u_{av}}{\alpha_2} \\ \frac{1-u_{av}}{\alpha_2} & 0 & \frac{1}{\alpha_1}\frac{1-u_{av}}{\alpha_2} & 0 \end{pmatrix} \frac{\partial H(x_d)}{\partial x_d}$$

$$+ \begin{pmatrix} 0 & 0 & 0 & 0 \\ 0 & 0 & 0 & 0 \\ 0 & 0 & 0 & 0 \\ 0 & 0 & 0 & -\frac{1}{\alpha_2^2 Q} \end{pmatrix} \frac{\partial H(x_d)}{\partial x_d} + \begin{pmatrix} 1 \\ 0 \\ 0 \\ 0 \end{pmatrix} - \begin{pmatrix} -R_1 & 0 & 0 & 0 \\ 0 & 0 & 0 & 0 \\ 0 & 0 & 0 & 0 \\ 0 & 0 & 0 & 0 \end{pmatrix} \begin{pmatrix} x_1 - x_{1d} \\ x_2 - x_{2d} \\ \alpha_1(x_3 - x_{3d}) \\ \alpha_2(x_4 - x_{4d}) \end{pmatrix}$$

with $R_1 > 0$, being a positive scalar, and

$$x_d = [x_{1d}, x_{2d}, x_{3d}, x_{4d}]^T, \quad \partial H(x_d)/\partial x_d = [x_{1d}, x_{2d}, \alpha_1 x_{3d}, \alpha_2 x_{4d}]^T$$

Define the following average normalized error variables, $e_i = x_i - x_{id}$, $i = 1, \ldots, 4$. We obtain the following tracking error dynamics:

$$\dot{e}_1 = -(1 - u_{av})(e_2 + e_4) - R_1 e_1$$
$$\dot{e}_2 = (1 - u_{av})e_1 - u_{av} e_3$$
$$\alpha_1 \dot{e}_3 = u_{av} e_2 - (1 - u_{av})e_4$$
$$\alpha_2 \dot{e}_4 = (1 - u_{av})(e_1 + e_3) - \frac{1}{Q} e_4$$

Consider now the following Lyapunov function candidate, defined in the trajectory tracking error space described by the coordinates, $e = (e_1, e_2, e_3, e_4)$,

$$H(e) = \frac{1}{2}\left(e_1^2 + e_2^2 + \alpha_1 e_3^2 + \alpha_2 e_4^2\right)$$

The time derivative of such a positive definitive function, along the controlled trajectories of the tracking error dynamics, yields

$$\dot{H}(e) = -R_1 e_1^2 - \frac{1}{Q} e_4^2 \leq 0$$

The set points in the tracking error space which satisfy $\dot{H}(e) = 0$ are given by the intersection of the hyper-planes, $e_1 = e_4 = 0$. This implies, from the tracking error dynamics, that also $e_3 = 0$ and $e_2 = 0$. According to LaSalle's theorem, the equilibrium point $e_i = 0$, $i = 1, \ldots, 4$ is a globally asymptotically stable equilibrium point for the controlled tracking error dynamics. This means that the average converter system trajectories, $x(t)$, and the auxiliary system trajectories, $x_d(t)$, asymptotically converge towards each other. It then suffices to fix a suitable trajectory, or constant reference value, (for the desired average inductor current x_{1d}) in the auxiliary system dynamics and define the control input in a corresponding fashion.

Letting, $x_{1d} = \bar{x}_1 = constant$, we obtain the following average dynamical output feedback controller expression for converter Sepic:

$$u_{av} = -\frac{1}{\xi_2 + \xi_4}[1 - (\xi_2 + \xi_4) + R_1(x_1 - \bar{x}_1)]$$
$$\dot{\xi}_2 = (1 - u_{av})\bar{x}_1 - u_{av}\xi_3$$
$$\alpha_1 \dot{\xi}_3 = u_{av}\xi_2 - (1 - u_{av})\xi_4$$
$$\alpha_2 \dot{\xi}_4 = (1 - u_{av})(\bar{x}_1 + \xi_3) - \frac{1}{Q}\xi_4$$

where the variables ξ_2, ξ_3 and ξ_4, acting as the dynamical controller states, have replaced the auxiliary state variables x_{2d}, x_{3d} and x_{4d}, respectively.

We propose to implement the average designed passivity based controller on the switched converter by means of a $\Sigma-\Delta$-modulator. As usual we have,

$$u = \frac{1}{2}\left[1 + \operatorname{sign} z\right], \qquad \dot{z} = e = u_{av} - u$$

$$u_{av} = -\frac{1}{\xi_2 + \xi_4}\left[1 - (\xi_2 + \xi_4) + R_1\left(x_1 - \overline{x}_1\right)\right]$$

$$\dot{\xi}_2 = (1 - u_{av})\overline{x}_1 - u_{av}\xi_3$$

$$\alpha_1 \dot{\xi}_3 = u_{av}\xi_2 - (1 - u_{av})\xi_4$$

$$\alpha_2 \dot{\xi}_4 = (1 - u_{av})\left(\overline{x}_1 + \xi_3\right) - \frac{1}{Q}\xi_4$$

Simulations

A typical Sepic converter circuit is characterized by the following parameters:

$$L_1 = 30 \text{ mH}, \qquad C_1 = 150 \text{ }\mu\text{F}, \qquad L_2 = 30 \text{ mH}, \qquad C_2 = 50 \text{ }\mu\text{F},$$

$$R = 10 \text{ }\Omega, \qquad E = 100 \text{ V}$$

We desire to regulate the system towards a steady state, non-normalized, voltage of value $\overline{v}_2 = 200$ V. The corresponding steady state variables yield: $\overline{i}_1 = 40$ A, $\overline{v}_1 = 100$ V and $\overline{i}_2 = 20$ A with $\overline{u}_{av} = 0.666$.

Figure 5.9 depicts the average Sepic system responses to the designed continuous, average passivity based controller synthesized via the energy shaping plus damping injection method.

Figure 5.10 shows the switched closed loop response of a Sepic power converter to a $\Sigma-\Delta$-modulator implementation of a passivity based stabilizing controller that uses energy shaping plus damping injection.

5.3.5 The Zeta Converter

Consider the average normalized model of a Zeta converter,

$$\dot{x}_1 = -(1 - u_{av})x_2 + u_{av}$$

$$\dot{x}_2 = (1 - u_{av})x_1 - u_{av}x_3$$

$$\alpha_1 \dot{x}_3 = u_{av}x_2 - x_4 + u_{av}$$

$$\alpha_2 \dot{x}_4 = x_3 - \frac{1}{Q}x_4$$

Given a constant average value of the output capacitor voltage $\overline{x}_4 = V_d$, the corresponding average equilibrium values for x_1, x_2, x_3 and u_{av} are given, respectively, by:

$$\overline{x}_1 = \frac{V_d^2}{Q}, \qquad \overline{x}_2 = V_d, \qquad \overline{x}_3 = \frac{V_d}{Q}, \qquad \overline{u}_{av} = \frac{V_d}{V_d + 1}$$

5.3 Passivity Based Control   275

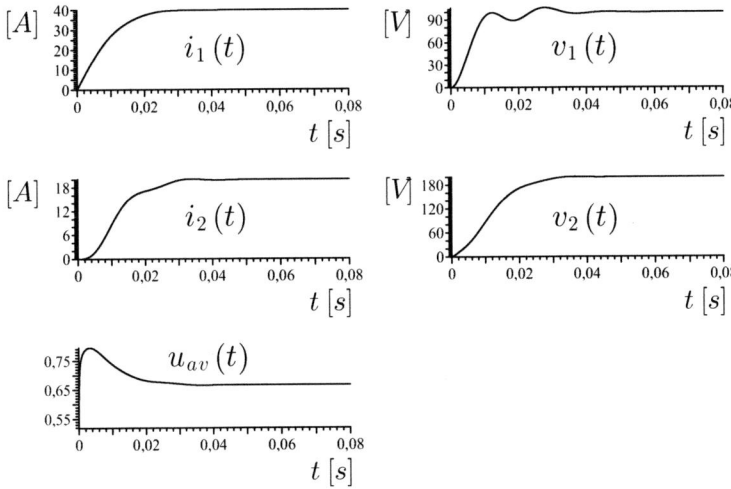

Fig. 5.9. Average responses of Sepic converter controlled by a passivity based controller.

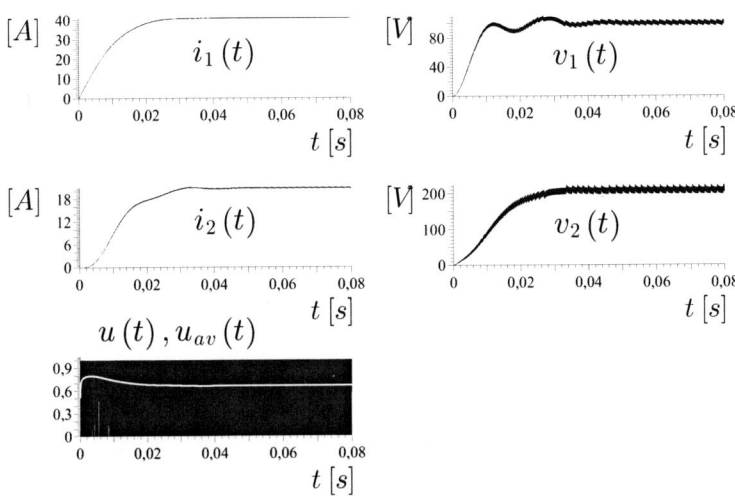

Fig. 5.10. Switched responses of Sepic converter controlled by an average passivity based controller implemented via a $\Sigma - \Delta$ modulator.

The average normalized output voltage x_4 is a *non-minimum phase output* while the average normalized inductor current x_1 is a *minimum phase output*. We proceed to indirectly control the capacitor output voltage x_4 through the regulation of x_1.

Consider the average normalized storage function,

$$H(x) = \frac{1}{2}\left(x_1^2 + x_2^2 + \alpha_1 x_3^2 + \alpha_2 x_4^2\right)$$

The time derivative of $H(x)$ along the controlled motions of the system is given by,

$$\dot{H}(x) = -\frac{1}{Q}x_4^2 + (x_1 + x_3)u_{av} \leq (x_1 + x_3)u_{av}$$

The system is then a passive system with a passive mapping existing between the control input u_{av} and the sum of the currents $(x_1 + x_3)$ acting as the output of the system. The integration of the time derivative of H, produces the passivity inequality,

$$H[x(t)] - H[x(0)] \leq \int_0^t [x_1(\sigma) + x_3(\sigma)]u_{av}(\sigma)\,d\sigma$$

Consider an auxiliary, *exogenous*, system, which is a copy of the original system, with injected damping represented by a term of the form: $R_1(x_1 - x_{1d})$, $R_1 > 0$. The injected damping affects only the copy of the average input inductor current dynamics, i.e., we consider,

$$\dot{x}_{1d} = -(1 - u_{av})x_{2d} + u_{av} + R_1(x_1 - x_{1d})$$
$$\dot{x}_{2d} = (1 - u_{av})x_{1d} - u_{av}x_{3d}$$
$$\alpha_1 \dot{x}_{3d} = u_{av}x_{2d} - x_{4d} + u_{av}$$
$$\alpha_2 \dot{x}_{4d} = x_{3d} - \frac{x_{4d}}{Q}$$

Define the following average normalized error variables, $e_i = x_i - x_{id}$, $i = 1,\ldots,4$. We then obtain the following tracking error dynamics:

$$\dot{e}_1 = -(1 - u_{av})e_2 - R_1 e_1$$
$$\dot{e}_2 = (1 - u_{av})e_1 - ue_3$$
$$\alpha_1 \dot{e}_3 = u_{av}e_2 - e_4$$
$$\alpha_2 \dot{e}_4 = e_3 - \frac{1}{Q}e_4$$

Consider now the following Lyapunov function candidate, defined in the trajectory tracking error space described by the coordinates, $e = (e_1, e_2, e_3, e_4)$,

$$H(e) = \frac{1}{2}\left(e_1^2 + e_2^2 + \alpha_1 e_3^2 + \alpha_2 e_4^2\right)$$

The time derivative of such a positive definitive function, along the controlled trajectories of the tracking error dynamics, yields

$$\dot{H}(e) = -R_1 e_1^2 - \frac{1}{Q}e_4^2 \leq 0$$

The set points in the tracking error space which satisfy $\dot{H}(e) = 0$ are given by the intersection of the hyper-planes, $e_1 = e_4 = 0$. This implies, from the tracking error dynamics, that also $e_3 = 0$ and $e_2 = 0$. According to LaSalle's theorem, the equilibrium point $e_i = 0$, $i = 1, \ldots, 4$ is a globally asymptotically stable equilibrium point for the controlled tracking error dynamics. This means that the average converter system trajectories, $x(t)$, and the auxiliary system trajectories, $x_d(t)$, asymptotically converge towards each other.

We fix a desired constant reference value for the average inductor current x_{1d} in the auxiliary system dynamics and solve from the control input from the first equation of the auxiliary dynamics. We denote, $x_{1d} = \overline{x}_1$, and obtain, a dynamic average feedback controller, u_{av}, of the form:

$$u_{av} = \frac{1}{1+\xi_2}[\xi_2 - R_1(x_1 - \overline{x}_1)]$$

$$\dot{\xi}_2 = (1 - u_{av})\overline{x}_1 - u_{av}\xi_3$$

$$\alpha_1 \dot{\xi}_3 = u_{av}\xi_2 - \xi_4 + u_{av}$$

$$\alpha_2 \dot{\xi}_4 = \xi_3 - \frac{1}{Q}\xi_4$$

where the variables ξ_2, ξ_3 and ξ_4, acting as the dynamical controller states, have replaced the auxiliary state variables x_{2d}, x_{3d} and x_{4d}, respectively. The signal u_{av} is, therefore, the output of the derived average controller. The only measurement required from the converter system is represented by the normalized average input inductor current x_1. The obtained controller is then truly a dynamic *output* feedback controller.

Simulations

A typical Zeta DC-to-DC power converter circuit is characterized by the following component parameter values: $L_1 = 600$ μH, $C_1 = 10$ μF, $L_2 = 10$ mH, $C_2 = 10$ μF, $R = 40$ Ω and $E = 100$ V.

We desired a steady state output voltage of value, $\overline{v}_2 = 200$ V, with the corresponding equilibrium values for the state variables given by,

$$\overline{i}_1 = 10 \text{ A}, \qquad \overline{v}_1 = 200 \text{ V}, \qquad \overline{i}_2 = 5 \text{ A}$$

and
$$\overline{u}_{av} = 0.666$$

Figure 5.11 shows the simulated average responses of the Zeta converter to the actions of the designed average passivity based feedback controller using the energy shaping plus damping injection method.

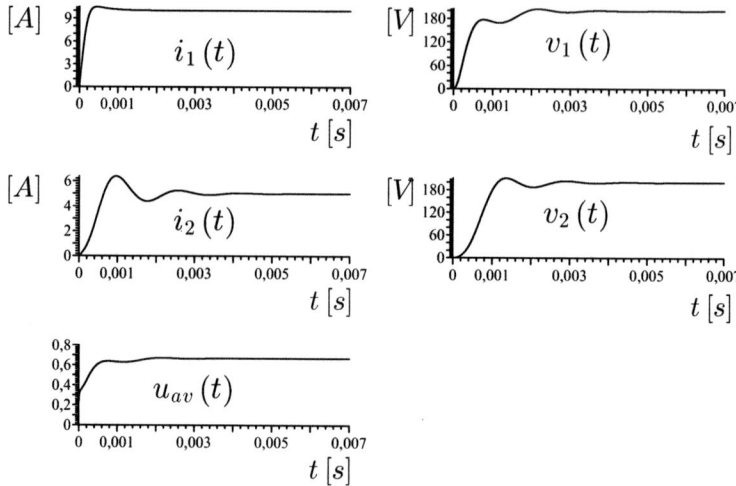

Fig. 5.11. Average responses of Zeta converter from a passivity based controller.

Figure 5.12 depicts the closed loop, switched, response of a Zeta power converter to a $\Sigma-\Delta$-modulator implementation of a passivity based stabilizing controller using the energy shaping plus damping injection design method.

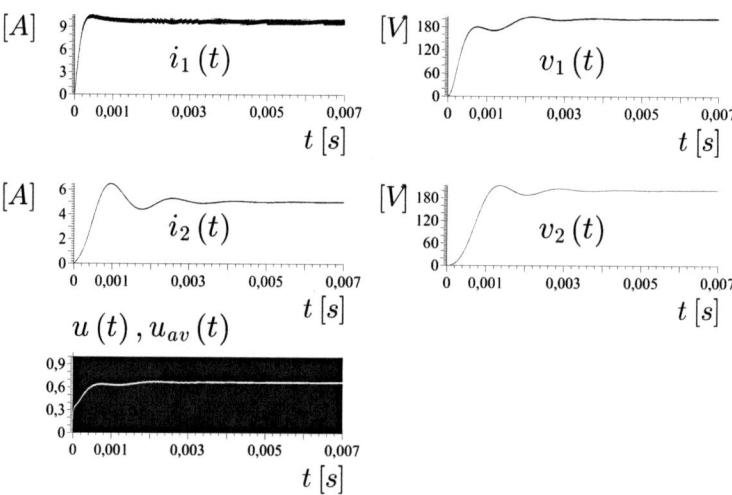

Fig. 5.12. Responses of a switched Zeta converter from a passivity based controller implemented via a $\Sigma - \Delta$ modulator.

5.3.6 The Quadratic Buck Converter

The average normalized model of the quadratic Buck converter is given by,

$$\dot{x}_1 = -x_2 + u_{av}$$
$$\dot{x}_2 = x_1 - u_{av}x_3$$
$$\alpha_1 \dot{x}_3 = u_{av}x_2 - x_4$$
$$\alpha_2 \dot{x}_4 = x_3 - \frac{1}{Q}x_4$$

The equilibrium point parameterized in terms of a constant output voltage $x_4 = \overline{x}_4$ for this converter is given by,

$$\overline{x}_1 = \frac{\sqrt{\overline{x}_4}^3}{Q}, \qquad \overline{x}_2 = \sqrt{\overline{x}_4}, \qquad \overline{x}_3 = \frac{\overline{x}_4}{Q}$$

In the average normalized quadratic Buck model, the output voltage x_4 is a *non-minimum phase output* and the inductor current x_1 is a *minimum phase output*. Regulation of the output voltage can be indirectly achieved by the regulation of the inductor current x_1 towards its equilibrium value.

The average normalized model of the quadratic Buck converter is found to represent a passive map between the average input u_{av} and the inductor current variable x_1. Consider the normalized average total stored energy of the system,

$$H(x) = \frac{1}{2}\left(x_1^2 + x_2^2 + \alpha_1 x_3^2 + \alpha_2 x_4^2\right)$$

The time derivative of $H(x)$ along the controlled motions of the system is given by,

$$\dot{H}(x) = -\frac{1}{Q}x_4^2 + x_1 u_{av} \leq x_1 u_{av}$$

This reveals a passive map between the average control input u_{av} and the current x_1. The integration of the time derivative of H, produce the following passivity inequality,

$$H[x(t)] - H[x(0)] \leq \int_0^t x_1(\sigma) u_{av}(\sigma) d\sigma$$

The auxiliary, *exogenous*, system, is written as a copy of the original system, with injected damping represented by a term of the form: $R_1(x_1 - x_{1d})$, $R_1 > 0$. We have,

$$\dot{x}_{1d} = -x_{2d} + u_{av} + R_1(x_1 - x_{1d})$$
$$\dot{x}_{2d} = x_{1d} - u_{av}x_{3d}$$
$$\alpha_1 \dot{x}_{3d} = u_{av}x_{2d} - x_{4d}$$
$$\alpha_2 \dot{x}_{4d} = x_{3d} - \frac{x_{4d}}{Q}$$

Define the following average normalized error variables, $e_i = x_i - x_{id}$, $i = 1, \ldots, 4$. We then obtain the following tracking error dynamics:

$$\dot{e}_1 = -e_2 - R_1 e_1$$
$$\dot{e}_2 = e_1 - u_{av} e_3$$
$$\alpha_1 \dot{e}_3 = u_{av} e_2 - e_4$$
$$\alpha_2 \dot{e}_4 = e_3 - \frac{1}{Q} e_4$$

Consider now the following Lyapunov function candidate, defined in the trajectory tracking error space described by the coordinates, $e = (e_1, e_2, e_3, e_4)$,

$$H(e) = \frac{1}{2} \left(e_1^2 + e_2^2 + \alpha_1 e_3^2 + \alpha_2 e_4^2 \right)$$

The time derivative of such a positive definitive function, along the controlled trajectories of the tracking error dynamics, yields

$$\dot{H}(e) = -R_1 e_1^2 - \frac{1}{Q} e_4^2 \leq 0$$

The set points in the tracking error space which satisfy $\dot{H}(e) = 0$ are given by the intersection of the hyper-planes, $e_1 = e_4 = 0$. This implies, from the tracking error dynamics, that also $e_3 = 0$ and $e_2 = 0$. According to LaSalle's theorem, the equilibrium point $e_i = 0$, $i = 1, \ldots, 4$ is a globally asymptotically stable equilibrium point for the controlled tracking error dynamics. This means that the average converter system trajectories, $x(t)$, and the auxiliary system trajectories, $x_d(t)$, asymptotically converge towards each other.

We fix a constant reference equilibrium value for the desired average inductor current x_{1d} in the auxiliary system dynamics and define the control input by solving from u_{av} from the first auxiliary dynamics equation. We obtain the following dynamic average feedback controller u_{av} for the system

$$u_{av} = \xi_2 - R_1 (x_1 - \overline{x}_1)$$
$$\dot{\xi}_2 = \overline{x}_1 - u_{av} \xi_3$$
$$\alpha_1 \dot{\xi}_3 = u_{av} \xi_2 - \xi_4$$
$$\alpha_2 \dot{\xi}_4 = \xi_3 - \frac{1}{Q} \xi_4$$

where the variables ξ_2, ξ_3 and ξ_4, representing the dynamical controller states, replace the auxiliary state variables x_{2d}, x_{3d} and x_{4d}, in the auxiliary dynamics model. We have also let $x_{1d} = \overline{x}_1$.

The signal u_{av} is, therefore, the output of the derived average dynamic feedback controller. The only measurement required from the converter system is represented by the normalized average input inductor current x_1. The obtained controller is then truly a dynamic *output* feedback controller.

A $\Sigma - \Delta$-modulator can be used for the switched implementation of the average feedback control law u_{av}.

Simulations

Consider a quadratic Buck converter characterized by the following parameter values:

$$L_1 = 600 \ \mu H, \quad C_1 = 10 \ \mu F, \quad L_2 = 600 \ \mu H, \quad C_2 = 10 \ \mu F,$$
$$R = 40 \ \Omega, \quad E = 100 \ V$$

These parameter values allow the computation of the normalized values:

$$Q = 5.164, \quad \sqrt{L_1 C_1} = 7.746 \times 10^{-5} \ s, \quad \alpha_1 = 1, \quad \alpha_2 = 1$$

We prescribe a desired steady state voltage of value, $\overline{v}_2 = 25$ V with the following corresponding steady state values for the rest of the state variables:

$$\overline{i}_1 = 0.3125 \ A, \quad \overline{v}_1 = 50 \ V, \quad \overline{i}_2 = 0.625 \ A$$

and $\overline{u}_{av} = 0.5$.

Figure 5.13 depicts the average quadratic Buck system response to the actions of an average passivity based controller designed on the basis of energy shaping plus damping injection.

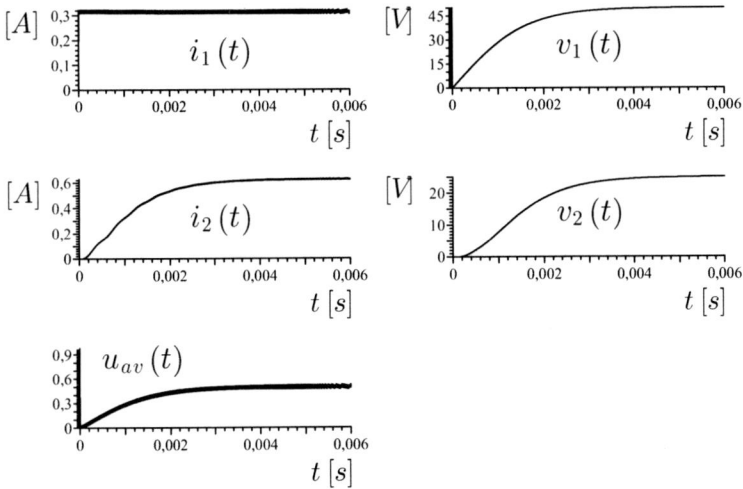

Fig. 5.13. Average feedback controlled responses of a quadratic Buck converter using a passivity based controller of the energy shaping plus damping injection type.

Figure 5.14 shows the closed loop response of a quadratic Buck power converter to a $\Sigma-\Delta$-modulator implementation of a passivity based stabilizing controller. The controller and the system parameters were chosen to be exactly the same as in the previous simulation.

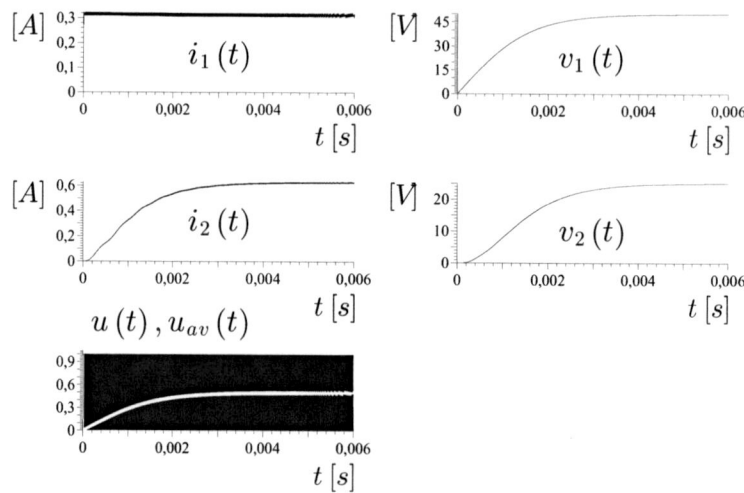

Fig. 5.14. Feedback controlled responses of a switched quadratic Buck converter using a $\Sigma - \Delta$ modulator implementation of a passivity based controller.

5.4 Exact Error Dynamics Passive Output Feedback Control

In this section we explore a rather direct approach to the feedback controller design for the stabilization of a large class of DC-to-DC power converters. The approach is based on first generating an exact dynamic model of the stabilization error of the average system model. Exploiting the energy managing structure of the error dynamics, which can be placed in Generalized Hamiltonian form, and identifying the passive output associated with this stabilization error dynamics, a simple linear, time invariant, feedback controller may be synthesized which renders the desired equilibrium point into a semi-globally asymptotically stable equilibrium for the closed loop system, provided a structural dissipation matching condition is satisfied.

5.4.1 A General Result

Consider the general model of a normalized DC-to-DC power converter, discussed in Chapter 2, written now in Generalized Hamiltonian canonical form (see [58]):

$$\dot{x} = \mathcal{J}(u_{av})\frac{\partial H}{\partial x} - \mathcal{R}\frac{\partial H}{\partial x} + bu_{av} + \mathcal{E} \qquad (5.12)$$

where $H(x)$ is the total stored energy given by the quadratic form $H(x) = \frac{1}{2}x^T x$, from where it is clear that the term $\partial H/\partial x = x$. The matrix $\mathcal{J}(u_{av})$ is skew-symmetric, $\mathcal{R}$ is symmetric and positive semi-definite, the vector b is constant and $\mathcal{E}$ represents non-switched constant external voltage sources.

5.4 Exact Error Dynamics Passive Output Feedback Control

For instance, in the normalized Boost converter, we have

$$\mathcal{J}(u_{av}) = \begin{bmatrix} 0 & -u_{av} \\ u_{av} & 0 \end{bmatrix}, \quad b = 0, \quad \mathcal{E} = \begin{bmatrix} 1 \\ 0 \end{bmatrix}$$

while in the quadratic Buck converter we have,

$$\mathcal{J}(u_{av}) = \begin{bmatrix} 0 & -1 & 0 & 0 \\ 1 & 0 & -u_{av} & 0 \\ 0 & u_{av} & 0 & -1 \\ 0 & 0 & 1 & 0 \end{bmatrix}, \quad b = \begin{bmatrix} 1 \\ 0 \\ 0 \\ 0 \end{bmatrix}, \quad \mathcal{E} = 0$$

It is quite straightforward to check that all the average models of the converters treated so far, conform to the Generalized Hamiltonian form model given above in (5.12).

We explore now fundamental properties of such model. The following assumptions are common to all nonlinear converters (except for the Buck converter, which is a linear system, and the matrix $\mathcal{J}(u_{av}) = \mathcal{J}$ is constant).

• The matrix $\mathcal{J}(u_{av})$, which is skew-symmetric, is *at most* affine in the average control input u_{av}. This means that $\mathcal{J}(u_{av})$ satisfies, for any constant $\bar{u}$, the following exact expansion property:

$$\mathcal{J}(u_{av}) = \mathcal{J}(\bar{u}) + \left.\frac{\partial \mathcal{J}(u_{av})}{\partial u_{av}}\right|_{u_{av}=\bar{u}} (u_{av} - \bar{u})$$

Because $\mathcal{J}(u_{av})$ is affine in u_{av} then the matrix $\partial \mathcal{J}(u_{av})/\partial u_{av}$ is a skew symmetric constant matrix.

• Under equilibrium conditions, the system equations read

$$0 = \mathcal{J}(\bar{u})\left.\frac{\partial H}{\partial x}\right|_{x=\bar{x}} - \mathcal{R}\left.\frac{\partial H}{\partial x}\right|_{x=\bar{x}} + b\bar{u} + \mathcal{E}$$

i.e.,

$$0 = \mathcal{J}(\bar{u})\bar{x} - \mathcal{R}\bar{x} + b\bar{u} + \mathcal{E}$$

where $\bar{x}$ is a constant average state equilibrium corresponding to the constant average control input $\bar{u}$ satisfying $\bar{u} \in [0,1]$.

Define the stabilization errors $e = x - \bar{x}$, $e_u = u_{av} - \bar{u}$. Recall that,

$$e = x - \bar{x} = \frac{\partial H(x)}{\partial x} - \frac{\partial H(\bar{x})}{\partial \bar{x}} = \frac{\partial H(e)}{\partial e}$$

and, clearly $\dot{e} = \dot{x}$.

We have the following proposition:

Proposition 5.1. *The stabilization error dynamics satisfies, without any approximations, the following dynamics:*

$$\dot{e} = \mathcal{J}(u_{av})\frac{\partial H(e)}{\partial e} - \mathcal{R}\frac{\partial H(e)}{\partial e} + be_u + \left[\frac{\partial \mathcal{J}(u_{av})}{\partial u_{av}}\left.\frac{\partial H}{\partial x}\right|_{x=\bar{x}}\right]e_u$$

or, in simpler terms

$$\dot{e} = \mathcal{J}(u_{av})e - \mathcal{R}e + \left[b + \frac{\partial \mathcal{J}(u_{av})}{\partial u_{av}}\bar{x}\right]e_u$$

The proof proceeds by direct computation, adding and subtracting the required equilibrium related quantities,

$$\dot{e} = \mathcal{J}(u_{av})\frac{\partial H(e)}{\partial e} - \mathcal{R}\frac{\partial H(e)}{\partial e} + be_u + \mathcal{E}$$
$$+ \mathcal{J}(u_{av})\left.\frac{\partial H}{\partial x}\right|_{x=\bar{x}} - \mathcal{R}\left.\frac{\partial H}{\partial x}\right|_{x=\bar{x}} + b\bar{u}$$

Using the equilibrium relations:

$$0 = \mathcal{E} + \mathcal{J}(\bar{u})\left.\frac{\partial H}{\partial x}\right|_{x=\bar{x}} - \mathcal{R}\left.\frac{\partial H}{\partial x}\right|_{x=\bar{x}} + b\bar{u}$$

we have that the error dynamics satisfies:

$$\dot{e} = \mathcal{J}(u_{av})\frac{\partial H(e)}{\partial e} - \mathcal{R}\frac{\partial H(e)}{\partial e} + be_u + [\mathcal{J}(u_{av}) - \mathcal{J}(\bar{u})]\left.\frac{\partial H}{\partial x}\right|_{x=\bar{x}}$$

where use of the exact expansion property of $\mathcal{J}(u_{av})$ around $\bar{u}$, we have:

$$\dot{e} = \mathcal{J}(u_{av})\frac{\partial H(e)}{\partial e} - \mathcal{R}\frac{\partial H(e)}{\partial e} + be_u + \frac{\partial \mathcal{J}(u_{av})}{\partial u_{av}}\left[\left.\frac{\partial H}{\partial x}\right|_{x=\bar{x}}\right]e_u$$

i.e.,

$$\dot{e} = \mathcal{J}(u_{av})e - \mathcal{R}e + be_u + \frac{\partial \mathcal{J}(u_{av})}{\partial u_{av}}\bar{x}e_u$$

which we rewrite as

$$\dot{e} = \mathcal{J}(u_{av})e - \mathcal{R}e + \left[b + \frac{\partial \mathcal{J}(u_{av})}{\partial u_{av}}\bar{x}\right]e_u$$

The crucial observations on this exact stabilization error dynamics are that:
- The term $\mathcal{J}(u_{av})e = \mathcal{J}(u_{av})\frac{\partial H(e)}{\partial e}$ is the only nonlinear term in the derived dynamics. This term happens to be conservative, i.e., for any u_{av}

$$e^T \mathcal{J}(u_{av})e = \frac{\partial H(e)}{\partial e^T}\mathcal{J}(u_{av})\frac{\partial H(e)}{\partial e} = 0, \quad \forall e$$

The conservative term, as expected, has no contribution in the stability properties of the closed loop system from the *incremental input* $e_u = u - \bar{u}$

5.4 Exact Error Dynamics Passive Output Feedback Control

- The term $-\mathcal{R}e + be_u + \frac{\partial \mathcal{J}(u_{av})}{\partial u_{av}}\overline{x}e_u$, representing the rest of the error dynamics, exactly coincides with the tangent linearization part of the dynamics which is independent of the matrix $\mathcal{J}(u_{av})$. In other words, note that the tangent linearization of the nonlinear dynamics

$$\dot{x} = \mathcal{J}(u_{av})\frac{\partial H}{\partial x} - \mathcal{R}\frac{\partial H}{\partial x} + bu_{av} + \mathcal{E}$$

around the equilibrium point $x = \overline{x}$, $u_{av} = \overline{u}$, given by:

$$\dot{x}_\delta = \mathcal{J}(\overline{u})x_\delta - \mathcal{R}x_\delta + bu_\delta + \frac{\mathcal{J}(u_{av})}{\partial u_{av}}\overline{x}u_\delta$$

exhibits, exactly, the same three last terms in the right hand side as the derived exact stabilization error dynamics. For this, of course, we agree in the validity of the equivalence of $x_\delta = x - \overline{x}$ with e and $u_\delta = u_{av} - \overline{u}$ with e_u.

We have the following theorem

Theorem 5.2. *A linear incremental feedback controller, deduced on the basis of the stabilization to zero of the tangent linearization average model of the converter around a desired equilibrium point, also stabilizes the nonlinear system to the desired equilibrium from any permissible initial condition. In other words, the linearized feedback control law, obtained from the tangent linearized model, makes the equilibrium point of the nonlinear converter semi-globally asymptotically stable.*

The proof, which is based on the previous developments, follows now quite easily. Indeed, Let the average linear incremental feedback control law

$$e_u = u_\delta = -k^T e = -k^T x_\delta$$

locally stabilize the nonlinear system thanks to the appropriate pole placement of the tangent linearization average model dynamics. Let k^T then be a row vector of gains feeding back the stabilization errors of the state. The closed loop system error dynamics is given by

$$\dot{e} = \mathcal{J}(u_{av})e - \mathcal{R}e - \left[b + \frac{\partial \mathcal{J}(u_{av})}{\partial u_{av}}\overline{x}\right]k^T e$$

i.e.,

$$\dot{e} = \mathcal{J}(u_{av})e - \left[\mathcal{R} + \left(b + \frac{\partial \mathcal{J}(u_{av})}{\partial u_{av}}\overline{x}\right)k^T\right]e$$

Let for simplicity,

$$\mathcal{M} = \left[\mathcal{R} + \left(b + \frac{\partial \mathcal{J}(u_{av})}{\partial u_{av}}\overline{x}\right)k^T\right]$$

Clearly, the matrix $\mathcal{M}$ has all its eigenvalues in the right portion of the complex plane.

Note that $\mathcal{M}$ is not symmetric or skew-symmetric, but it can, nevertheless, be written as:
$$\mathcal{M} = \mathcal{J}_M + \mathcal{R}_M$$
where $\mathcal{J}_M$ is skew-symmetric and $\mathcal{R}_M$ is symmetric and positive definite (i.e., $-\mathcal{R}_M$ is negative definite). Indeed,
$$\mathcal{M} = \frac{1}{2}\left[\mathcal{M} - \mathcal{M}^T\right] + \frac{1}{2}\left[\mathcal{M} + \mathcal{M}^T\right]$$

The closed loop system is then of the form
$$\dot{e} = \left[\mathcal{J}(u_{av}) - \mathcal{J}_M\right]e - \left[\mathcal{R} + \mathcal{R}_M\right]e$$

The semi-global stability of the closed loop system is obvious from the skew-symmetry of the matrix $\mathcal{J}(u_{av}) - \mathcal{J}_M$ for any u_{av}, and the positive definite nature of the symmetric matrix $\mathcal{R} + \mathcal{R}_M$.

This theorem has evident implications in the stability of nonlinear average converters using feedback of the passive incremental output. This simple linear feedback also semi-globally stabilizes the nonlinear average converter models.

5.4.2 The Boost Converter

Consider the normalized average model of the Boost DC-to-DC power converter
$$\dot{x}_1 = -u_{av}x_2 + 1$$
$$\dot{x}_2 = u_{av}x_1 - \frac{1}{Q}x_2$$
$$y = x_2$$

It is desired to regulate the system trajectories towards a constant average state equilibrium point characterized, in terms of the desired output equilibrium voltage $\bar{x}_2 = V_d$, by
$$\bar{x}_1 = \frac{V_d^2}{Q}, \qquad \bar{x}_2 = V_d, \qquad \bar{u}_{av} = \frac{1}{V_d}$$

A translation of the state coordinates to the stabilization error space $e_1 = x_1 - V_d^2/Q$, $e_2 = x_2 - V_d$ yields,
$$\dot{e}_1 = -u_{av}e_2 + 1 - u_{av}V_d$$
$$\dot{e}_2 = u_{av}e_1 - \frac{1}{Q}e_2 + u_{av}\frac{V_d^2}{Q} - \frac{1}{Q}V_d$$

which written in matrix form yields,

5.4 Exact Error Dynamics Passive Output Feedback Control

$$\dot{e} = \begin{bmatrix} 0 & -u_{av} \\ u_{av} & 0 \end{bmatrix} e - \begin{bmatrix} 0 & 0 \\ 0 & \frac{1}{Q} \end{bmatrix} e + \begin{bmatrix} -V_d \\ \frac{V_d^2}{Q} \end{bmatrix} u_{av} + \begin{bmatrix} 1 \\ -\frac{V_d}{Q} \end{bmatrix}$$

Consider the following error energy function candidate,

$$H(e) = \frac{1}{2}\left[\left(x_1 - \frac{V_d}{Q}\right)^2 + (x_2 - V_d)^2\right] = \frac{1}{2}[e_1^2 + e_2^2]$$

Since $\partial H(e)/\partial e = e$ we may write the average stabilization error system in classical Hamiltonian form:

$$\dot{e} = \begin{bmatrix} 0 & -u_{av} \\ u_{av} & 0 \end{bmatrix} \frac{\partial H(e)}{\partial e} - \begin{bmatrix} 0 & 0 \\ 0 & \frac{1}{Q} \end{bmatrix} \frac{\partial H(e)}{\partial e} + \begin{bmatrix} -V_d \\ \frac{V_d^2}{Q} \end{bmatrix} u_{av} + \begin{bmatrix} 1 \\ -\frac{V_d}{Q} \end{bmatrix}$$

which we conveniently rewrite as

$$\dot{e} = \begin{bmatrix} 0 & -u_{av} \\ u_{av} & 0 \end{bmatrix} \frac{\partial H(e)}{\partial e} - \begin{bmatrix} 0 & 0 \\ 0 & \frac{1}{Q} \end{bmatrix} \frac{\partial H(e)}{\partial e} + \begin{bmatrix} -V_d \\ \frac{V_d^2}{Q} \end{bmatrix} e_{u_{av}}$$

where

$$e_{u_{av}} = u_{av} - \frac{1}{V_d}$$

The error dynamics is then of the form

$$\dot{e} = \mathcal{J}\frac{\partial H(e)}{\partial e} - \mathcal{R}\frac{\partial H(e)}{\partial e} + be_{u_{av}}$$

The passive output corresponding to this Hamiltonian stabilization error representation is given by

$$e_y = -V_d e_1 + \frac{V_d^2}{Q} e_2$$

The dissipation matching condition is evidently satisfied since

$$\mathcal{R} + \gamma bb^T = \begin{bmatrix} 0 & 0 \\ 0 & \frac{1}{Q} \end{bmatrix} + \gamma \begin{bmatrix} V_d^2 & -\frac{V_d^2}{Q} \\ -\frac{V_d^3}{Q} & \frac{V_d^4}{Q^2} \end{bmatrix} = \begin{bmatrix} \gamma V_d^2 & -\gamma \frac{V_d^2}{Q} \\ -\gamma \frac{V_d^2}{Q} & \frac{1}{Q} + \gamma \frac{V_d^4}{Q^2} \end{bmatrix} > 0$$

A feedback controller which makes the equilibrium point globally asymptotically stable is just

$$e_{u_{av}} = -\gamma e_y = -\gamma\left[-V_d e_1 + \frac{V_d^2}{Q} e_2\right] = -\gamma\left[-V_d\left(x_1 - \frac{V_d^2}{Q}\right) + \frac{V_d^2}{Q}(x_2 - V_d)\right]$$

The average stabilizing feedback control, based on passive output feedback, is then given by

$$u_{av} = \frac{1}{V_d} - \gamma\left[-V_d\left(x_1 - \frac{V_d^2}{Q}\right) + \frac{V_d^2}{Q}(x_2 - V_d)\right]$$

this expression can be simplified to the form:

$$u_{av} = \frac{1}{V_d} + \gamma\left(x_1 - \frac{V_d}{Q}x_2\right) V_d \tag{5.13}$$

288   5 Nonlinear Methods

Simulations

Figure 5.15 depicts the normalized switched controlled response of the Boost converter with the average control input designed on the basis of static exact stabilization error dynamics passive output feedback.

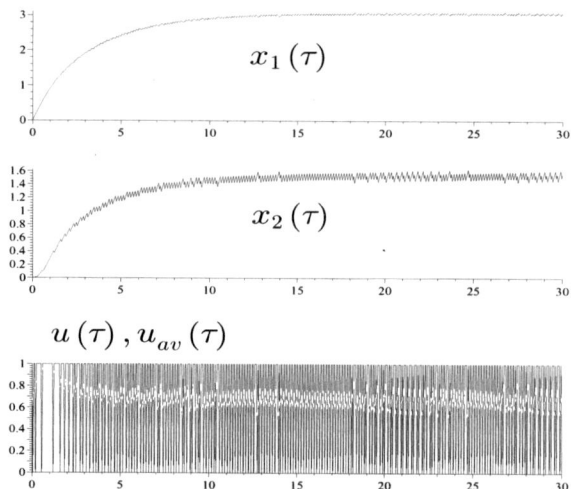

Fig. 5.15. Switch controlled responses of Boost converter to linear static passive feedback.

5.4.3 Experimental Implementation

In order to evaluate the validity of the proposed static linear passivity based control, which was determined in normalized form by Equation 5.13, this controller is implemented and tested on the experimental *Boost system* developed in Chapter 2 with the help of a $\Sigma - \Delta$-*modulator circuit* already described also in Chapter 3.

In Figure 5.16 we show the functional block diagram of the Boost converter. It illustrates all the components of the system that we built with its respective control block already inserted. A $\Sigma - \Delta$ modulator with a limiting circuit is used in connection with the static linear passivity based average feedback control law implementation.

Control Block

In this block the average control static passivity based control strategy is implemented for the Boost converter. The inductor current and the output voltage signals (i and v, respectively) are received from the *Boost system* block.

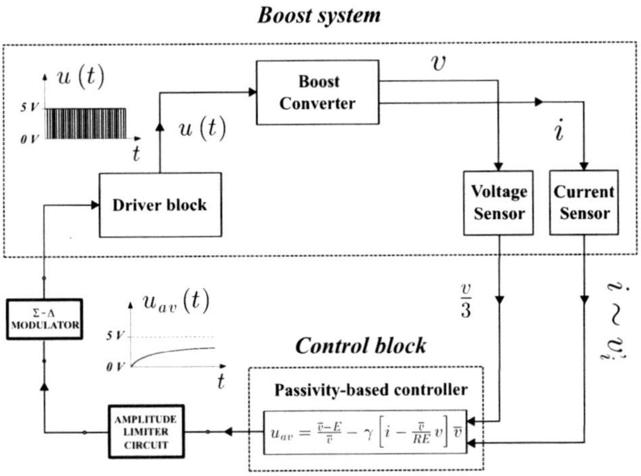

5.4 Exact Error Dynamics Passive Output Feedback Control

Fig. 5.16. Functional block diagram of the Boost power converter to a $\Sigma - \Delta$-modulator implementation of a passivity based stabilizing controller.

The average feedback control strategy (5.13) is implemented using analog electronics by noticing that, using

$$x_1 = \frac{1}{E}\sqrt{\frac{L}{C}}i, \quad x_2 = \frac{v}{E}, \quad Q = R\sqrt{\frac{C}{L}}$$

we can rewrite (5.13) in non-normalized form as:

$$u_{av} = \frac{E}{\overline{v}} + \gamma_{actual}\left[i - \frac{\overline{v}}{RE}v\right]\overline{v} \qquad (5.14)$$

for the alternative model of the Boost converter (see Figure 3.1), or

$$u_{av} = \frac{\overline{v} - E}{\overline{v}} - \gamma_{actual}\left[i - \frac{\overline{v}}{RE}v\right]\overline{v} \qquad (5.15)$$

for the model given by Equation 2.15 (see Figure 2.11).

Figure 5.17 shows the actual *control* circuit block. It also shows the transfer functions that realize the op-amps for achieving the actual implementation of the designed linear static passivity based controller, expressed in non-normalized form by (5.15).

Experimental Results

Figure 5.18 depicts the experimental results portraying the closed loop controlled responses of the *Boost system* when the average linear static passivity based stabilizing controller is implemented through the designed $\Sigma - \Delta$ modulator. The controller and the system parameters were chosen to be:

290   5 Nonlinear Methods

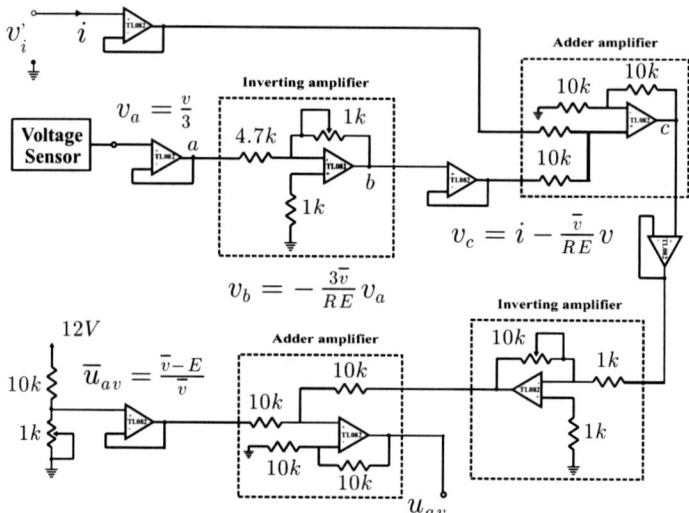

Fig. 5.17. Control circuit structure implemented for the passivity based stabilizing controller.

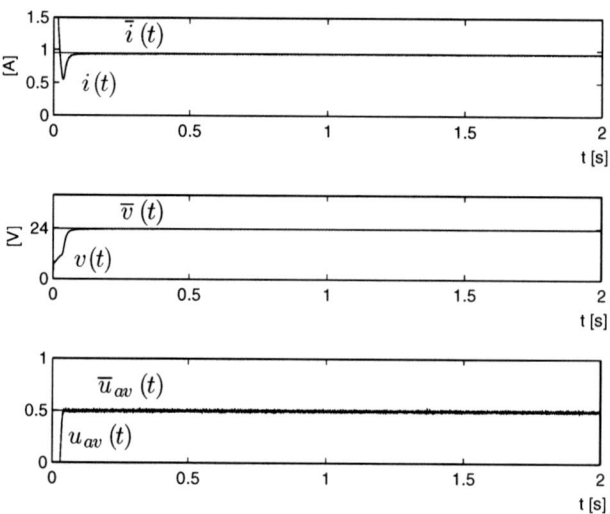

Fig. 5.18. Experimental closed loop response of the Boost DC-to-DC power converter to a $\Sigma - \Delta$-modulator implementation of a static linear passivity based stabilizing controller.

$$L = 15.91 \text{ mH}, \quad C = 50 \text{ μF}, \quad R = 52 \text{ Ω}, \quad E = 12 \text{ V}$$

with $\gamma_{actual} = 0.1$. We take the steady state output voltage value $\bar{v} = 24$ V as the desired output voltage. The corresponding steady state current is given by $\bar{i} = 0.923$ A, and $\bar{u}_{av} = 0.5$.

5.4.4 The Buck-Boost Converter

Consider the average normalized model of the Buck-Boost DC-to-DC power converter

$$\dot{x}_1 = u_{av}x_2 + (1 - u_{av})$$
$$\dot{x}_2 = -u_{av}x_1 - \frac{1}{Q}x_2$$
$$y = x_2$$

It is desired to regulate the system trajectories towards a constant average state equilibrium point characterized, in terms of the desired output equilibrium voltage $\bar{x}_2 = V_d$, by

$$\bar{x}_1 = -\frac{V_d(1 - V_d)}{Q}, \quad \bar{x}_2 = V_d, \quad \bar{u}_{av} = \frac{1}{1 - V_d}$$

A translation of the state coordinates to the stabilization error space $e_1 = x_1 + V_d(1 - V_d)/Q$, $e_2 = x_2 - V_d$ yields,

$$\dot{e}_1 = u_{av}e_2 - (1 - V_d)\left(u_{av} - \frac{1}{1 - V_d}\right)$$
$$\dot{e}_2 = -u_{av}e_1 - \frac{1}{Q}e_2 + \frac{V_d(1 - V_d)}{Q}\left(u_{av} - \frac{1}{1 - V_d}\right)$$

which, written in matrix form, yields

$$\dot{e} = \begin{bmatrix} 0 & u_{av} \\ -u_{av} & 0 \end{bmatrix} e - \begin{bmatrix} 0 & 0 \\ 0 & \frac{1}{Q} \end{bmatrix} e + \begin{bmatrix} 1 - V_d \\ \frac{V_d(1-V_d)}{Q} \end{bmatrix}\left(u_{av} - \frac{1}{1 - V_d}\right)$$

The error energy function candidate,

$$H(e) = \frac{1}{2}\left[\left(x_1 - \frac{V_d}{Q}\right)^2 + (x_2 - V_d)^2\right] = \frac{1}{2}[e_1^2 + e_2^2]$$

allows, by virtue of the fact that $\partial H(e)/\partial e = e$, to write the system in classical Hamiltonian form:

$$\dot{e} = \begin{bmatrix} 0 & u_{av} \\ -u_{av} & 0 \end{bmatrix}\frac{\partial H(e)}{\partial e} - \begin{bmatrix} 0 & 0 \\ 0 & \frac{1}{Q} \end{bmatrix}\frac{\partial H(e)}{\partial e} + \begin{bmatrix} -(1 - V_d) \\ \frac{V_d(1-V_d)}{Q} \end{bmatrix} e_{u_{av}}$$

where

5 Nonlinear Methods

$$e_{u_{av}} = u_{av} - \frac{1}{1-V_d}$$

The passive output, corresponding to this exact Hamiltonian stabilization error dynamics representation, is given by

$$e_y = -(1-V_d)e_1 + \frac{V_d(1-V_d)}{Q}e_2$$

The dissipation matching condition is evidently satisfied since

$$\mathcal{R} + \gamma bb^T = \begin{bmatrix} 0 & 0 \\ 0 & \frac{1}{Q} \end{bmatrix} + \gamma \begin{bmatrix} (1-V_d)^2 & -\frac{V_d(1-V_d)^2}{Q} \\ -\frac{V_d(1-V_d)^2}{Q} & \frac{V_d^2(1-V_d)^2}{Q^2} \end{bmatrix}$$

$$= \begin{bmatrix} \gamma(1-V_d)^2 & -\gamma\frac{V_d(1-V_d)^2}{Q} \\ -\gamma\frac{V_d(1-V_d)^2}{Q} & \frac{1}{Q}+\gamma\frac{V_d^2(1-V_d)^2}{Q^2} \end{bmatrix} > 0$$

A feedback controller which makes the equilibrium point globally asymptotically stable is just,

$$e_{u_{av}} = -\gamma e_y = -\gamma\left[-(1-V_d)e_1 + \frac{V_d(1-V_d)}{Q}e_2\right]$$

$$= \gamma\left[(1-V_d)\left(x_1 + \frac{V_d(1-V_d)}{Q}\right) - \frac{V_d(1-V_d)}{Q}(x_2 - V_d)\right]$$

The average stabilizing feedback control, based on passive output feedback, is then given by

$$u_{av} = \frac{1}{1-V_d} + \gamma\left[(1-V_d)\left(x_1 + \frac{V_d(1-V_d)}{Q}\right) - \frac{V_d(1-V_d)}{Q}(x_2 - V_d)\right]$$

This is, precisely, the linear feedback controller obtained by tangent linearization and feedback of the linearized passive output given by the Equation 4.23.

Figure 4.46 depicts the response of the nonlinear system to the proposed average static passivity based controller implemented through a $\Sigma - \Delta$ modulator.

Note that the error vector coordinates have the origin as a semi-global asymptotically stable equilibrium with the static incremental linear passivity based control just proposed.

Indeed, the time derivative of the stabilization error energy, along the closed loop trajectories of the error system is given by,

$$\dot{H}(e) = \frac{\partial H(e)}{\partial e^T}\mathcal{J}(u)\frac{\partial H(e)}{\partial e} - \frac{\partial H(e)}{\partial e^T}\left[\mathcal{R}+\gamma bb^T\right]\frac{\partial H(e)}{\partial e}$$

$$= 0 - \frac{\partial H(e)}{\partial e^T}\left[\mathcal{R}+\gamma bb^T\right]\frac{\partial H(e)}{\partial e} < 0$$

The stability of the closed loop system does not depend on the nonlinear part of the system, which is conservative. Only the linear part is relevant.

5.4.5 The Cúk Converter

Consider the average normalized Cúk converter model

$$\dot{x}_1 = -(1 - u_{av})x_2 + 1$$
$$\dot{x}_2 = (1 - u_{av})x_1 + u_{av}x_3$$
$$\alpha_1 \dot{x}_3 = -u_{av}x_2 - x_4$$
$$\alpha_2 \dot{x}_4 = x_3 - \frac{1}{Q}x_4$$

Let us express the system in state stabilization error coordinates with respect to the average equilibrium point corresponding to a desired output voltage, $\bar{x}_4 = V_d$. We have,

$$e_1 = x_1 - \frac{V_d^2}{Q}, \quad e_2 = x_2 - (1 - V_d), \quad e_3 = x_3 - \frac{V_d}{Q}, \quad e_4 = x_4 - V_d$$

The average input error $e_{u_{av}}$ is defined as

$$e_{u_{av}} = u_{av} + \frac{V_d}{1 - V_d}$$

We obtain the following expression for the transformed system

$$\dot{e}_1 = -(1 - u_{av})e_2 + (1 - V_d)\left(u_{av} + \frac{V_d}{1 - V_d}\right)$$
$$\dot{e}_2 = (1 - u_{av})e_1 + u_{av}\,e_3 + \frac{V_d(1 - V_d)}{Q}\left(u_{av} + \frac{V_d}{1 - V_d}\right)$$
$$\alpha_1 \dot{e}_3 = -u_{av}e_2 - e_4 - (1 - V_d)\left(u_{av} + \frac{V_d}{1 - V_d}\right)$$
$$\alpha_2 \dot{e}_4 = e_3 - \frac{1}{Q}e_4$$

Taking the average error energy function $H(e)$ as

$$H(e) = \frac{1}{2}\left[e_1^2 + e_2^2 + e_3^2 + e_4^2\right]$$

we have that $\partial H(e)/\partial e = [e_1,\ e_2,\ e_3,\ e_4]^T$.

Writing the stabilization error in Generalized Hamiltonian form we obtain

$$\mathcal{A}\dot{e} = \begin{bmatrix} 0 & -(1-u_{av}) & 0 & 0 \\ (1-u_{av}) & 0 & u_{av} & 0 \\ 0 & -u_{av} & 0 & -1 \\ 0 & 0 & 1 & 0 \end{bmatrix}\frac{\partial H}{\partial e} - \begin{bmatrix} 0 & 0 & 0 & 0 \\ 0 & 0 & 0 & 0 \\ 0 & 0 & 0 & 0 \\ 0 & 0 & 0 & \frac{1}{Q} \end{bmatrix}\frac{\partial H}{\partial e} + \begin{bmatrix} (1-V_d) \\ \frac{V_d(1-V_d)}{Q} \\ -(1-V_d) \\ 0 \end{bmatrix} e_{u_{av}}$$

where $\mathcal{A} = \mathrm{diag}\,[1, 1, \alpha_1, \alpha_2]$.

The passive output of the stabilization error dynamics is given by

$$e_y = y - V_d = b^T \frac{\partial H(e)}{\partial e} = (1 - V_d)e_1 + \frac{V_d(1-V_d)}{Q}e_2 - (1-V_d)e_3$$

The dissipation matching condition is not strictly satisfied and reads

$$\mathcal{R} + \gamma bb^T = \begin{bmatrix} \gamma(1-V_d)^2 & \gamma\frac{V_d(1-V_d)^2}{Q} & -\gamma(1-V_d)^2 & 0 \\ \gamma\frac{V_d(1-V_d)^2}{Q} & \gamma\frac{V_d^2(1-V_d)^2}{Q^2} & -\gamma\frac{V_d(1-V_d)^2}{Q} & 0 \\ -\gamma(1-V_d)^2 & -\gamma\frac{V_d(1-V_d)^2}{Q} & \gamma(1-V_d)^2 & 0 \\ 0 & 0 & 0 & \frac{1}{Q} \end{bmatrix} \geq 0$$

Again, the nonlinear system is stabilized by a feedback control law using the passive output of the exact stabilization error dynamics with an average equilibrium input feed-forward

$$u_{av} = -\frac{V_d}{1-V_d} - \gamma e_y \qquad (5.16)$$

This control law renders the origin of the error space as an asymptotically stable equilibrium point by virtue of LaSalle's theorem.

It is important to remark that (5.16) is, precisely, the linear feedback controller obtained by tangent linearization and feedback of the linearized passive output given by Equation 4.25.

Figure 5.19 depicts the response of the switched Cúk converter model, to the passivity based average feedback controller implemented through a $\Sigma - \Delta$ modulator. The system parameters were chosen to be exactly the same as in the Section 4.6.2, with $\gamma = 1$.

5.4.6 The Sepic Converter

Consider the normalized average model of the Sepic DC-to-DC power converter

$$\dot{x}_1 = -(1 - u_{av})(x_2 + x_4) + 1$$
$$\dot{x}_2 = (1 - u_{av})x_1 - u_{av}x_3$$
$$\alpha_1\dot{x}_3 = u_{av}x_2 - (1 - u_{av})x_4$$
$$\alpha_2\dot{x}_4 = (1 - u_{av})(x_1 + x_3) - \frac{1}{Q}x_4$$

It is desired to regulate the system trajectories towards a constant average state equilibrium point characterized, in terms of the desired output equilibrium voltage, $\bar{x}_4 = V_d$, by

$$\bar{x}_1 = \frac{V_d^2}{Q}, \quad \bar{x}_2 = 1, \quad \bar{x}_3 = \frac{V_d}{Q}, \quad \bar{u}_{av} = \frac{V_d}{1+V_d}$$

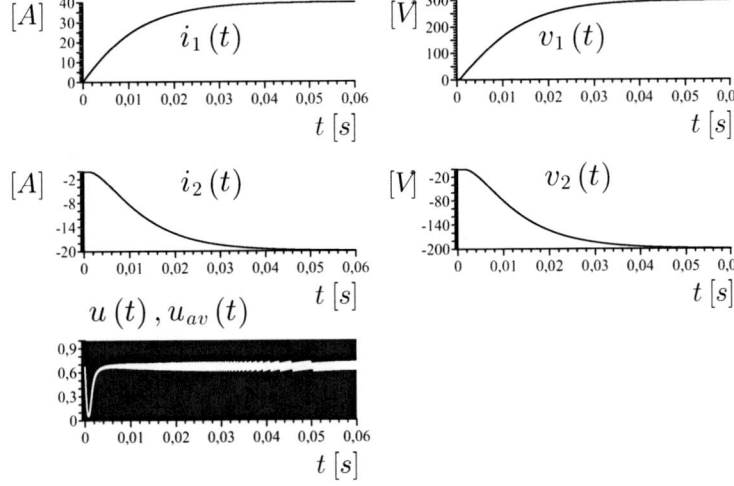

Fig. 5.19. Responses of the switched Cúk converter based on passive output feedback.

Define

$$e_1 = x_1 - \frac{V_d^2}{Q}, \quad e_2 = x_2 - 1, \quad e_3 = x_3 - \frac{V_d}{Q}, \quad e_4 = x_4 - V_d$$

and define also the average input error $e_{u_{av}}$, as:

$$e_{u_{av}} = u_{av} - \frac{V_d}{1 + V_d}$$

We transform the average normalized system into state stabilization error coordinates with respect to the average equilibrium point corresponding to the prescribed output reference voltage, $\bar{x}_4 = V_d$. We obtain the following expression for the transformed system:

$$\dot{e}_1 = -(1 - u_{av})(e_2 + e_4) + (1 + V_d)\left(u_{av} - \frac{V_d}{1 + V_d}\right)$$

$$\dot{e}_2 = (1 - u_{av})e_1 - ue_3 - \frac{V_d(1 + V_d)}{Q}\left(u_{av} - \frac{V_d}{1 + V_d}\right)$$

$$\alpha_1 \dot{e}_3 = u_{av}e_2 - (1 - u_{av})e_4 + (1 + V_d)\left(u_{av} - \frac{V_d}{1 + V_d}\right)$$

$$\alpha_2 \dot{e}_4 = (1 - u_{av})(e_1 + e_3) - \frac{1}{Q}e_4 - \frac{V_d(1 + V_d)}{Q}\left(u_{av} - \frac{V_d}{1 + V_d}\right)$$

Consider the average stabilization error energy function $H(e)$ as

296   5 Nonlinear Methods

$$H(e) = \frac{1}{2}\left[\left(x_1 - \frac{V_d^2}{Q}\right)^2 + (x_2 - 1)^2 + \alpha_1\left(x_3 - \frac{V_d}{Q}\right)^2 + \alpha_2(x_4 - V_d)^2\right]$$

$$= \frac{1}{2}e^T \mathcal{A} e = \frac{1}{2}\left[e_1^2 + e_2^2 + \alpha_1 e_3^2 + \alpha_2 e_4^2\right]$$

where

$$\mathcal{A} = \mathcal{A}^T = \mathrm{diag}\,(1, 1, \alpha_1, \alpha_2), \qquad e = [e_1, e_2, e_3, e_4]^T$$

The exact stabilization error dynamics can be placed in Generalized Hamiltonian form. We obtain,

$$\begin{bmatrix}\dot{e}_1\\ \dot{e}_2\\ \dot{e}_3\\ \dot{e}_4\end{bmatrix} = \begin{bmatrix} 0 & -(1-u_{av}) & 0 & -\frac{1}{\alpha_2}(1-u_{av})\\ (1-u_{av}) & 0 & -\frac{1}{\alpha_1}u_{av} & 0\\ 0 & \frac{1}{\alpha_1}u_{av} & 0 & -\frac{1}{\alpha_1\alpha_2}(1-u_{av})\\ \frac{1}{\alpha_2}(1-u_{av}) & 0 & \frac{1}{\alpha_1\alpha_2}(1-u_{av}) & 0 \end{bmatrix}\frac{\partial H}{\partial e}$$

$$-\begin{bmatrix}0 & 0 & 0 & 0\\ 0 & 0 & 0 & 0\\ 0 & 0 & 0 & 0\\ 0 & 0 & 0 & \frac{1}{\alpha_2^2}\frac{1}{Q}\end{bmatrix}\frac{\partial H}{\partial e} + \begin{bmatrix}(1+V_d)\\ -\frac{V_d(1+V_d)}{Q}\\ \frac{1}{\alpha_1}(1+V_d)\\ -\frac{1}{\alpha_2}\frac{V_d(1+V_d)}{Q}\end{bmatrix}e_{u_{av}}$$

The passive output associated with the exact stabilization error dynamics is given by

$$e_y = b^T \frac{\partial H(e)}{\partial e} = e_y = (1+V_d)\left[e_1 - \frac{V_d}{Q}e_2 + e_3 - \frac{V_d}{Q}e_4\right]$$

We may clearly choose the average feedback control input as an output feedback control law of the form:

$$e_{u_{av}} = -\gamma b^T \frac{\partial H(e)}{\partial e} = -\gamma e_y$$

where γ is a positive scalar quantity.

The total time derivative of the energy function $H(e) > 0$ is given by

$$\dot{H}(e) = -\frac{\partial H(e)}{\partial e^T}\left[\mathcal{R} + \gamma bb^T\right]\frac{\partial H(e)}{\partial e} = -e^T \mathcal{A}\left[\mathcal{R} + \gamma bb^T\right]\mathcal{A} e$$

and the dissipation matching condition is not strictly satisfied and takes the form

$$\mathcal{A}\left[\mathcal{R} + \gamma bb^T\right]\mathcal{A} = \gamma(1+V_d)^2\begin{bmatrix}1 & -\frac{V_d}{Q} & 1 & -\frac{V_d}{Q}\\ -\frac{V_d}{Q} & \frac{V_d^2}{Q^2} & -\frac{V_d}{Q} & \frac{V_d^2}{Q^2}\\ 1 & -\frac{V_d}{Q} & 1 & -\frac{V_d}{Q}\\ -\frac{V_d}{Q} & \frac{V_d^2}{Q^2} & -\frac{V_d}{Q} & \frac{1}{\gamma(1+V_d)^2 Q} + \frac{V_d^2}{Q^2}\end{bmatrix} \geq 0$$

The set of vectors which are in the null space of the preceding matrix, are of the form: $z = \begin{bmatrix}e_1 & e_2 & e_3 & 0\end{bmatrix}$ such that $\xi_\delta = e_1 - \frac{V_d}{Q}e_2 + e_3 - \frac{V_d}{Q}e_4 = 0$, i.e., they

5.4 Exact Error Dynamics Passive Output Feedback Control

lay in a subspace of R^4 and correspond to $e_y = (1 + V_d) \xi_\delta$. This means the nonlinear system is controlled by the equilibrium input $\overline{u}_{av} = V_d/(1+V_d)$, i.e., the error system is controlled by $e_{u_{av}} = 0$. The only trajectory of the error system with $e_4 = 0$ and $e_{u_{av}} = 0$ corresponds to the origin.

This is compatible with the fact that in order for the closed loop incremental average system to have the origin as an asymptotically stable equilibrium, the trajectories of the system should have no other equilibrium than the origin itself. The origin of the average output feedback controlled system is, hence, an asymptotically stable equilibrium.

The output feedback control law, with design parameter $\gamma > 0$, is given by

$$e_{u_{av}} = -\gamma e_y = -\gamma (1 + V_d) \left[e_1 - \frac{V_d}{Q} e_2 + e_3 - \frac{V_d}{Q} e_4 \right]$$

The nonlinear average normalized Sepic converter model is stabilized by the feedback of the passive output associated with the exact stabilization error dynamics, complemented with average equilibrium input feed-forward.

The average feedback control law to be implemented is synthesized as

$$u_{av} = \frac{V_d}{1 + V_d} - \gamma(1 + V_d) \left[\left(x_1 - \frac{V_d^2}{Q} \right) - \frac{V_d}{Q} (x_2 - 1) \right.$$
$$\left. + \left(x_3 - \frac{V_d}{Q} \right) - \frac{V_d}{Q} (x_4 - V_d) \right]$$

This control law renders the origin of the error space as a semi-global asymptotically stable equilibrium point by virtue of LaSalle's theorem.

The above expression can be rewritten, in a simpler form, as:

$$u_{av} = \frac{V_d}{1 + V_d} - \gamma(1 + V_d) \left[x_1 + x_3 - \frac{V_d}{Q} (x_2 + x_4) \right] \quad (5.17)$$

Simulations

Consider a Sepic converter circuit with parameter values:

$$L_1 = 30 \text{ mH}, \quad C_1 = 150 \text{ μF}, \quad L_2 = 30 \text{ mH}, \quad C_2 = 50 \text{ μF},$$
$$R = 10 \text{ Ω}, \quad E = 100 \text{ V}$$

with $\gamma = 1$.

We take the steady state output voltage value $\overline{v}_2 = 200$ V as the desired output voltage with the following corresponding steady state values for the rest of the state variables:

$$\overline{i}_1 = 40 \text{ A}, \quad \overline{v}_1 = 100 \text{ V}, \quad \overline{i}_2 = 20 \text{ A}$$

and

$$\overline{u}_{av} = 0.666$$

Figure 5.20 depicts the actual average state variables responses of the switched controlled system accomplishing the required stabilization task.

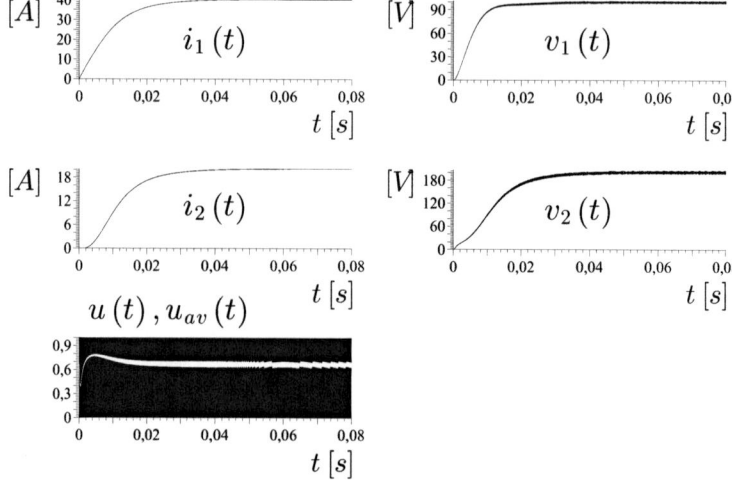

Fig. 5.20. Switched controlled responses of Sepic power converter implemented via a $\Sigma - \Delta$ modulator.

5.4.7 The Zeta Converter

Consider the normalized average model of the Zeta DC-to-DC power converter

$$\dot{x}_1 = -(1 - u_{av})x_2 + u_{av}$$
$$\dot{x}_2 = (1 - u_{av})x_1 - u_{av}x_3$$
$$\alpha_1 \dot{x}_3 = u_{av}x_2 - x_4 + u_{av}$$
$$\alpha_2 \dot{x}_4 = x_3 - \frac{1}{Q}x_4$$

It is desired to regulate the system trajectories towards a constant average state equilibrium point characterized, in terms of the desired output equilibrium voltage $\overline{x}_4 = V_d$, by

$$\overline{x}_1 = \frac{V_d^2}{Q}, \quad \overline{x}_2 = V_d, \quad \overline{x}_3 = \frac{V_d}{Q}, \quad \overline{u}_{av} = \frac{V_d}{1 + V_d}$$

A state coordinates transformation, of the translation type, to the stabilization error space: $e_1 = x_1 - V_d^2/Q$, $e_2 = x_2 - V_d$, $e_3 = x_3 - V_d/Q$, $e_4 = x_4 - V_d$ yields,

5.4 Exact Error Dynamics Passive Output Feedback Control

$$\dot{e}_1 = -(1 - u_{av})e_2 + (1 + V_d)\left(u_{av} - \frac{V_d}{1 + V_d}\right)$$

$$\dot{e}_2 = (1 - u_{av})e_1 - u_{av}e_3 - \frac{V_d(1 + V_d)}{Q}\left(u_{av} - \frac{V_d}{1 + V_d}\right)$$

$$\alpha_1 \dot{e}_3 = u_{av}e_2 - e_4 + (1 + V_d)\left(u_{av} - \frac{V_d}{1 + V_d}\right)$$

$$\alpha_2 \dot{e}_4 = e_3 - \frac{1}{Q}e_4$$

Consider the following stabilization error energy function candidate,

$$H(e) = \frac{1}{2}e^T \mathcal{A} e = \frac{1}{2}\left[e_1^2 + e_2^2 + \alpha_1 e_3^2 + \alpha_2 e_4^2\right]$$

where

$$\mathcal{A} = \mathcal{A}^T = \text{diag}(1, 1, \alpha_1, \alpha_2), \qquad e = [e_1, e_2, e_3, e_4]^T$$

Since

$$\frac{\partial H(e)}{\partial e} = \mathcal{A}e = [e_1, e_2, \alpha_1 e_3, \alpha_2 e_4]^T$$

we may write the system in Generalized Hamiltonian form:

$$\begin{bmatrix}\dot{e}_1\\\dot{e}_2\\\dot{e}_3\\\dot{e}_4\end{bmatrix} = \begin{bmatrix}0 & -(1-u_{av}) & 0 & 0\\(1-u_{av}) & 0 & -\frac{1}{\alpha_1}u_{av} & 0\\0 & \frac{1}{\alpha_1}u_{av} & 0 & -\frac{1}{\alpha_1\alpha_2}\\0 & 0 & \frac{1}{\alpha_1\alpha_2} & 0\end{bmatrix}\frac{\partial H(e)}{\partial e}$$

$$-\begin{bmatrix}0&0&0&0\\0&0&0&0\\0&0&0&0\\0&0&0&\frac{1}{\alpha_2^2 Q}\end{bmatrix}\frac{\partial H(e)}{\partial e} + \begin{bmatrix}(1+V_d)\\-\frac{V_d(1+V_d)}{Q}\\\frac{1}{\alpha_1}(1+V_d)\\0\end{bmatrix}e_{u_{av}}$$

where $e_{u_{av}} = u_{av} - \frac{V_d}{1+V_d}$.

The passive output corresponding to this Hamiltonian stabilization error representation is given by

$$e_y = b^T \frac{\partial H(e)}{\partial e} = (1 + V_d)\left[e_1 - \frac{V_d}{Q}e_2 + e_3\right]$$

The dissipation matching condition is not strictly satisfied and reads

$$\mathcal{A}\left[\mathcal{R} + \gamma b b^T\right]\mathcal{A} = \gamma(1+V_d)^2 \begin{bmatrix}1 & -\frac{V_d}{Q} & 1 & 0\\-\frac{V_d}{Q} & \frac{V_d^2}{Q^2} & -\frac{V_d}{Q} & 0\\1 & -\frac{V_d}{Q} & 1 & 0\\0 & 0 & 0 & \frac{1}{\gamma(1+V_d)^2 Q}\end{bmatrix} \geq 0$$

The set of vectors which are in the null space of the matrix, preceding matrix are of the form: $z = \begin{bmatrix}e_1 & e_2 & e_3 & 0\end{bmatrix}$ such that $\xi_\delta = e_1 - \frac{V_d}{Q}e_2 + e_3 = 0$, i.e., they lay in a subspace of R^4 and corresponds to $e_y = (1 + V_d) \xi_\delta$. This means the nonlinear system is controlled by the equilibrium input $\bar{u}_{av} = V_d/(1 + V_d)$, i.e., the error system is controlled by $e_{u_{av}} = 0$. The only trajectory of the error system with $e_4 = 0$ and $e_{u_{av}} = 0$ corresponds to the origin.

This is compatible with the fact that in order for the closed loop incremental average system to have the origin as an asymptotically stable equilibrium, the trajectories of the error system should have no other equilibrium than the origin itself. The origin of the average output feedback controlled system is, hence, an semi-globally asymptotically stable equilibrium.

A feedback controller which makes the equilibrium point asymptotically stable, by virtue of LaSalle's theorem, is given by the feedback of the passive output associated with the exact stabilization error dynamics,

$$e_{u_{av}} = -\gamma e_y = -\gamma (1 + V_d) \left[e_1 - \frac{V_d}{Q} e_2 + e_3 \right]$$

The average stabilizing feedback control, based on linear, static, passive output feedback, is then given by

$$u_{av} = \frac{V_d}{1 + V_d} - \gamma (1 + V_d) \left[\left(x_1 - \frac{V_d^2}{Q} \right) - \frac{V_d}{Q} (x_2 - V_d) + \left(x_3 - \frac{V_d}{Q} \right) \right]$$

Simulations

A set of typical parameters for a Zeta power converter circuit is given by:

$$L_1 = 600 \ \mu H, \qquad C_1 = 10 \ \mu F, \qquad L_2 = 600 \ \mu H, \qquad C_2 = 10 \ \mu F,$$

$$R = 40 \ \Omega, \qquad E = 100 \ V$$

These parameter values allow the computation of the quantities:

$$Q = 5.164, \qquad \sqrt{L_1 C_1} = 7.746 \times 10^{-5} \ s, \qquad \alpha_1 = 1, \qquad \alpha_2 = 1$$

We have set the controller design parameter γ to be 1, and we specify a desired steady state equilibrium voltage of value $\bar{v}_2 = 200$ V with the corresponding steady state equilibrium values for the rest of the original circuit variables:

$$\bar{i}_1 = 10 \ A, \qquad \bar{v}_1 = 200 \ V, \qquad \bar{i}_2 = 5 \ A$$

and $\bar{u}_{av} = 0.666$.

Figure 5.21 shows the switched responses of Zeta converter to the action of the static linear feedback controller based on the passive output of the exact stabilization error dynamics implemented through a $\Sigma - \Delta$-modulator.

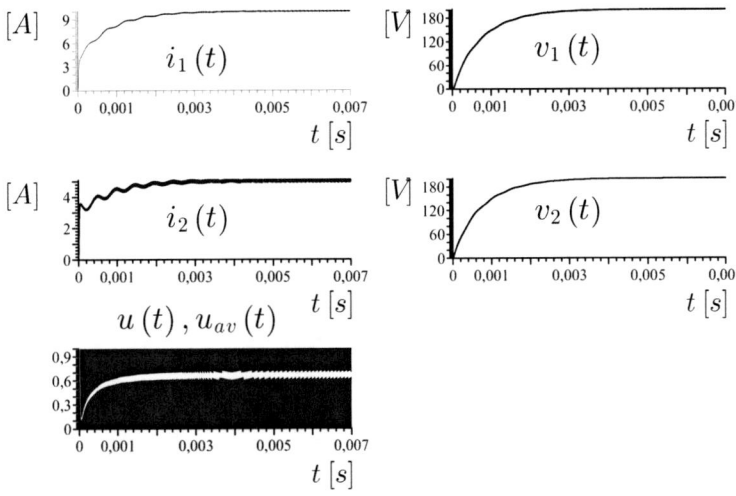

Fig. 5.21. Switched stabilization response of Zeta converter.

5.4.8 The Quadratic Buck Converter

Consider the normalized average model of the quadratic Buck DC-to-DC power converter

$$\dot{x}_1 = -x_2 + u_{av}$$
$$\dot{x}_2 = x_1 - u_{av}x_3$$
$$\alpha_1 \dot{x}_3 = u_{av}x_2 - x_4$$
$$\alpha_2 \dot{x}_4 = x_3 - \frac{1}{Q}x_4$$

It is desired to regulate the system trajectories towards a constant average state equilibrium point characterized, in terms of the desired output equilibrium voltage, $\bar{x}_4 = V_d$, by

$$\bar{x}_1 = \frac{(V_d)^{3/2}}{Q}, \quad \bar{x}_2 = \sqrt{V_d}, \quad \bar{x}_3 = \frac{V_d}{Q}, \quad \bar{u}_{av} = \sqrt{V_d}$$

A translation of the state coordinates to the stabilization error space defined, by

$$e_1 = x_1 - \frac{(V_d)^{3/2}}{Q}, \quad e_2 = x_2 - \sqrt{V_d}, \quad e_3 = x_3 - \frac{V_d}{Q}, \quad e_4 = x_4 - V_d$$

yields the following relations:

$$\dot{e}_1 = -e_2 + \left(u_{av} - \sqrt{V_d}\right)$$

$$\dot{e}_2 = e_1 - u_{av}e_3 - \frac{V_d}{Q}\left(u_{av} - \sqrt{V_d}\right)$$

$$\alpha_1\dot{e}_3 = u_{av}e_2 - e_4 + \sqrt{V_d}\left(u_{av} - \sqrt{V_d}\right)$$

$$\alpha_2\dot{e}_4 = e_3 - \frac{1}{Q}e_4$$

The error energy function candidate,

$$H(e) = \frac{1}{2}e^T \mathcal{A}e = \frac{1}{2}\left[e_1^2 + e_2^2 + \alpha_1 e_3^2 + \alpha_2 e_4^2\right]$$

where

$$\mathcal{A} = \mathcal{A}^T = \text{diag}(1, 1, \alpha_1, \alpha_2), \qquad e = [e_1, e_2, e_3, e_4]^T$$

allows, by virtue of the fact that

$$\frac{\partial H(e)}{\partial e} = \mathcal{A}e = [e_1, e_2, \alpha_1 e_3, \alpha_2 e_4]^T$$

to write the (open loop) stabilization error system in Generalized Hamiltonian form:

$$\dot{e} = \begin{bmatrix} 0 & -1 & 0 & 0 \\ 1 & 0 & -\frac{1}{\alpha_1}u_{av} & 0 \\ 0 & \frac{1}{\alpha_1}u_{av} & 0 & -\frac{1}{\alpha_1\alpha_2} \\ 0 & 0 & \frac{1}{\alpha_1\alpha_2} & 0 \end{bmatrix}\frac{\partial H(e)}{\partial e} - \begin{bmatrix} 0 & 0 & 0 & 0 \\ 0 & 0 & 0 & 0 \\ 0 & 0 & 0 & 0 \\ 0 & 0 & 0 & \frac{1}{\alpha_2^2 Q} \end{bmatrix}\frac{\partial H(e)}{\partial e} + \begin{bmatrix} 1 \\ -\frac{V_d}{Q} \\ \frac{\sqrt{V_d}}{\alpha_1} \\ 0 \end{bmatrix} e_{u_{av}}$$

where $e_{u_{av}} = u_{av} - \sqrt{V_d}$.

The passive output corresponding to this Hamiltonian stabilization error representation is given by

$$e_y = b^T \frac{\partial H(e)}{\partial e} = e_1 - \frac{V_d}{Q}e_2 + \sqrt{V_d}e_3$$

The dissipation matching condition is not strictly satisfied since

$$\dot{H}(e) = -\frac{\partial H(e)}{\partial e^T}\left[\mathcal{R} + \gamma bb^T\right]\frac{\partial H(e)}{\partial e} = -e^T \mathcal{A}\left[\mathcal{R} + \gamma bb^T\right]\mathcal{A}e \leq 0$$

i.e.,

$$\mathcal{A}\left[\mathcal{R} + \gamma bb^T\right]\mathcal{A} = \begin{bmatrix} \gamma & -\gamma\frac{V_d}{Q} & \gamma\sqrt{V_d} & 0 \\ -\gamma\frac{V_d}{Q} & \gamma\frac{V_d^2}{Q^2} & -\gamma\frac{(V_d)^{3/2}}{Q} & 0 \\ \gamma\sqrt{V_d} & -\gamma\frac{(V_d)^{3/2}}{Q} & \gamma V_d & 0 \\ 0 & 0 & 0 & \frac{1}{Q} \end{bmatrix} \geq 0$$

The set of vectors which are in the null space of the above matrix are of the form: $z = \begin{bmatrix} e_1 & e_2 & e_3 & 0 \end{bmatrix}$ such that $\xi_\delta = e_1 - \frac{V_d}{Q}e_2 + \sqrt{V_d}e_3 = 0$, i.e.,

5.4 Exact Error Dynamics Passive Output Feedback Control

they lay in a subspace of R^4 that corresponds to $e_y = \xi_\delta$. This means that in such a subspace, the nonlinear system is controlled by the equilibrium input $\overline{u}_{av} = \sqrt{V_d}$, i.e., the error system is controlled by $e_{u_{av}} = 0$. The only trajectory of the error system with $e_4 = 0$ and $e_{u_{av}} = 0$ corresponds to the origin.

According to LaSalle's theorem for asymptotic stability, in order for the average closed loop error system to have the origin as an asymptotically stable equilibrium, the trajectories of the system taking place in the set $\{e \mid \dot{H}(e) = 0\}$ should have no other equilibrium than the origin itself. The origin of the average feedback controlled error system is, hence, an asymptotically stable equilibrium.

A feedback controller which makes the equilibrium point asymptotically stable, is just

$$e_{u_{av}} = -\gamma e_y = -\gamma \left[e_1 - \frac{V_d}{Q} e_2 + \sqrt{V_d} e_3 \right]$$

The average stabilizing feedback control, based on passive output feedback, is then given by

$$u_{av} = \sqrt{V_d} - \gamma \left[\left(x_1 - \frac{(V_d)^{3/2}}{Q} \right) - \frac{V_d}{Q} \left(x_2 - \sqrt{V_d} \right) + \sqrt{V_d} \left(x_3 - \frac{V_d}{Q} \right) \right]$$

This is, precisely, the linear feedback controller obtained by tangent linearization and feedback of the incremental passive output.

Simulations

A typical set of parameter values for the quadratic Buck converter circuit is given by:

$$L_1 = 1.5 \text{ H}, \quad C_1 = 10 \text{ }\mu\text{F}, \quad L_2 = 600 \text{ }\mu\text{H}, \quad C_2 = 10 \text{ }\mu\text{F},$$

$$R = 40 \text{ }\Omega, \quad E = 100 \text{ V}$$

For the simulations we have chosen $\gamma = 1$, and a constant reference output equilibrium voltage of value $\overline{v}_2 = 25$ V with the actual corresponding steady state variables specified by:

$$\overline{i}_1 = 0.3125 \text{ A}, \quad \overline{v}_1 = 50 \text{ V}, \quad \overline{i}_2 = 0.625 \text{ A}$$

and $\overline{u}_{av} = 0.5$.

Figure 5.22 shows the switched system responses to the static linear passivity based controller processing the passive output of the exact stabilization error dynamics for the quadratic Buck DC-to-DC power converter.

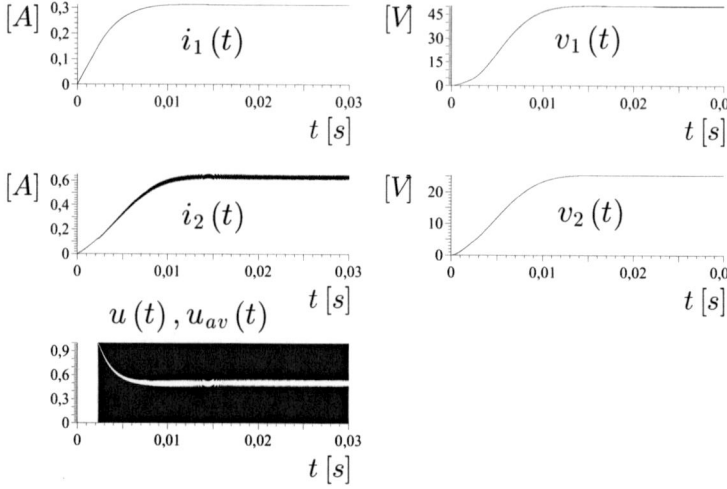

Fig. 5.22. Switched stabilization response of quadratic Buck converter controlled via a static linear passivity based feedback controller.

5.4.9 The Boost-Boost Converter

Consider now the average normalized model of a multi-variable Boost-Boost converter, with the following simplification: $\alpha_1 = \alpha_2 = 1$, we have

$$\dot{x}_1 = -u_{1av}x_2 + 1$$
$$\dot{x}_2 = u_{1av}x_1 - \frac{1}{Q_1}x_2 - x_3$$
$$\dot{x}_3 = x_2 - u_{2av}x_4$$
$$\dot{x}_4 = u_{2av}x_3 - \frac{1}{Q_L}x_4$$

The equilibrium point of the system, for a desired set of output average equilibrium voltages $\overline{x}_2 = V_{2d}$ and $\overline{x}_4 = V_{4d}$, is given by,

$$\overline{x}_1 = \frac{V_{2d}^2}{Q_1} + \frac{V_{4d}^2}{Q_L}, \qquad \overline{x}_2 = V_{2d}, \qquad \overline{x}_3 = \frac{V_{4d}^2}{Q_L V_{2d}}, \qquad \overline{x}_4 = V_{4d}$$

$$\overline{u}_{1av} = \frac{1}{V_{2d}}, \qquad \overline{u}_{2av} = \frac{V_{2d}}{V_{4d}}$$

Transforming the system into average stabilization state error variables yields,

5.4 Exact Error Dynamics Passive Output Feedback Control

$$\dot{e}_1 = -u_{1av}e_2 - V_{2d}\left(u_{1av} - \frac{1}{V_{2d}}\right)$$

$$\dot{e}_2 = u_{1av}e_1 - \frac{e_2}{Q_1} - e_3 + \left(\frac{V_{2d}^2}{Q_1} + \frac{V_{4d}^2}{Q_L}\right)\left(u_{1av} - \frac{1}{V_{2d}}\right)$$

$$\dot{e}_3 = e_2 - u_{2av}e_4 - V_{4d}\left(u_{2av} - \frac{V_{2d}}{V_{4d}}\right)$$

$$\dot{e}_4 = u_{2av}e_3 - \frac{e_4}{Q_L} + \frac{V_{4d}^2}{Q_L V_{2d}}\left(u_{2av} - \frac{V_{2d}}{V_{4d}}\right)$$

The system may be written in Hamiltonian canonical form

$$\dot{e} = \begin{bmatrix} 0 & -u_{1av} & 0 & 0 \\ u_{1av} & 0 & -1 & 0 \\ 0 & 1 & 0 & -u_{2av} \\ 0 & 0 & u_{2av} & 0 \end{bmatrix}\frac{\partial H(e)}{\partial e} - \begin{bmatrix} 0 & 0 & 0 & 0 \\ 0 & \frac{1}{Q_1} & 0 & 0 \\ 0 & 0 & 0 & 0 \\ 0 & 0 & 0 & \frac{1}{Q_L} \end{bmatrix}\frac{\partial H(e)}{\partial e}$$

$$+ \begin{bmatrix} -V_{2d} & 0 \\ \left(\frac{V_{2d}^2}{Q_1} + \frac{V_{4d}^2}{Q_L}\right) & 0 \\ 0 & -V_{4d} \\ 0 & \frac{V_{4d}^2}{V_{2d}Q_L} \end{bmatrix}\begin{bmatrix} e_{u_{1av}} \\ e_{u_{2av}} \end{bmatrix}$$

We note that the nonlinear part of the system, which includes the interaction between the stages, does not intervene in the average system stability and the dissipative linear incremental part, which is decoupled, plays the important role in stability. As a consequence, a linear decoupled controller of the average passive outputs is all that is required for semi-global asymptotic stability of the desired equilibrium state.

Note that by choosing the Γ matrix in a diagonal form: $\Gamma = \text{diag}[\gamma_1, \gamma_2]$, with $\gamma_1, \gamma_2 > 0$, the dissipation matching condition takes the following natural block-decoupled form:

$$\mathcal{R} + B\Gamma B^T =$$
$$\begin{bmatrix} \gamma_1 V_{2d}^2 & -\gamma_1 V_{2d}\left(\frac{V_{2d}^2}{Q_1} + \frac{V_{4d}^2}{Q_L}\right) & 0 & 0 \\ -\gamma_1 V_{2d}\left(\frac{V_{2d}^2}{Q_1} + \frac{V_{4d}^2}{Q_L}\right) & \frac{1}{Q_1} + \gamma_1\left(\frac{V_{2d}^2}{Q_1} + \frac{V_{4d}^2}{Q_L}\right)^2 & 0 & 0 \\ 0 & 0 & \gamma_2 V_{4d}^2 & -\gamma_2\frac{V_{4d}^3}{Q_L V_{2d}} \\ 0 & 0 & -\gamma_2\frac{V_{4d}^3}{Q_L V_{2d}} & \frac{1}{Q_L} + \gamma_2\frac{V_{4d}^4}{Q_L^2 V_{2d}^2} \end{bmatrix} > 0$$

The passive outputs of the stabilization error dynamics are given by

$$e_y = B^T\frac{\partial H(e)}{\partial e}$$

which, in explicit form yields:

$$y_1 = -V_{2d}e_1 + \left(\frac{V_{2d}^2}{Q_1} + \frac{V_{4d}^2}{Q_L}\right)e_2$$

$$y_2 = -V_{4d}e_3 + \frac{V_{4d}^2}{QV_{2d}}e_4$$

Each passive output involves state variables which are ascribed to his own converter block. The average passive output feedback control policy can be proposed to be decoupled, as follows

$$e_{u_{1av}} = \gamma_1 V_{2d}e_1 - \gamma_1\left(\frac{V_{2d}^2}{Q_1} + \frac{V_{4d}^2}{Q_L}\right)e_2$$

$$e_{u_{2av}} = \gamma_2 V_{4d}e_3 - \gamma_2 \frac{V_{4d}^2}{Q_L V_{2d}}e_4$$

5.4.10 The Double Buck-Boost Converter

Consider now the average normalized model of a multi-variable double Buck-Boost converter:

$$\dot{x}_1 = (1 - u_{1av})x_2 + u_{1av}$$

$$\dot{x}_2 = -(1 - u_{1av})x_1 - \frac{1}{Q_1}x_2 - u_{2av}x_3$$

$$\alpha_1 \dot{x}_3 = u_{2av}x_2 + (1 - u_{2av})x_4$$

$$\alpha_2 \dot{x}_4 = -(1 - u_{2av})x_3 - \frac{1}{Q_L}x_4$$

The equilibrium point of the average normalized system, for a desired set of output average equilibrium voltages $\bar{x}_2 = V_{2d}$ and $\bar{x}_4 = V_{4d}$, is given by,

$$\bar{x}_1 = -\left(\frac{V_{2d}^2}{Q_1} + \frac{V_{4d}^2}{Q_L}\right)\left(\frac{1 - V_{2d}}{V_{2d}}\right), \quad \bar{x}_2 = V_{2d}, \quad \bar{x}_3 = \frac{V_{4d}}{Q_L}\left(\frac{V_{4d}}{V_{2d}} - 1\right), \quad \bar{x}_4 = V_{4d}$$

$$\bar{u}_{1av} = -\frac{V_{2d}}{1 - V_{2d}}, \quad \bar{u}_{2av} = \frac{V_{4d}}{V_{4d} - V_{2d}}$$

We express the system in state error coordinates with respect to the average equilibrium point corresponding to a desired set of output voltages $\bar{x}_2 = V_{2d}$ and $\bar{x}_4 = V_{4d}$,

$$e_1 = x_1 + \left(\frac{V_{2d}^2}{Q_1} + \frac{V_{4d}^2}{Q_L}\right)\left(\frac{1 - V_{2d}}{V_{2d}}\right), \qquad e_2 = x_2 - V_{2d}$$

$$e_3 = x_3 - \frac{V_{4d}}{Q_L}\left(\frac{V_{4d}}{V_{2d}} - 1\right), \qquad e_4 = x_4 - V_{4d}$$

The average components of the input error vector $e_{u_{av}}$ are defined as

5.4 Exact Error Dynamics Passive Output Feedback Control

$$e_{u_{1av}} = u_{1av} + \frac{V_{2d}}{1 - V_{2d}}, \quad e_{u_{2av}} = u_{2av} - \frac{V_{4d}}{V_{4d} - V_{2d}}$$

Transforming the system into average state error variables we obtain the following expressions

$$\dot{e}_1 = (1 - u_{1av})e_2 + (1 - V_{2d})\left(u_{1av} + \frac{V_{2d}}{1 - V_{2d}}\right)$$

$$\dot{e}_2 = -(1 - u_{1av})e_1 - \frac{1}{Q_1}e_2 - u_{2av}e_3 + \beta_1 e_{u_{1av}} + \beta_2 e_{u_{2av}}$$

$$\alpha_1 \dot{e}_3 = u_{2av}e_2 + (1 - u_{2av})e_4 - (V_{4d} - V_{2d})\left(u_{2av} - \frac{V_{4d}}{V_{4d} - V_{2d}}\right)$$

$$\alpha_2 \dot{e}_4 = -(1 - u_{2av})e_3 - \frac{1}{Q_L}e_4 - \beta_2\left(u_{2av} - \frac{V_{4d}}{V_{4d} - V_{2d}}\right)$$

where:

$$\beta_1 = -\left(\frac{V_{2d}}{Q_1} + \frac{V_{4d}^2}{V_{2d}Q_L}\right)(1 - V_{2d}), \quad \beta_2 = -\frac{V_{4d}}{V_{2d}Q_L}(V_{4d} - V_{2d})$$

Consider the average stabilization error energy function $H(e)$ as

$$H(e) = \frac{1}{2}e^T \mathcal{A} e = \frac{1}{2}\left[e_1^2 + e_2^2 + \alpha_1 e_3^2 + \alpha_2 e_4^2\right]$$

where

$$\mathcal{A} = \mathcal{A}^T = \text{diag}(1, 1, \alpha_1, \alpha_2), \quad e = [e_1, e_2, e_3, e_4]^T$$

Since

$$\frac{\partial H(e)}{\partial e} = \mathcal{A}e = [e_1, e_2, \alpha_1 e_3, \alpha_2 e_4]^T$$

we may write the average normalized model of the double Buck-Boost system in Generalized Hamiltonian canonical form:

$$\dot{e} = \begin{bmatrix} 0 & (1 - u_{1av}) & 0 & 0 \\ -(1 - u_{1av}) & 0 & -\frac{1}{\alpha_1}u_{2av} & 0 \\ 0 & \frac{1}{\alpha_1}u_{2av} & 0 & \frac{1}{\alpha_1 \alpha_2}(1 - u_{2av}) \\ 0 & 0 & -\frac{1}{\alpha_1 \alpha_2}(1 - u_{2av}) & 0 \end{bmatrix} \frac{\partial H(e)}{\partial e}$$

$$- \begin{bmatrix} 0 & 0 & 0 & 0 \\ 0 & \frac{1}{Q_1} & 0 & 0 \\ 0 & 0 & 0 & 0 \\ 0 & 0 & 0 & \frac{1}{\alpha_2^2 Q_L} \end{bmatrix} \frac{\partial H(e)}{\partial e} + \begin{bmatrix} (1 - V_{2d}) & 0 \\ \beta_1 & \beta_2 \\ 0 & -\frac{1}{\alpha_1}(V_{4d} - V_{2d}) \\ 0 & -\frac{1}{\alpha_2}\beta_2 \end{bmatrix} \begin{bmatrix} e_{u_{1av}} \\ e_{u_{2av}} \end{bmatrix}$$

Note that by choosing the Γ matrix in a diagonal form: $\Gamma = \text{diag}[\gamma_1, \gamma_2]$, with $\gamma_1, \gamma_2 > 0$, the dissipation matching condition takes the following form:

$$\mathcal{A}\left[\mathcal{R}+B\Gamma B^T\right]\mathcal{A} = \begin{bmatrix} \gamma_1\delta_1^2 & \gamma_1\beta_1\delta_1 & 0 & 0 \\ \gamma_1\beta_1\delta_1 & \frac{1}{Q_1}+\gamma_1\beta_1^2+\gamma_2\beta_2^2 & -\gamma_2\beta_2\delta_2 & -\gamma_2\beta_2^2 \\ 0 & -\gamma_2\beta_2\delta_2 & \gamma_2\delta_2^2 & \gamma_2\beta_2\delta_2 \\ 0 & -\gamma_2\beta_2^2 & \gamma_2\beta_2\delta_2 & \gamma_2\beta_2^2+\frac{1}{Q_L} \end{bmatrix}$$

where:
$$\delta_1 = (1-V_{2d}), \qquad \delta_2 = (V_{4d}-V_{2d})$$

Sylvester's test to the above matrix, $\mathcal{A}\left[\mathcal{R}+B\Gamma B^T\right]\mathcal{A}$, yields principal minor determinants given by,

$$0 < \left|\gamma_1(1-V_{2d})^2\right|$$

$$0 < \begin{vmatrix} \gamma_1(1-V_{2d})^2 & \gamma_1\beta_1(1-V_{2d}) \\ \gamma_1\beta_1(1-V_{2d}) & \frac{1}{Q_1}+\gamma_1\beta_1^2+\gamma_2\beta_2^2 \end{vmatrix} = \eta_1$$

$$0 < \begin{vmatrix} \gamma_1(1-V_{2d})^2 & \gamma_1\beta_1(1-V_{2d}) & 0 \\ \gamma_1\beta_1(1-V_{2d}) & \frac{1}{Q_1}+\gamma_1\beta_1^2+\gamma_2\beta_2^2 & -\gamma_2\beta_2(V_{4d}-V_{2d}) \\ 0 & -\gamma_2\beta_2(V_{4d}-V_{2d}) & \gamma_2(V_{4d}-V_{2d})^2 \end{vmatrix} = \eta_2$$

$$0 < \left|\mathcal{A}\left[\mathcal{R}+B\Gamma B^T\right]\mathcal{A}\right| = \eta_3$$

with

$$\eta_1 = \gamma_1\frac{(1-V_{2d})^2\left(1+\gamma_2\beta_2^2 Q_1\right)}{Q_1}$$

$$\eta_2 = \gamma_1\gamma_2\frac{(1-V_{2d})^2(V_{4d}-V_{2d})^2}{Q_1}$$

$$\eta_3 = \gamma_1\gamma_2\frac{(1-V_{2d})^2(V_{4d}-V_{2d})^2}{Q_1 Q_L}$$

which are all strictly positive, due to the fact that the parameters: Q_1, Q_L and the design gains γ_1, γ_2 are all positive and the average equilibrium point corresponding to a desired set of output voltages is such that $V_{2d} < 0$ and $V_{4d} > 0$, respectively. Thus the matrix $\mathcal{A}\left[\mathcal{R}+B\Gamma B^T\right]\mathcal{A}$ is positive definite. Therefore $\dot{H}(e) = -e^T\mathcal{A}\left[\mathcal{R}+B\Gamma B^T\right]\mathcal{A}e$ is negative definite.

The passive outputs are given by

$$e_y = B^T\frac{\partial H(e)}{\partial e}$$

which, in explicit form read as:

$$e_y = \begin{bmatrix} e_{y_1} \\ e_{y_2} \end{bmatrix} = \begin{bmatrix} (1-V_{2d})e_1+\beta_1 e_2 \\ \beta_2 e_2 - (V_{4d}-V_{2d})e_3 - \beta_2 e_4 \end{bmatrix}$$

In this case, the passive outputs involves the sharing of information, among the two system controllers, of the the error variable e_2. A feedback controller which makes the equilibrium point globally asymptotically stable is just the linear controller: $e_{u_{av}} = -\Gamma e_y$, hence:

$$e_{u_{av}} = \begin{bmatrix} e_{u_{1av}} \\ e_{u_{2av}} \end{bmatrix} = \begin{bmatrix} -\gamma_1 \left[(1 - V_{2d}) e_1 + \beta_1 e_2 \right] \\ -\gamma_2 \left[\beta_2 e_2 - (V_{4d} - V_{2d}) e_3 - \beta_2 e_4 \right] \end{bmatrix}$$

The average stabilizing feedback control, based on passive output feedback, is then given by

$$u_{av} = \begin{bmatrix} u_{1av} \\ u_{2av} \end{bmatrix} = \begin{bmatrix} -\frac{V_{2d}}{1-V_{2d}} - \gamma_1 \left[(1 - V_{2d}) e_1 + \beta_1 e_2 \right] \\ \frac{V_{4d}}{V_{4d}-V_{2d}} - \gamma_2 \left[\beta_2 e_2 - (V_{4d} - V_{2d}) e_3 - \beta_2 e_4 \right] \end{bmatrix}$$

This is, precisely, the linear feedback controller obtained by tangent linearization and feedback of the linearized incremental passive output.

Note that the error vector has the origin as a semi-global asymptotically stable equilibrium with the incremental control proposed above.

Simulations

Simulations were carried out with the following design parameter values:

$$L_1 = 20 \text{ mH}, \quad C_1 = 20 \text{ }\mu\text{F}, \quad L_2 = 20 \text{ mH}, \quad C_2 = 20 \text{ }\mu\text{F},$$

$$R_1 = 30 \text{ }\Omega, \quad R_L = 30 \text{ }\Omega, \quad E = 15 \text{ V}$$

with $\gamma_1 = 0.1$ and $\gamma_2 = 0.1$. The corresponding values for the normalized parameter turned out to be:

$$Q_1 = Q_L = 0.9487, \quad \sqrt{L_1 C_1} = 632.46 \text{ }\mu\text{s}, \quad \alpha_1 = \alpha_2 = 1$$

It is desired to regulate the voltage variables to the reference equilibrium values: $\overline{v}_1 = -22.5$ V and $\overline{v}_2 = 22.5$ V. The corresponding control inputs and the equilibrium currents are given, respectively, by: $\overline{u}_{1av} = 0.6$ and $\overline{u}_{2av} = 0.5$, while $\overline{i}_1 = 3.75$ A and $\overline{i}_2 = -1.5$ A.

Figure 5.23 depicts the average response of the switched system to a passivity based controller.

5.5 Trajectory Tracking via Error Dynamics Passive Output Feedback

A variant of the previous result is obtained when we consider trajectory tracking problems in terms of the exact nonlinear average tracking error model. We present the general formulation for all DC-to-DC power converters studied thus far.

A general average model, discussed in Chapter 2, of the SISO DC-to-DC power converters is of the form

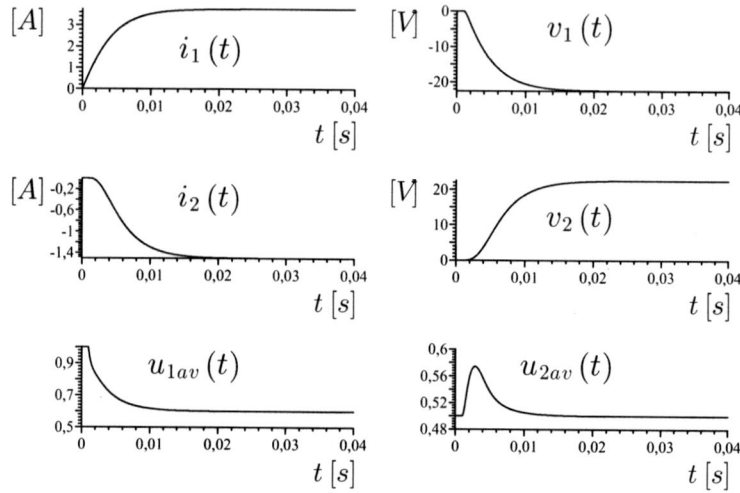

Fig. 5.23. Average responses of double Buck-Boost converter to an exact stabilization error passive output feedback controller.

$$\mathcal{A}\dot{x} = \mathcal{J}(u_{av})x - \mathcal{R}x + bu_{av} + \mathcal{E}$$

Whenever we need to emphasize the Hamiltonian character of the system we simply use the fact that with $H(x) = \frac{1}{2}x^T\mathcal{A}x$, then $\frac{\partial H}{\partial x}$ is simply equals to $\mathcal{A}x$. Multiplying out the system equation by $\mathcal{A}^{-1}$ and replacing the vector x by the equivalent expression $\mathcal{A}^{-1}\mathcal{A}x$ we obtain

$$\dot{x} = \mathcal{A}^{-1}\mathcal{J}(u_{av})\mathcal{A}^{-1}\mathcal{A}x - \mathcal{A}^{-1}\mathcal{R}\mathcal{A}^{-1}\mathcal{A}x + \mathcal{A}^{-1}bu_{av} + \mathcal{A}^{-1}\mathcal{E}$$

Defining now $\tilde{b} = \mathcal{A}^{-1}b$ and $\mathcal{A}^{-1}\mathcal{E} = \tilde{\mathcal{E}}$, we see from the fact that the conservative character of the term $\mathcal{J}(u_{av})x$ and the dissipative of $\mathcal{R}x$ is preserved, respectively, in their new forms $\mathcal{A}^{-1}\mathcal{J}(u_{av})\mathcal{A}^{-1} = \tilde{\mathcal{J}}(u_{av})$ and $\mathcal{A}^{-1}\mathcal{R}\mathcal{A}^{-1} = \tilde{\mathcal{R}}$, we have:

$$\dot{x} = \tilde{\mathcal{J}}(u_{av})\frac{\partial H}{\partial x} - \tilde{\mathcal{R}}\frac{\partial H}{\partial x} + \tilde{b}u_{av} + \tilde{\mathcal{E}}$$

From the above developments we consider in this section, when needed, and without loss of generality, the Generalized Hamiltonian models of the form:

$$\dot{x} = \mathcal{J}(u_{av})\frac{\partial H}{\partial x} - \mathcal{R}\frac{\partial H}{\partial x} + bu_{av} + \mathcal{E}$$

Let $x^*(t)$ be a desired state trajectory which can be effectively accomplished by means of the nominal average control input $u_{av}^*(t)$. We thus have

$$\dot{x}^*(t) = \mathcal{J}(u_{av}^*)x^* - \mathcal{R}x^*(t) + bu_{av}^*(t) + \mathcal{E}$$

5.5 Error Dynamics Passive Output Feedback

Therefore, defining $e = x - x^*(t)$, and $e_{u_{av}} = u_{av} - u^*$, we have, using the last relation,

$$\dot{e} = \mathcal{J}(u_{av})(x - x^*) - \mathcal{R}(x - x^*) + b(u_{av} - u^*) + \mathcal{E}$$
$$+ \mathcal{J}(u_{av})x^* - \mathcal{R}x^* + bu^* - \dot{x}^*$$
$$= \mathcal{J}(u_{av})e - \mathcal{R}e + be_{u_{av}} + \mathcal{J}(u_{av})x^* - \mathcal{R}x^* + bu^*_{av}$$
$$- \mathcal{J}(u^*_{av})x^* + \mathcal{R}x^*(t) - bu^*_{av}(t) - \mathcal{E}$$
$$= \mathcal{J}(u_{av})e - \mathcal{R}e + be_{u_{av}} + [\mathcal{J}(u_{av}) - \mathcal{J}(u^*_{av})]x^*$$

Using the affine nature of $\mathcal{J}(u_{av})$ on the average control input u_{av}, we have the following exact model for the tracking error dynamics:

$$\dot{e} = \mathcal{J}(u_{av})e - \mathcal{R}e + be_{u_{av}} + \frac{\partial \mathcal{J}(u_{av})}{\partial u_{av}} x^* e_{u_{av}}$$

It is interesting to note that the nonlinear part of the error system dynamics is conservative and that the control input vector is now a time-varying vector depending upon the desired state trajectory. We rewrite the tracking error dynamics in Hamiltonian form, with $H(e) = \frac{1}{2}e^T e$, as

$$\dot{e} = \mathcal{J}(u_{av})\frac{\partial H(e)}{\partial e} - \mathcal{R}\frac{\partial H(e)}{\partial e} + \left[b + \frac{\partial \mathcal{J}(u_{av})}{\partial u_{av}} x^*(t)\right] e_{u_{av}}$$

The passive output tracking error is just given by

$$e_y = y - y^* = \left[b + \frac{\partial \mathcal{J}(u_{av})}{\partial u_{av}} x^*(t)\right]^T \frac{\partial H(e)}{\partial e}$$

A linear time-varying average incremental passive output feedback controller is simply given by

$$e_{u_{av}} = -\gamma e_y = -\gamma \left[b + \frac{\partial \mathcal{J}(u_{av})}{\partial u_{av}} x^*(t)\right]^T \frac{\partial H(e)}{\partial e}$$

which produces the closed loop system given by

$$\dot{e} = \mathcal{J}(u_{av})\frac{\partial H(e)}{\partial e}$$
$$- \left(\mathcal{R} + \gamma \left[b + \frac{\partial \mathcal{J}(u_{av})}{\partial u_{av}} x^*(t)\right] \left[b + \frac{\partial \mathcal{J}(u_{av})}{\partial u_{av}} x^*(t)\right]^T\right)\frac{\partial H(e)}{\partial e}$$

The time derivative of the positive definite tracking error energy function $H(e)$ is given by:

$$\dot{H}(e) = -\frac{\partial H(e)}{\partial e^T}\left(\mathcal{R} + \gamma \left[b + \frac{\partial \mathcal{J}(u_{av})}{\partial u_{av}} x^*(t)\right] \left[b + \frac{\partial \mathcal{J}(u_{av})}{\partial u_{av}} x^*(t)\right]^T\right)\frac{\partial H(e)}{\partial e}$$

The dissipation matching condition adopts the following time varying form:

$$\left(\mathcal{R} + \gamma \left[b + \frac{\partial \mathcal{J}(u_{av})}{\partial u_{av}} x^*(t) \right] \left[b + \frac{\partial \mathcal{J}(u_{av})}{\partial u_{av}} x^*(t) \right]^T \right) > 0$$

We assume that the dissipation matching condition is strictly satisfied. Otherwise, if the dissipation matching condition is not strictly satisfied as in:

$$\left(\mathcal{R} + \gamma \left[b + \frac{\partial \mathcal{J}(u_{av})}{\partial u_{av}} x^*(t) \right] \left[b + \frac{\partial \mathcal{J}(u_{av})}{\partial u_{av}} x^*(t) \right]^T \right) \geq 0$$

we have to resort to LaSalle's theorem to establish semi-global asymptotic stability of the origin of the tracking error space.

The nature of the average passive output feedback control law is that of a linear time-varying average incremental state feedback law:

$$e_{u_{av}} = -\gamma e_y = -\gamma \left[b + \frac{\partial \mathcal{J}(u_{av})}{\partial u_{av}} x^*(t) \right]^T \frac{\partial H(e)}{\partial e}$$

The average control input is therefore synthesized as

$$u_{av} = u_{av}^*(t) - \gamma \left[b + \frac{\partial \mathcal{J}(u_{av})}{\partial u_{av}} x^*(t) \right]^T (x - x^*(t))$$

Obtaining the average nominal state trajectory $x^*(t)$, in an explicit manner, may prove to be a difficult task sometimes (as in the Buck-Boost converter, for instance). For obtaining an explicit expression of the nominal state trajectory is sometimes useful resorting to the converter's flatness property.

Recall that the matrix $\partial \mathcal{J}(u_{av})/\partial u_{av}$ is assumed to be constant, that we denote by $\mathcal{J}_1$, we denote by $b^*(t)$ the vector

$$b^*(t) = [\, b + \mathcal{J} x^*(t) \,]$$

5.5.1 The Boost Converter

Recall the normalized average model of the Boost DC-to-DC power converter

$$\dot{x}_1 = -u_{av} x_2 + 1$$
$$\dot{x}_2 = u_{av} x_1 - \frac{1}{Q} x_2$$

It is desired to regulate the system trajectories between two average state equilibrium points characterized, in terms of the desired output equilibrium voltage $\bar{x}_2(\tau_1) = V_{d1}$, and $\bar{x}_2(\tau_2) = V_{d2}$ while following a corresponding compatible state trajectory $x^*(\tau)$.

A time-varying translation of the state coordinates to the tracking error space $e_1 = x_1 - x_1^*$, $e_2 = x_2 - x_2^*$ yields,

$$\dot{e}_1 = -u_{av}e_2 + 1 - u_{av}x_2^*(\tau) - \dot{x}_1^*(\tau)$$
$$\dot{e}_2 = u_{av}e_1 - \frac{1}{Q}e_2 + u_{av}x_1^*(\tau) - \frac{1}{Q}x_2^* - \dot{x}_2^*(\tau)$$

where

$$\dot{x}_1^*(\tau) = -u_{av}^*(\tau)x_2^*(\tau) + 1$$
$$\dot{x}_2^*(\tau) = u_{av}^*(\tau)x_1^*(\tau) - \frac{1}{Q}x_2^*(\tau)$$

We then have,

$$\dot{e}_1 = -u_{av}e_2 - x_2^*(\tau)(u_{av} - u_{av}^*(\tau))$$
$$\dot{e}_2 = u_{av}e_1 - \frac{1}{Q}e_2 + x_1^*(\tau)(u_{av} - u_{av}^*(\tau))$$

or

$$\dot{e}_1 = -u_{av}e_2 - x_2^*(\tau)e_{u_{av}}$$
$$\dot{e}_2 = u_{av}e_1 - \frac{1}{Q}e_2 + x_1^*(\tau)e_{u_{av}}$$

The passive average output is given by

$$e_y = -x_2^*(\tau)e_1 + x_1^*(\tau)e_2$$

The dissipation matching condition adopts the form,

$$\begin{bmatrix} \gamma(x_2^*(\tau))^2 & -\gamma x_1^*(\tau)x_2^*(\tau) \\ -\gamma x_1^*(\tau)x_2^*(\tau) & \frac{1}{Q} + \gamma(x_1^*(\tau))^2 \end{bmatrix} > 0$$

and the average passive output linear feedback control law for the average input error is given by:

$$e_{u_{av}} = -\gamma e_y = \gamma x_2^*(\tau)(x_1 - x_1^*(\tau)) - \gamma x_1^*(\tau)(x_2 - x_2^*(\tau))$$

The average control input is readily established to be

$$u_{av} = u^*(\tau) + \gamma[x_2^*(t)x_1 - x_1^*(\tau)x_2] \tag{5.18}$$

Nominal Trajectory Generation

The normalized Boost converter model is differentially flat, with flat output given by the normalized total stored energy

$$F = \frac{1}{2}[x_1^2 + x_2^2]$$

Indeed, all variables are parameterizable in terms of F and a finite number of its time derivatives

$$x_1 = -\frac{Q}{2} + \sqrt{\frac{Q^2}{4} + Q\dot{F} + 2F}$$

$$x_2 = \sqrt{2F - \left(-\frac{Q}{2} + \sqrt{\frac{Q^2}{4} + Q\dot{F} + 2F}\right)^2}$$

$$u_{av} = \frac{\left(1 + \frac{2}{Q^2}x_2^2\right) - \ddot{F}}{x_2\left(1 + \frac{2}{Q}x_1\right)}$$

Recall it is desired to regulate the system trajectories between two average state equilibrium points characterized, in terms of the desired normalized output equilibrium voltages $\bar{x}_2(\tau_{init}) = V_{d1}$, and $\bar{x}_2(\tau_{final}) = V_{d2}$, while following a corresponding state trajectory $x^*(\tau)$.

Given an average normalized equilibrium, say $\bar{x}_2(\tau_{init}) = V_{d1}$ for the normalized average output capacitor voltage, the corresponding equilibrium of the average normalized inductor current is given by

$$\bar{x}_1(t_{init}) = \frac{V_{d1}^2}{Q}$$

The corresponding equilibria of the average normalized flat output are

$$\bar{F}_{init} = \frac{1}{2}\left[\frac{V_{d1}^4}{Q^2} + V_{d1}^2\right], \quad \bar{F}_{final} = \frac{1}{2}\left[\frac{V_{d2}^4}{Q^2} + V_{d2}^2\right]$$

We prescribe a nominal trajectory for the flat output that smoothly interpolates between $\bar{F}_{init}, \bar{F}_{final}$ in a reasonable time interval $[\tau_{init}, \tau_{final}]$.

$$F^*(\tau) = \bar{F}_{init} + (\bar{F}_{final} - \bar{F}_{init})\varphi(\tau, \tau_{init}, \tau_{final})$$

with

$$\varphi(\tau, \tau_{init}, \tau_{final}) = \begin{cases} 0 & \text{for } \tau \leq \tau_{init} \\ \Delta_\tau^8\left[r_1 - r_2\Delta_\tau + \cdots - r_8\Delta_\tau^7 + r_9\Delta_\tau^8\right] & \\ & \text{for } \tau \in [\tau_{init}, \tau_{final}] \\ 1 & \text{for } \tau \geq \tau_{final} \end{cases}$$

where

$$\Delta_t = \left[\frac{\tau - \tau_{init}}{\tau_{final} - \tau_{init}}\right]$$

and

$r_1 = 12870, \quad r_2 = 91520, \quad r_3 = 288288, \quad r_4 = 524160, \quad r_5 = 600600,$

$r_6 = 443520, \quad r_7 = 205920, \quad r_8 = 54912, \quad r_9 = 6435$

This type of polynomial interpolation functions are addressed as Bezier polynomials

Simulations

We used a Boost converter model with the following parameters:

$$L = 20 \text{ [mH]}, \quad C = 20 \text{ [}\mu\text{F]}, \quad R = 30 \text{ [}\Omega\text{]}, \quad E = 15 \text{ [V]}$$

We set $t_1 = 6.32$ [ms], $t_2 = 37.94$ [ms], with $v(t_{init}) = 22.36$ [V], and $v(t_{final}) = 45.0$ [V]. We have chosen $\gamma = 0.5$ as the controller gain parameter.

Figure 5.24 depicts the simulated responses of the non-normalized switched Boost converter circuit to the exact tracking error passive output feedback control. The average designed feedback controller is implemented via a $\Sigma - \Delta$ modulator.

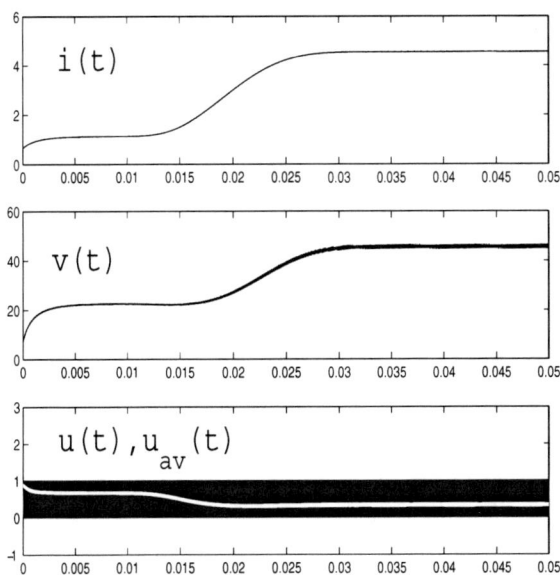

Fig. 5.24. Trajectory tracking response of the switched Boost converter circuit.

5.5.2 Experimental Results

The corresponding non-normalized controller (5.18) is given by

$$u_{av}(t) = u_{av}^*(t) + \gamma_{actual} \left[v^*(t) i - i^*(t) v \right] \quad (5.19)$$

We implemented this controller using the PCI-6025E National Instruments card, in connection with the MATLAB®-Simulink® program. A nominal desired output energy profile, exhibiting a rather smooth start for the Boost converter, was specified using an interpolating Bezier polynomial of tenth order, defined by:

$$F^*(t) = \overline{F}(t_1) + \left[\overline{F}(t_2) - \overline{F}(t_1)\right]\varphi(t, t_1, t_2) \tag{5.20}$$

where $\varphi(t, t_1, t_2)$ is a piecewise polynomial function interpolating between the values of 0 and 1. This function is of the following form:

$$\varphi(t, t_1, t_2) = \begin{cases} 0 & \text{for } t \leq t_1 \\ \left(\frac{t-t_1}{t_2-t_1}\right)^5 \left[252 - 1050\left(\frac{t-t_1}{t_2-t_1}\right) + 1800\left(\frac{t-t_1}{t_2-t_1}\right)^2 \right. \\ \left. - 1575\left(\frac{t-t_1}{t_2-t_1}\right)^3 + 700\left(\frac{t-t_1}{t_2-t_1}\right)^4 - 126\left(\frac{t-t_1}{t_2-t_1}\right)^5\right] \\ & \text{for } t \in (t_1, t_2) \\ 1 & \text{for } t \geq t_2 \end{cases} \tag{5.21}$$

We have used: $t_1 = 0.5$ s, $t_2 = 1$ s, i.e., $t_2 - t_1 = 0.5$ s, and $\gamma_{actual} = 0.1$.
It is desired to transfer the system from the initial equilibrium point,

$$\left[\overline{i}(t_1), \overline{v}(t_1)\right] = \left[\frac{\overline{v}_{init}^2}{RE}, \overline{v}_{init}\right] = [360.58 \text{ mA}, 15 \text{ V}]$$

towards the final equilibrium:

$$\left[\overline{i}(t_2), \overline{v}(t_2)\right] = \left[\frac{\overline{v}_{final}^2}{RE}, \overline{v}_{final}\right] = [923.08 \text{ mA}, 24 \text{ V}]$$

during an interval of time $[t_1, t_2]$. The corresponding average total stored energy values are given by

$$\overline{F}(t_1) = \frac{1}{2}\left[\frac{L}{R^2E^2}\overline{v}_{init}^4 + C\overline{v}_{init}^2\right], \quad \overline{F}(t_2) = \frac{1}{2}\left[\frac{L}{R^2E^2}\overline{v}_{final}^4 + C\overline{v}_{final}^2\right]$$

Finally, the corresponding average control input signal generated by the linear feedback controller of the Boost converter varies between the initial and final values, respectively, $\overline{u}_{av}(t_1) = 0.8$ and $\overline{u}_{av}(t_2) = 0.5$. The system parameters were chosen to be exactly the same as in the corresponding experimental results at the end of Section 5.4.3.

Figure 5.25 depicts the experimental results which achieve the demanded rest to rest task.

5.6 Controller Design via Fliess' Generalized Canonical Form

In this section, dynamic controllers for DC-to-DC power converters are designed by means of state feedback using Fliess' generalized observability canonical form (GOCF). We examine this control synthesis methodology in the

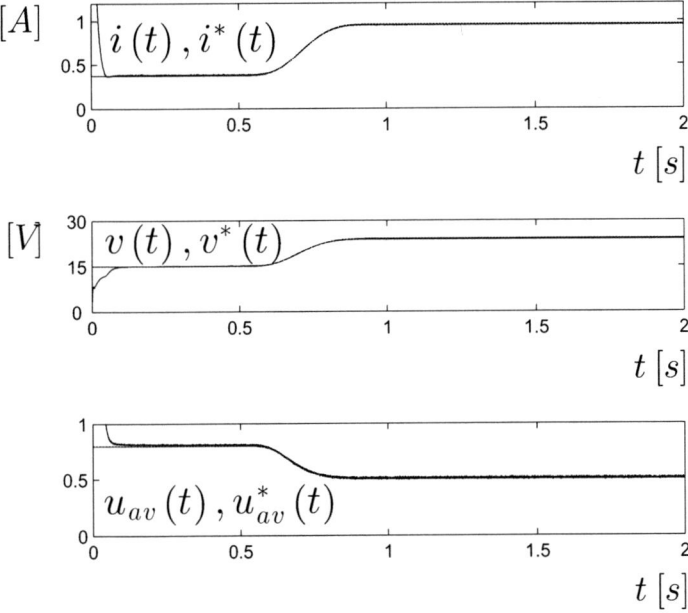

Fig. 5.25. Experimental trajectory tracking response of the switched Boost DC-to-DC converter.

cases of the following converter topologies: Boost, Buck-Boost and quadratic Buck. Also, we carry out the closed loop stability analysis of the nonlinear zero dynamics corresponding to the considered output of the system (normalized currents and/or voltages, respectively). The algebraic theory of nonlinear systems, basically *differential algebra*, which supports the development of Fliess' GOCF is contained in several articles by Prof. Michel Fliess. The reader is advised to browse through [15] for a rather rigorous account of the theory.

5.6.1 The Boost Converter

The average normalized dynamics of the Boost converter is given by:

$$\dot{x}_1 = -(1 - u_{av})x_2 + 1$$
$$\dot{x}_2 = (1 - u_{av})x_1 - \frac{x_2}{Q}$$

System Output: The Input Current

Consider $y = h(x) = x_1$ to be the output of the average normalized Boost system. This output is clearly relative degree one. Consider then the input-dependent state coordinate transformation:

$$\begin{bmatrix} z_1 \\ z_2 \end{bmatrix} = \begin{bmatrix} h(x) \\ \dot{h}(x) \end{bmatrix} = \begin{bmatrix} x_1 \\ -(1-u_{av})x_2 + 1 \end{bmatrix}$$

The corresponding inverse transformation can be obtained, by inspection, as:

$$x_1 = z_1$$
$$x_2 = \frac{1-z_2}{1-u_{av}}$$

We obtain then the following generalized observability canonical form for the Boost converter,

$$\dot{z}_1 = z_2$$
$$\dot{z}_2 = \frac{\dot{u}_{av}}{1-u_{av}}(1-z_2) - (1-u_{av})^2 z_1 + \frac{1-z_2}{Q}$$
$$y = z_1 \qquad (5.22)$$

The zero dynamics, which is now a nonlinear differential equation relating u_{av} and $\dot{u}_{av}$, is found to be stable around a desired output equilibrium point. Substituting $y = \bar{z}_1$ in (5.22) yields to the following relation:

$$\dot{u}_{av} = (1-u_{av})\left[(1-u_{av})^2 \bar{z}_1 - \frac{1}{Q}\right] \qquad (5.23)$$

The following are possible equilibrium points for the zero dynamics:

$$u_{av} = 1, \qquad u_{av} = 1 - \sqrt{\frac{1}{\bar{z}_1 Q}}, \qquad u_{av} = 1 + \sqrt{\frac{1}{\bar{z}_1 Q}}$$

The solution $u_{av} = 1 - \sqrt{1/\bar{z}_1 Q}$ has physical and control theoretic significance. The phase diagram of the zero dynamics (5.23) shown in Figure 5.26, graphically illustrates that this equilibrium value is locally stable. We verify, once more, that the average model of the Boost converter with average inductor current z_1 taken as the system output is of *minimum phase* nature.

Dynamic Feedback Controller Design

Consider the auxiliary input variable v_{av} defined as

$$v_{av} = \frac{\dot{u}_{av}}{1-u_{av}}(1-z_2) - (1-u_{av})^2 z_1 + \frac{1-z_2}{Q}$$

The system can be then be written as the following linear system

$$\dot{z}_1 = z_2$$
$$\dot{z}_2 = v_{av}$$
$$y = z_1$$

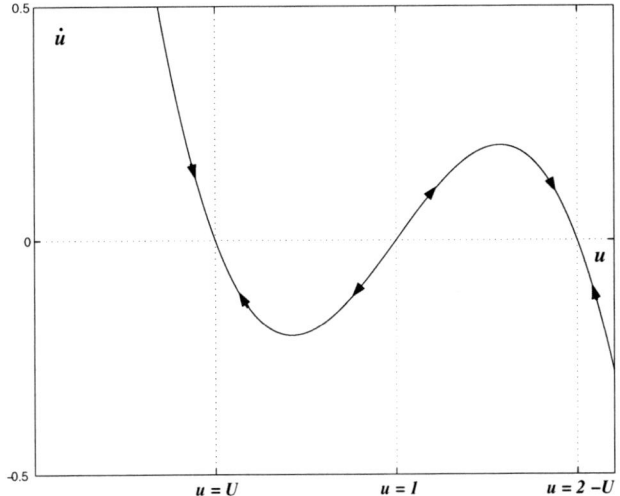

Fig. 5.26. Phase diagram of the inductor current zero dynamics for the Boost converter.

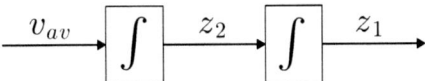

Fig. 5.27. Block diagram of the linearized system.

In fact, the system simply reduces to a set of two integrators arranged in cascade, as shown in Figure 5.27.

It is evident that the problem of stabilizing the average converter current output to a desired reference equilibrium value $\overline{z}_1$, appears now to be trivial. Indeed, the control law

$$v_{av} = -2\xi\omega_n z_2 - \omega_n^2 (z_1 - \overline{z}_1)$$

produces a linear closed loop system with assignable transient features by proper choices of $\xi > 0$ and $\omega_n > 0$,

$$\dot{z}_1 = z_2$$
$$\dot{z}_2 = -2\xi\omega_n z_2 - \omega_n^2 (z_1 - \overline{z}_1)$$
$$y = z_1$$

The trajectories of the closed loop system asymptotically converge to the desired reference equilibrium point, $z_1 = \overline{z}_1$, $\overline{z}_2 = 0$.

The dynamic feedback controller required for synthesizing the auxiliary input v_{av} can be obtained from the relations

$$\frac{\dot{u}_{av}}{1-u_{av}}(1-z_2)-(1-u_{av})^2 z_1+\frac{1-z_2}{Q}=\underbrace{-2\xi\omega_n z_2-\omega_n^2(z_1-\overline{z}_1)}_{=:v_{av}}$$

that results in

$$\dot{u}_{av}=\frac{1-u_{av}}{(1-z_2)}\left[v_{av}+(1-u_{av})^2 z_1-\frac{1-z_2}{Q}\right] \quad (5.24)$$

The proposed controller (5.24) constitutes a *dynamic* nonlinear state feedback control law. We require that the restriction: $z_2 \neq 1$ be valid. This can be guaranteed since on the one hand: $z_2 = 1$ implies that $x_2 = 0$ and, on the other hand, under non-saturated average control, in the case of $y = \overline{x}_1 = \overline{z}_1$, the amplifying condition $x_2 > 1$ of the converter is fulfilled.

Finally, the average dynamic feedback control law may be expressed in terms of the original average normalized variables, x_1 and x_2, resulting in,

$$\dot{u}_{av}=\frac{1}{x_2}\Bigg[-2\xi\omega_n\left[-(1-u_{av})x_2+1\right]-\omega_n^2(x_1-\overline{x}_1)$$
$$+(1-u_{av})^2 x_1-\frac{(1-u_{av})x_2}{Q}\Bigg]$$

Simulations

We use the following component and design parameters for simulating the Boost converter:

$$L=20\text{ mH},\quad C=20\text{ }\mu\text{F},\quad R=30\text{ }\Omega,\quad E=15\text{ V}$$

These parameters result in a quality factor of $Q = 0.9487$ and a time normalization factor of $\sqrt{LC}=6.3246\times 10^{-4}$ s. Furthermore, the desired normalized voltage is enforced to be: $\overline{x}_2 = 2$, in steady state conditions (corresponding to $\overline{v} = 30$ V). These values correspond to a normalized equilibrium current of value, $\overline{x}_1 = 4.2164$ (corresponding to $\overline{i} = 2$ A). The equilibrium value of the average input is $\overline{u}_{av} = 0.5$.

Figure 5.28 shows the closed loop response of the system using the average dynamic controller based on Fliess' GOCF, the purpose of which is regulating the output voltage in an indirect manner employing current control, using a $\Sigma - \Delta$ modulator.

System Output: The Output Capacitor Voltage

We take now the normalized voltage x_2 as the system output output $y = h(x) = x_2$. The relative degree of this output is equal to 1.

Define the following input dependent state coordinate transformation:

$$\begin{bmatrix}z_1\\z_2\end{bmatrix}=\begin{bmatrix}h(x)\\\dot{h}(x)\end{bmatrix}=\begin{bmatrix}x_2\\(1-u_{av})x_1-\frac{x_2}{Q}\end{bmatrix}$$

5.6 Control via Fliess' Generalized Canonical Form

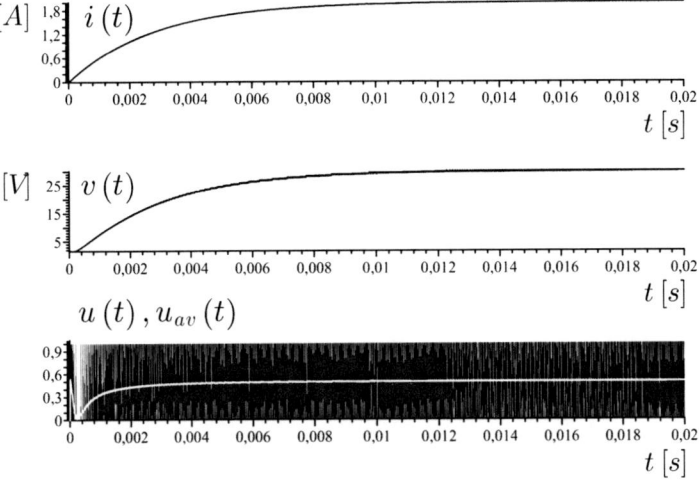

Fig. 5.28. Switched closed loop responses of a Boost power converter to a $\Sigma - \Delta$ modulator implementation of a linearizing controller based on Fliess's GOCF.

The inverse transformation is readily found to be,

$$x_1 = \frac{z_1 + Qz_2}{(1 - u_{av})Q}$$
$$x_2 = z_1$$

Thus, the Fliess' GOCF for the Boost converter, assuming $y = x_2$, is given by

$$\dot{z}_1 = z_2$$
$$\dot{z}_2 = -\frac{\dot{u}_{av}}{(1 - u_{av})} \frac{(z_1 + Qz_2)}{Q} - (1 - u_{av})^2 z_1 + (1 - u_{av}) - \frac{z_2}{Q}$$
$$y = z_1$$

The zero dynamics associated to an equilibrium point of the average output voltage $z_2 = \overline{z}_2$ and corresponding average equilibrium current, $z_1 = \overline{z}_1$, is represented by the following nonlinear differential equation in u_{av},

$$\dot{u}_{av} = -\frac{Q}{\overline{z}_1}(1 - u_{av})^2 \left[(1 - u_{av})\overline{z}_1 - 1\right] \qquad (5.25)$$

This zero dynamics has the following equilibrium points

$$u_{av} = 1, \qquad u_{av} = 1 - \frac{1}{\overline{z}_1}$$

only the second one has a physical meaning since $\overline{z}_1 = \overline{x}_2 > 1$.

322   5 Nonlinear Methods

The phase diagram of the zero dynamics (5.25) is shown in Figure 5.29. Clearly the feasible equilibrium point is unstable. We conclude that the average model of the Boost converter, with output $y = x_2$, is a *non-minimum phase* system.

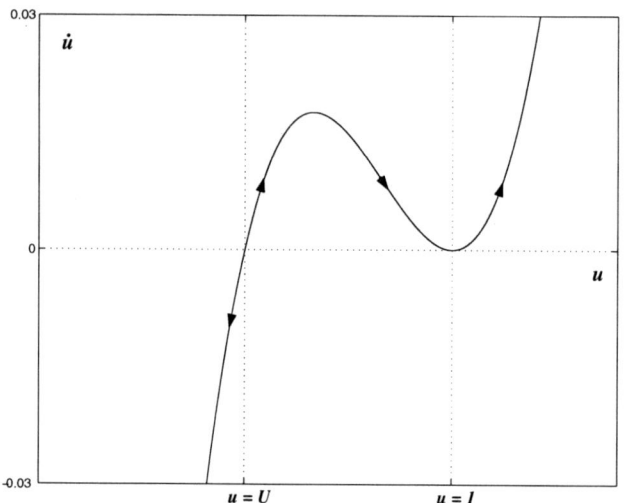

Fig. 5.29. Phase diagram of the zero dynamics of the output capacitor voltage of a Boost converter.

5.6.2 The Buck-Boost Converter

In this section, using Fliess' GOCF for the average normalized Buck-Boost converter, we establish the main features of the stability of the zero dynamics associated to the possible output variables, x_1 and x_2 around given reference equilibrium points and derive a feasible stabilizing dynamical feedback controller for the indirect regulation of the non-minimum phase output x_2.

The average normalized dynamics of the Buck-Boost is given by:

$$\dot{x}_1 = (1 - u_{av})x_2 + u_{av}$$
$$\dot{x}_2 = -(1 - u_{av})x_1 - \frac{1}{Q}x_2$$

System Output: The Input Inductor Current

Consider the average normalized model of the Buck-Boost converter circuit and let $y = x_1$ to be the output of the system. The output $y = x_1$ is, clearly, relative degree 1.

5.6 Control via Fliess' Generalized Canonical Form

Define the following input-dependent state coordinates transformation:

$$\begin{bmatrix} z_1 \\ z_2 \end{bmatrix} = \begin{bmatrix} h(x) \\ \dot{h}(x) \end{bmatrix} = \begin{bmatrix} x_1 \\ (1-u_{av})x_2 + u_{av} \end{bmatrix} \quad (5.26)$$

with inverse transformation given by,

$$\begin{bmatrix} x_1 \\ x_2 \end{bmatrix} = \begin{bmatrix} z_1 \\ \frac{z_2 - u_{av}}{1 - u_{av}} \end{bmatrix}$$

The GOCF of the average normalized Buck-Boost converter system, with output $y = z_1 = x_1$ is then given by,

$$\begin{aligned} \dot{z}_1 &= z_2 \\ \dot{z}_2 &= \left(\frac{1 - z_2}{1 - u_{av}} \right) \dot{u}_{av} - (1 - u_{av})^2 z_1 + \frac{u_{av} - z_2}{Q} \\ y &= z_1 \end{aligned} \quad (5.27)$$

The zero dynamics of the system (5.27), associated with an average output equilibrium point $y = \bar{z}_1 = \bar{x}_1$, can be expressed as

$$\dot{u}_{av} = (1 - u_{av}) \left[(1 - u_{av})^2 \bar{z}_1 - \frac{u_{av}}{Q} \right]$$

The equilibrium points of the zero dynamics are given by

$$u_{av} = 1, \quad u_{av} = 1 + \frac{1}{2Q\bar{z}_1} + \sqrt{\left(\frac{1}{2Q\bar{z}_1} \right)^2 + \frac{1}{Q\bar{z}_1}},$$

$$u_{av} = 1 + \frac{1}{2Q\bar{z}_1} - \sqrt{\left(\frac{1}{2Q\bar{z}_1} \right)^2 + \frac{1}{Q\bar{z}_1}}$$

The first two equilibrium points of the zero dynamics are unstable. The remaining equilibrium, which is the only one with physical meaning, is locally stable. This fact is illustrated in Figure 5.30.

Dynamic Feedback Controller Design

We propose an average dynamical feedback control law for regulating the Buck-Boost converter in the vicinity of the equilibrium point using Fliess's GOCF of the average normalized converter model,

$$\begin{aligned} \dot{z}_1 &= z_2 \\ \dot{z}_2 &= \left(\frac{1 - z_2}{1 - u_{av}} \right) \dot{u}_{av} - (1 - u_{av})^2 z_1 + \frac{u_{av} - z_2}{Q} \\ y &= z_1 \end{aligned} \quad (5.28)$$

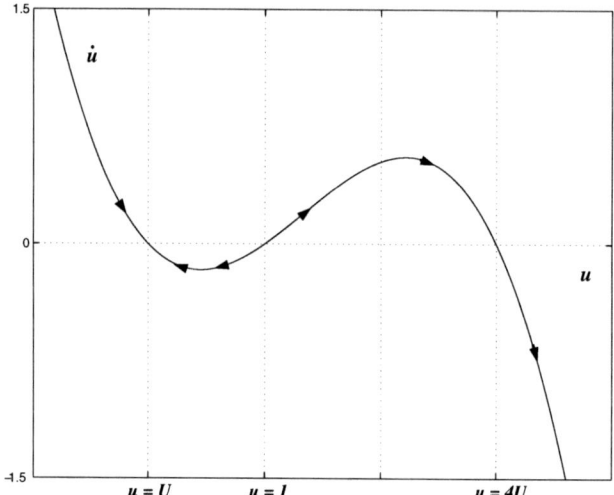

Fig. 5.30. Phase diagram of the output current zero dynamics for the Buck-Boost converter.

Define an auxiliary input v_{av} given by

$$v_{av} = \left(\frac{1-z_2}{1-u_{av}}\right)\dot{u}_{av} - (1-u_{av})^2 z_1 + \frac{u_{av} - z_2}{Q}$$

The input transformed system results in the following linear system:

$$\dot{z}_1 = z_2$$
$$\dot{z}_2 = v_{av}$$
$$y = z_1$$

The problem of stabilizing the average normalized inductor current to a constant value $\overline{z}_1 = \overline{x}_1$, from the auxiliary input v_{av} can be carried out by means of a linear feedback control law. Indeed, the average state feedback control law

$$v_{av} = -2\xi\omega_n z_2 - \omega_n^2 (z_1 - \overline{z}_1)$$

yields a closed loop system of the form:

$$\dot{z}_1 = z_2$$
$$\dot{z}_2 = -2\xi\omega_n z_2 - \omega_n^2 (z_1 - \overline{z}_1)$$
$$y = z_1$$

The trajectories of the average controlled system can be asymptotically stabilized to the desired equilibrium point $z_1 = \overline{z}_1$, $\overline{z}_2 = 0$ by suitable choice of the design parameters $\xi > 0$ and $\omega_n > 0$.

The nonlinear average dynamic state feedback controller for the transformed system (5.28) is given

$$\dot{u}_{av} = \frac{1 - u_{av}}{1 - z_2} \left[-2\xi\omega_n z_2 - \omega_n^2 (z_1 - \overline{z}_1) + (1 - u_{av})^2 z_1 - \frac{u_{av} - z_2}{Q} \right] \quad (5.29)$$

It is required, however, that the condition $z_2 \neq 1$ remains valid throughout the transient performance of the system. The condition is actually a saturation condition thanks to the negative amplifying character of the Buck-Boost converter.

Substituting the coordinate transformation (5.26) in (5.29) we obtain the expression for the average dynamic feedback controller based on Fliess' GOCF of the Buck-Boost converter circuit in the original, normalized state variables.

Simulations

Figure 5.31 illustrates the indirect output voltage regulation for a typical average Buck-Boost converter circuit model with parameter values: $L = 20$ mH, $C = 20$ μF, $R = 30$ Ω, $E = 15$ V. We set as a desired steady state output voltage the value $\overline{x}_2 = -1.5$, then $\overline{x}_1 = 3.95$. This value corresponds to $\overline{v} = -22.5$ V and $\overline{i} = 1.875$ A. The parameter values yield the normalized parameter: $Q = 0.9487$ and the time normalization factor is found to be $\sqrt{LC} = 6.3246 \times 10^{-4}$[s]. The average control input equilibrium value is $\overline{u}_{av} = 0.6$.

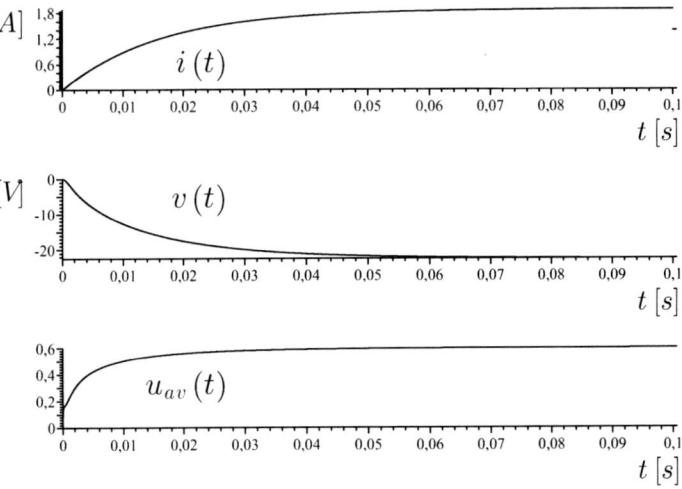

Fig. 5.31. Average responses of a Buck-Boost converter to dynamic linearizing state feedback controller based on Fliess' GOCF.

System Output: The Output Capacitor Voltage

Considering the customary average normalized model of the Buck-Boost converter with output variable $y = x_2$ representing the average normalized output capacitor voltage. Clearly, the relative degree of this output is equal to 1. We define the following input-dependent state coordinates transformation:

$$\begin{bmatrix} z_1 \\ z_2 \end{bmatrix} = \begin{bmatrix} h(x) \\ \dot{h}(x) \end{bmatrix} = \begin{bmatrix} x_2 \\ -(1 - u_{av})x_1 - \frac{x_2}{Q} \end{bmatrix}$$

The inverse transformation can be obtained directly from solving for x_1 and x_2, that is

$$\begin{bmatrix} x_1 \\ x_2 \end{bmatrix} = \begin{bmatrix} -\frac{z_1 + Q z_2}{(1 - u_{av})Q} \\ z_1 \end{bmatrix}$$

Fliess' GOCF for the Buck-Boost converter, assuming $y = z_1 = x_2$, is determined by

$$\dot{z}_1 = z_2$$
$$\dot{z}_2 = v_{av} = -\frac{\dot{u}_{av}}{(1 - u_{av})} \frac{z_1 + Q z_2}{Q} - (1 - u_{av})\left[(1 - u_{av}) z_1 + u_{av}\right] - \frac{z_2}{Q}$$
$$y = z_1$$

The resulting zero dynamics under equilibrium conditions for the output, $y = \overline{z}_1$ results in,

$$\dot{u}_{av} = -\frac{Q}{\overline{z}_1}(1 - u_{av})^2 \left[(1 - u_{av})\overline{z}_1 + u_{av}\right] \tag{5.30}$$

The equilibrium points of the zero dynamics are given by

$$u_{av} = 1, \qquad u_{av} = \frac{\overline{z}_1}{\overline{z}_1 - 1}$$

Among these equilibrium points, $u_{av} = \frac{\overline{z}_1}{\overline{z}_1 - 1}$, is the only one that has a physical meaning, since $\overline{z}_1 = \overline{x}_2 < 0$. The phase diagram of (5.30) is shown in Figure 5.32. It is clear that this equilibrium point is unstable. Hence, the output $y = x_2 = z_1$ representing the output capacitor voltage is a *non-minimum phase* output.

5.6.3 The Quadratic Buck Converter

In this section we obtain Fliess' GOCF of the quadratic Buck converter only for the case when the output of the system is the normalized current x_1. Using Fliess' GOCF for the quadratic Buck converter, we design an indirect dynamic state feedback control for the regulation of the output capacitor voltage which is based on the regulation of the inductor current.

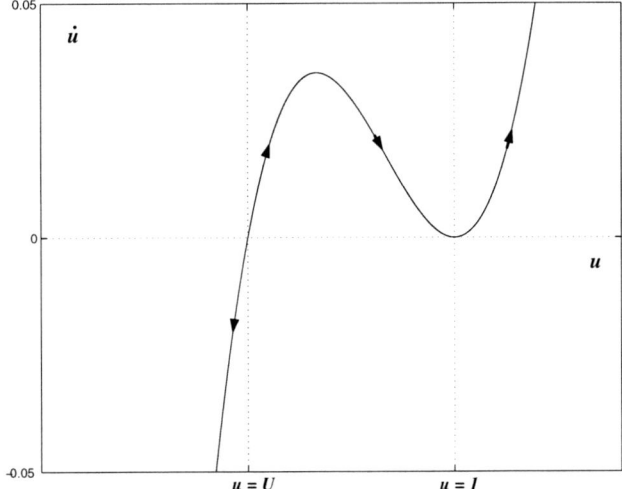

Fig. 5.32. Phase diagram of the zero dynamics of the average output voltage for a Buck-Boost converter.

Consider the normalized average model of the quadratic Buck converter, i.e.,

$$\dot{x}_1 = -x_2 + u_{av}$$
$$\dot{x}_2 = x_1 - u_{av}x_3$$
$$\alpha_1\dot{x}_3 = u_{av}x_2 - x_4$$
$$\alpha_2\dot{x}_4 = x_3 - \frac{x_4}{Q}$$
$$y = x_1$$

The relative degree of the system output is equal to 1, as it can be easily established. The following invertible input-dependent state coordinates transformation

$$\begin{bmatrix} z_1 \\ z_2 \\ z_3 \\ z_4 \end{bmatrix} = \phi(x) = \begin{bmatrix} h(x) \\ \dot{h}(x) \\ \ddot{h}(x) \\ h^{(3)}(x) \end{bmatrix} = \begin{bmatrix} x_1 \\ u_{av} - x_2 \\ \dot{u}_{av} + u_{av}x_3 - x_1 \\ \ddot{u}_{av} + x_3\dot{u}_{av} + \frac{1}{\alpha_1}(u_{av}x_2 - x_4 - \alpha_1)u_{av} + x_2 \end{bmatrix}$$
(5.31)

leads to the Fliess' GOCF for the system that we do not write for the sake of brevity.

The inverse transformation associated to (5.31) is

$$\begin{bmatrix} x_1 \\ x_2 \\ x_3 \\ x_4 \end{bmatrix} = \phi^{-1}(z) = \begin{bmatrix} z_1 \\ u_{av} - z_2 \\ -(\dot{u}_{av} - z_1 - z_3)\frac{1}{u_{av}} \\ \alpha_1 \frac{\ddot{u}_{av}}{u_{av}} - \alpha_1 \frac{(\dot{u}_{av} - z_1 - z_3)}{u_{av}^2} \dot{u}_{av} + (u_{av} - z_2) u_{av} - \alpha_1 \frac{(z_2+z_4)}{u_{av}} \end{bmatrix}$$
(5.32)

The zero dynamics of the Fliess' GOCF for the quadratic Buck converter around a desired average reference equilibrium value for the output, can be shown to be given by:

$$u_{av}^{(3)} + \left(\frac{\bar{z}_1}{u_{av}} - \frac{3}{u_{av}} \dot{u}_{av} + \frac{1}{\alpha_2 Q}\right) \ddot{u}_{av} + \frac{2}{u_{av}^2} \dot{u}_{av}^3 - \left(\frac{2\bar{z}_1}{u_{av}^2} + \frac{1}{\alpha_2 u_{av} Q}\right) \dot{u}_{av}^2$$
$$+ \left(2\frac{u_{av}^2}{\alpha_1} + \frac{1}{\alpha_1 \alpha_2} + \frac{\bar{z}_1}{\alpha_2 u_{av} Q}\right) \dot{u}_{av} + \frac{1}{\alpha_1 \alpha_2} \left(\frac{u_{av}^3}{Q} - \bar{z}_1\right) = 0 \quad (5.33)$$

The equilibrium points of the zero dynamics (5.33) are the roots of the polynomial

$$p(u_{av}) = \frac{1}{\alpha_1 \alpha_2} \left(\frac{u_{av}^3}{Q} - \bar{z}_1\right)$$

that is

$$u_{av} = \sqrt[3]{Q\bar{z}_1}, \quad u_{av} = \frac{1}{2} \sqrt[3]{Q\bar{z}_1} \left(-1 \pm \sqrt{3}i\right)$$

Among these equilibrium points only $u_{av} = \sqrt[3]{Q\bar{z}_1}$ has a physical meaning. Since it is not easy to check the stability of the zero dynamics via a phase diagram, we resort to the construction of a respective Lyapunov function given in Chapter 3. We conclude that the zero dynamics (5.33) is asymptotically stable.

Dynamic Feedback Controller Design

The model of quadratic Buck converter in Fliess' GOCF is given by

$$\dot{z}_1 = z_2$$
$$\dot{z}_2 = z_3$$
$$\dot{z}_3 = z_4$$
$$\dot{z}_4 = u_{av}^{(3)} + x_3 \ddot{u}_{av} + \frac{3u_{av}x_2 - 2x_4 - \alpha_1}{\alpha_1} \dot{u}_{av}$$
$$+ \frac{(u_{av}^2 + \alpha_1)(x_1 - u_{av}x_3)}{\alpha_1} + \frac{x_4 - Qx_3}{\alpha_1 \alpha_2 Q} u_{av}$$
$$y = z_1$$

with the state vector (x_1, x_2, x_3, x_4) defined by the corresponding inverse transformation (5.32).

The linearizing average dynamic feedback controller results from equating the right hand side of $\dot{z}_4$ to an auxiliary input variable v_{av} to obtain:

$$\dot{z}_1 = z_2$$
$$\dot{z}_2 = z_3$$
$$\dot{z}_3 = z_4$$
$$\dot{z}_4 = v_{av}$$
$$y = z_1$$

where

$$v_{av} = u_{av}^{(3)} + x_3 \ddot{u}_{av} + \frac{3u_{av}x_2 - 2x_4 - \alpha_1}{\alpha_1}\dot{u}_{av}$$
$$+ \frac{(u_{av}^2 + \alpha_1)(x_1 - u_{av}x_3)}{\alpha_1} + \frac{x_4 - Qx_3}{\alpha_1\alpha_2 Q}u_{av} \quad (5.34)$$

The equilibrium point of this system is $\overline{z} = (\overline{z}_1, 0, 0, 0)$. Thus, if it is desired that the trajectories of the system to converge to the equilibrium point $\overline{z}$ the auxiliary variable v_{av} may be chosen as

$$v_{av} = -\beta_4 z_4 - \beta_3 z_3 - \beta_2 z_2 - \beta_1 (z_1 - \overline{z}_1) \quad (5.35)$$

which forces that the average system, in closed loop, takes the following form:

$$\begin{pmatrix} \dot{z}_1 \\ \dot{z}_2 \\ \dot{z}_3 \\ \dot{z}_4 \end{pmatrix} = \begin{pmatrix} 0 & 1 & 0 & 0 \\ 0 & 0 & 1 & 0 \\ 0 & 0 & 0 & 1 \\ -\beta_1 & -\beta_2 & -\beta_3 & -\beta_4 \end{pmatrix} \begin{pmatrix} z_1 - \overline{z}_1 \\ z_2 - \overline{z}_2 \\ z_3 - \overline{z}_3 \\ z_4 - \overline{z}_4 \end{pmatrix}$$

Obviously, the characteristic polynomial of this system is

$$p(s) = s^4 + \beta_4 s^3 + \beta_3 s^2 + \beta_2 s + \beta_1 \quad (5.36)$$

and guaranteed to be Hurwitz by means of an appropriate choice of a desired polynomial $p_d(s)$. Hence, we calculate the coefficients of $p_d(s)$ according to stable roots located in the left semi-plane of the complex plane. A desired appropriate characteristic polynomial is the choice

$$p_d(s) = \left(s^2 + 2\xi_1 \omega_{n1} s + \omega_{n1}^2\right)\left(s^2 + 2\xi_2 \omega_{n2} s + \omega_{n2}^2\right)$$
$$= s^4 + 2\left(\xi_1 \omega_{n1} + \xi_2 \omega_{n2}\right) s^3 + \left(\omega_{n1}^2 + 4\xi_1\xi_2\omega_{n1}\omega_{n2} + \omega_{n2}^2\right) s^2$$
$$+ 2\left(\xi_1 \omega_{n1}\omega_{n2}^2 + \xi_2 \omega_{n1}^2 \omega_{n2}\right) s + \omega_{n1}^2 \omega_{n2}^2 \quad (5.37)$$

Equating the coefficients of the polynomials (5.36) and (5.37) we obtain the feedback gains β_1, β_2, β_3, and β_4:

$$\beta_1 = \omega_{n1}^2 \omega_{n2}^2$$
$$\beta_2 = 2\left(\xi_1 \omega_{n1}\omega_{n2}^2 + \xi_2 \omega_{n1}^2 \omega_{n2}\right)$$
$$\beta_3 = \left(\omega_{n1}^2 + 4\xi_1\xi_2\omega_{n1}\omega_{n2} + \omega_{n2}^2\right)$$
$$\beta_4 = 2\left(\xi_1 \omega_{n1} + \xi_2 \omega_{n2}\right)$$

Equating (5.34) and (5.35) we determine the feedback law of the nonlinear dynamic controller, which ensures the stabilization of the system by means of a linearization scheme in the closed loop. Hence,

$$u_{av}^{(3)} = -\beta_4 z_4 - \beta_3 z_3 - \beta_2 z_2 - \beta_1 (z_1 - \bar{z}_1) - x_3 \ddot{u}_{av} - \frac{3 u_{av} x_2 - 2 x_4 - \alpha_1}{\alpha_1} \dot{u}_{av}$$

$$- \frac{(u_{av}^2 + \alpha_1)(x_1 - u_{av} x_3)}{\alpha_1} - \frac{x_4 - Q x_3}{\alpha_1 \alpha_2 Q} u_{av} \quad (5.38)$$

For the simulation we define the state variables

$$\mu_1 = u_{av}, \qquad \mu_2 = \dot{u}_{av}, \qquad \mu_3 = \ddot{u}_{av}$$

The dynamic controller (5.38), expressed in these state variables, exhibits the following form:

$$\dot{\mu}_1 = \mu_2$$
$$\dot{\mu}_2 = \mu_3$$
$$\dot{\mu}_3 = -\beta_4 z_4 - \beta_3 z_3 - \beta_2 z_2 - \beta_1 (z_1 - \bar{z}_1) - x_3 \mu_3 - \frac{3 \mu_1 x_2 - 2 x_4 - \alpha_1}{\alpha_1} \mu_2$$

$$- \frac{(\mu_1^2 + \alpha_1)(x_1 - \mu_1 x_3)}{\alpha_1} - \frac{x_4 - Q x_3}{\alpha_1 \alpha_2 Q} \mu_1 \quad (5.39)$$

Carrying out the corresponding substitutions in (5.31), the transformed state variables z_1, z_2, z_3 and z_4, result in

$$z_1 = x_1$$
$$z_2 = \mu_1 - x_2$$
$$z_3 = \mu_2 + \mu_1 x_3 - x_1$$
$$z_4 = \mu_3 + \mu_2 x_3 + \frac{1}{\alpha_1} (\mu_1 x_2 - x_4 - \alpha_1) \mu_1 + x_2$$

Simulations

Figure 5.33 depicts computer simulations which resemble the closed loop response of the system. The controller design based on Fliess' GOCF is implemented using a $\Sigma - \Delta$ modulator. It illustrates the indirect output voltage regulation for the quadratic Buck converter whose parameters are:

$$L_1 = 600 \ \mu\text{H}, \qquad C_1 = 10 \ \mu\text{F}, \qquad L_2 = 600 \ \mu\text{H}, \qquad C_2 = 10 \ \mu\text{F},$$

$$R = 40 \ \Omega, \qquad E = 100 \ \text{V}$$

with a desired steady state voltage of $\bar{x}_4 = 0.25$, (corresponding to $\bar{v}_2 = 25$ V), and $\bar{x}_1 = 2.4206 \times 10^{-2}$ (corresponding to $\bar{i}_1 = 0.3125$ A). Moreover, in this converter we have $\bar{x}_2 = \bar{u}_{av} = 0.5$ and $\bar{x}_3 = 4.8412 \times 10^{-2}$ (corresponding to $\bar{v}_1 = 50$ V and $\bar{i}_2 = 0.625$ A, respectively). These parameter values yield $Q = 5.164$ and the time normalization factor is $\sqrt{L_1 C_1} = 7.746 \times 10^{-5}$ s.

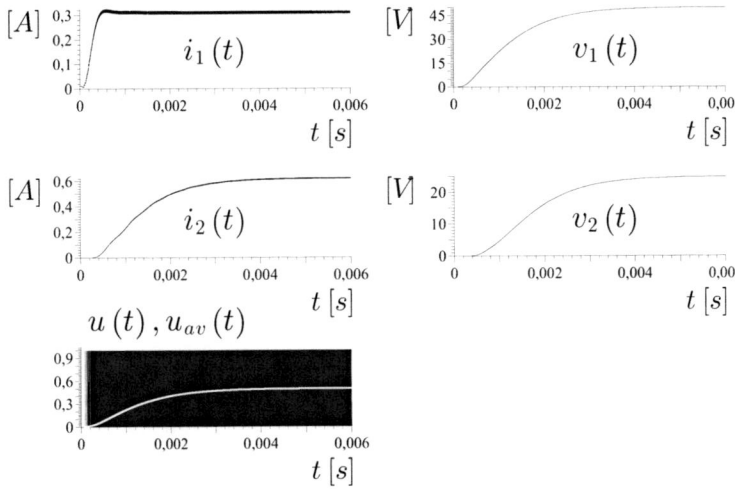

Fig. 5.33. Closed loop response of a quadratic Buck converter to a $\Sigma-\Delta$ modulator implementation of a stabilizing controller based on Fliess' GOCF.

5.7 Nonlinear Observer Design for DC-to-DC Power Converters

The topic of nonlinear observers has been of sustained interest over the last twenty years in the automatic control literature. Several books exist on the topic with various levels of theoretical exposition. The reader is referred to the work of Gauthier and Kupka [28]. A book, containing a rather concise and rigorous treatment of nonlinear observers and the many variations in the direction of adaptive schemes, is that of Marino [41]. Some recent developments on nonlinear observers have been gathered in the edited volume by Nijmeijer and Fossen [46]. New interesting directions about *algebraic observers*, which may have important implications in power electronics devices, has been published by Fliess and Sira-Ramírez [24], [25] and [64]. For a parallel development of nonlinear observers from the standpoint of Generalized Hamiltonian Systems, the reader is referred to [71].

5.7.1 Full Order Observers

Consider the rather general average model of DC-to-DC power converters, which may well be a multi-input system

$$\mathcal{A}\dot{x} = \mathcal{J}(u_{av})x - \mathcal{R}x + \mathcal{B}u_{av} + \mathcal{E}$$
$$y = \mathcal{C}^T x$$

where $x \in R^n$, $u \in R^m$ and $y \in R^p$

We assume the system to be observable, i.e., there exists a differential parametrization of the states in terms of the inputs u and the outputs y.

The nonlinear observer:

$$\mathcal{A}\dot{\hat{x}} = \mathcal{J}(u_{av})\hat{x} - \mathcal{R}\hat{x} + Bu_{av} + \mathcal{E} + \mathcal{K}(y - \hat{y})$$
$$y = C^T \hat{x}$$

yields the following input dependent error system:

$$\mathcal{A}\dot{e} = \mathcal{J}(u_{av})e - \mathcal{R}e - \mathcal{K}e_y$$
$$e_y = C^T e$$

where e is the state estimation error defined to be $e = x - \hat{x}$, e_y is the output estimation error, $e_y = y - \hat{y}$.

Note that if the observer gain vector $\mathcal{K}$ is chosen to be the column vector (or matrix) ΓC, with the $p \times n$ matrix Γ being a strictly positive matrix $\Gamma > 0$, then the output injected estimation error dynamics is given by

$$\mathcal{A}\dot{e} = \mathcal{J}(u_{av})e - \left[\mathcal{R} + \gamma C \Gamma C^T\right] e$$
$$e_y = C^T e$$

The *dual dissipation matching condition* establishes that if the symmetric matrix:

$$\left[\mathcal{R} + C\Gamma C^T\right] \tag{5.40}$$

is positive definite, then the estimation error vector e asymptotically semi-globally converges to zero.

If the matrix (5.40) is only positive semi-definite, then the origin of the estimation error space may still represent an asymptotically stable equilibrium point for the error dynamics, provided that the set of error trajectories e satisfy the following condition:

$$\{e \mid e^T \left[\mathcal{R} + C\Gamma C^T\right] e = 0\} = \{0\}$$

This result follows directly from LaSalle's theorem.

Interestingly enough, if the output vector of the system is given by the system's *passive* output $y = B^T x$, then the input dependent observer enjoys a passive output injection stability property, provided the dissipation matching condition

$$\left[\mathcal{R} + B\Gamma B^T\right] \tag{5.41}$$

is satisfied.

5.7.2 The Boost Converter

Consider the normalized average model of the Boost DC-to-DC power converter.

$$\dot{x}_1 = -u_{av}x_2 + 1$$
$$\dot{x}_2 = u_{av}x_1 - \frac{1}{Q}x_2$$
$$y = x_2$$

where x_1 represents the average normalized inductor current, x_2 is the average normalized output voltage. The function u_{av} denotes the average control input.

It is desired to asymptotically obtain the average inductor current x_1 on the basis of a full order observer based on the measured average input u_{av} and the average output y.

We propose the following full order Luenberger type of observer:

$$\dot{\hat{x}}_1 = -u_{av}\hat{x}_2 + 1 + \lambda_1(y - \hat{x}_2)$$
$$\dot{\hat{x}}_2 = u_{av}\hat{x}_1 - \frac{1}{Q}\hat{x}_2 + \lambda_2(y - \hat{x}_2)$$
$$y = x_2$$

The estimation errors, $e_1 = x_1 - \hat{x}_1$, and, $e_2 = x_2 - \hat{x}_2$, evolve according to the following input dependent dynamics

$$\begin{bmatrix} \dot{e}_1 \\ \dot{e}_2 \end{bmatrix} = \begin{bmatrix} 0 & -u_{av} \\ u_{av} & 0 \end{bmatrix} \begin{bmatrix} e_1 \\ e_2 \end{bmatrix} - \begin{bmatrix} 0 & 0 \\ 0 & \frac{1}{Q} \end{bmatrix} \begin{bmatrix} e_1 \\ e_2 \end{bmatrix} - \begin{bmatrix} \lambda_1 e_2 \\ -\lambda_2 e_2 \end{bmatrix}$$

Note that the influence of the average control input u_{av} is centered on the conservative forces of the estimation error dynamics and, hence, it does not affect the stability of the observer error.

Consider the following Lyapunov function candidate for assessing the stability of the estimation error dynamics

$$V(e) = \frac{1}{2}[e_1^2 + e_2^2] > 0$$

The time derivative of this average estimation energy function, along the trajectories of the system, is given by

$$\dot{V}(e) = -\lambda_1 e_1 e_2 - \left(\lambda_2 + \frac{1}{Q}\right)e_2^2$$

Setting $\lambda_1 = 0$, we obtain that $\dot{V}(e) \leq 0$. The set of error vectors where $\dot{V}(e)$ is zero coincides then with $e_2 = 0$. But $e_2 = 0$ implies $u_{av}e_1 = 0$. Since u_{av} is not identically zero when the system is being controlled, then $e_1 = 0$ is the only controlled trajectory of the system which is compatible with the largest invariant set.

The estimation error dynamics has the origin as an asymptotically stable equilibrium point. The estimated average inductor current converges to the actual average current.

Simulations

We tested the performance of the full order observer with a controller derived on the basis of the passivity based control methodology using energy shaping plus damping injection previously explained in this chapter.

$$u_{av} = \frac{-\dot{x}_1^*(t) + 1 + R_I(\widehat{x}_1 - x_1^*(t))}{\zeta}$$

$$\dot{\zeta} = u_{av} x_1^*(t) - \frac{1}{Q}\zeta$$

where $x_1^*(t)$ is the desired reference trajectory for x_1. We set as $x_1^*(t)$ the constant desired equilibrium value $\overline{x}_1 = \frac{V_d^2}{Q}$ with $V_d = 1.5$. We set the controller gain R_I to the value of 1. The observer gain λ_2 was also set to 1. The value of Q was also set to 1.

Figure 5.34 illustrates the response of an average Boost converter model to a passivity based controller, synthesized with the help of a full order observer $(Q = 1)$. The corresponding switched controlled trajectories, implemented via a $\Sigma - \Delta$ modulator are shown in Figure 5.35.

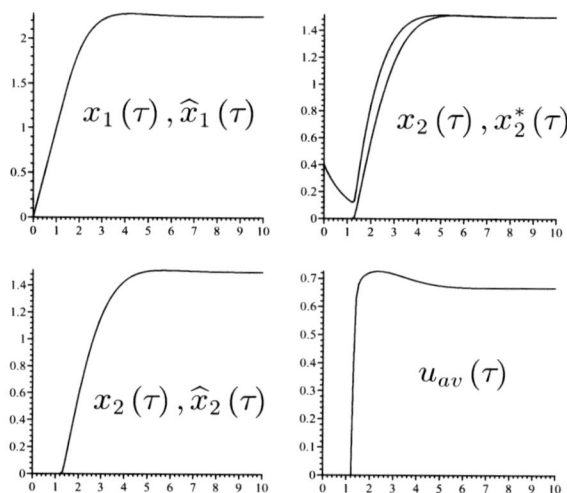

Fig. 5.34. Average responses of Boost converter controlled via passivity based control and a full order observer.

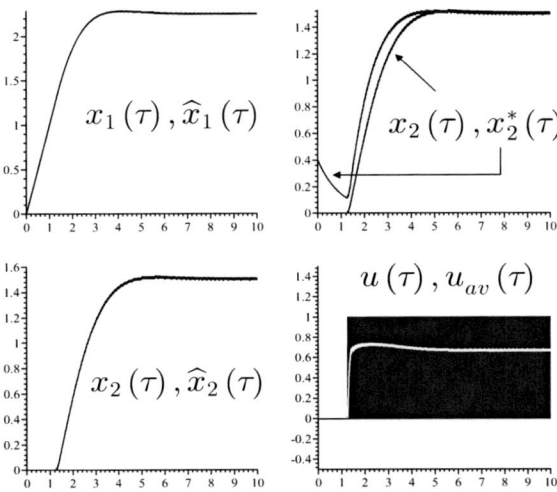

Fig. 5.35. Responses of switched Boost converter controlled via passivity based control, a full order observer and a $\Sigma - \Delta$ modulator.

5.7.3 The Buck-Boost Converter

Consider the normalized average model of the Buck-Boost DC-to-DC power converter.

$$\dot{x}_1 = u_{av}x_2 + (1 - u_{av})$$
$$\dot{x}_2 = -u_{av}x_1 - \frac{1}{Q}x_2$$
$$y = x_2$$

where x_1 is the normalized inductor current, x_2 stands for the normalized output voltage and u_{av} is the average control input or average switch position function.

It is desired to asymptotically obtain the average inductor current x_1 on the basis of a full order observer based on the measured average input u_{av} and the average output y.

We propose as a full order observer the Luenberger type:

$$\dot{\widehat{x}}_1 = u_{av}\widehat{x}_2 + 1 - u_{av} + \lambda_1(y - \widehat{x}_2)$$
$$\dot{\widehat{x}}_2 = -u_{av}\widehat{x}_1 - \frac{1}{Q}\widehat{x}_2 + \lambda_2(y - \widehat{x}_2)$$
$$y = x_2$$

The estimation errors, $e_1 = x_1 - \widehat{x}_1$, and, $e_2 = x_2 - \widehat{x}_2$, evolve according to the following input dependent dynamics

$$\begin{bmatrix} \dot{e}_1 \\ \dot{e}_2 \end{bmatrix} = \begin{bmatrix} 0 & u_{av} \\ -u_{av} & 0 \end{bmatrix} \begin{bmatrix} e_1 \\ e_2 \end{bmatrix} - \begin{bmatrix} 0 & 0 \\ 0 & \frac{1}{Q} \end{bmatrix} \begin{bmatrix} e_1 \\ e_2 \end{bmatrix} - \begin{bmatrix} \lambda_1 e_2 \\ -\lambda_2 e_2 \end{bmatrix}$$

Note that the influence of the average control input u_{av} is ascribed to the conservative forces of the estimation error dynamics and, hence, it does not affect the stability of the origin of coordinates of the observation error.

Consider the following Lyapunov function candidate for assessing the stability of the estimation error dynamics

$$V(e) = \frac{1}{2}\left[e_1^2 + e_2^2\right]$$

The time derivative of this energy function, along the trajectories of the system, is given by

$$\dot{V}(e) = -\lambda_1 e_1 e_2 - \left(\lambda_2 + \frac{1}{Q}\right)e_2^2$$

Setting $\lambda_1 = 0$, we obtain that $\dot{V}(e) \leq 0$. The set of error vectors where $\dot{V}(e)$ is zero coincides then with $e_2 = 0$. But $e_2 = 0$ implies $u_{av}e_1 = 0$. Since u_{av} is not identically zero, then $e_1 = 0$ is the only trajectory of the system which is compatible with the largest invariant set.

The estimation error dynamics has the origin as a semi-globally asymptotically stable equilibrium point. The estimated average inductor current converges to the actual average current.

Simulations

We tested the performance of the full order observer with a controller derived on the basis of the passivity based control methodology that uses energy shaping plus damping injection.

$$u_{av} = \frac{-x_1^*(t) - 1 + u_{av} + R_I(\widehat{x}_1 - x_1^*(t))}{\zeta}$$

$$\dot{\zeta} = -u_{av}x_1^*(t) - \frac{1}{Q}\zeta$$

where $x_1^*(t)$ is the desired trajectory for x_1. We set $x_1^*(t)$ to be the constant desired equilibrium value $\overline{x}_1 = -\frac{(1-V_d)V_d}{Q}$ with $V_d = -1.5$. We set the controller gain R_I to the value 1. The observer gain λ_2 was also set to 1 and $Q = 1$.

The figure 5.36 depicts the response of the switched Buck-Boost normalized converter model to a passivity based controller synthesized with the help of a full order observer and a $\Sigma - \Delta$ modulator.

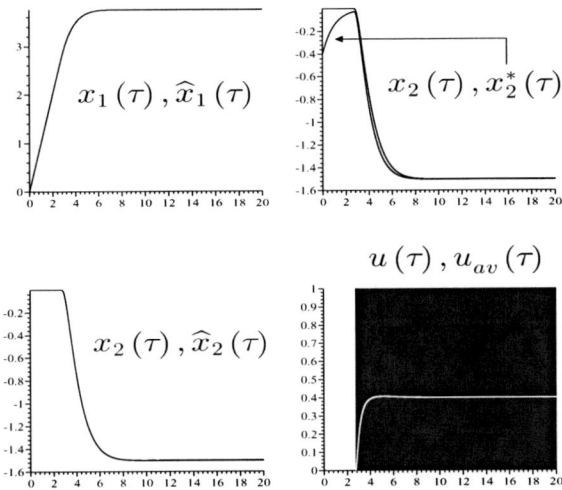

Fig. 5.36. Switched response of Buck-Boost converter controlled via a passivity based control, a full order observer and $\Sigma - \Delta$ modulation.

5.8 Reduced Order Observers with Input Dependent Error

5.8.1 The Boost Converter

Consider the normalized average model of the Boost DC-to-DC power converter

$$\dot{x}_1 = -u_{av}x_2 + 1$$
$$\dot{x}_2 = u_{av}x_1 - \frac{1}{Q}x_2$$
$$y = x_2$$

It is desired to asymptotically obtain the average inductor current x_1 on the basis of a reduced order observer which uses only the measured average input u_{av} and the average output y.

Recall that the average control input u_{av} is a signal constrained to take values in the open interval $(0, 1)$ of the real line. Also, note that the signal $u_{av}x_1$ can be expressed as follows in terms of the measured output y

$$u_{av}x_1 = \dot{y} + \frac{1}{Q}y$$

We thus propose the following reduced order observer:

$$\dot{\hat{x}}_1 = -u_{av}y + 1 + \lambda(u_{av}x_1 - u_{av}\hat{x}_1)$$

The estimation error $e = x_1 - \hat{x}_1$ evolves according to the input dependent linear dynamics

$$\dot{e} = -\lambda u_{av} e$$

which, by virtue of the strictly positive character of the average input u_{av} results in a dynamics with the origin as an asymptotically stable equilibrium point provided the input trajectory $u_{av}(t)$ is bounded away from zero by a strictly negative real constant.

Under these circumstances, letting, $d\rho = u_{av}(\tau)d\tau$, we find that the estimation error, in transformed time scale satisfies the asymptotically stable linear time-invariant dynamics

$$\frac{d}{d\rho} e(\rho) = -\lambda e(\rho)$$

Substituting the expression for $u_{av} x_1$ in the observer dynamics we obtain:

$$\dot{\hat{x}}_1 = -u_{av} y + 1 + \lambda(\dot{y} + \frac{1}{Q} y - u_{av} \hat{x}_1)$$

Defining $\zeta = \hat{x}_1 - \lambda y$ yields the following reduced order nonlinear observer:

$$\dot{\zeta} = -\lambda u_{av} \zeta + \left[\frac{\lambda}{Q} - u_{av}(1 + \lambda^2)\right] y + 1$$
$$\hat{x}_1 = \zeta + \lambda y$$

Note that if the initial states are at rest at the origin and the observer state is initially set to zero, then the estimation error is identically zero for all t. Such a singular situation must and can be avoided in practise.

Simulations

We test the performance of the proposed controller with several feedback controllers.

First, consider the traditional sliding mode controller

$$u = \frac{1}{2}[1 + \text{sign}(s)]$$

with an estimated sliding surface of the form: $s = \hat{x}_1 - \frac{V_d^2}{Q}$.

It is desired to drive the output voltage to the average equilibrium $V_d = 1.5$. This requires to indirectly control the inductor current to the value V_d^2/Q. We show the simulations corresponding to zero initial conditions of the converter and the observer when $Q = 1$.

The control is initially saturated to the value of zero, while the inductor charges from the external voltage source. The estimator dynamics corresponds with $\dot{\zeta} = 1$ which is exactly the same as that for the inductor current $\dot{x}_1 = 1$.

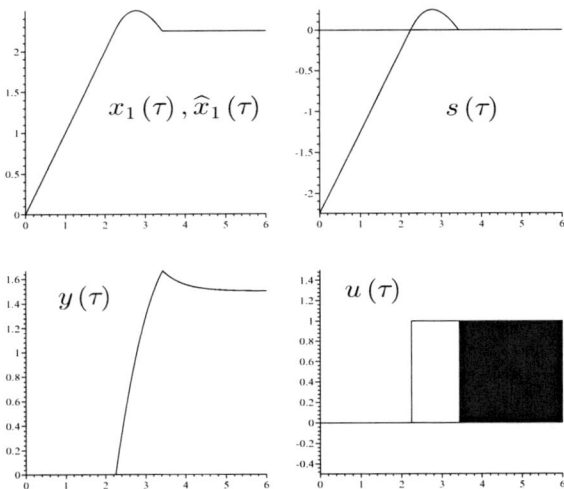

Fig. 5.37. Sliding mode controller responses of a Boost converter using a reduced order observer.

Since both the variable x_1 and ζ start from zero, the responses are identical. When the current overshoots the desired equilibrium value the switched control input changes and saturates to $u = 1$, the observer is activated and perfectly tracks the state due to the identical values up to that activation moment. The sliding mode creation is imminent. This controlled behavior of the Boost converter is shown in Figure 5.37

If the initial states of the system and of the estimator are not equal, then due to the initial saturation of the control input to zero, the inductor current and the reduced order observer state variables both evolve as straight parallel lines with a finite difference error. Once the desired current is overshot the observer is activated and a sliding regime is induced. The controlled behavior for the considered Boost converter is depicted in Figure 5.38.

Consider now an average controller based on the linearized system model and the exact tracking error dynamics passive output feedback scheme. Such a controller was shown to have interesting semi-global stability features (see, respectively, Sections 4.4.9 and 5.4.2). The controller is recalled as given by,

$$u_{av} = \frac{1}{V_d} - \gamma \left[-V_d \left(x_1 - \frac{V_d^2}{Q} \right) + \frac{V_d^2}{Q} (x_2 - V_d) \right]$$

We use following observer based controller using the previously introduced reduced order nonlinear observer,

$$u_{av} = \frac{1}{V_d} - \gamma \left[-V_d \left(\widehat{x}_1 - \frac{V_d^2}{Q} \right) + \frac{V_d^2}{Q} (y - V_d) \right]$$

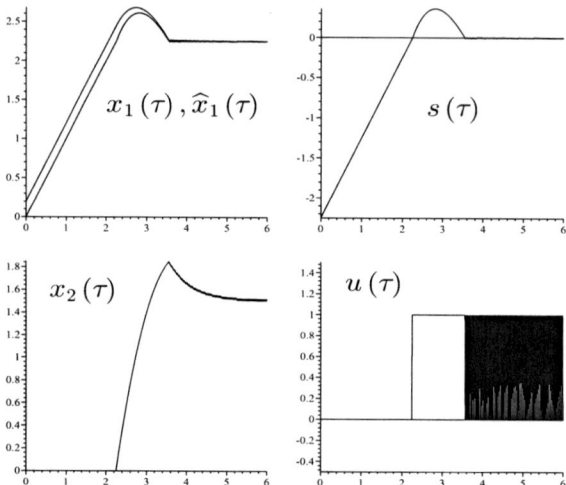

Fig. 5.38. Sliding mode controller responses of a Boost converter using a reduced order observer.

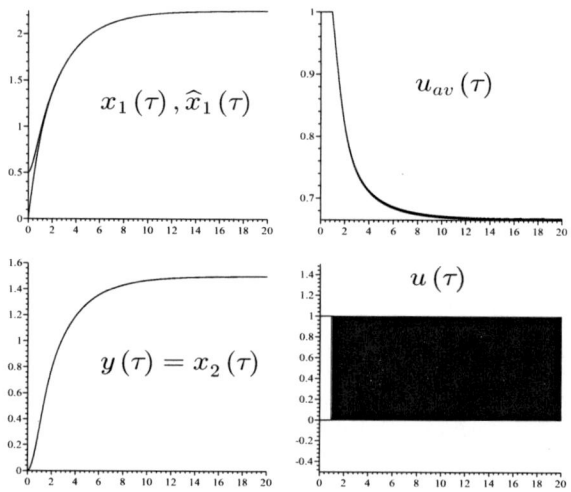

Fig. 5.39. Switched responses of Boost converter to an exact stabilization error passive output feedback controller using a reduced order observer and $\Sigma - \Delta$ modulation.

5.8.2 The Buck-Boost Converter

Consider the normalized average model of the Buck-Boost DC-to-DC power converter.

$$\dot{x}_1 = u_{av}x_2 + (1 - u_{av})$$
$$\dot{x}_2 = -u_{av}x_1 - \frac{1}{Q}x_2$$

with x_1 being the normalized inductor current, x_2 represents the normalized output voltage and u_{av} stands for the average control input.

It is desired to asymptotically obtain the average inductor current x_1 on the basis of a reduced order observer based on the measured average input u_{av} and the average output y

As in most of the studied converters, the average control input u_{av} is a signal constrained to take values in the interval $[0, 1]$ of the real line. Also, note that the signal $u_{av}x_1$ can be expressed as follows, in terms of the measured output y

$$u_{av}x_1 = -\dot{y} - \frac{1}{Q}y$$

We thus propose the following reduced order observer:

$$\dot{\hat{x}}_1 = u_{av}y + 1 - u_{av} + \lambda(u_{av}x_1 - u_{av}\hat{x}_1)$$

The estimation error $e = x_1 - \hat{x}_1$ evolves according to the input dependent linear dynamics:

$$\dot{e} = -\lambda u_{av} e$$

which, by virtue of the strictly positive character of the average input u_{av} exhibited when the system is being controlled, results in a dynamics with the origin as an asymptotically stable equilibrium point.

Substituting the expression for $u_{av}x_1$ in the observer dynamics, we obtain, after some algebraic manipulations which involve the definition of the variable ζ as $\zeta = \hat{x}_1 - \lambda y$:

$$d\zeta = -\lambda u_{av}\zeta + \left[(1 + \lambda^2)u_{av} - \frac{\lambda}{Q}\right] y + (1 - u_{av})$$
$$z = \zeta - \lambda y$$

Note that, as in the previous case, if the system initial states are at rest at zero and the observer is initially set to zero, then the estimation error is identically zero for all t.

Simulations

We test the performance of the proposed reduced order observer with several average feedback controllers.

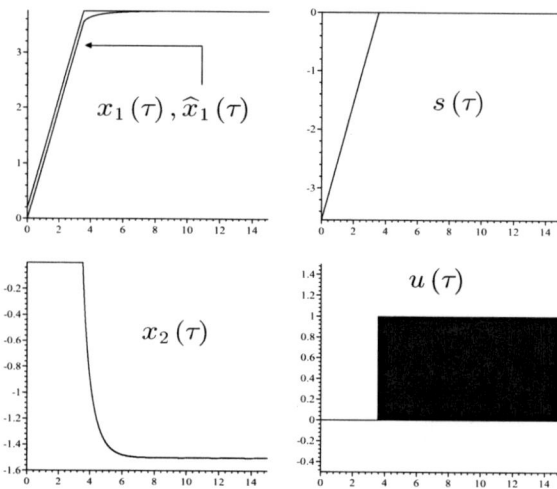

Fig. 5.40. Sliding mode controlled responses of a Buck-Boost converter using a reduced order observer.

First, consider the traditional sliding mode controller

$$u = \frac{1}{2}[1 + \text{sign}(s)]$$

with an estimated sliding surface of the form: $s = \hat{x}_1 + \frac{(1-V_d)V_d}{Q}$.

It is desired to drive the output voltage to the average equilibrium value: $V_d = -1.5$. This requires to indirectly control the inductor current to the value: $\overline{x}_1 = -(1 - V_d)V_d/Q$. We show the simulations corresponding to non zero initial conditions for the converter and for the normalized resistor value, $Q = 1$.

The control is initially saturated to the value of zero, while the inductor charges from the external voltage source. The estimator dynamics is blocked and it corresponds with, $\dot{\zeta} = 1$, which is exactly the same as that for the closed loop inductor current $\dot{x}_1 = 1$. Since both the variable x_1 and ζ start from different initial conditions, the responses differ only by a constant. When the current overshoots the desired equilibrium value, the control input changes from zero to one and it remains saturated to $u = 1$ while the sliding surface is newly reached. The observer being activated tracks the state. The sliding mode controlling the system is thus created. The controlled behavior of the system is depicted in Figure 5.40.

Consider now an average feedback controller based on the linearized system model and the static passive output linear feedback scheme (see, respectively, Sections 4.5.10 and 5.4.4). The controller is recalled as given by,

$$u_{av} = \frac{1}{1-V_d} + \gamma \left[(1-V_d)\left(x_1 + \frac{(1-V_d)V_d}{Q}\right) - \frac{(1-V_d)V_d}{Q}(x_2 - V_d)\right]$$

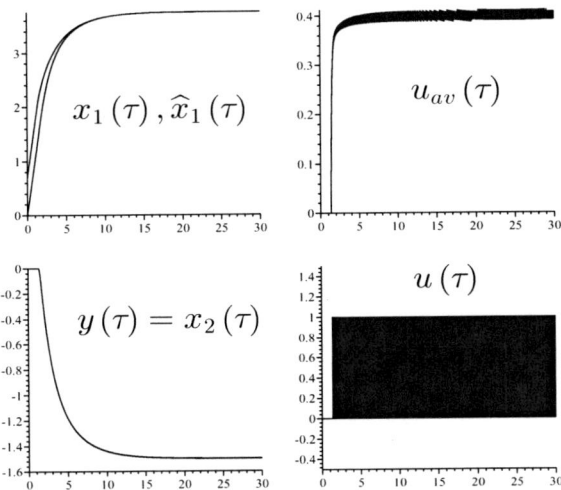

Fig. 5.41. Switched controlled responses of a Buck-Boost converter using an incremental static passive controller, a reduced order observer and a $\Sigma - \Delta$ modulator.

We use following controller based on the previously introduced nonlinear observer,

$$u_{av} = \frac{1}{1-V_d} + \gamma \left[(1-V_d)\left(\widehat{x}_1 + \frac{(1-V_d)V_d}{Q}\right) - \frac{(1-V_d)V_d}{Q}(y - V_d) \right]$$

5.9 GPI Sliding Mode Control of DC-to-DC Power Converters

In this section we propose some additional sliding mode feedback control options for the Buck, the Boost and the Buck-Boost converter circuits. The proposed schemes are based on the idea of *Integral State Reconstructors*, which has resulted in a far reaching generalization of classical PID control and is addressed as *Generalized PI controllers* (GPI) (See Fliess et al. [20], [21], Fliess and Márquez [19], and Fliess [16]). The GPI control technique side-steps the need for asymptotic observers, or for on-line calculations based on samplings and time-discretizations, in the feedback regulation of observable linear dynamic systems. The extension of the integral reconstructor-based feedback control technique to the nonlinear arena, and in particular to switched systems, is here accomplished in the context of the sliding mode regulation of DC-to-DC power converter circuits, of the Buck, Boost and Buck-Boost types, operating in continuous conduction mode. The results concerning the GPI control of the Boost converter can also be found in an article by Sira-Ramírez et al. [69].

The integral reconstructor-based sliding mode schemes for the treated power converters are shown to exhibit similar stabilizing features as the traditional sliding mode controllers, but they turn out to be *vastly superior* as far as robustness, with respect to un-modelled load resistance parameter variations, is concerned. The proposed feedback control schemes gives traditional "op-amps", and modern integrated analog circuits, a renewed importance in the feedback regulation of power electronics circuits. The direct economic consequence is then a substantial lowering of implementation costs.

5.9.1 The Buck Converter

Consider the average normalized model of the Buck converter

$$\dot{x}_1 = -x_2 + u$$
$$\dot{x}_2 = x_1 - \frac{z_2}{Q} \tag{5.42}$$

with output signal represented by the normalized output capacitor voltage $y = x_2$. The system is clearly observable from this output variable and, hence, reconstruction of the unmeasured state variable x_1 is possible (see Fliess et al., [20]).

An integral input output parametrization, or an integral resconstructor, of the average normalized inductor current, $x_1(\tau)$, is directly obtained from the first of the system equations (5.42) by simple integration,

$$\widehat{x}_1(\tau) = \int_0^\tau (u(\rho) - y(\rho)) \, d\rho \tag{5.43}$$

The integral reconstructor of x_1 may be considered to be a "open loop estimate" of the normalized inductor current x_1 which is biased by an unknown constant value, represented by the initial condition $x_1(0)$.

It is clear that the relation linking the estimated value $\widehat{x}_1$ of x_1 to its actual value, is just given by

$$x_1(\tau) = \widehat{x}_1(\tau) + x_1(0) \tag{5.44}$$

We use the estimate (5.43) of the inductor current, x_1, in the traditional sliding surface definition

$$S = \{\, x \in R^2 \mid \sigma(x) = x_1 - \overline{x}_1 = x_1 - V_d/Q = 0 \,\} \tag{5.45}$$

and proceed to complement the expression with an integral control action, computed on the basis of the output voltage stabilization error, $y - V_d$.

Consider then the following integral reconstructor-based sliding mode controller,

$$u_{av} = \begin{cases} 1 \text{ for } \widehat{\sigma}(y, u), \xi) < 0 \\ 0 \text{ for } \widehat{\sigma}(y, u, \xi) > 0 \end{cases} \tag{5.46}$$

5.9 GPI Sliding Mode Control

$$\widehat{\sigma}(y, u), \xi) = \int_0^\tau (u(\rho) - y(\rho)) \, d\rho - \frac{V_d}{Q} + k_0\xi \tag{5.47}$$

$$\dot{\xi} = y(\tau) - V_d, \quad \xi(0) = 0 \tag{5.48}$$

with k_0 being a strictly positive design constant to be chosen later.

The modified sliding surface coordinate function, $\widehat{\sigma}$, can also be equivalently written in terms of the, non-measured, actual state x_1 as,

$$\widehat{\sigma}(x_1, \xi) = x_1 - \frac{V_d}{Q} - x_1(0) + k_0\xi \tag{5.49}$$

In spite of the unknown constant value of $x_1(0)$, the expression (5.49) is found to be useful for our analysis purposes.

The time derivative of any of the two equivalent expressions of the modified sliding surface coordinate function (5.49), or (5.47), is given by

$$\dot{\widehat{\sigma}}(y, u, \xi) = u - y + k_0(y - V_d) \tag{5.50}$$

Note that on $\widehat{\sigma} = 0$ the inductor current, x_1, is given by the expression $x_1 = \frac{V_d}{Q} + x_1(0) - k_0\xi$.

The equivalent control, corresponding to the modified sliding surface coordinate function is now given by

$$u_{eq} = y - k_0(y - V_d) \tag{5.51}$$

The ideal sliding dynamics, obtained from the *invariance conditions*, $\widehat{\sigma} = 0$, $\dot{\widehat{\sigma}} = 0$, is obtained as

$$\begin{aligned} \dot{x}_1 &= -k_0(y - V_d) \\ \dot{y} &= x_1 - \frac{y}{Q} \\ \dot{\xi} &= y - V_d \end{aligned} \tag{5.52}$$

Thus, the ideal closed loop behavior of the normalized output capacitor voltage is governed by the second order differential equation

$$\ddot{y} + \frac{1}{Q}\dot{y} + k_0(y - V_d) = 0 \tag{5.53}$$

Given the strictly positive character of Q and k_0, the ideal sliding behavior of the output signal y exponentially asymptotically converges towards the desired equilibrium value $\overline{y} = V_d$. The corresponding average equilibrium point of the normalized inductor current is then given by $\overline{x}_1 = V_d/Q$.

The only constant equilibrium point, $(\overline{y}, \overline{x}_1, \overline{\xi})$, of the ideal closed loop sliding dynamics, according to (5.49) and (5.52), is given by

$$\overline{x}_1 = \frac{V_d}{Q}, \quad \overline{y} = V_d, \quad \overline{\xi} = \frac{1}{k_0} x_1(0), \tag{5.54}$$

A sliding regime locally exists on $\hat{\sigma}(y, u, \xi) = 0$ whenever the following *existence condition*, $0 < u_{eq} < 1$, is satisfied (see Sira-Ramírez [66]):

$$0 < (1 - k_0)y + k_0 V_d < 1 \tag{5.55}$$

which, under the prevailing physical considerations, is equivalent to the following set of inequalities

$$\begin{cases} \max\left\{0, \frac{1-k_0 V_d}{k_0 - 1}\right\} < y < \min\left\{1, \frac{k_0 V_d}{k_0 - 1}\right\} & \text{for } k_0 > 1 \\ 0 < y < 1 \leq \frac{1-k_0 V_d}{1-k_0}, & \text{for } 0 < k_0 < 1 \end{cases} \tag{5.56}$$

Thus, the set of values for k_0 that guarantees a larger region of existence of a sliding regime, compatible with the physical limitations, corresponds to the condition, $k_0 \in (0, 1)$. The following choice of, k_0, as a strictly positive constant, within the interval:

$$0 < k_0 < V_d < 1 \tag{5.57}$$

clearly guarantees the non-empty character of the region of existence of sliding motions and it will prove to be most convenient to assure sliding surface reachability from the origin of the system's state space. We thus assume that condition 5.57 remains valid throughout.

The local reachability of the sliding surface, $\hat{\sigma} = 0$, from an arbitrary, though physically compatible, initial state value, is established by the well known condition, $\hat{\sigma}\dot{\hat{\sigma}} < 0$, to be verified in a neighborhood of the modified sliding surface. Suppose that (5.57) is valid. Let $\hat{\sigma} < 0$, then, according to (5.46), the control is set to $u = 1$. The time derivative of the modified sliding surface coordinate is given by $\dot{\hat{\sigma}} = 1 + (k_0 - 1)y - k_0 V_d$. Then for all $y < (1 - k_0 V_d)/(1 - k_0)$, which is always the case since $y < 1$ and $k_0 \in (0, V_d)$, the time derivative, $\dot{\hat{\sigma}}$ is positive and the product $\hat{\sigma}\dot{\hat{\sigma}}$ is negative. Suppose now that $\hat{\sigma}$ is positive, then, the control input is given by $u = 0$. The time derivative of the sliding surface coordinate is $\dot{\hat{\sigma}} = (k_0 - 1)y - k_0 V_d$. Thus, for all $y > 0$, the product $\hat{\sigma}\dot{\hat{\sigma}}$ is, again, negative. We conclude that the modified sliding surface, $\hat{S} = \{ (y, u, \xi) \mid \sigma(y, u, \xi) = 0\}$, is reachable in finite time, by means of the proposed discontinuous control law (5.46). Due to the physical restrictions on the state of the system $y, x_1 > 0$ and $y < 1$, we say that the sliding surface is *semi-globally reachable* in finite time.

For the reachability of the sliding surface from the origin, suppose the system is initially resting at the zero state, $x_1(0) = 0, x_2(0) = 0, \xi(0) = 0$, then, the initial value of the modified sliding surface is negative, $\hat{\sigma}(x_1(0), 0) = \hat{\sigma}(0,0) = -V_d/Q < 0$, and u is set to 1. The initial value of the product, $\hat{\sigma}\dot{\hat{\sigma}}$, is given by:

$$\hat{\sigma}(0,0)\dot{\hat{\sigma}}(0,0) = -\frac{V_d}{Q}(1 - k_0 V_d)$$

The modified sliding surface $\hat{\sigma}$, thus, starts increasing towards zero from the given zero initial condition, provided k_0 is chosen within the prescribed interval, $k_0 \in (0, V_d) \subset (0, 1)$. In light of the semi-global reachability property shown above, we have that, in particular, the sliding surface is always reachable, by means of the proposed switched control strategy (5.46), from the origin of the state space. Note that the origin is commonly regarded as a starting point for the operation of DC-to-DC power converters. We summarize the proven result in the following proposition.

Proposition 5.3. *Consider a Buck converter in normalized form in which it is desired to stabilize the measured output variable, $y = x_2$, towards the given constant value, $V_d > 0$. Suppose that the control input, u, is also available for measurement. Then, the following integral reconstructor-based sliding mode controller, using only input-output, information:*

$$u = \begin{cases} 1 \text{ for } \hat{\sigma}(y, u, \xi) < 0 \\ 0 \text{ for } \hat{\sigma}(y, u, \xi) > 0 \end{cases}$$

$$\hat{\sigma}(y, u, \xi) = \int_0^\tau (u(\rho) - y(\rho))\, d\rho - \frac{V_d}{Q} + k_0 \xi$$

$$\dot{\xi} = y(\tau) - V_d, \quad \xi(0) = 0, \quad 0 < k_0 < V_d$$

(5.58)

yields a permanent sliding motion on the surface:

$$\hat{S} = \{ (y, u, \xi) \mid \hat{\sigma}(y, u, \xi) = 0 \}$$
$$= \{ (x, \xi) \mid x_1 - \frac{V_d}{Q} - x_1(0) + k_0 \xi = 0 \} \quad (5.59)$$

which is reachable from the origin of coordinates in finite time. The induced ideal sliding motions on the sliding manifold, $\hat{S}$ exponentially asymptotically stabilize the trajectories of the variables x_1, x_2 and ξ towards the unique equilibrium point:

$$\bar{x}_1 = \frac{V_d}{Q}, \quad \bar{x}_2 = V_d, \quad \bar{\xi} = \frac{1}{k_0} x_1(0)$$

where $x_1(0)$ is the unknown initial state of the normalized inductor current variable x_1. The sliding motions globally exist on the physically significant domain of $\hat{S}$, provided the design constant k_0 is chosen so that $0 < k_0 < 1$.

Figure 5.42 depicts the integral reconstructor-based indirect sliding mode feedback control scheme for the stabilization of the normalized Boost converter circuit.

Remark 5.4. The sliding mode controller (5.46)-(5.48), based on integral reconstruction of the normalized inductor current, x_1, exhibits a fundamental limitation in the injection of appropriate damping to the closed loop dynamics of the output capacitor voltage $y = x_2$, as it follows from Equation 5.53. The following alternative sliding surface

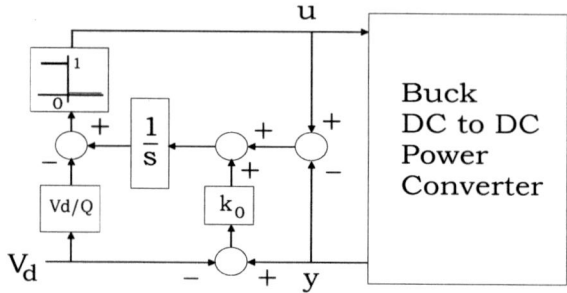

Fig. 5.42. Integral reconstructor-based indirect sliding mode control scheme for the stabilization of the Buck converter circuit.

$$\hat{\sigma}(y, u, \xi) = \int_0^\tau (u(\rho) - y(\rho)) \, d\rho - \frac{V_d}{Q} + k_0 \xi + k_1 y$$
$$\dot{\xi} = y - V_d$$

results in an ideal sliding dynamics for y governed by

$$\ddot{y} + (\frac{1}{Q} + k_1)\dot{y} + k_0(y - V_d) = 0$$

which has complete command over the damping and the natural frequency of the closed loop ideal sliding dynamics. The existence of a sliding regime and the reachability of the sliding surface follows from a similar analysis as that carried out for the previous sliding mode controller. The closed loop average dynamics for y clearly shows that the scheme is also robust with respect to load, or quality parameter, step variations since these only imply a change, always within strictly positive limits, in the damping factor but not in the achieved steady state equilibrium value.

Simulation Results

Simulations were performed on a typical Buck converter circuit with the same parameter values used before. It was desired to bring the state trajectories from the origin towards the final desired value of $\bar{z}_2 = 7.5$ [V], with corresponding $\bar{z}_1 = 0.2488$ [A]. The simulations, shown in Figure 5.43, depict the performance of the proposed sliding mode plus integral reconstructor-based feedback control scheme on the behavior of the considered DC-to-DC Buck converter circuit. The value of the design constant k_0 was set to be $k_0 = 1/(4Q^2)$.

Robustness to Load Variations

In order to test the robustness of the proposed GPI sliding mode control scheme, we let the load resistor R undergo a sudden un-modelled and *permanent* variation of 100% of its nominal value of 30Ω. This variation took place,

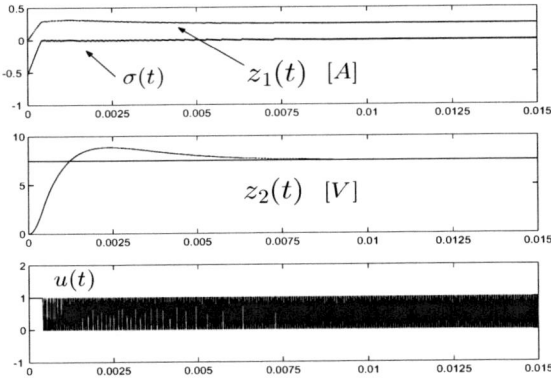

Fig. 5.43. Integral reconstruction based sliding mode controlled Buck converter performance.

approximately, at time, $t = 0.01586$ [s], while the system was already stabilized to the desired voltage value. Figure 5.44 shows the excellent recovering features of the proposed controller to the imposed load variation.

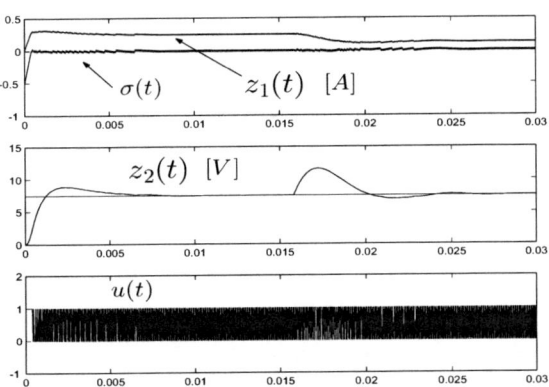

Fig. 5.44. Robust performance of integral reconstructor based sliding mode control of the Buck converter subject to un-modelled load variations of 100%.

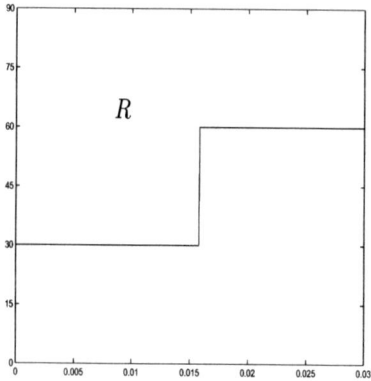

Fig. 5.45. Robust performance of integral reconstructor based sliding mode control of the Buck converter subject to un-modelled load variations of 100%.

5.9.2 The Boost Converter

The normalized Boost system

$$\begin{aligned} \dot{x}_1 &= -ux_2 + 1 \\ \dot{x}_2 &= ux_1 - \frac{x_2}{Q} \end{aligned} \quad (5.60)$$

is observable, in an average sense, from the measured normalized output variable $y = x_2$. This is easily verified since the "observability" matrix:

$$\frac{\partial(y, \dot{y})}{\partial x} = \begin{bmatrix} 0 & 1 \\ u & -\frac{1}{Q} \end{bmatrix} \quad (5.61)$$

is rank 2 for all average values of u which are not identically equal to zero. Since the average value of the input, under ideal sliding mode conditions, is $u_{eq} = 1/V_d > 0$, the observability condition is clearly met.

An integral state resconstructor for x_1 may be obtained in this case as

$$\widehat{x}_1(\tau) = \int_0^\tau (1 - u(\rho)y(\rho))\, d\rho \quad (5.62)$$

The relation linking the estimated value $\widehat{x}_1$ of x_1 to its actual value, is clearly given by

$$x_1(\tau) = \widehat{x}_1(\tau) + x_1(0) \quad (5.63)$$

We propose using the estimate (5.62) of the inductor current, x_1, in the following traditional sliding surface definition

$$S = \{\, x \in R^2 \mid \sigma(x) = x_1 - \overline{x}_1 = x_1 - V_d^2/Q = 0 \,\} \quad (5.64)$$

and, as before, proceed to complement the expression with an integral output error control action based on the stabilization error, $y - V_d$.

Consider then the following integral reconstructor-based sliding mode controller,

$$u = \begin{cases} 1 \text{ for } \hat{\sigma}(y,u,\xi) > 0 \\ 0 \text{ for } \hat{\sigma}(y,u,\xi) < 0 \end{cases} \quad (5.65)$$

$$\hat{\sigma}(y,u,\xi) = \int_0^\tau (1 - u(\rho)y(\rho))\, d\rho - \frac{V_d^2}{Q} + k_0 \xi \quad (5.66)$$

$$\dot{\xi} = y(\tau) - V_d, \quad \xi(0) = 0 \quad (5.67)$$

with k_0 a strictly positive design constant to be chosen later.

The modified sliding surface coordinate function, $\hat{\sigma}$, can be written in terms of the actual state x_1 as,

$$\hat{\sigma}(x_1, \xi) = x_1 - \frac{V_d^2}{Q} - x_1(0) + k_0 \xi \quad (5.68)$$

In spite of the unknown value of $x_1(0)$, the expression (5.68) is useful for analysis purposes.

The time derivative of any of the modified sliding surface coordinate function (5.68), or (5.66), is given by

$$\dot{\hat{\sigma}}(y,u,\xi) = 1 - uy + k_0(y - V_d) \quad (5.69)$$

Note that on $\hat{\sigma} = 0$ the inductor current, x_1, is given by the expression $x_1 = \frac{V_d^2}{Q} + x_1(0) - k_0 \xi$. The equivalent control, corresponding to the modified sliding surface coordinate function is now given by

$$u_{eq} = \frac{1 + k_0(y - V_d)}{y} \quad (5.70)$$

A sliding regime locally exists on $\hat{\sigma}(y,u,\xi) = 0$ whenever the condition, $0 < u_{eq} < 1$, is satisfied:

$$0 < 1 + k_0(y - V_d) < y \quad (5.71)$$

which is equivalent to the following set of inequalities

$$\begin{cases} V_d - \frac{1}{k_0} < y < V_d + \frac{1-V_d}{1-k_0} & \text{for } k_0 > 1 \\ y > V_d - \min\left\{\frac{1}{k_0}, \frac{V_d - 1}{1 - k_0}\right\} & \text{for } 0 < k_0 < 1 \end{cases} \quad (5.72)$$

Thus, the set of values for k_0 that guarantees a larger region of existence of a sliding regime corresponds to the condition, $k_0 \in (0,1)$. The following choice of, k_0, as a strictly positive constant, within the interval:

$$0 < k_0 < \frac{1}{V_d} < 1 \tag{5.73}$$

clearly guarantees the non-empty character of the region of existence of sliding motions and it will prove to be the most convenient to assure reachability from the origin of state coordinates.

The ideal sliding dynamics, obtained from the *invariance conditions*, $\hat{\sigma} = 0$, $\dot{\hat{\sigma}} = 0$, is now obtained as

$$\dot{x}_1 = -k_0(y - V_d)$$
$$\dot{y} = \frac{1 + k_0(y - V_d)}{y}\left[\frac{V_d^2}{Q} + x_1(0) - k_0\xi\right] - \frac{y}{Q}$$
$$\dot{\xi} = y - V_d \tag{5.74}$$

where the output signal y is assumed to satisfy the non-singularity condition, $y > 1 > 0$.

The only constant equilibrium point, $(\bar{y}, \bar{\xi})$, of the ideal closed loop sliding dynamics (5.74) is given by

$$\bar{y} = V_d, \quad \bar{\xi} = \frac{1}{k_0} x_1(0), \tag{5.75}$$

The reachability of the sliding surface, $\hat{\sigma} = 0$, from a given initial state value, is established by the well known condition, $\hat{\sigma}\dot{\hat{\sigma}} < 0$, to be verified in a local neighborhood of the modified sliding surface. Suppose, then, that (5.73) is valid. Let $\hat{\sigma} < 0$, then, according to (5.65), the control is set to $u = 0$. The time derivative of the modified sliding surface coordinate is given by $\dot{\hat{\sigma}} = 1 + k_0(y - V_d)$. Then for all $y > V_d - 1/k_0$, the time derivative, $\dot{\hat{\sigma}}$ is positive and the product $\hat{\sigma}\dot{\hat{\sigma}}$ is negative. Suppose now that $\hat{\sigma}$ is positive, then, the control input is given by $u = 1$. The time derivative of the sliding surface coordinate is $\dot{\hat{\sigma}} = 1 - y + k_0(y - V_d)$. Thus, for all $y > \frac{1 - k_0 V_d}{1 - k_0} = V_d - \frac{V_d - 1}{1 - k_0}$, the product $\hat{\sigma}\dot{\hat{\sigma}}$ is, again, negative. Due to the constrained physically plausible values of the system states, we conclude that a sliding regime *semi-globally* exists on the modified sliding surface, $\widehat{S} = \{ (y, u, \xi) \mid \sigma(y, u, \xi) = 0\}$, which is also reachable in finite time, by means of the proposed discontinuous control law (5.65).

For the reachability of the sliding surface from the origin, suppose the system is initially resting at the zero state, $x_1(0) = 0, x_2(0) = 0, \xi(0) = 0$, then, the initial value of the modified sliding surface is negative, $\hat{\sigma}(x_1(0), 0) = \hat{\sigma}(0,0) = -V_d^2/Q < 0$, and the initial value of the product, $\hat{\sigma}\dot{\hat{\sigma}}$, is given by:

$$\hat{\sigma}(0,0)\dot{\hat{\sigma}}(0,0) = -\frac{V_d^2}{Q}(1 - k_0 V_d)$$

The modified sliding surface $\hat{\sigma}$, thus, starts increasing towards zero from the given zero initial condition, provided k_0 is chosen within the prescribed interval. According to the previously demonstrated semi-global attractiveness of the sliding surface, the sliding surface, is, therefore, always reachable from the origin by the proposed switched control strategy (5.65).

It remains to be proved the nature of the stability of the average equilibrium point for ideal sliding trajectories starting on the sliding surface $\widehat{S}$. Due to the physical limitations of the variables only local existence of the sliding motions may be guaranteed. The equilibrium point is thus clearly *not* attractive from every point of the sliding surface. We prove then local asymptotic stability, which suffices for our purposes, by resorting to tangent linearization of the ideal sliding dynamics.

The tangent linearization of the ideal sliding dynamics (5.74) is given by

$$\dot{\xi}_\delta = y_\delta$$
$$\dot{y}_\delta = -\frac{k_0}{V_d}\xi_\delta - \frac{2 - k_0 V_d}{Q} y_\delta \quad (5.76)$$

where $\xi_\delta = \xi - x_1(0)/k_0$ and $y_\delta = y - V_d$. Since $0 < k_0 < 1/V_d$, the linearized system (5.76) is asymptotically stable to zero. The result follows.

Note that a small value of the design parameter, k_0, not only increases the damping in the linearized average version of the closed loop system, but it also lowers the corresponding natural frequency. This results, generally speaking, in a slower convergence of the controlled motions towards the origin of the incremental variables and, hence, a slower convergence of the nonlinear controlled system output towards the desired constant equilibrium.

We summarize the proven result in the following proposition.

Proposition 5.5. *Consider a Boost converter, represented in normalized form, in which it is desired to stabilize the measured output variable, $y = x_2$, towards the given constant value, $V_d > 0$. Suppose that the control input, u, is also available for measurement. Then, the following integral reconstructor-based sliding mode controller, using only input-output, information:*

$$u = \begin{cases} 1 \text{ for } \widehat{\sigma}(y,u) > 0 \\ 0 \text{ for } \widehat{\sigma}(y,u) < 0 \end{cases}$$

$$\widehat{\sigma}(y,u,\xi) = \int_0^\tau (1 - u(\rho)y(\rho))\,d\rho - \frac{V_d^2}{Q} + k_0 \xi$$

$$\dot{\xi} = y(\tau) - V_d, \quad \xi(0) = 0, \quad 0 < k_0 < \frac{1}{V_d}$$

(5.77)

yields a permanent sliding motion on the surface:

$$\widehat{S} = \{\,(y,u,\xi) \mid \widehat{\sigma}(y,u,\xi) = 0\,\}$$
$$= \{\,(x,\xi) \mid x_1 - \frac{V_d^2}{Q} - x_1(0) + k_0 \xi = 0\,\} \quad (5.78)$$

which is reachable from the origin in finite time. The induced sliding motions on the sliding manifold, $\widehat{S}$, ideally, locally asymptotically stabilize the trajectories of the circuit variables x_1, x_2 and ξ towards the equilibrium values:

$$\bar{x}_1 = \frac{V_d^2}{Q}, \quad \bar{x}_2 = V_d, \quad \bar{\xi} = \frac{1}{k_0} x_1(0)$$

where $x_1(0)$ is the unknown initial state of the normalized inductor current variable x_1. The sliding motions exist on, $\widehat{S}$, whenever the regulated values of the output, y, satisfy the inequality:

$$y > V_d - \min\left\{\frac{1}{k_0}, \frac{V_d - 1}{1 - k_0}\right\} \qquad (5.79)$$

□

Figure 5.46 depicts the integral reconstructor-based sliding mode feedback control scheme for the stabilization of the normalized Boost converter circuit.

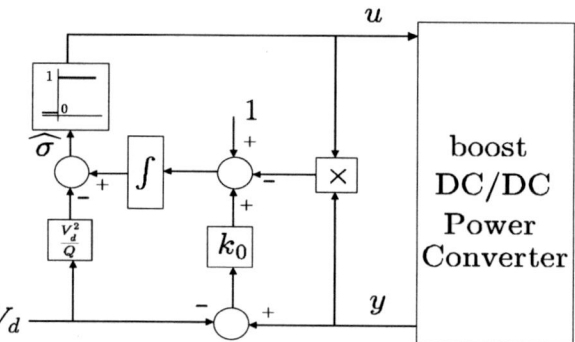

Fig. 5.46. Integral reconstructor-based sliding mode control scheme for the stabilization of the Boost converter circuit.

Simulations

Simulations were performed on a typical Boost converter circuit with parameter values given by

$$L = 20 \text{ [mH]}, \quad C = 20 \text{ [}\mu\text{F]}, \quad R = 30 \text{ [}\Omega\text{]}, \quad E = 15 \text{ [V]}$$

This parameter values yield a value of Q given by $Q = 0.9486$ and a time normalization factor given by $t = 6.32 \times 10^{-4} \tau$.

It was desired to bring the Boost converter trajectories from unknown initial conditions (taken to be, for the simulation purposes, $x_1(0) = 0.5$ and $x_2(0) = 0.8$) towards the final desired value of $\bar{z}_2 = 30$ [V], with corresponding $\bar{z}_1 = 2$ [A]. The simulations, shown in Figure 5.47, depict the performance of the proposed sliding mode plus integral reconstructor-based feedback control scheme on the behavior of the considered DC-to-DC Boost converter circuit. The value of the design constant k_0 was set to be $k_0 = 0.1 < 1/Vd = 0.5$.

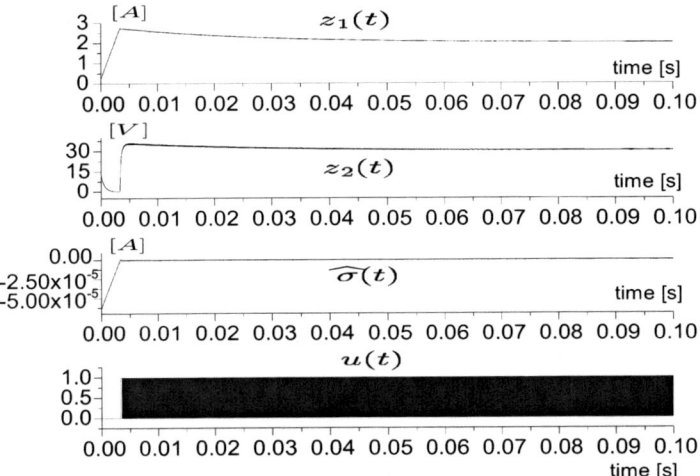

Fig. 5.47. Integral reconstructor based sliding mode controlled Boost converter performance.

Robustness to Load Variations

In order to test the robustness of the proposed GPI sliding mode control scheme, we let the load resistor R undergo a sudden un-modelled and *permanent* variation of 400% of its nominal value of 30Ω. This variation took place, approximately, at time, $t = 0.0633$ [s], while the system was not yet stabilized to the desired voltage value. Figure 5.48 shows the excellent recovering features of the proposed controller to the imposed load variation.

5.9.3 The Buck-Boost Converter

The normalized model of the system is given by (where "·" stands for $\frac{d}{d\tau}$ and $Q = R\sqrt{C/L}$):

$$\dot{x}_1 = (1-u)x_2 + u$$
$$\dot{x}_2 = -(1-u)x_1 - \frac{1}{Q}x_2$$

Consider the following integral reconstructor of the normalized inductor current for the Buck-Boost converter

$$\hat{x}_1 = \int_0^\tau [u(\rho) + (1 - u(\rho))\, y(\rho)]\, d\rho$$

The relation between the integral reconstructor of x_1 and its actual value is given by

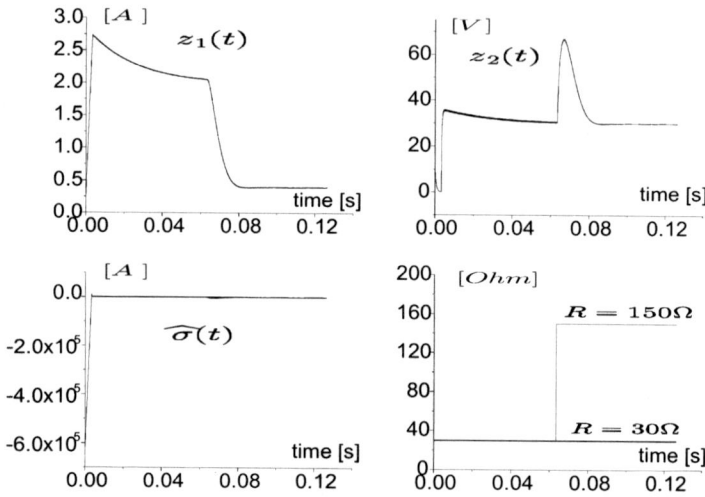

Fig. 5.48. Robust performance of integral reconstructor based sliding mode control of the Boost converter subject to un-modelled load variations of 400%.

$$x_1(\tau) = \widehat{x_1}(\tau) + x_1(0)$$

We consider, then, the modified sliding surface

$$\widehat{\sigma}(x) = \widehat{x}_1 - \frac{V_d(1+V_d)}{Q} - k_0 \int_0^\tau (y + V_d)\, d\rho$$

The invariance conditions $\widehat{\sigma} = \dot{\widehat{\sigma}} = 0$ result in the ideal sliding dynamics

$$\dot{x}_1 = k_0\,(y + V_d)$$
$$\dot{y} = -\frac{1}{1-y}\left[\frac{V_d(1+V_d)}{Q} - x_1(0) + k_0\xi\right] - \frac{y}{Q}$$
$$\dot{\xi} = y + V_d$$

which exhibits the physically meaningful equilibrium point,

$$\overline{x}_1 = \frac{V_d(1+V_d)}{Q},\quad \overline{y} = -V_d,\quad \overline{\xi} = \frac{x_1(0)}{k_0}$$

Figure 5.49 depicts the robustness of the integral reconstructor based sliding mode controller when the resistance load undergoes an un-modelled variation of 200% above its nominal value.

In this section we have pointed to some of the basic limitations of traditional indirect sliding mode control when applied to the regulation of switched power converters. These limitations generally refer to: 1) complete system state availability and 2) a lack of robustness with respect to un-modelled load

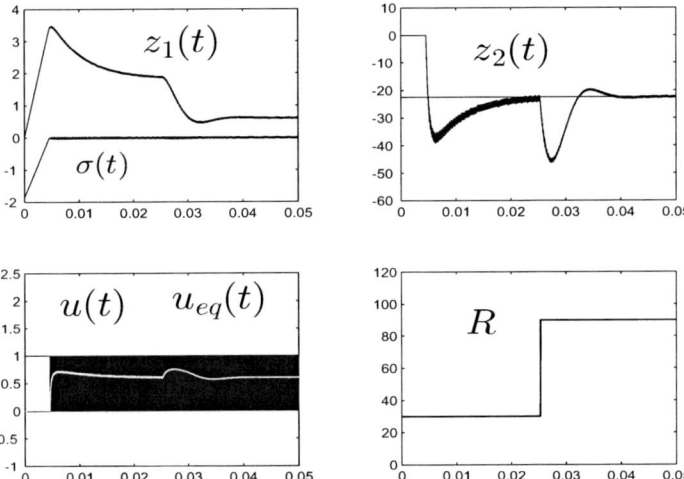

Fig. 5.49. Robust performance of integral reconstructor based sliding mode control of the Buck-Boost converter subject to un-modelled load variations of 200%.

resistance variations. We have proposed a direct use of the GPI control technique, based on integral reconstructors, to the realm of sliding mode control within the context of three specific physical examples of wide interest in the Power Electronics area. Integral reconstructors with suitable integral output error compensation provide asymptotically stabilizing sliding mode controllers which only require measurements of the output voltage of the converter (i.e., of the non-minimum phase state variable) and the availability of the input signal. The integral reconstructor-based controller may be motivated by the usual indirect design of the traditional sliding surface coordinate function in terms of the normalized inductor current variable stabilization around a constant value. Alternatively, in some cases, the technique allows for a closed loop average linearization approach. An integral reconstructor of the normalized inductor current variable, exhibiting a constant "off-set" error, is synthesized in terms of an integral of simple algebraic functions of the available input and the measured output signals. The sliding surface synthesis uses this "open loop" estimate of the inductor current in combination with a suitable integral output error compensation term. The integral input-output parameterized sliding surface is shown to be semi-globally reachable and, once a sliding regime is established on the sliding manifold, an asymptotically stable ideal sliding dynamics is obtained which converges to the desired equilibrium values.

Through computer simulations, the proposed control schemes were shown to be remarkably robust with respect to unusually large, un-modelled, load parameter step variations. Note that for an extremely large load variation it is possible that the inductor current signal drops to the zero value, saturating the controller action to yield a fixed switch position and a consequent, temporary or permanent, loss of feedback. Strategies to efficiently emerge from, or avoid, such situations, are known as *operation in discontinuous conduction mode*. These are the object of sustained studies in the current power electronics systems literature. Our approach, while being quite robust in this respect, is not devised to entirely avoid such adverse possibility. An on-line algebraic identification scheme combined with GPI control has been proposed to effectively deal with the problem (see Fliess and Sira-Ramírez [23] and Sira-Ramírez et al. [65])

The same integral reconstructor-based sliding mode control technique is readily applicable to un-interruptible power supplies and, possibly, to the Cúk converter. An interesting topic for further study is represented by the integral reconstructor-based AC voltage generation problem using traditional DC-to-DC Power Converters (see Sira-Ramírez, [60]).

Part III

Applications

6
DC-to-AC Power Conversion

6.1 Introduction

Traditional approaches to DC-to-AC power conversion include, among other options: PWM commanded switch based *inverters*, feeded by un-interruptible power supplies or rectified voltage sources; various combinations of series-resonant DC-to-AC inverters and, more recently, the so called zero-voltage-switching (ZVS) PWM commutation cells linking constant voltage sources and Buck converters (see, among an immense wealth of articles in these areas, the works of Mendes de Seixas [42], [43], García and Barbi [27], Jung and Tzou [32], Hsieh *et al.* [30] and the many references therein).

DC-to-AC power conversion using the traditional DC-to-DC power converter topologies constitutes a relatively recent sub-area of the power electronics field which has proven to constitute a challenging area from the feedback controller design viewpoint. This is specially so for DC-to-AC conversion schemes using converters other than the step down, Buck, converter (see the many articles published in the yearly Power Electronics Specialist Conference (PESC)). Our work in this chapter, however, is motivated by that of Cáceres and Barbi [6] and the article by Zinober *et al.* [85]). In [6], a sliding mode controller is proposed for a set of coupled Boost converters, viewing each converters AC output capacitor voltage as a bounded, unknown, perturbation for the other converters AC signal tracking task. In [85] two approaches are proposed. The first one reduces the tracking task to the Fourier series solution of an Abel type of differential equation. The second approach proposes a *back-stepping* controller for the tracking task. Some of the developments in this chapter are taken from [60]. We base our considerations on some of the more traditional DC-to-DC power converter topologies, basically, the Buck, the Boost and the Buck-Boost converters, as working examples.

Generally speaking, DC-to-AC power conversion is a special class of trajectory tracking problems for the output voltage variable of the converter system. As such, the control synthesis problem based on total, or partial, system inversion presents severe difficulties in all converter topologies except for the Buck converter. The reason being that in those topologies the output capacitor voltage is a non-minimum phase output variable. The solution to the trajectory tracking problem for this class of systems is usually found by exercising an indirect control approach (see Benvenuti, Di Benedetto and Grizzle [2]). In other words, design efforts are primarily placed on synthesizing a feedback controller for the induced trajectory tracking problem, described now in terms of a corresponding desired trajectory for a minimum phase output variable, such as the inductor current or the total stored energy.

In this chapter, we consider a particular but important aspect of the DC-to-AC power conversion problem viewed as a particular trajectory tracking problem. Namely, that of nominal trajectory generation. The details of the specific feedback controller design for the trajectory tracking problem have already been studied in previous chapters from a variety of methodological viewpoints. The feedback control techniques that may be used for DC-to-AC power conversion trajectory tracking task include: approximate linearization, differential flatness based solutions, indirect sliding mode control and a Lyapunov control approach based on the exact tracking error dynamics passive output feedback (ETEDPOF). We remark that the presented approaches, however, can also be successfully extended to include the Cúk converter and many others DC-to-DC power conversion topologies.

We base great part of our considerations on differential flatness, which is an important structural property of many physical nonlinear systems in general and of some DC-to-DC power converters in particular (see the seminal work of Fliess *et al.* [18], [17], for the underlying theoretical considerations and see [63] for the potential of this technique in applications). As it is known, the Buck converter is linear and controllable, hence flat. For this converter, the flat output is given by the output capacitor voltage. Therefore, there is no special difficulty in generating the nominal current and input trajectories used in the already studied feedback control design schemes solving the underlying trajectory tracking problem. The differential flatness of the Boost converter is characterized by the fact that the total stored energy completely parameterizes all system variables. A similar statement may be made for the Buck-Boost converter. In DC-to-AC power conversion tasks, however, it is not trivial, nor intuitive, to assess the waveform of this type of flat output. Flatness is shown to allow for the development of an efficient, rapidly convergent, iterative, off-line, computational scheme which yields, in an approximate manner, a suitable finite differential parametrization of the inductor current reference trajectory in terms of the desired capacitor voltage AC reference signal. This approximate finite dimensional parametrization simply means that the inductor current reference trajectory is approximately expressible as a function of the desired capacitor voltage and a finite number of its time derivatives.

6.2 Nominal Trajectories in DC-to-AC Power Conversion

In many of the feedback control schemes presented in this chapter, it will be required to know the off-line nominal state and input trajectories corresponding to a DC-to-AC power conversion task on the controlled behavior of a particular switch-mode power converter topology. For this reason, in this part of the chapter we devote attention to the problem of finding such nominal trajectories based on the concept of differential flatness. We shall concentrate, mainly, on three basic topologies: The Buck converter, the Boost converter and the Buck-Boost converter. The fundamental problem for finding such nominal state and input trajectories, at least in the Boost and Buck-Boost converters, is that by fixing the output capacitor voltage to be a biased sinusoid, it is not straightforward to find out what the corresponding input inductor current and the associated average control input should be. This difficulty is due to the non-minimum phase character of the output voltage variable. As it will be shown, the Buck converter offers no difficulty whatsoever.

6.2.1 The Buck Converter

It is not difficult to realize that a traditional Buck converter circuit is capable of producing only biased sinusoidal average output voltage signals whose values are never to become negative. The reason being that negative output reference voltages saturate the average switch position function which takes values on the set $[0, 1]$. For this reason, the Buck converter must be slightly modified to include the possibilities of producing negative average output voltage values. This is achieved by a suitable input source polarity reversal at those moments when the sinusoidal tracking requires the generation of negative output values. The net result is equivalent to expanding the set of possible control inputs to the discrete set $\{-1, 0, 1\}$. This particular modification may be realized by means of a double bridge circuit changing the polarity of the voltage source as required. The consequences of this extended switch position function on the average model of the double bridge Buck converter is that the closed interval of existence for the average control input is now the interval $[-1, 1]$ of the real line. The average model of the double bridge Buck converter is therefore the same as the traditional Buck converter but with the average control input taking values in the closed interval $[-1, 1]$ of the real line. The switched implementation of the average feedback control law requires now of a two sided $\Sigma - \Delta$ modulation as explained at the end of this section.

The normalized average model of a double bridge Buck converter (see Figure 6.1) is given by

$$\dot{x}_1 = -x_2 + u_{av}$$
$$\dot{x}_2 = x_1 - \frac{x_2}{Q} \qquad (6.1)$$

where x_1 is the average normalized input inductor current, x_2 is the average normalized output capacitor voltage. The distinctive feature of this converter is that the actual control input u takes value in the discrete set $\{-1, 0, 1\}$. Thus, the average control input u_{av} is assumed to take values in the closed set $[-1, 1]$ of the real line.

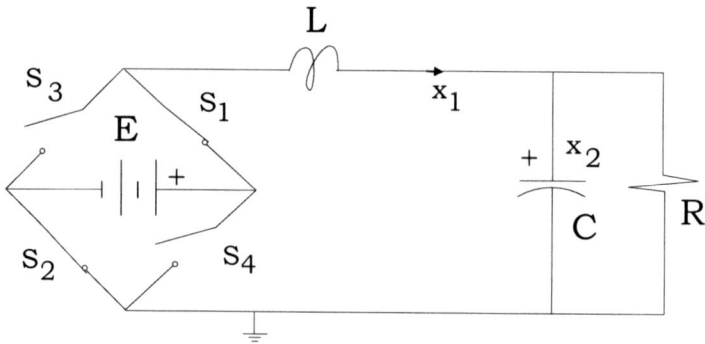

Fig. 6.1. The double bridge Buck converter.

The average normalized double bridge Buck converter model (6.1) is flat, with flat output given by the average normalized output voltage $F = x_2$. Indeed, all system variables are parameterizable in terms of F and a finite number of its time derivatives,

$$x_1 = \dot{F} + \frac{1}{Q}F$$
$$x_2 = F$$
$$u_{av} = \ddot{F} + \frac{1}{Q}\dot{F} + F \qquad (6.2)$$

We are interested in generating a sinusoidal signal at the output voltage of the average normalized system. Let such a desired output voltage trajectory be specified by:

$$x_2^*(\tau) = F^*(\tau) = A \sin(\omega_0 \tau) \qquad (6.3)$$

where τ is the dimensionless normalized time scale $\tau = t/\sqrt{LC}$.

The differential parametrization (6.2) yields the following corresponding values of the average normalized inductor current $x_1^*(\tau)$ and the nominal average control input $u_{av}^*(\tau)$.

$$x_1^*(\tau) = \frac{A}{Q}\left[\sqrt{1 + Q^2\omega_0^2}\right]\sin(\omega_0\tau + \arctan[\omega_0 Q])$$
$$u_{av}^*(\tau) = \frac{A}{Q}\left[\sqrt{Q^2(1-\omega_0^2)^2 + \omega_0^2}\right]\sin\left(\omega_0\tau + \arctan\left[\frac{\omega_0}{Q(1-\omega_0^2)}\right]\right)$$
$$(6.4)$$

6.2 Nominal Trajectories in DC-to-AC Power Conversion

The last expression in Equation 6.4 reveals the limitations, naturally imposed by the bounded nature of the average control input, u_{av}, on the amplitude and frequencies of sinusoidal signals that can be demanded on the normalized average output voltages. Indeed the fact that $u_{av} \in [-1, 1]$ yields:

$$A \leq \frac{Q}{\sqrt{Q^2(1-\omega_0^2)^2 + \omega_0^2}} = \frac{1}{\sqrt{(1-\omega_0^2)^2 + \left[\frac{\omega_0}{Q}\right]^2}} \quad (6.5)$$

This expression reveals that the amplitude and the angular frequency of the normalized average output sinusoidal voltage cannot be independently chosen. The critical values of the right hand side, as a function of the angular normalized frequency ω_0, occur at the values: $\omega_0 = 0$, which has no meaning from the DC-to-AC conversion task viewpoint, and at $\omega_0 = \sqrt{2 - \frac{1}{Q^2}}$. Hence, the normalized load value should be larger than $Q = \sqrt{2}/2$. The sinusoidal frequency ω_0 is thus limited to normalized values which are bounded below by $\sqrt{2}$.

6.2.2 Two-Sided $\Sigma - \Delta$ Modulation

It is possible to extend the $\Sigma - \Delta$ modulation scheme to deal with control inputs taking values in the discrete set $\{-1, 0, 1\}$. To this category of switched systems can be reduced the great majority of switched power electronics devices provided with "double bridges".

Suppose that the average input signal $u_{av}(t)$ takes values on the closed interval $[-1, 1]$ of the real line. We propose the following two sided $\Sigma - \Delta$ modulation system governed by the following equations

$$\dot{e} = u_{av}(t) - u, \quad u = \begin{cases} \frac{1}{2}(1 + \text{sign } e) & \text{for } u_{av} > 0 \\ -\frac{1}{2}(1 - \text{sign } e) & \text{for } u_{av} < 0 \end{cases} \quad (6.6)$$

A sliding regime exists on $e = 0$ in any of the two cases ($u_{av} > 0$ and $u_{av} < 0$), as it may be easily verified. The double sided $\Sigma - \Delta$ modulation equations may be summarized as follows:

$$\dot{e} = u_{av}(t) - u, \quad u = \frac{1}{2}[\text{sign } u_{av}(t) + \text{sign } e] \quad (6.7)$$

Figure 6.2 shows a typical response of a two sided $\Sigma - \Delta$ modulator to an input signal $u_{av}(t)$ of varying polarity, like that represented by a sinusoid function. The switchings actively commute between 0 and 1 when the input signal $u_{av}(t)$ is positive and between 0 and -1 when the input signal $u_{av}(t)$ is negative.

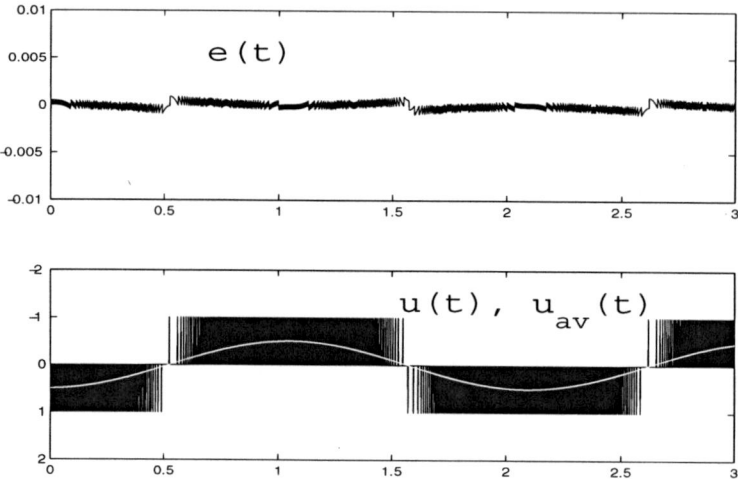

Fig. 6.2. Typical two sided $\Sigma - \Delta$ modulation-based switched implementation of a bounded input signal, $u_{av}(t)$, of changing polarity.

6.2.3 The Boost Converter

The Boost converter circuit cannot produce sinusoidal signals of varying polarity around the origin. One cannot easily modify the Boost converter, as it was done in the Buck case, in order to produce zero mean value sinusoids. In fact, due to the need of having output voltages higher than the constant input source voltage, the Boost converter will only produce (positively) polarized sinusoidal voltages whose minimum values must be uniformly bounded below by the constant value of the input source voltage. In normalized average terms, the nominal sinusoidal output voltages that can be demanded from a traditional Boost converter are of the form

$$x_2^*(\tau) = B + A\sin(\omega_0\tau) \tag{6.8}$$

where $B - A > 1$, i.e., $B > 1 + A$.

Consider now the average normalized model of the Boost converter.

$$\dot{x}_1 = -u_{av}x_2 + 1$$
$$\dot{x}_2 = u_{av}x_1 - \frac{x_2}{Q}$$
$$y = x_2 \tag{6.9}$$

where the "dot" notation stands for derivation with respect to the dimensionless normalized time variable τ. The flat output of the average normalized Boost converter is given by the average normalized total stored energy,

$$F = \frac{1}{2}\left(x_1^2 + x_2^2\right) \tag{6.10}$$

6.2 Nominal Trajectories in DC-to-AC Power Conversion

Note that the time derivative of F, with respect to the normalized time τ, is given by

$$\dot{F} = x_1 - \frac{x_2^2}{Q} \qquad (6.11)$$

From (6.10), (6.11) we obtain, by elementary algebraic manipulations, the differential parameterizations of the average normalized inductor current, x_1, and the average normalized output capacitor voltage, x_2, as follows:

$$x_1 = -\frac{Q}{2} + \sqrt{\frac{Q^2}{4} + \left(Q\dot{F} + 2F\right)}$$

$$x_2 = \sqrt{Q\left[\left(-\frac{Q}{2} + \sqrt{\frac{Q^2}{4} + \left(Q\dot{F} + 2F\right)} - \dot{F}\right)\right]} \qquad (6.12)$$

where we have chosen the positive average normalized current solution for natural physical reasons.

The differential parametrization of the average control input, u_{av}, follows from the following relation, obtained directly from the time derivative of Expression 6.11.

$$u_{av} = \frac{\left[\left(1 + \frac{2}{Q^2}x_2^2\right) - \ddot{F}\right]}{x_2\left(1 + \frac{2x_1}{Q}\right)} \qquad (6.13)$$

The problem with the flat output based differential parametrization resides in the fact that it is rather difficult to "guess" what the flat output variable trajectory, $F^*(\tau)$, should be in order to have $x_2^*(\tau)$ exactly coincide with a desired sinusoidal signal of pre-specified amplitude and frequency. Besides, if one uses the differential parametrization for x_2 in (6.12), one obtains a differential equation for F^* in terms of the desired x_2^*. Unfortunately, this differential equation yields unstable solutions when x_2 is particularized to be $x_2^*(\tau) = B + A\sin(\omega_0\tau)$.

We propose a functional iterative scheme in order to obtain suitable approximations to the nominal average normalized inductor current trajectory, $x_1^*(\tau)$, on the basis of equations (6.10) and (6.11).

Consider,

$$x_{1,k+1}^*(\tau) = \dot{F}_k(\tau) + \frac{x_2^*(\tau)}{Q}$$

$$F_{k+1}(\tau) = \frac{1}{2}\left[(x_{1,k}^*)^2(\tau) + (x_2^*(\tau))^2\right] \qquad (6.14)$$

Note that this algorithm produces entire trajectories for $x_1^*(\tau)$ and $F^*(\tau)$ at each iteration. As such, the recursive formula represents an operator mapping the space of smooth functions into itself. The operator is, evidently, an unbounded nonlinear operator, thanks to the presence of the time derivative operator. The convergence properties of this algorithm are quite difficult to establish. We can only state that if this iterative functional algorithm converges, it converges to the $F^*(\tau)$ trajectory that corresponds with the nominal trajectories $x_2^*(\tau)$ and $x_1^*(\tau)$.

Suppose we started with a rather "wild" guess for $F^*(\tau)$, by setting $F_0^* = constant$. We obtain the following sequence of trajectories candidates for $x_{1,k}^*(\tau)$ and $F_k^*(\tau)$.

$$F_0^*(\tau) = constant$$

$$x_{1,1}^*(\tau) = \dot{F}_0(\tau) + \frac{[x_2^*(\tau)]^2}{Q} = \frac{[x_2^*(\tau)]^2}{Q}$$

$$F_1^*(\tau) = \frac{1}{2}\left([x_{1,1}^*(\tau)]^2 + [x_2^*(\tau)]^2\right) = \frac{[x_2^*(\tau)]^2}{2}\left(1 + \frac{[x_2^*(\tau)]^2}{Q^2}\right)$$

$$x_{1,2}^*(\tau) = \dot{F}_1^*(\tau) + \frac{[x_2^*(\tau)]^2}{Q} = x_2^*(\tau)\dot{x}_2^*(\tau)\left(1 + \frac{x_2^*(\tau)}{Q^2}\right)$$

$$F_2^*(\tau) = \frac{1}{2}\left\{[x_2^*(\tau)]^2[\dot{x}_2^*(\tau)]^2\left(1 + \frac{x_2^*(\tau)}{Q^2}\right)^2 + [x_2^*(\tau)]^2\right\}$$

$$x_{1,3}^*(\tau) = \ldots$$

(6.15)

Simulations

We present below some typical simulations on the outcome of the iterative functional algorithm and its rapid convergent features.

In Figure 6.3 a run is depicted for the off-line iterative computation of the non-normalized inductor current reference trajectories $x_{1,1}^*(t)$, $x_{1,2}^*(t)$, for a Boost converter with the following data:

$$L = 20 \text{ mH}, \quad C = 20 \text{ μF}, \quad Q = 0.3535$$

The reference normalized output voltage $x_2^*(\tau)$ was set to be $B + A\sin(w_0\tau)$ with $B = 1.5$, $A = 0.4$, with normalized angular frequency: $w_0 = 0.02$. The candidate non-normalized reference trajectory $x_{1,3}^*(t)$ for the inductor current, not shown in Figure, is practically coincident with $x_{1,2}^*(t)$.

Figure 6.4 depicts the tracking performance of the switched Boost converter in response to the linear time-varying controller synthesized on the basis of exact tracking error dynamics passive output feedback. The data for the Boost converter used in these simulations are the same as in the above example except that a larger frequency was used for the normalized sinusoidal output voltage reference $x_2^*(\tau)$.

6.2 Nominal Trajectories in DC-to-AC Power Conversion

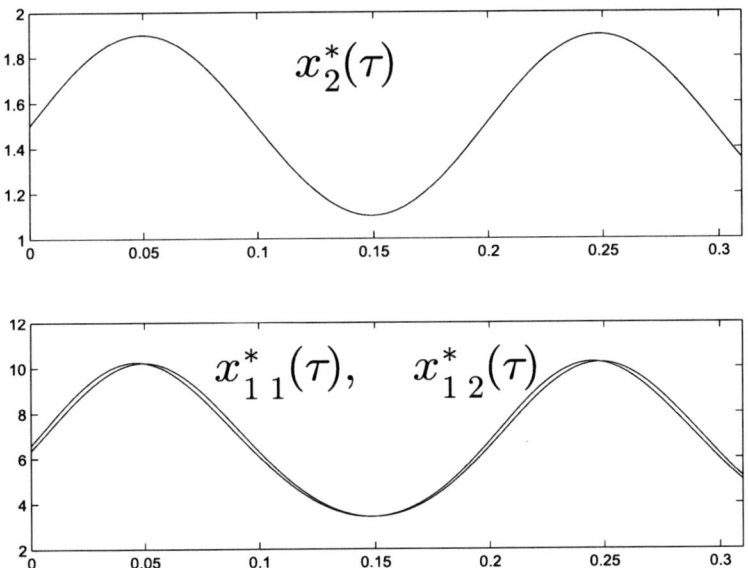

Fig. 6.3. Convergence of functional iterative algorithm in reference trajectory generation for inductor current for a Boost converter.

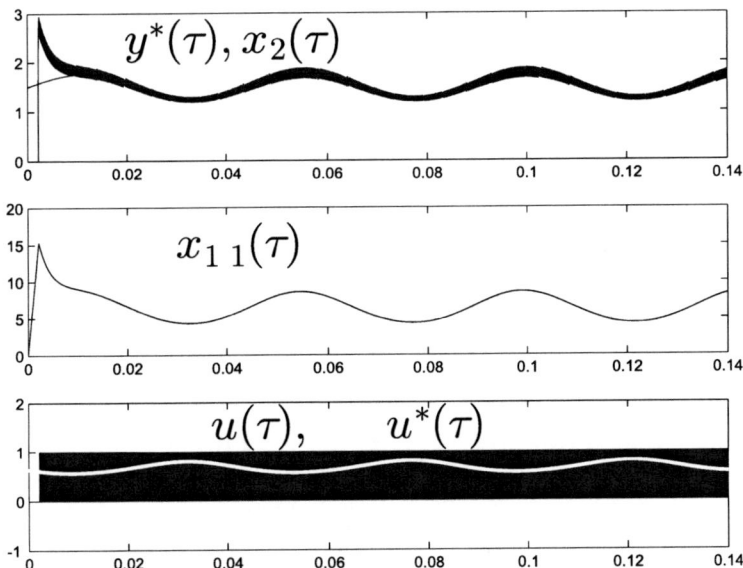

Fig. 6.4. Performance of switched Boost converter to a DC-to-AC tracking task accomplished via time-varying linear passivity based control.

6.2.4 The Buck-Boost Converter

The same remarks made for the Boost converter, concerning the difficulties for system inversion due to its non-minimum phase character, can be reproduced for the Buck-Boost converter modulo the voltage inversion performed by this circuit. Thus, the sinusoidal voltages available from the Buck-Boost converter are negatively polarized in such a manner that the maximum voltage value is upper bounded by the normalized value of -1. As a result, the nominal normalized output sinusoidal voltage of a Buck-Boost converter is, necessarily, of the form $x_2^*(\tau) = B + A\sin(\omega_0 \tau)$ with $B < 0$ and $B + A < -1$, i.e., $B < -(1+A)$.

Consider now the average normalized model of the Buck-Boost converter

$$\frac{dx_1}{d\tau} = u_{av}x_2 + (1 - u_{av})$$
$$\frac{dx_2}{d\tau} = -u_{av}x_1 - \frac{x_2}{Q} \tag{6.16}$$

The following state coordinate transformation significantly simplifies the considerations and the algebraic manipulations,

$$\xi_1 = x_1$$
$$\xi_2 = x_2 - 1 \tag{6.17}$$

The system is now written as

$$\frac{d\xi_1}{d\tau} = u_{av}\xi_2 + 1$$
$$\frac{d\xi_2}{d\tau} = -u_{av}\xi_1 - \frac{1+\xi_2}{Q} \tag{6.18}$$

System 6.18 is rather similar to the normalized model of the Boost converter given in Equation 6.9.

The flat output of the transformed system (6.18) is given by

$$F = \frac{1}{2}\left(\xi_1^2 + \xi_2^2\right) \tag{6.19}$$

The time derivative of the flat output is given by

$$\dot{F} = \xi_1 - \frac{\xi_2(\xi_2 + 1)}{Q} \tag{6.20}$$

and, after some tedious but straightforward algebraic manipulations, one obtains a differential parametrization of ξ_1 and ξ_2 in terms of F and $\dot{F}$. Rather than proceeding in that fashion, we directly resort to an unbounded operator iterative scheme for the approximation of the nominal average normalized current $x_1^*(\tau)$ in terms of $x_2^*(\tau)$, the desired output voltage sinusoidal signal.

$$\dot{\xi}_{1,k}(\tau) = \dot{F}_k(\tau) + \frac{\xi_2^*(\tau)(\xi_2^*(\tau)+1)}{Q}$$
$$F_{k+1}(\tau) = \frac{1}{2}\left(\xi_{1,k}^2 + [\xi_2^*(\tau)]^2\right) \qquad (6.21)$$

or, in terms of the original normalized state variables

$$\dot{x}_{1,k}(\tau) = \dot{F}_k(\tau) + \frac{[x_2^*(\tau)-1]\,x_2^*(\tau)}{Q}$$
$$F_{k+1}(\tau) = \frac{1}{2}\left(x_{1,k}^2 + [x_2^*(\tau)-1]^2\right) \qquad (6.22)$$

At each iteration stage, an approximate flat output trajectory is obtained for the indirect solution of the underlying non-minimum phase output trajectory tracking problem involved in the DC-to-AC power conversion task using the Buck-Boost converter.

Since the formulae are rather close to those derived for the Boost converter, we leave it to the reader to perform by himself some simulation runs to assess the rapid convergent features of the above off-line functional iterative algorithm for suitable flat output reference trajectory generation.

6.3 An Approximate Linearization Approach

We briefly summarize the possibilities of regulating the output voltages of the studied DC-to-DC power converter topologies to track a sinusoidal signal. The controller design methodology to be used is that of approximate linearization around the desired state and input trajectory corresponding to the required output voltage sinusoidal waveform. The linearization of the average normalized nonlinear DC-to-DC power converter dynamics, around the nominal state and input trajectories, invariably results in a time-varying linearized system dynamics. The feedback controller task consists in driving to zero the incremental state variables by means of appropriately bounded control inputs. Since the double bridge Buck converter average model is linear, an approximate linearization approach does not make much sense. We thus present only the Boost and Buck-Boost cases.

6.3.1 The Boost Converter

The normalized average model of the Boost converter is given by,

$$\dot{x}_1 = -u_{av}x_2 + 1$$
$$\dot{x}_2 = u_{av}x_1 - \frac{x_2}{Q} \qquad (6.23)$$

The tangent linearization of the nonlinear average model around the nominal state and average control input trajectories $x_1^*(\tau)$, $x_2^*(\tau)$, $u^*(\tau)$, is given by

$$\dot{x}_{1\delta} = -u_{av}^*(\tau)x_{2\delta} - x_2^*(\tau)u_\delta$$
$$\dot{x}_{2\delta} = u_{av}^*(\tau)x_{1\delta} - \frac{1}{Q}x_{2\delta} + x_1^*(\tau)u_\delta \qquad (6.24)$$

where $x_{1\delta} = x_1 - x_1^*(\tau)$, $x_{2\delta} = x_2 - x_2^*(\tau)$ and $u_\delta = u_{av} - u_{av}^*(\tau)$ are the incremental states and incremental inputs of the linearized time-varying system.

To deduce a stabilizing time-varying feedback controller we adopt a Lyapunov design approach. For this, consider the total stored normalized incremental energy, defined to be the following positive definite function:

$$H_\delta = \frac{1}{2}\left[x_{1\delta}^2 + x_{2\delta}^2\right]$$

The time derivative of H_δ along the controlled solutions of the linearized system is given by

$$\dot{H}_\delta = -\frac{1}{Q}x_{2\delta}^2 + \left[-x_2^*(\tau)x_{1\delta} + x_1^*(\tau)x_{2\delta}\right]u_\delta$$

Let γ be a strictly positive scalar parameter. The choice of u_δ as the following time varying feedback control law

$$u_\delta = -\gamma\left[-x_2^*(\tau)x_{1\delta} + x_1^*(\tau)x_{2\delta}\right] = -\gamma\left[-x_2^*(\tau)x_1 + x_1^*(\tau)x_2\right]$$

leads to the following negative definite evaluation of the closed loop time derivative of the Lyapunov function H_δ,

$$\dot{H}_\delta = -\frac{1}{Q}x_{2\delta}^2 - \gamma\left[-x_2^*(\tau)x_{1\delta} + x_1^*(\tau)x_{2\delta}\right]^2$$
$$= -[x_{1\delta}\ x_{2\delta}]\begin{bmatrix} \gamma[x_2^*(\tau)]^2 & -\gamma x_1^*(\tau)x_2^*(\tau) \\ -\gamma_\gamma x_1^*(\tau)x_2^*(\tau) & \frac{1}{Q} + \gamma[x_1^*(\tau)]^2 \end{bmatrix}\begin{bmatrix} x_{1\delta} \\ x_{2\delta} \end{bmatrix} < 0$$

We have proven the following result:

Theorem 6.1. *Given a sinusoidal average normalized output reference signal $x_2^*(\tau)$ of the Boost converter model (6.23), to which it corresponds the nominal average normalized current, $x_1^*(\tau)$, and the nominal average control input trajectory, $u_{av}^*(\tau)$, then the linear time-varying state feedback controller:*

$$u_{av} = u_{av}^*(\tau) - \gamma\left[-x_2^*(\tau)x_1 + x_1^*(\tau)x_2\right],\ \gamma > 0$$

locally asymptotically stabilize the closed loop state trajectories of the average normalized Boost converter model towards the nominal reference state trajectories.

It will be shown further ahead in this chapter, that this linear controller actually semi-globally asymptotically stabilizes the state trajectories towards the nominal reference trajectories.

The previous theorem nevertheless prescribes knowledge of the nominal average normalized states and control input trajectory compatible with the sinusoidal average normalized reference output voltage trajectory. A task that, as it was already seen, is possible only in an approximate manner.

6.3.2 The Buck-Boost Converter

The tangent linearization of the average normalized Buck-Boost converter model:

$$\frac{dx_1}{d\tau} = u_{av}x_2 + (1 - u_{av})$$
$$\frac{dx_2}{d\tau} = -u_{av}x_1 - \frac{x_2}{Q} \quad (6.25)$$

is given by

$$\dot{x}_{1\delta} = u_{av}^*(\tau)x_{2\delta} + (x_2^*(\tau) - 1)u_\delta$$
$$\dot{x}_{2\delta} = -u_{av}^*(\tau)x_{1\delta} - \frac{x_{2\delta}}{Q} - x_1^*(\tau)u_\delta \quad (6.26)$$

where $x_{1\delta} = x_1 - x_1^*(\tau)$, $x_{2\delta} = x_2 - x_2^*(\tau)$ and $u_\delta = u_{av} - u_{av}^*(\tau)$.

A linear incremental state feedback controller can be synthesized with the help of the Lyapunov function candidate

$$H_\delta = \frac{1}{2}\left[x_{1\delta}^2 + x_{2\delta}^2\right]$$

Indeed, the time derivative of this incremental average normalized total stored energy is given by

$$\dot{H}_\delta = -\frac{1}{Q}x_{2\delta}^2 + [x_{1\delta}(x_2^*(\tau) - 1) - x_{2\delta}x_1^*(\tau)]\,u_\delta$$

thus suggesting the following incremental average normalized state feedback controller:

$$u_\delta = -\gamma\left[x_{1\delta}(x_2^*(\tau) - 1) - x_{2\delta}x_1^*(\tau)\right], \quad \gamma > 0$$

The closed loop evaluation of the time derivative of the Lyapunov function H_δ may be written as

$$\dot{H}_\delta = -\begin{bmatrix} x_{1\delta} & x_{2\delta} \end{bmatrix}\begin{bmatrix} \gamma(x_2^*(\tau) - 1)^2 & -\gamma x_1^*(\tau)(x_2^*(\tau) - 1) \\ -\gamma x_1^*(\tau)(x_2^*(\tau) - 1) & \frac{1}{Q} + \gamma[x_1^*(\tau)]^2 \end{bmatrix}\begin{bmatrix} x_{1\delta} \\ x_{2\delta} \end{bmatrix} < 0$$

thus demonstrating the local asymptotic stability of the origin of the incremental error state space for the closed loop system. We have the following theorem:

Theorem 6.2. *Given a sinusoidal average normalized output reference signal $x_2^*(\tau)$ of the Buck-Boost converter model (6.25), to which it corresponds the nominal average normalized current, $x_1^*(\tau)$, and the nominal average control input trajectory, $u_{av}^*(\tau)$, then the linear time-varying state feedback controller:*

$$u_{av} = u_{av}^*(\tau) - \gamma\left[x_{1\delta}(x_2^*(\tau) - 1) - x_{2\delta}x_1^*(\tau)\right], \quad \gamma > 0$$

locally asymptotically stabilize the closed loop state trajectories of the average normalized Buck-Boost converter model towards the nominal reference state trajectories.

6.4 A Flatness Based Approach

6.4.1 The Double Bridge Buck Converter

Under the assumption of full state availability, the differential parametrization (6.2), for the average input variable, immediately suggests a flatness based linearizing tracking error feedback controller, with integral action, given by:

$$u_{av} = v_{av} + \frac{1}{Q}\dot{F} + F$$

$$v_{av} = \ddot{F}^*(\tau) - k_2(\dot{F} - \dot{F}^*(\tau)) - k_1(F - F^*(\tau)) - k_0 \int_0^\tau (F - F^*(\sigma))d\sigma \tag{6.27}$$

We remark that given a sinusoidal reference trajectory for the normalized average output voltage, $x_2^*(\tau)$, the corresponding average normalized nominal inductor current trajectory $x_1^*(\tau)$ and the average nominal control input trajectory $u_{av}^*(\tau)$, can only be approximately computed.

The closed loop tracking error system, with $e = F - F^*(\tau)$, is given by

$$\ddot{e} + k_2\dot{e} + k_1 e + k_0 \int_0^\tau e(\sigma)d\sigma = 0 \tag{6.28}$$

This integro-differential system has a characteristic polynomial expressed as,

$$p(s) = s^3 + k_2 s^2 + k_1 s + k_0 \tag{6.29}$$

A suitable choice of the controller design parameters k_2, k_1 and k_0 is obtained by equating the characteristic polynomial in (6.29) to a desired polynomial with pre-specified stable roots.

$$\begin{aligned} p_d(s) &= (s+p)(s^2 + 2\zeta\omega_n s + \omega_n^2) \\ &= s^3 + (p + 2\zeta\omega_n)s^2 + (2p\zeta\omega_n + \omega_n^2)s + \omega_n^2 p, \\ p &> 0, \quad \zeta > 0, \quad \omega_n > 0 \end{aligned} \tag{6.30}$$

We set then

$$k_2 = p + 2\zeta\omega_n, \quad k_1 = 2p\zeta\omega_n + \omega_n^2, \quad k_0 = \omega_n^2 p$$

The designed feedback controller exponentially asymptotically drives the tracking error system e towards zero.

The feedback controller (6.27) requires the flat output F and its first order time derivative, $\dot{F}$. These two variables may be directly obtained, from the system equations, in terms of the state variables, which are here assumed to be measurable,

$$F = x_2$$
$$\dot{F} = x_1 - \frac{x_2}{Q} \tag{6.31}$$

Hence, the state feedback controller is readily found, from (6.27), to be given by

$$u_{av} = v_{av} + \frac{1}{Q}\left(x_1 - \frac{x_2}{Q}\right) + x_2$$
$$v_{av} = \ddot{F}^*(\tau) - k_2\left[x_1 - \frac{x_2}{Q} - \dot{F}^*(\tau)\right] - k_1(x_2 - F^*(\tau))$$
$$- k_0 \int_0^\tau (x_2 - F^*(\sigma))d\sigma \tag{6.32}$$

Simplifying the controller expression, we find,

$$u_{av} = u_{av}^*(\tau) + \left(\frac{1}{Q} - k_2\right)\left[x_1 - \frac{x_2}{Q} - \dot{F}^*(\tau)\right] + (1 - k_1)(x_2 - F^*(\tau))$$
$$- k_0 \int_0^\tau (x_2 - F^*(\sigma))d\sigma \tag{6.33}$$

where $u_{av}^*(\tau) = \ddot{F}^* + \frac{1}{Q}\dot{F}^* + F^*$.

6.4.2 The Boost Converter

Consider the average normalized model of the Boost converter

$$\dot{x}_1 = -u_{av}x_2 + 1$$
$$\dot{x}_2 = u_{av}x_1 - \frac{x_2}{Q}$$

The flat output of this system, as already determined in previous sections and chapters, is the total average normalized energy, defined as

$$F = \frac{1}{2}\left(x_1^2 + x_2^2\right)$$

The differential parametrization of the system variables in terms of the flat output F is readily obtained to be

$$x_1 = \frac{1}{2}\left[-Q + \sqrt{Q^2 + 4\left(Q\dot{F} + 2F\right)}\right]$$

$$x_2 = \sqrt{-Q\dot{F} + \frac{1}{2}\left[-Q^2 + Q\sqrt{Q^2 + 4\left(Q\dot{F} + 2F\right)}\right]}$$

$$u_{av} = \frac{1 + \frac{2}{Q^2}x_2^2 - \ddot{F}}{x_2\left(1 + \frac{2}{Q}x_1\right)} \tag{6.34}$$

Note that the singular points $x_2 = 0$ and $x_1 = -Q/2$ are naturally excluded from consideration thanks to the underlying tracking problem defined on the bases or a positively biased sinusoid for the output voltage reference trajectory.

The nonlinear state dependent input coordinate transformation

$$u_{av} = \frac{1 + \frac{2}{Q^2} x_2^2 - v_{av}}{x_2 \left(1 + \frac{2}{Q} x_1\right)}$$

where v_{av} is the new, auxiliary, average control input leads to the exact second order integration dynamics

$$\ddot{F} = v_{av}$$

A trajectory tracking linear feedback controller, with integral action, is readily designed to be

$$v_{av} = \ddot{F}^*(\tau) - k_2 \left(\dot{F} - \dot{F}^*(\tau)\right) - k_1 \left(F - F^*(\tau)\right) - k_0 \int_0^\tau (F - F^*(\sigma))\, d\sigma$$

The choice of the design parameters k_2, k_1 and k_0 readily follows from the fact that the exactly linearized closed loop system exhibits a characteristic polynomial given by

$$p(s) = s^3 + k_2 s^2 + k_1 s + k_0$$

As it is usual with the flatness based approach, the required time derivatives of the flat output can always be placed back in terms of nonlinear functions of the measured states variables. In this case, we have:

$$F = \frac{1}{2}\left(x_1^2 + x_2^2\right), \qquad \dot{F} = x_1 - \frac{x_2^2}{Q}$$

The average nominal trajectory for F, denoted by $F^*(\tau)$ may be obtained, as a function of the desired biased sinusoidal output $x_1^*(\tau)$, from the functional iterative procedure previously explained in this chapter.

6.4.3 The Buck-Boost Converter

The average model of the transformed Buck-Boost converter given in Equation 6.18 is flat with flat output given by 6.19, recalled here for convenience:

$$F = \frac{1}{2}\left(\xi_1^2 + \xi_2^2\right) \tag{6.35}$$

A differential parametrization for the transformed system average state variables ξ_1, ξ_2, is readily obtained from the Expression 6.35 and the time derivative of F given by

$$\dot{F} = \xi_1 - \frac{\xi_2(1+\xi_2)}{Q} \tag{6.36}$$

We obtain:

$$\xi_2 = -\frac{1}{2(1+Q)} - \sqrt{\frac{1}{4(1+Q)^2} - \frac{Q}{1+Q}\left(\dot{F} - 2F\right)}$$

$$\xi_1 = \frac{\xi_2(1+\xi_2)}{Q} + \dot{F} \tag{6.37}$$

The parametrization for the average control input is obtained from $\ddot{F}$ as follows:

$$u_{av} = \frac{Q(\ddot{F}-1) - (1+\xi_2)(1+2\xi_2)}{Q\xi_2 + \xi_1(1+2\xi_2)} \tag{6.38}$$

The state dependent input coordinate transformation:

$$u_{av} = \frac{Q(v_{av}-1) - (1+\xi_2)(1+2\xi_2)}{Q\xi_2 + \xi_1(1+2\xi_2)} \tag{6.39}$$

where v_{av} is a new control input leads to the exact linearization of the flat output dynamics

$$\ddot{F} = v_{av}$$

As in the previous case, a trajectory tracking linear feedback controller, with integral action, is readily designed to be

$$v_{av} = \ddot{F}^*(\tau) - k_2\left(\dot{F} - \dot{F}^*(\tau)\right) - k_1\left(F - F^*(\tau)\right) - k_0 \int_0^\tau (F - F^*(\sigma))\,d\sigma$$

The prescription of the roots of the corresponding closed loop characteristic polynomial:

$$p(s) = s^3 + k_2 s^2 + k_1 s + k_0$$

by means of the design gains k_2, k_1 and k_0 completes the flatness based feedback controller design.

The fact that necessarily the restriction, $u_{av}(\tau) \in [0,1]$ must be uniformly valid for a given biased sinusoidal transformed output voltage of the form:

$$\xi_2^*(\tau) = (B-1) + A\sin(\omega_0 \tau)$$

with $A > 0$, $B < 0$ and $B < -(1+A)$, leads to important nonlinear amplitude frequency tradeoffs whose derivation details are left for the reader.

6.5 A Sliding Mode Control Approach

As already remarked, the solution of the AC voltage generation problem, as a trajectory tracking problem, the internal instability issue seems to be unavoidable when inversion or partial input-output inversion of the system is invoked. The main reason being that a direct solution to the output tracking problem results in an unfeasible, internally unstable, closed loop, behavior of the system due to the non-minimum phase properties of the output capacitor voltage variable (see Sira-Lischinsky [68]).

In the sliding mode control approach off-line computed candidates for the inductor current reference signals are then used to devise time-varying sliding surfaces for the converter dynamics on which the sliding mode existence conditions must be inspected. The frequency and amplitude limitations for the desired AC output voltage signal naturally emerge as a consequence of the well-known sliding mode existence conditions (see Utkin [75]).

6.5.1 The Boost Converter

Consider the normalized model of the Boost converter:

$$\dot{x}_1 = -ux_2 + 1 \tag{6.40}$$

$$\dot{x}_2 = ux_1 - \frac{x_2}{Q} \tag{6.41}$$

The normalized total stored energy, here denoted by F, is given by

$$F = \frac{1}{2}\left(x_1^2 + x_2^2\right) \tag{6.42}$$

It is desired to devise a discontinuous feedback control law for u, such that the normalized capacitor voltage, x_2, tracks a given desired voltage reference signal $x_2^*(\tau)$. This signal is assumed to be bounded and sufficiently differentiable. In fact, we assume that $x_2^*(\tau)$ is smooth, i.e., infinitely differentiable. Specifically, we are interested in generating a normalized output voltage of the form $x_2(t) = A + (B/2)\sin \omega\tau$ with A, $\omega > 0$ and B being a constant of arbitrary sign.

6.5.2 A Feasible Indirect Input Current Tracking Approach

The idea that circumvents the underlying non-minimum phase output control problem is to indirectly generate the desired capacitor voltage signal, $x_2^*(\tau)$, on the basis of tracking a suitable corresponding inductor current signal $x_1^*(\tau)$. The difficulty, as previously remarked, resides in finding such a suitable inductor current reference signal $x_1^*(\tau)$. This issue has been treated, using a plausible approximation scheme, in the previous sections.

Suppose that a suitable smooth inductor current reference signal is given as $x_1^*(\tau)$, whose time derivative is, of course, also bounded. A discontinuous feedback controller which reaches and sustains a sliding motion on the time-varying surface defined as:

$$\sigma = x_1 - x_1^*(\tau)$$

is given by

$$u = 0.5(1 + \text{sign}\,\sigma) \qquad (6.43)$$

Indeed, starting from zero initial conditions for x_1 and x_2, we have that initially $x_1(\tau)$ is smaller than $x_1^*(\tau)$ (i.e., $\sigma < 0$). The switching strategy (6.43) sets $u = 0$ and the normalized inductor current x_1 grows with slope equals to 1, while x_2 remains at zero. The sliding surface reaching condition is thus satisfied from "below", provided the reference signal $x_1^*(\tau)$ is designed with a time derivative which is bounded above by 1. Clearly, under such assumptions, the quantity $\sigma\dot{\sigma}$ is negative and given by $\sigma(1 - \dot{x}_1^*(\tau)) < 0$. When the sliding surface is reached and slightly overshot, the controller (6.43) starts to inject large positive current pulses to the output RC filter by letting $u = 1$. As a consequence, x_2 immediately starts to grow from zero, rapidly reaching the converters amplifying mode $x_2 > 1$. Thus, while σ is positive, its time derivative, $\dot{\sigma} = -x_2 + 1$, becomes negative. Hence, the sliding surface reaching condition $\sigma\dot{\sigma} < 0$ is also satisfied from "above" after the circuit is found in its amplifying mode.

The corresponding "equivalent control" is now obtained as

$$u_{eq} = \frac{1 - \dot{x}_1^*(\tau)}{x_2} \qquad (6.44)$$

The necessary and sufficient conditions for the existence of a sliding regime, given by

$$0 < u_{eq} < 1 \qquad (6.45)$$

imply that, at each instant, the following set of inequalities must be satisfied,

$$0 < 1 - \dot{x}_1^*(\tau) < x_2 \qquad (6.46)$$

The restriction $\dot{x}_1^*(\tau) < 1$ implies, roughly speaking, a limitation on the amplitude and frequency of the desired reference signal. Specific tracking limitations of the sliding mode control approach have to be worked out, in detail, for each particular given reference signal waveform $x_1^*(\tau)$.

The ideal sliding dynamics corresponding to the sliding surface $\sigma = x_1 - x_1^*(\tau)$ is given by the following stable time-varying nonlinear dynamics,

$$\dot{x}_2 = \left(\frac{1 - \dot{x}_1^*(\tau)}{x_2}\right) x_1^*(\tau) - \frac{x_2}{Q} \qquad (6.47)$$

In order to establish the stability of (6.47) we define the variable $\rho = x_2^2$ which is easily seen to satisfy the following stable linear differential equation subject to bounded perturbations input signals,

$$\dot{\rho} = -\frac{2}{Q}\left[\rho - Q\left(1 - \dot{x}_1^*(\tau)\right)x_1^*(\tau)\right] \qquad (6.48)$$

This linear time-invariant forced system is clearly asymptotically stable due to the negativity of its only constant eigenvalue. The forcing signal is clearly bounded. The result follows.

6.6 Exact Tracking Error Dynamics Passive Output Feedback Control

6.6.1 The Double Bridge Buck Converter

Consider the normalized average model of a double bridge Buck converter:

$$\begin{aligned}\dot{x}_1 &= -x_2 + u_{av} \\ \dot{x}_2 &= x_1 - \frac{x_2}{Q}\end{aligned} \qquad (6.49)$$

Consider a smooth nominal state trajectory of the system

$$t \to (x_1^*(\tau), x_2^*(\tau), u^*(\tau))$$

This trajectory actually represents a solution of the average normalized dynamics,

$$\begin{aligned}\dot{x}_1^* &= -x_2^* + u_{av}^* \\ \dot{x}_2^* &= x_1^* - \frac{x_2^*}{Q}\end{aligned} \qquad (6.50)$$

The exact tracking error dynamics is readily obtained to be

$$\begin{aligned}\dot{e}_1 &= -e_2 + e_u \\ \dot{e}_2 &= e_1 - \frac{e_2}{Q}\end{aligned} \qquad (6.51)$$

where $e_1 = x_1 - x_1^*(\tau)$, $e_2 = x_2 - x_2^*(\tau)$ and $e_u = u_{av} - u_{av}^*(\tau)$.

Note that the passive output error, corresponding to the dynamics (6.51) is represented by $e_z = e_1$.

A Lyapunov function candidate of the form

$$V(e) = \frac{1}{2}\left[e_1^2 + e_2^2\right] \qquad (6.52)$$

has as a time derivative, along the controlled solutions of the controlled exact tracking error dynamics (6.51), the following expression:

$$\dot{V}(e) = e_1 e_u - \frac{e_2^2}{Q} \qquad (6.53)$$

6.6 Exact Tracking Error Dynamics Passive Output Feedback Control

Let $\gamma > 0$. Then, the passive output feedback controller expressed in terms of the input error:

$$e_u = -\gamma e_1 = -\gamma(x_1 - x_1^*(\tau)) \qquad (6.54)$$

yields,

$$\dot{V}(e) = -\gamma e_1^2 - \frac{e_2^2}{Q} \leq 2\min\left\{\gamma, \frac{1}{Q}\right\} V(e)$$

i.e., the origin of the error space, if there are no control input limitations, is a globally exponentially asymptotic equilibrium point for the closed loop error dynamics. Due to the average control input limitations, represented by $u_{av} \in [-1, 1]$ the origin of the error space will be, generally speaking, only a semi-globally exponentially asymptotic equilibrium point for the closed loop error dynamics.

The average, full state, linear feedback controller may be rewritten as,

$$u_{av} = u_{av}^*(\tau) - \gamma(x_1 - x_1^*(\tau)) \qquad (6.55)$$

For the DC-to-AC power conversion task, the average reference signal, $x_1^*(\tau)$, required by the feedback controller (6.55), is obtained from Equation 6.4, derived in the first section of this chapter.

6.6.2 The Boost Converter

Consider the average normalized model of the Boost converter

$$\dot{x}_1 = -u_{av} x_2 + 1$$
$$\dot{x}_2 = u_{av} x_1 - \frac{x_2}{Q}$$

Take a smooth average normalized reference state trajectory

$$t \to (x_1^*(\tau), x_2^*(\tau), u^*(\tau))$$

i.e., a smooth solution of the average normalized dynamics

$$\dot{x}_1^* = -u_{av}^* x_2^* + 1$$
$$\dot{x}_2^* = u_{av}^* x_1^* - \frac{x_2^*}{Q}$$

Define the tracking error state as $e_1 = x_1 - x_1^*(\tau)$, $e_2 = x_2 - x_2^*(\tau)$ while the control input error is given by $e_u = u_{av} - u_{av}^*(\tau)$. Rather straightforward algebraic manipulations lead to the following open loop error dynamics, that we address as exact tracking error dynamics (ETED),

$$\dot{e}_1 = -u_{av} e_2 - x_2^* e_u$$
$$\dot{e}_2 = u_{av} e_1 - \frac{e_2}{Q} + x_1^* e_u \qquad (6.56)$$

A Lyapunov function of the form

$$V(e) = \frac{1}{2}\left(e_1^2 + e_2^2\right)$$

exhibits a time derivative along the solutions of the controlled ETED given by

$$\dot{V}(e) = [-x_2^*(\tau)e_1 + x_1^*(\tau)e_2]\,e_u - \frac{e_2^2}{Q} \tag{6.57}$$

A natural choice for the average feedback control input error, e_u, is given by:

$$e_u = -\gamma\left[-x_2^*(\tau)e_1 + x_1^*(\tau)e_2\right], \quad \gamma > 0 \tag{6.58}$$

i.e., after using the definition of the tracking error variables, we have,

$$u_{av} = u_{av}^*(\tau) + \gamma\left[x_2^*(\tau)x_1 - x_1^*(\tau)x_2\right] \tag{6.59}$$

The proposed controller (6.59) renders, for all strictly positive values of γ, a negative definite time derivative for $V(e)$. If the average control input u_{av} were not constrained to an interval of the real line, then the origin of the error space (e_1, e_2) would indeed be a globally asymptotically stable equilibrium point for the closed loop ETED.

Note that the time-varying linear tracking error controller simply amounts to a static feedback of the passive output of the ETED. Indeed, the ETED, (6.56), has as its passive output the following time-varying linear combination of state error variables:

$$e_z = -x_2^*(\tau)e_1 + x_1^*(\tau)e_2 = -x_2^*(\tau)x_1 + x_1^*(\tau)x_2 \tag{6.60}$$

and the proposed feedback controller is simply of the form $e_u = -\gamma e_z$. This justifies the name of Exact Tracking Error Dynamics Passive Output Feedback (ETEDPOF) to this type of feedback controller. The feedback controller (6.59), when faced with the natural limitations of u_{av} to lie within the interval $[0, 1]$, may not necessarily stabilize all initial state errors towards the origin due to possible control input saturations. This implies that the proposed ETEDPOF controller renders the origin of the error space only as a semi-globally asymptotically stable equilibrium point.

A set of approximate trajectories $x_1^*(\tau)$, $u^*(\tau)$, generated from the desired AC biased sinusoidal voltage $x_2^*(\tau)$, as explained in the first section of this chapter, may plausibly integrate the time-varying part of the proposed controller (6.59). The reader is invited to perform simulations to assess the validity of the previous statement.

6.6.3 The Buck-Boost Converter

Consider the average normalized model of the Boost converter

$$\dot{x}_1 = u_{av}x_2 + 1 - u_{av}$$
$$\dot{x}_2 = -u_{av}x_1 - \frac{x_2}{Q}$$

As before, consider a smooth average normalized reference state trajectory,

$$t \to (x_1^*(\tau), x_2^*(\tau), u^*(\tau))$$

which is indeed a smooth solution of the average normalized dynamics:

$$\dot{x}_1^* = u_{av}^* x_2^* + 1 - u_{av}^*$$
$$\dot{x}_2^* = -u_{av}^* x_1^* - \frac{x_2^*}{Q}$$

Let $e_1 = x_1 - x_1^*(\tau)$, $e_2 = x_2 - x_2^*(\tau)$ and $e_u = u_{av} - u_{av}^*(\tau)$. We obtain the following open loop error dynamics,

$$\dot{e}_1 = u_{av}e_2 + [x_2^* - 1]e_u$$
$$\dot{e}_2 = -u_{av}e_1 - \frac{e_2}{Q} - x_1^* e_u \quad (6.61)$$

A Lyapunov function of the form

$$V(e) = \frac{1}{2}\left(e_1^2 + e_2^2\right)$$

exhibits a time derivative along the solutions of the controlled ETED given by

$$\dot{V}(e) = [e_1(x_2^*(\tau) - 1) - x_1^*(\tau)e_2]e_u - \frac{e_2^2}{Q} \quad (6.62)$$

A natural choice for the average feedback control input error, e_u, is given by:

$$e_u = -\gamma[e_1(x_2^*(\tau) - 1) - x_1^*(\tau)e_2], \quad \gamma > 0 \quad (6.63)$$

i.e., after using the definition of the tracking error variables, we have,

$$u_{av} = u_{av}^*(\tau) - \gamma[x_1(x_2^*(\tau) - 1) - x_2 x_1^*(\tau) + x_1^*(\tau)] \quad (6.64)$$

The proposed controller (6.64) renders, for all strictly positive values of γ, a negative definite time derivative for $V(e)$. It follows that the origin of the error space (e_1, e_2) would indeed be a globally asymptotically stable equilibrium point for the closed loop ETED provided the average control input u_{av} were not constrained to a closed interval of the real line.

Note that, as in the previous case, the time-varying linear tracking error controller simply amounts to a static feedback of the passive output of the

ETED. Indeed, the ETED, (6.61), has as its passive output the following time-varying linear combination of state error variables:

$$e_z = (x_2^*(\tau) - 1)e_1 - x_1^*(\tau)e_2 = (x_2^*(\tau) - 1)x_1 + x_1^*(\tau) - x_2(\tau)x_1^*(\tau) \quad (6.65)$$

and the proposed feedback controller is simply of the form $e_u = -\gamma e_z$. The proposed ETEDPOF controller renders the origin of the error space only as a semi-globally asymptotically stable equilibrium point due to the effect of possible controller saturation for some initial states.

7
AC Rectifiers

7.1 Introduction

In this chapter we will address the control of two Boost AC rectifiers: The monophasic Boost rectifier and the three phase Boost rectifier. In both cases the control objectives are twofold: the enhancement and regulation of the DC component of the output voltage and the alignment of input line currents with the input voltages. The first task represents the desired rectifying features of the controlled converter while the second task constitutes a desirable operation feature bestowing a betterment of the converter system power factor. Both tasks are suitably accounted for in the proposed feedback control strategies which combine the passivity based control with the differential flatness properties of the system. At the end of this chapter we also present two specific applications dealing with a three phase rectifier dc-motor combination.

Here, we pretend to be rather tutorial in nature. For this reason, this chapter presents a somewhat idealized version of the underlying challenging problem of efficient control of the switched AC rectifiers based on average normalized models. In order to introduce the reader to the fundamentals of average based feedback controller design we purposefully avoid a rather crucial issue in AC rectifier system control. Namely, that the control must be robust with respect to a lack of knowledge of the value of the constant loads. Load uncertainty is quite a common feature in this field. It has most frequently been handled from an adaptive feedback control viewpoint. We feel that with the advent of recent developments in the are of *algebraic identification* techniques, the problem of on-line load estimation largely overcomes the need for asymptotic adaptive control techniques in Power Electronics. The reader is referred to the work of Fliess and Sira-Ramírez [23] for an introduction to the subject of on-line parameter identification from an algebraic approach.

A brief, necessarily incomplete, survey of some contributions in this field of controlling AC rectifiers is in order just to provide some guide to the interested reader in his search for some other fundamental features of switched control of AC rectifiers. A sliding mode control approach for the unity power factor rectifier of the Boost type has been addressed in the work of Silva [55] and also in the article by Morici et al. [45]. A, so called, three-dimensional pulse width modulated control scheme for the four wire version of the three phase rectifier is the subject of the work in Wong et al. [78]. A geometric approach exploiting the input output linearization possibilities in a three phase rectifier are addressed in the work of Lee [39]. The feedback linearization features of three phase rectifiers were established and exploited in Lee et al. [40]. Interesting experimental tests, and precise comparisons among several adaptive feedback control schemes based on passivity considerations, are reported in Karagiannis et al. [34] for mono-phasic Boost DC-AC converters with unit power factor. An experimental implementation of advanced nonlinear feedback control techniques is reported in the article by Yacoubi et al. [81]. The references of Wu, Dewand and Slemon [80] and that of Blasko and Kaura [3] are quite useful in many respects regarding the modelling and control of AC rectifiers from alternative viewpoints.

7.2 Boost Unit Power Factor Rectifier

Power factor rectifiers also receive the name of Power Factor Pre-compensators to emphasize one of the basic tasks in the switched regulation of AC rectifiers. Namely, that of enhancing the power factor of the inserted rectifier and achieving an alignment of input currents and voltages. This natural demand poses interesting limitations to the underlying output voltage DC component regulation to a desirable constant level. In the next section we examine this issue. The results in this section stem, and are largely motivated, from the work by Escobar et al. [12], where an adaptive control viewpoint is also adopted for the control of a unity power factor rectifier of the Boost type with unknown loads. We closely follow their analytic developments except for the controller design, which in our case turns out to be a linear feedback controller.

7.2.1 Model of the Monophasic Boost Rectifier

Consider the Boost type unit power pre-compensator, or rectifier, shown in Figure 7.1

The system is described by the following set of differential equations:

$$L\frac{di}{dt} = -uv + E\sin(\omega t)$$
$$C\frac{dv}{dt} = ui - \frac{1}{R}v$$

where i is the inductor current, v is the output capacitor voltage and u is the switch position function taking values in the discrete set $\{-1, 0, 1\}$.

While the input voltage is positive, the input u takes values in the set $\{0, 1\}$ and when it becomes negative, it takes values in the set $\{-1, 0\}$.

7.2 Boost Unit Power Factor Rectifier

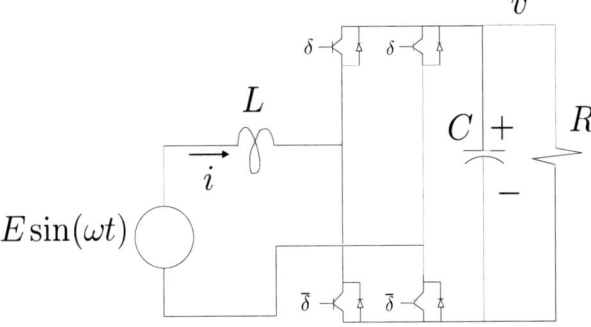

Fig. 7.1. Monophasic Boost power factor rectifier.

The normalization of the system equations is carried out according to the following state and time coordinates transformation

$$x_1 = \frac{i}{E}\sqrt{\frac{L}{C}}, \quad x_2 = \frac{v}{E}, \quad Q = R\sqrt{\frac{C}{L}}, \quad \tau = \frac{t}{\sqrt{LC}}$$

The normalized average system equations are then obtained in the following form

$$\dot{x}_1 = -ux_2 + \sin(\omega_0 \tau)$$
$$\dot{x}_2 = ux_1 - \frac{x_2}{Q}$$

7.2.2 The Control Objectives

The control objectives for the average monophasic Boost rectifier system are twofold:
- It is desired to have the average normalized inductor current track a sinusoidal signal of the same angular frequency ω_0 and an amplitude A to be determined. This guarantees a unit power factor
- It is desired that the DC component of the average normalized voltage v stabilizes to a constant desired value V_d.

7.2.3 Steady State Considerations

The total stored average normalized energy of the system is given by

$$H = \frac{1}{2}\left[x_1^2 + x_2^2\right]$$

The total power is given by the time derivative of H,

$$\frac{dH}{dt} = x_1 \sin(\omega_0 \tau) - \frac{x_2^2}{Q}$$

where the first summand corresponds with the input power and the second term corresponds to the delivered power at the load. The steady state value of the DC component of the total power should balance to zero, since the system is lossless. We have then the following steady state power balance condition:

$$\langle \overline{x}_1 \sin(\omega_0 \tau) \rangle_{dc} = \left\langle \frac{\overline{x}_2^2}{Q} \right\rangle_{dc}$$

where the "over-line" stands for steady state value of the involved variable.

Using the desired values as the steady state value we obtain the following relationship

$$\langle A \sin^2(\omega_0 \tau) \rangle_{dc} = \frac{V_d^2}{Q}$$

From where it is immediate to obtain:

$$A = \frac{2V_d^2}{Q}$$

This relation will be quite useful in the sequel.

The fact that the inductor current amplitude A and the desired DC component of the output voltage satisfy the above relation is sometimes addressed as the *solvability condition*. When inductor resistances are considered, the obtained condition further reveals a natural limitation of the reachable output voltages.

7.2.4 Exact Open Loop Tracking Error Dynamics and Controller Design

In the particular case of the Boost based unity power factor pre-compensator, we have that the tracking error dynamic system is given by

$$\dot{e}_1 = -ue_2 - x_2^*(\tau)e_u$$
$$\dot{e}_2 = ue_1 - \frac{e_2}{Q} + x_1^*(\tau)e_u$$
$$e_y = e_2$$

The proposed linear time varying feedback control law reads

$$e_u = -\gamma[-x_2^* e_1 + x_1^*(\tau)e_2]$$

and the dissipation matching condition takes the form:

$$\begin{bmatrix} \gamma[x_2^*]^2 & -\gamma x_2^* x_1^* \\ -\gamma x_2^* x_1^* & \frac{1}{Q} + \gamma[x_1^*]^2 \end{bmatrix} > 0$$

which is certainly valid for any non-zero state trajectory.

The average linear time varying controller

$$u = u^*(\tau) + \gamma\left[x_2^*(x_1 - x_1^*) - x_1^*(x_2 - x_2^*)\right]$$
$$= u^*(\tau) + \gamma\left[x_2^* x_1 - x_1^* x_2\right]$$

is the desired feedback control law.

One possibility for specifying the feedback control law consists in using as nominal reference trajectories the desired values. For instance, We may use the steady state value of the inductor current which ideally guarantees a power factor of 1 and for the output capacitor voltage, the constant steady state DC component. In other words, we may set:

$$x_1^*(\tau) = A\sin(\omega_0 \tau), \qquad x_2^*(\tau) = V_d$$

with the nominal control input, $u^*(\tau)$, computed from the first equation of the average normalized model as follows:

$$u^* = \frac{(\sin(\omega\tau) - \dot{x}_1^*)}{x_2^*(\tau)} = \frac{\sin(\omega\tau) - A\omega\cos(\omega\tau)}{V_d}$$

7.2.5 Simulations

We considered the following Boost type Power Factor Pre-Compensator, characterized by:

$$L = 1 \text{ mH}, \qquad C = 2 \text{ mF}, \qquad R = 2.4 \text{ } \Omega$$

These values yield normalized parameter values given by

$$\omega_0 = 2\pi f_0, \qquad f_0 = 60\sqrt{LC} = 0.08472,$$

$$Q = R\sqrt{\frac{C}{L}} = 33.94, \qquad \tau = t/0.001414$$

We set the controller gain to $\gamma = 0.5$ and $V_d = 1.5$.

Figure 7.2 depicts the computer simulations of the closed loop response of the Boost monophasic rectifier circuit when the static passivity based controller is used.

7.2.6 The Use of the Differential Flatness Property in the Passive Controller Design

A problem with this response lies in the fact that the quality of the transient and the precision of the tracking is never guaranteed due to the fact that the adopted nominal trajectories are not really trajectories of the system.

The main task becomes then one of specifying the nominal state and input trajectories in accordance with the control objectives which are actual system

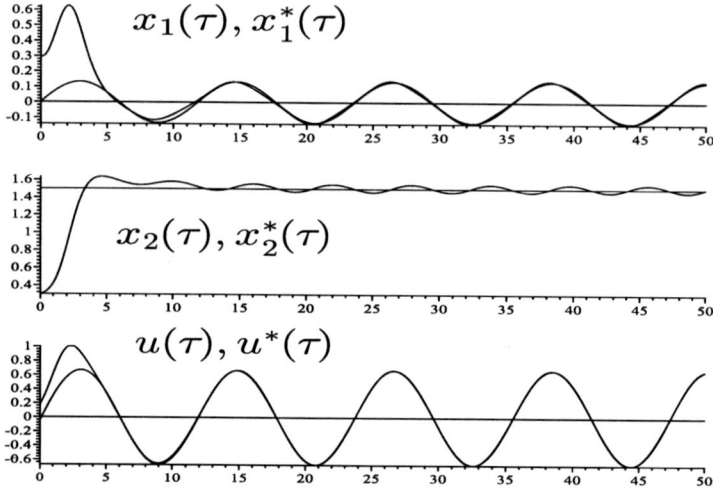

Fig. 7.2. Closed loop response of Boost unit power factor rectifier.

trajectories. The problem is by no means a trivial one unless one resorts to flatness of the original system. We have the following property of the Boost based unity power factor pre-compensator system.

The system

$$\dot{x}_1 = -ux_2 + \sin(\omega_0 \tau)$$
$$\dot{x}_2 = ux_1 - \frac{x_2}{Q}$$
$$y = x_2$$

is flat, with flat output given by the total stored energy:

$$F = \frac{1}{2}\left[x_1^2 + x_2^2\right]$$

and its time derivative is given by

$$\dot{F} = x_1 \sin(\omega_0 \tau) - \frac{x_2^2}{Q}$$

Eliminating x_2 from the last two relations we obtain a quadratic equation for x_1 in terms of F and $\dot{F}$.

$$x_1^2 + [Q\sin(\omega_0 \tau)]x_1 - (Q\dot{F} + 2F) = 0$$

and the positive solution for the differential parametrization of the average normalized current is readily obtained as

$$x_1 = -\frac{Q}{2}\sin(\omega_0 \tau) + \sqrt{\frac{Q^2}{4}\sin^2(\omega_0 \tau) + (Q\dot{F} + 2F)}$$

7.2 Boost Unit Power Factor Rectifier

Using the obtained parametrization for x_1 one obtains, from the system equations, the corresponding parametrization for x_2

$$x_2 = \sqrt{Q\left[-\frac{Q}{2}\sin^2(\omega_0\tau) + \sin(\omega_0\tau)\sqrt{\frac{Q^2}{4}\sin^2(\omega_0\tau) + (Q\dot{F} + 2F)} - \dot{F}\right]}$$

The average control input signal u is also differentially parameterized in terms of F, $\dot{F}$ and $\ddot{F}$ using, for instance, the relation obtained from the average normalized inductor current equation:

$$u = \frac{\sin(\omega_0\tau) - \dot{x}_1}{x_2}$$

The previous differential parameterizations allow us to compute the nominal state trajectories and the nominal control input associated with a nominal trajectory of the flat output which is compatible with the control objectives.

Let the unit power factor desired nominal value of $x_1(\tau)$ be given by the signal $x_1^*(\tau) = A\sin(\omega_0\tau)$ then the differential parametrization of x_1 leads to the following (stable) differential equation for F^*:

$$\dot{F}^* = -\frac{2}{Q}F^* + \left[\frac{A(A+Q)}{Q}\right]\sin^2(\omega_0\tau)$$

In terms of the desired steady state constant average output voltage $\langle\bar{x}_2\rangle_{dc} = V_d$ the differential equation satisfied by the flat output (average total stored energy) is obtained by using the relation $A = 2V_d^2/Q$. We get:

$$\dot{F}^* = -\frac{2}{Q}F^* + \frac{2V_d^2}{Q}\left(1 + \frac{2V_d^2}{Q^2}\right)\sin^2(\omega_0\tau)$$

The DC component of the steady state solution of the above differential equation is computed to be

$$\langle\overline{F}\rangle_{dc} = \frac{V_d^2}{2}\left(1 + \frac{2V_d^2}{Q^2}\right)$$

which precisely coincides with the value obtained from the flat output definition

$$\langle\overline{F}\rangle_{dc} = \frac{1}{2}[\langle\bar{x}_1^2\rangle_{dc} + \langle\bar{x}_2^2\rangle_{dc}] = \frac{1}{2}[\langle A^2\sin^2(\omega_0\tau)\rangle_{dc} + \langle V_d^2\rangle_{dc}]$$
$$= \frac{1}{2}\left[\frac{A^2}{2} + V_d^2\right] = \frac{V_d^2}{2}\left[1 + \frac{2V_d^2}{Q^2}\right]$$

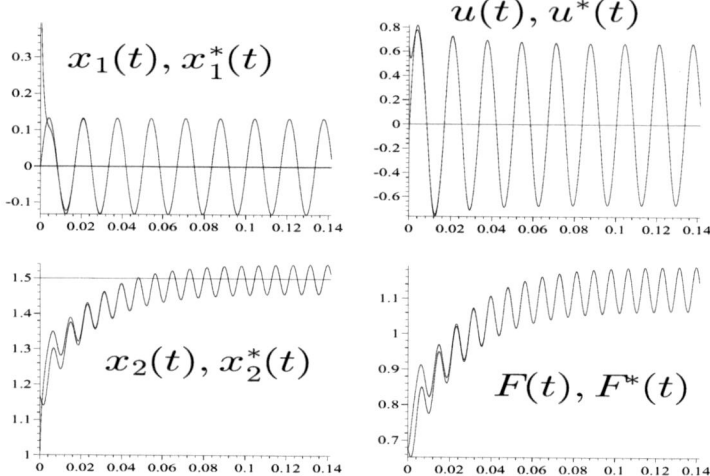

Fig. 7.3. Closed loop response of unit power factor Boost rectifier.

7.2.7 Simulations

Figure 7.3 depicts the computer simulations of the controlled monophasic Boost rectifier controlled by a static passivity based controller.

The controller used for the simulations is the linear time-varying controller with $x_1^*(\tau) = A\sin(\omega_0\tau)$, and $x_2^*(\tau)$, as given by the differential parametrization involving F^* and $\dot{F}^*$. The nominal control input, u^*, was computed from the first system equation using the computed nominal states. We have set the controller gain to $\gamma = 1$, in this case.

7.3 Three Phase Boost Rectifier

In this section, we propose a linear, time-varying, feedback controller for the uniform semi-global stabilization of the output voltage in a three phase switched Boost rectifier. The approach combines differential flatness, linear static modified output tracking error feedback and $\Sigma - \Delta$ modulation. The passive output considerations of the exact tracking error model allows for a simple linear state feedback which requires the nominal state trajectories and control inputs as data. The nominal state and inputs trajectories are related to the desire of having constant flat outputs for the reduced average normalized balanced rectifier model. The nominal state and input trajectories are thus planned on the basis of the flat outputs ideal constant desired behavior. We specify these flat output trajectories by imposing ideal behaviors which conveniently imply: 1) unit power factor for each line, as well as 2) perfect balancing conditions on the rectifier. The designed average control input signals are then feed into independent $\Sigma - \Delta$ modulators for an efficient sliding mode type of controller signal implementation. The outputs of the $\Sigma - \Delta$ modulators are, in fact, switched output signals acting as the actual control inputs. These switched inputs cause average responses which represent the ideal sliding features of the underlying sliding regime taking place on the error space of the $\Sigma - \Delta$ modulators. The designed ideal closed loop average features are thus efficiently recovered in the switched implementation.

7.3.1 The Three Phase Boost Rectifier Average Model

Consider the following average model of a Boost type three phase rectifier (see Figure 7.4)

$$L\dot{x}_1 = -u_{1,av}x_4 - Rx_1 + V_1$$
$$L\dot{x}_2 = -u_{2,av}x_4 - Rx_2 + V_2$$
$$L\dot{x}_3 = -u_{3,av}x_4 - Rx_3 + V_3$$
$$C\dot{x}_4 = u_{1,av}x_1 + u_{2,av}x_2 + u_{3,av}x_3 - \frac{x_4}{R_L}$$

where $V_1 = V\cos(\omega t)$, $V_2 = V\cos\left(\omega t - \frac{2\pi}{3}\right)$, $V_3 = V\cos\left(\omega t + \frac{2\pi}{3}\right)$, represent the balanced external AC voltages. The average inputs, representing the switching actions, satisfy $u_{i,av} \in [-1, 1]\ \forall\ i$. R is the line resistance.

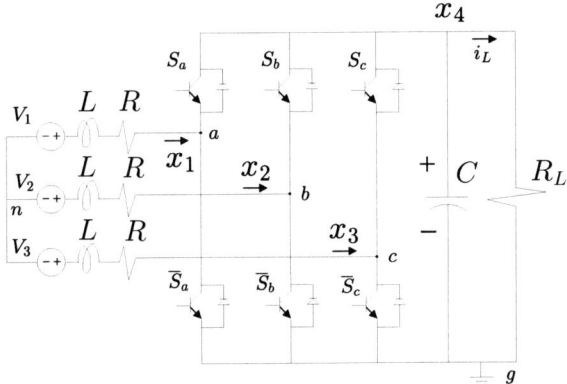

Fig. 7.4. Three phase Boost rectifier.

The state coordinate and time scale transformation

$$z_i = \left(\frac{1}{V}\sqrt{\frac{L}{C}}\right)x_i,\ i = 1, 2, 3. \qquad z_4 = \frac{x_4}{V}, \qquad \tau = \frac{t}{\sqrt{LC}}$$

yields the following normalized average model of the system:

$$\dot{z}_1 = -u_{1,av}z_4 - Qz_1 + \cos(\omega_n \tau)$$

$$\dot{z}_2 = -u_{2,av}z_4 - Qz_2 + \cos\left(\omega\tau - \frac{2\pi}{3}\right)$$

$$\dot{z}_3 = -u_{3,av}z_4 - Qz_3 + \cos\left(\omega\tau + \frac{2\pi}{3}\right)$$

$$\dot{z}_4 = u_{1,av}z_1 + u_{2,av}z_2 + u_{3,av}z_3 - \frac{z_4}{Q_L}$$

where

$$Q = R\sqrt{\frac{C}{L}}, \quad Q_L = R_L\sqrt{\frac{C}{L}}, \quad \omega_n = \omega\sqrt{LC}$$

We rewrite the normalized average system in the following "energy management" form: $\mathcal{A}\dot{z} = \mathcal{J}(u_{av})z - \mathcal{R}z + \mathcal{E}(t)$,

$$\frac{d}{d\tau}\begin{bmatrix} z_1 \\ z_2 \\ z_3 \\ z_4 \end{bmatrix} = \begin{bmatrix} 0 & 0 & 0 & -u_{1,av} \\ 0 & 0 & 0 & -u_{2,av} \\ 0 & 0 & 0 & -u_{3,av} \\ u_{1,av} & u_{2,av} & u_{3,av} & 0 \end{bmatrix}\begin{bmatrix} z_1 \\ z_2 \\ z_3 \\ z_4 \end{bmatrix}$$

$$-\begin{bmatrix} Q & 0 & 0 & 0 \\ 0 & Q & 0 & 0 \\ 0 & 0 & Q & 0 \\ 0 & 0 & 0 & \frac{1}{Q_L} \end{bmatrix}\begin{bmatrix} z_1 \\ z_2 \\ z_3 \\ z_4 \end{bmatrix} + \begin{bmatrix} \cos(\omega_n\tau) \\ \cos(\omega_n\tau - \frac{2\pi}{3}) \\ \cos(\omega_n\tau + \frac{2\pi}{3}) \\ 0 \end{bmatrix}$$

i.e., according to our previous notations; $\mathcal{A} = I$, $\mathcal{B} = 0$ and $\mathcal{E}(\tau)$ represents the vector of unit amplitude, normalized, line voltages with zero as the last component. (We have abusively used the "dot" notation "$\dot{z}$" to mean $\frac{dz}{d\tau}$).

The matrix $\mathcal{B}^*(\tau)$ is, in this case, given by

$$\mathcal{B}^*(\tau) = \begin{bmatrix} -z_4^*(\tau) & 0 & 0 \\ 0 & -z_4^*(\tau) & 0 \\ 0 & 0 & -z_4^*(\tau) \\ z_1^*(\tau) & z_2^*(\tau) & z_3^*(\tau) \end{bmatrix}$$

We choose the matrix Γ to be diagonal of the form $\Gamma = \text{diag}[\gamma_1, \gamma_2, \gamma_3]$ with $\gamma_i > 0 \ \forall \ i$. It is easy to verify that the dissipation matching condition is strongly uniformly satisfied in this case[1].

---

[1]

$$\mathcal{R} + \mathcal{B}^*(\tau)\Gamma[\mathcal{B}^*(\tau)]^T =$$
$$\begin{bmatrix} Q + \gamma_1[z_4^*(\tau)]^2 & 0 & 0 & \gamma_1 z_4^*(\tau)z_1^*(\tau) \\ 0 & Q + \gamma_2[z_4^*(\tau)]^2 & 0 & \gamma_2 z_4^*(\tau)z_2^*(\tau) \\ 0 & 0 & Q + \gamma_3[z_4^*(\tau)]^2 & \gamma_3 z_4^*(\tau)z_3^*(\tau) \\ \gamma_1 z_4^*(\tau)z_1^*(\tau) & \gamma_2 z_4^*(\tau)z_2^*(\tau) & \gamma_3 z_4^*(\tau)z_3^*(\tau) & \frac{1}{Q_L} + \gamma_1[z_1^*(\tau)]^2 + \gamma_2[z_2^*(\tau)]^2 + \gamma_3[z_3^*(\tau)]^2 \end{bmatrix} > 0$$

7.3.2 A Static Passivity Based Controller

The average linear tracking error feedback controller, based on static passive output feedback, is readily obtained as

$$u_{i\,av} = u^*_{i\,av}(\tau) - \gamma_i\left[-z^*_4(\tau)(z_i - z^*_i) + z^*_i(z_4 - z^*_4)\right]$$
$$= u^*_{i\,av}(\tau) - \gamma_i\left(-z^*_4 z_i + z^*_i z_4\right), \quad i = 1, 2, 3$$

Contrary to what is customary in many publications in the control of these devices, we do not resort to $d-q$ transformations nor do we impose the *balance conditions* on the original model. Rather, we have used the actual average normalized current and voltage model to obtain the average linear, passive output tracking error, feedback controller. We will now resort to flatness in order to specify the required nominal state and input trajectories. In specifying such desired nominal trajectories, we shall use the perfect balance conditions. In other words, we shall force the system to track the response of an ideal balanced system whose line currents exhibit a unit power factor.

7.3.3 Trajectory Planning

Consider the nominal average normalized system

$$\dot{z}^*_1 = -u^*_{1,av} z^*_4 - Qz^*_1 + \cos(\omega_n \tau)$$
$$\dot{z}^*_2 = -u^*_{2,av} z^*_4 - Qz^*_2 + \cos\left(\omega\tau - \frac{2\pi}{3}\right)$$
$$\dot{z}^*_3 = -u^*_{3,av} z^*_4 - Qz^*_3 + \cos\left(\omega\tau + \frac{2\pi}{3}\right)$$
$$\dot{z}^*_4 = u^*_{1,av} z^*_1 + u^*_{2,av} z^*_2 + u^*_{3,av} z^*_3 - \frac{z^*_4}{Q_L}$$

along with the normalized average current balance condition:

$$z^*_1 + z^*_2 + z^*_3 = 0$$

This condition implies that one of the average nominal line current variables, say z^*_3 does not qualify as a state variable due to its (linear) dependance on the first two states. Moreover, adding the first three equations and imposing the balance conditions yields the relation $(u^*_1 + u^*_2 + u^*_3)z^*_4 = 0$ for all z^*_4. Evidently, the nominal output voltage should not be identically zero. Hence, we also have that the average nominal control inputs $\{u^*_1, u^*_2, u^*_3\}$ are not independent and in fact,

$$u^*_1 + u^*_2 + u^*_3 = 0$$

Similarly, under the perfect balance conditions, one of the control inputs, say, u^*_3 can always be expressed in terms of the two other control inputs. Using

$z_3^* = -(z_1^* + z_2^*)$ and $u_3^* = -(u_1^* + u_2^*)$. The normalized reduced nominal average balanced system is readily expressed as:

$$\dot{z}_1^* = -u_{1,av}^* z_4^* - Q z_1^* + \cos(\omega_n \tau)$$

$$\dot{z}_2^* = -u_{2,av}^* z_4^* - Q z_2^* + \cos\left(\omega \tau - \frac{2\pi}{3}\right)$$

$$\dot{z}_4^* = u_{1,av}^* (2z_1^* + z_2^*) + u_{2,av}^* (z_1^* + 2z_2^*) - \frac{z_4^*}{Q_L}$$

The normalized reduced nominal average balanced system is flat, with several interesting possible choices for the flat outputs. We choose

$$F^* = [z_1^*]^2 + [z_1^*][z_2^*] + [z_2^*]^2 + \frac{1}{2}[z_4^*]^2, \qquad L^* = z_4^*$$

After some tedious but straightforward computations we obtain:

$$\dot{F}^* = -2QF^* + \left(Q - \frac{1}{Q_L}\right) L^2 + z_1^* \left[\frac{3}{2}\cos(\omega_n \tau) + \frac{\sqrt{3}}{2}\sin\left(\omega_n \tau - \frac{2\pi}{3}\right)\right]$$

$$+ \sqrt{3} z_2^* \sin(\omega_n \tau)$$

We impose on the nominal system the following nominal values for z_1^* and z_2^*

$$z_1^* = N \cos(\omega_n \tau), \qquad z_2^* = N \cos\left(\omega_n \tau - \frac{2\pi}{3}\right)$$

where N is a constant amplitude to be determined. These nominal values guarantee unit power factor on the current lines z_1^* and z_2^*. Indeed, from the balance condition $z_3^* = -(z_1^* + z_2^*)$ one readily obtains:

$$z_3^* = N \cos\left(\omega_n \tau + \frac{2\pi}{3}\right)$$

We also adopt a constant nominal value for second flat output, L^*, representing the nominal average normalized output voltage. We let this value be expressed by $\overline{L}^*$. The adopted nominal average values of the line currents and the output voltage yield a constant value of the flat output F^* given by

$$\overline{F}^* = \frac{3}{4} N^2 + \frac{1}{2} \overline{L}^2$$

This, in turn, yields a zero value for the time derivative of the nominal flat output $\dot{F}^*$. One obtains, after further manipulations on the expression, $\dot{F}^* = 0$, the following relation:

$$\overline{L}^2 = \frac{3}{2} N (1 - QN) Q_L \qquad (7.1)$$

7.3 Three Phase Boost Rectifier

Thus, given a desired steady state constant average normalized output voltage $\overline{L}$, the normalized average line currents amplitude N is found to be

$$N = \frac{1}{2Q} \pm \sqrt{\frac{1}{4Q^2} - \frac{2}{3}\frac{\overline{L}^2}{QQ_L}}$$

We choose the minus sign to obtain a smaller amplitude line current. The last relation also yields a natural limitation on the achievable constant output voltage. This condition stems from the real (as opposed to complex) character of the line current amplitude N

$$[\overline{z}_4^*]^2 = \overline{L}^2 < \frac{3}{8}\left(\frac{Q_L}{Q}\right)$$

The average nominal control inputs are easily computed from the system relations:

$$u_{1,av}^* = \frac{-\dot{z}_1^* - Qz_1^* + \cos(\omega_n \tau)}{z_4^*}$$

$$u_{2,av}^* = \frac{-\dot{z}_2^* - Qz_2^* + \cos(\omega_n \tau - \frac{2\pi}{3})}{z_4^*}$$

and the balance condition $u_{3\ av}^* = -(u_{1\ av}^* + u_{2\ av}^*)$, as:

$$u_{3,av}^* = \left[\frac{\dot{z}_1^* + \dot{z}_2^* + Q(z_1^* + z_2^*) + \cos(\omega_n \tau + \frac{2\pi}{3})}{z_4^*}\right]$$

$$= \left[\frac{-\dot{z}_3^* - Q(z_3^*) + \cos(\omega_n \tau + \frac{2\pi}{3})}{z_4^*}\right]$$

The particular steady state values of the average nominal input signals, in view of the adopted values of the line currents and the steady state normalized average output voltage, result in:

$$u_{1,av}^* = \left[\frac{\sqrt{(N\omega_n)^2 + (1 - QN)^2}}{\overline{z}_4^*}\right] \sin(\omega_n \tau + \phi_1),$$

$$\phi_1 = \arctan\left[\frac{1 - QN}{N\omega_n}\right]$$

$$u_{2,av}^* = \left[\frac{\sqrt{(N\omega_n)^2 + (1 - QN)^2}}{\overline{z}_4^*}\right] \sin(\omega_n \tau + \phi_2),$$

$$\phi_2 = -\frac{2\pi}{3} + \arctan\left[\frac{1 - QN}{N\omega_n}\right]$$

$$u_{3,av}^* = \left[\frac{\sqrt{(N\omega_n)^2 + (1 - QN)^2}}{\overline{z}_4^*}\right] \sin(\omega_n \tau + \phi_3),$$

$$\phi_3 = \frac{2\pi}{3} + \arctan\left[\frac{1 - QN}{N\omega_n}\right]$$

The common amplitude of the steady state average control inputs must be bounded within the closed interval $[-1, 1]$ of the real line. This imposes a second limitation on the achievable normalized steady state value of the average output voltage $\bar{z}_4^*$. Indeed, we have

$$\left[\frac{\sqrt{(N\omega_n)^2 + (1 - QN)^2}}{\bar{z}_4^*}\right] < 1$$

Joining this restriction to the previously found one we obtain that the feasible average normalized output voltages satisfy the sector condition

$$\sqrt{(N\omega_n)^2 + (1 - QN)^2} < \bar{z}_4^* < \sqrt{\frac{3}{8}\left(\frac{Q_L}{Q}\right)}$$

The previous steady state and equilibrium formulae also allow us to carry out an efficient trajectory planning for, say, smoothly rising the output average normalized load voltage z_4 of the three phase rectifier from an initial equilibrium value, say $z_4^*(\tau_1)$, towards a final equilibrium value $z_4^*(\tau_2)$, within a finite interval of normalized time $[\tau_1, \tau_2]$. Once the initial and final equilibrium points of $z_4^*(\tau)$ are decided upon, we may determine the nominal corresponding constant values of the normalized amplitude parameter N, valid before τ_1 and after τ_2. We may then use a Bézier polynomial function for smoothly interpolating between the initial and the final value of N, thus obtaining a time-varying average normalized amplitude $N(\tau)$ for the line currents. This, in turn, determines the corresponding smooth trajectory for smoothly increasing the amplitude of the reference average normalized line currents. Since, in such a case, N becomes a time-varying function, then the nominal currents for the transition maneuver are given by: $z_1^*(\tau) = N(\tau)\cos(\omega_n\tau)$, $z_2^*(\tau) = N(\tau)\cos(\omega_n\tau - \frac{2\pi}{3})$, etc. These expressions must be taken into account for computing the nominal average control input signals: $u_{1,av}^*(\tau)$, $u_{2,av}^*(\tau)$ and $u_{3,av}^*(\tau)$. Computation of the required time derivative $\dot{N}(\tau)$ is clearly trivial.

7.3.4 Switched Implementation of the Average Design

The implementation of the average feedback control laws, as switched control actions, is easily accomplished by resorting to $\Sigma - \Delta$ modulation (See Sira-Ramírez [62]). The $\Sigma - \Delta$ devices accurately translating the average bounded control signals $u_{i\,av} \in [-1, 1]$ into switched signals $u_i \in \{1, -1\}$ are described by,

$$u_i = \text{sign}(e_i), \quad \frac{de_i}{d\tau} = u_{i\,av} - u_i, \quad i = 1, 2, 3 \qquad (7.2)$$

For further details and mathematical proofs associated with this implementation issue, the reader is invited to see [62].

7.3.5 Simulations

Simulations were performed on the given system with the following data, where, for simplicity, we have taken the line resistances to be ideally zero.

$$L = 2 \text{ mH}, \quad C = 2 \text{ mF}, \quad R_L = 5.9 \text{ }\Omega,$$
$$V = 230\sqrt{2} \text{ V}, \quad R = 0 \text{ }\Omega$$

It was desired to rise the rectified output voltage, x_4, from an initial steady state value of 500 V to a new voltage of 1106 V in approximately 0.04 s. This corresponds to planning a smooth transition for N from the initial value of 0.25 to the final value of 1.2. Since we have set, just for simplicity, $Q = 0$, we have no limitations in the achievable values of the output voltage ($0 < \overline{z}_4^* < +\infty$).

Figure 7.5 shows the response of the rectifier system, quickly achieving balanced line currents and steady state stabilization. Then it undergoes a smooth amplitude increase to the new desired steady state value in a finite time interval. The figure also shows the switching actions along with the average control input for one of the controls inputs (u_1, the rest being rather similar). The average control signal is obtained from the evolution of the proposed linear feedback controller and, finally, Figure 7.5 also shows the load voltage equilibrium to equilibrium transfer, as desired, in a pre-specified amount of time. This type of trajectory tracking possibility is important in the control of some electro-mechanical systems, such as DC motors, when they operate connected, as loads, to a controlled three phase rectifier.

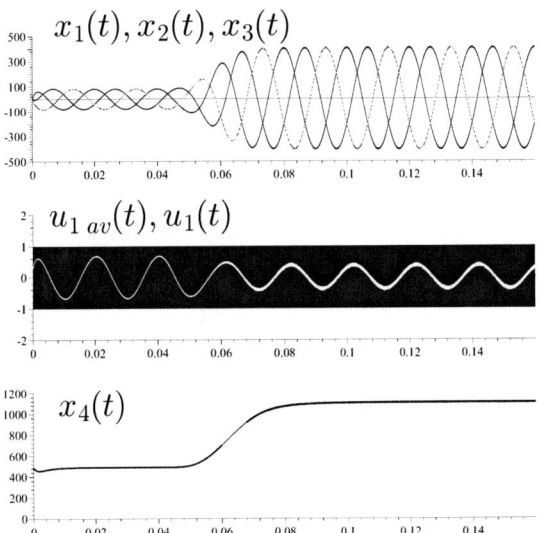

Fig. 7.5. Switched controlled responses of three phase Boost rectifier.

7.4 A Unit Power Factor Rectifier-DC Motor System

7.4.1 The Combined Rectifier-DC Motor Model

Consider the combination of a Boost type unit power factor rectifier and a DC motor connected in tandem, as shown in Figure 7.6.

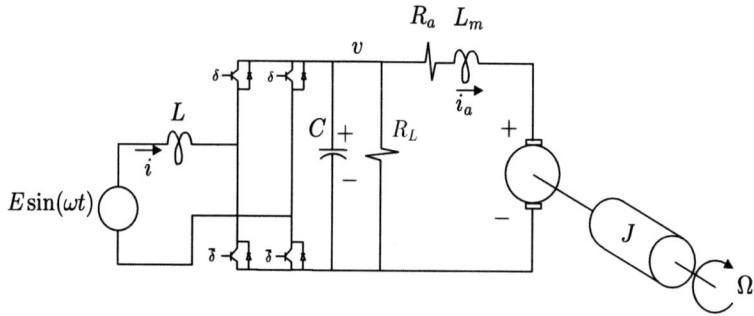

Fig. 7.6. Boost converter-DC motor system.

The system is described by the following set of differential equations:

$$L\frac{di}{dt} = -uv + E\sin(\omega t)$$
$$C\frac{dv}{dt} = ui - \frac{1}{R_L}v - i_a$$
$$L_m\frac{di_a}{dt} = -R_a i_a - K\Omega + v$$
$$J\frac{d\Omega}{dt} = Ki_a - B\Omega$$

where i is the inductor current, v is the output capacitor voltage and u is the switch position function taking values in the discrete set $\{-1, 1\}$. i_a is the armature current and Ω represents the motor shaft angular velocity. The line resistance was assumed to be negligible.

We consider the state average system equations by simply replacing the actual states of the system description by average states, while letting the control input continuously take values in the closed interval $[-1, 1]$ of the real line. In other words, we consider

7.4 A Unit Power Factor Rectifier-DC Motor System

$$L\frac{dI}{dt} = -u_{av}V + E\sin(\omega t)$$

$$C\frac{dV}{dt} = u_{av}I - \frac{V}{R_L} - I_a$$

$$L_m\frac{dI_a}{dt} = -R_a I_a - K\Omega + v$$

$$J\frac{d\Omega}{dt} = KI_a - B\Omega$$

to be the state average model of the original system, with u_{av} being now a continuous scalar signal taking values in the compact set $[-1,1]$.

The normalization of the average system equations is carried out according to the following state and time coordinates transformation:

$$x_1 = \frac{I}{E}\sqrt{\frac{L}{C}}, \quad x_2 = \frac{V}{E}, \quad x_3 = \frac{I_a}{E}\sqrt{\frac{L}{C}}, \quad x_4 = \Omega\sqrt{LC}$$

$$\omega_n = \omega\sqrt{LC}, \quad \tau = \frac{t}{\sqrt{LC}}$$

The normalized average system equations are then obtained in the following form

$$\dot{x}_1 = -u_{av}x_2 + \sin(\omega_n\tau)$$
$$\dot{x}_2 = u_{av}x_1 - \frac{x_2}{Q_L} - x_3$$
$$\alpha\dot{x}_3 = -Q_a x_3 - \gamma x_4 + x_2$$
$$\beta\dot{x}_4 = -Q_B x_4 + \gamma x_3$$

The normalized parameters are thus defined as:

$$Q_L = R_L\sqrt{\frac{C}{L}}, \quad Q_a = R_a\sqrt{\frac{C}{L}}, \quad Q_B = \frac{B}{E^2\sqrt{LC}}$$

$$\alpha = \frac{L_m}{L}, \quad \beta = \frac{J}{E^2 C^2 L}, \quad \gamma = \frac{K}{E\sqrt{LC}}$$

Note that, as usual, we have abusively used the "dot" notation to actually mean derivation with respect to the normalized time coordinate τ.

The control objectives for the average system are twofold:

- It is primordially desired that the normalized angular velocity of the motor shaft stabilizes to a constant desired value denoted by $\bar{x}_4$. Later on, we will also tackle the problem of tracking a desired normalized angular velocity trajectory.
- It is also desired to have the average normalized inductor current track a sinusoidal signal of constant amplitude A, yet to be determined, and of the same angular frequency ω_n as the input source. This objective guarantees, as before, a unit power factor.

The system is not differentially flat. Nevertheless, the normalized motor shaft angular velocity is capable of differentially parameterizing three of the four state variables. Indeed if we denote by F the angular velocity we obtain

$$x_4 = F$$
$$x_3 = \frac{1}{\gamma}\left[\beta \dot{F} + Q_B F\right]$$
$$x_2 = \left[\frac{\beta\alpha}{\gamma}\ddot{F} + \left(\frac{Q_B\alpha}{\gamma} + \frac{Q_a\beta}{\gamma}\right)\dot{F} + \left(\frac{Q_a Q_B}{\gamma} + \gamma\right)F\right]$$

The average equilibrium values of these three variables are found to be,

$$\overline{x}_4 = \overline{F}, \qquad \overline{x}_3 = \frac{Q_B}{\gamma}\overline{F}, \qquad \overline{x}_2 = \left[\frac{Q_a Q_B}{\gamma} + \gamma\right]\overline{F}$$

The total stored average normalized energy of the system is given by:

$$H = \frac{1}{2}\left[x_1^2 + x_2^2 + \alpha x_3^2 + \beta x_4^2\right]$$

The total power is then given by the time derivative of H,

$$\frac{dH}{d\tau} = x_1 \sin(\omega_n \tau) - \frac{x_2^2}{Q_L} - Q_a x_3^2 - Q_B x_4^2$$

where the first summand corresponds with the average normalized input power and the rest of the summands correspond to the average normalized delivered power at the electric and mechanical loads.

The steady state value of the DC component of the total power should balance to zero, since the system is lossless. We have then the following steady state power balance condition:

$$\langle x_1 \sin(\omega_n \tau)\rangle_{dc} = \left\langle \frac{\overline{x}_2^2}{Q_L} + Q_a \overline{x}_3^2 + Q_B \overline{x}_4^2\right\rangle_{dc}$$
$$= \left[\frac{1}{Q_L}\left(\frac{Q_a Q_B}{\gamma} + \gamma\right)^2 + Q_a \frac{Q_B^2}{\gamma^2} + Q_B\right]\overline{F}^2$$

where the "overline" stands for equilibrium (i.e., "steady state") value of the involved variable. We then have, using $x_1 = A\sin(\omega_n \tau)$, that the average inductor current amplitude is given by:

$$A = 2\left(\frac{Q_a Q_B}{\gamma} + \gamma\right)\left[\frac{1}{Q_L}\left(\frac{Q_a Q_B}{\gamma} + \gamma\right) + \frac{Q_B}{\gamma}\right]\overline{F}^2$$

This last relation will be quite useful below.

The fact that the steady state inductor current amplitude A and the desired equilibrium value of the average motor shaft angular velocity, $\overline{F} = \overline{x}_4$, satisfy the above relation, will be also addressed as the solvability condition.

7.4 A Unit Power Factor Rectifier-DC Motor System

When the line resistances are considered, acting in series with the inductor, the obtained corresponding solvability condition further reveals a natural limitation of the reachable set of output voltages and feasible angular velocities.

7.4.2 The Exact Tracking Error Dynamics Passive Output Feedback Controller

The exact open loop tracking error dynamics is easily found to be given by:

$$\dot{e}_1 = -u_{av}e_2 - x_2^*(\tau)e_u$$
$$\dot{e}_2 = u_{av}e_1 - \frac{e_2}{Q_L} - e_3 + x_1^*(\tau)e_u$$
$$\alpha\dot{e}_3 = -Q_a e_3 - \gamma e_4 + e_2$$
$$\beta\dot{e}_4 = -Q_B e_4 + \gamma e_3$$

The proposed linear time varying feedback control law reads as follows:

$$e_u = -\Gamma\left[-x_2^*(\tau)e_1 + x_1^*(\tau)e_2\right]$$

and the dissipation matching condition takes the form:

$$\begin{bmatrix} \Gamma[x_2^*(\tau)]^2 & -\Gamma x_2^*(\tau)x_1^*(\tau) & 0 & 0 \\ -\Gamma x_2^*(\tau)x_1^*(\tau) & \frac{1}{Q_L} + \Gamma[x_1^*(\tau)]^2 & 0 & 0 \\ 0 & 0 & Q_a & 0 \\ 0 & 0 & 0 & Q_B \end{bmatrix} > 0$$

which is indeed uniformly valid for any non-zero state trajectory.

The average linear time varying controller:

$$u_{av} = u^*(\tau) + \Gamma\left[x_2^*(\tau)(x_1 - x_1^*(\tau)) - x_1^*(\tau)(x_2 - x_2^*(\tau))\right]$$
$$= u^*(\tau) + \Gamma\left[x_2^*(\tau)x_1 - x_1^*(\tau)x_2\right]$$

is the required trajectory tracking state feedback control law.

7.4.3 Trajectory Generation

The main problem now becomes one of specifying the nominal state and input trajectories, $(x^*(\tau), u^*(\tau))$, in complete accordance with the announced control objectives. The problem is by no means a trivial one, unless one resorts to the partial flatness of the original composite system and the off-line planning of an auxiliary endogenous variable: the total average stored energy.

Consider the total stored energy of the system,

$$H = \frac{1}{2}\left[x_1^2 + x_2^2 + \alpha x_3^2 + \beta x_4^2\right]$$

The steady state value of H, denoted by $\langle H \rangle_{dc}$, can be expressed in terms of the system parameters and the steady state (equilibrium) value of the average motor shaft angular velocity $\overline{F}$. We obtain:

$$\langle H \rangle_{dc} = \frac{A^2}{4} + \frac{1}{2}\left[\left(\frac{Q_a Q_B}{\gamma} + \gamma\right)^2 + \alpha\left(\frac{Q_B}{\gamma}\right)^2 + \beta\right]\overline{F}^2$$

i.e.,

$$\langle H \rangle_{dc} = \left(\frac{Q_a Q_B}{\gamma} + \gamma\right)^2 \left[\frac{1}{Q_L}\left(\frac{Q_a Q_B}{\gamma} + \gamma\right) + \frac{Q_B}{\gamma}\right]^2 \overline{F}^4$$

$$+ \frac{1}{2}\left[\left(\frac{Q_a Q_B}{\gamma} + \gamma\right)^2 + \alpha\left(\frac{Q_B}{\gamma}\right)^2 + \beta\right]\overline{F}^2$$

Given an equilibrium value of the average angular velocity of the DC motor, denoted by $\overline{F}_{init}$ and valid up to time τ_{init}, and a desired final equilibrium value of this angular velocity, $\overline{F}_{final}$ valid only after τ_{final}, with $\tau_{final} > \tau_{init}$, we can then specify a nominal trajectory for the total steady state stored energy transfer, here denoted by $\overline{H}(\tau)$. Such a trajectory is specified to smoothly interpolate between the corresponding values $\langle H^*(\tau_{init}) \rangle_{dc} = \overline{H}_{init}$ and $\langle H^*(\tau_{final}) \rangle_{dc} = \overline{H}_{final}$ with the transfer taking place on the time interval $[\tau_{init}, \tau_{final}]$. i.e.,

$$\langle H^*(\tau) \rangle_{dc} = \begin{cases} \overline{H}_{init} & \text{for } \tau < \tau_{init} \\ \overline{H}(\tau) & \text{for } \tau \in [\tau_{init}, t_{final}] \\ \overline{H}_{final} & \text{for } \tau > \tau_{final} \end{cases}$$

This procedure allows us to off-line plan the nominal trajectory for the remaining state variable x_1, $x_1^*(\tau)$ and the control input u_{av}^*.

The nominal trajectory planning for the state variable x_1 is intimately related to the initial and final values of the average steady state total stored energy. This off line energy planning induces a corresponding nominal trajectory for the unit power factor line current amplitude $A(\tau)$, as follows:

$$A(\tau) = 2\sqrt{\left\{\langle H^*(\tau) \rangle_{dc} - \frac{1}{2}\left[\left(\frac{Q_a Q_B}{\gamma} + \gamma\right)^2 + \alpha\left(\frac{Q_B}{\gamma}\right)^2 + \beta\right][F(\tau)]^2\right\}}$$

Thus

$$x_1^*(\tau) = A(\tau)\sin(\omega_n \tau)$$

The nominal average control input $u^*(\tau)$ is computed in full compatibility with the average normalized (line) inductor current trajectory $x_1^*(\tau)$.

We obtain from the normalized average system equations,

$$u^*(\tau) = \frac{-\dot{x}_1^*(\tau) + \sin(\omega_n\tau)}{x_2^*(\tau)}$$

$$= \frac{\left[\sqrt{\left(1 - \frac{dA(\tau)}{d\tau}\right)^2 + (A(\tau)\omega_n)^2}\right]\sin(\omega_n\tau + \phi(\tau))}{x_2^*(\tau)}$$

$$\phi(\tau) = \arctan\left(\frac{A(\tau)\omega_n}{1 - \frac{dA(\tau)}{d\tau}}\right)$$

with

$$x_2^*(\tau) = \left[\frac{\beta\alpha}{\gamma}\dddot{F}^*(\tau) + \left(\frac{Q_B\alpha}{\gamma} + \frac{Q_a\beta}{\gamma}\right)\dot{F}^*(\tau) + \left(\frac{Q_aQ_B}{\gamma} + \gamma\right)F^*(\tau)\right]$$

where $F^*(\tau)$ represents the rest-to-rest desired trajectory for the motor's average angular velocity.

All the design elements required for the synthesis of the time varying linear feedback controller, achieving the rest to rest angular velocity tracking, have therefore been completely computed.

7.4.4 Simulations

We considered a Boost unit power factor rectifier with the same data considered before in the previous example

$$L = 6 \text{ mH}, \quad C = 2.2 \text{ mF}, \quad R_L = 60 \text{ }\Omega, \quad E = 150 \text{ V}, \quad \omega_n = 2\pi 60 \text{ rad/s}$$

The DC motor data was set to be

$$J = 1.625 \times 10^{-5} \text{ N.m.rad/s}^2, \quad L_m = 2.5 \text{ mH},$$

$$R_a = 600 \text{ }\Omega, \quad K = 9 \text{ V.s/rad}, \quad B = 0 \text{ N.m.s/rad}$$

The first task was that of stabilization towards a desired equilibrium point for the motor angular velocity. We set

$$\Omega^* = 27.52 \text{ rad/s}$$

To this equilibrium it corresponds a steady state input current amplitude of $A = 13.63$ A, an rectified converter output steady state voltage of $V = 247.71$ V and an equilibrium value of 0 A for the armature current. The control gain Γ was set to be equal to 2.

Figure 7.7 depicts the actual average state variables responses of the controlled system to the stabilization task. The graphs are shown in a non-normalized fashion in the actual time scale.

The switched implementation of the feedback controller is accomplished by means of a small variant of the $\Sigma - \Delta$ modulation scheme:

406   7 AC Rectifiers

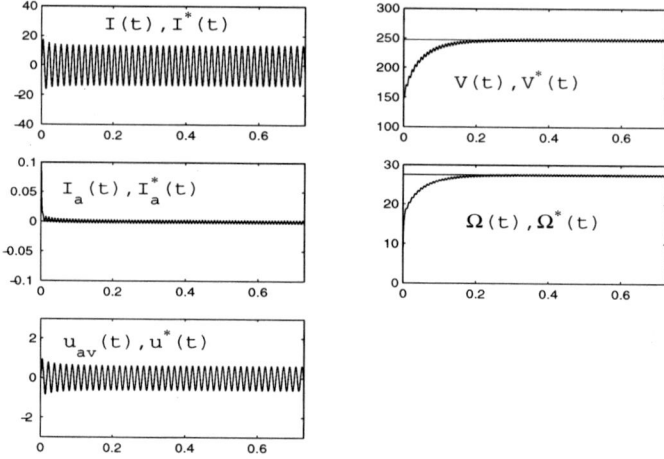

Fig. 7.7. Average stabilization response of Boost rectifier-DC motor system.

$$u = \text{sign}(e)$$
$$\frac{de}{dt} = u_{av} - u$$

i.e., the control input takes values in the discrete set $\{-1, +1\}$ while the average control input takes values in the compact set $[-1, 1]$. Figure 7.8 depicts the actual average state variables responses of the switched controlled system accomplishing the required stabilization task.

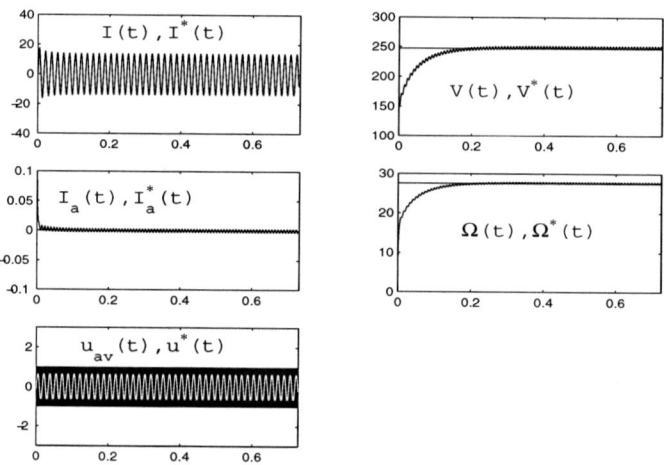

Fig. 7.8. Switched stabilization response of Boost rectifier-DC motor system.

7.4 A Unit Power Factor Rectifier-DC Motor System

A rest-to-rest tracking task is also presented which entitles a maneuver of the angular velocity from an initial equilibrium value towards a final equilibrium value. The prescribed angular velocity trajectories allows for the off-line computation of the armature current and the rectified output voltage.

As a second possibility for the planning of the inductor current, we proceed by computing the steady state sinusoidal inductor current amplitudes corresponding to the initial and final velocity equilibrium conditions. With these two equilibrium values for the current amplitudes, we prescribe a nominal current amplitude trajectory, $A^*(t)$, that smoothly joins these two steady state values. We prescribe then an inductor current reference trajectory of the form:

$$x_1^*(\tau) = A^*(\tau)\sin(\omega_n \tau)$$

This procedure, guarantees the invariance of the unit power factor condition during the rest to rest maneuver. The control task becomes now one of tracking this current reference trajectory. The nominal control input is computed as indicated before. Figure 7.9 shows the switched control regulation of the demanded rest to rest maneuver between initial and final equilibrium values of the angular velocity.

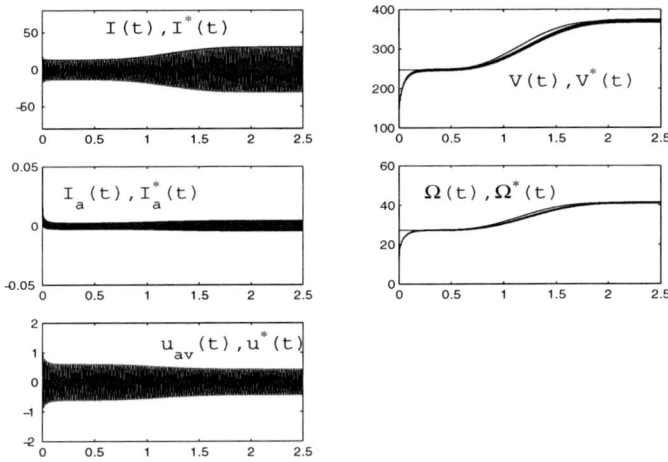

Fig. 7.9. Average rest-to-rest trajectory tracking of Boost rectifier-DC motor system.

7.5 A Three Phase Rectifier-DC Motor System

7.5.1 The Combined Three Phase Rectifier DC Motor Model

Consider the following model of a Boost type rectifier feeding a DC motor shown in Figure 7.10:

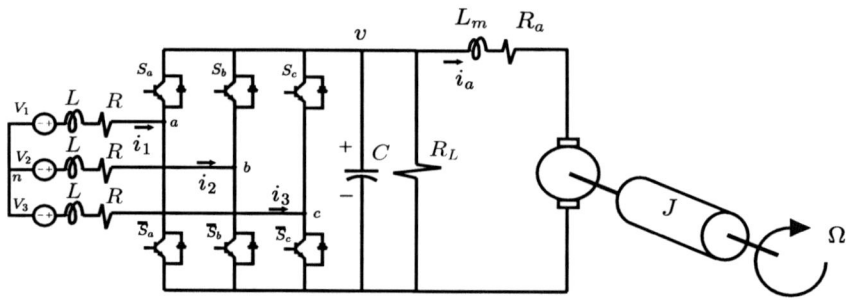

Fig. 7.10. A three phase Boost rectifier-DC motor system.

The dynamic model of the composite system is obtained as:

$$L\frac{d}{dt}i_1 = -u_1 v - R i_1 + V_m \cos(\omega t)$$

$$L\frac{d}{dt}i_2 = -u_2 v - R i_2 + V_m \cos(\omega t - 2\pi/2)$$

$$L\frac{d}{dt}i_3 = -u_3 v - R i_3 + V_m \cos(\omega t + 2\pi/3)$$

$$C\frac{d}{dt}v = u_1 i_1 + u_2 i_2 + u_3 i_3 - \frac{v}{R_L} - i_a$$

$$L_m \frac{d}{dt}i_a = -R_a i_a - K\Omega + v$$

$$J\frac{d}{dt}\Omega = K i_a - B\Omega$$

It is desired to rise the motor angular velocity from an initial equilibrium value towards a final equilibrium value within a feasible finite time interval.

We solve the problem by resorting to an average normalized model of the switched system.

$$z_j = \left(\frac{1}{V_m}\sqrt{\frac{L}{C}}\right) i_j, \ j = 1, 2, 3. \qquad z_4 = \frac{v}{V_m},$$

$$z_5 = \left(\frac{1}{V_m}\sqrt{\frac{L}{C}}\right) i_a, \qquad z_6 = \Omega\sqrt{LC}, \qquad \omega_n = \omega\sqrt{LC}, \qquad \tau = \frac{t}{\sqrt{LC}}$$

We define

$$\alpha = \frac{L_m}{L}, \quad \beta = \frac{J}{V_m^2 C^2 L}, \quad Q = R\sqrt{\frac{C}{L}}, \quad Q_L = R_L\sqrt{\frac{C}{L}},$$

$$Q_B = \frac{B}{V_m^2 \sqrt{LC}}, \quad \gamma = \frac{K}{V_m \sqrt{LC}}$$

We obtain the following average normalized system

$$\frac{d}{d\tau} z_1 = -u_{1\,av} z_4 - Q z_1 + \cos(\omega_n \tau)$$

$$\frac{d}{d\tau} z_2 = -u_{2\,av} z_4 - Q z_2 + \cos(\omega_n \tau - 2\pi/2)$$

$$\frac{d}{d\tau} z_3 = -u_{3\,av} z_4 - Q z_3 + \cos(\omega_n \tau + 2\pi/3)$$

$$\frac{d}{d\tau} z_4 = u_{1\,av} z_1 + u_{2\,av} z_2 + u_{3\,av} z_3 - \frac{z_4}{Q_L} - z_5$$

$$\alpha \frac{d}{d\tau} z_5 = -Q_a z_5 - \gamma z_6 + z_4$$

$$\beta \frac{d}{d\tau} z_6 = \gamma z_5 - Q_B z_6$$

The average normalized system is of the form

$$\mathcal{A}\dot{z} = \mathcal{J}(u_{av})z - \mathcal{R}z + \mathcal{E}(\tau)$$

where

$$\mathcal{A} = \mathrm{diag}[1,1,1,1,\alpha,\beta],$$
$$\mathcal{R} = \mathrm{diag}[Q,Q,Q,1/Q_L,Q_a,Q_B]$$
$$\mathcal{E}^T(\tau) = [\cos(\omega_n \tau)\ \cos(\omega_n \tau - 2\pi/3)\ \cos(\omega_n \tau + 2\pi/3)\ 0\ 0\ 0]$$

$$\mathcal{J}(u_{av}) = \begin{bmatrix} 0 & 0 & 0 & -u_{1\,av} & 0 & 0 \\ 0 & 0 & 0 & -u_{2\,av} & 0 & 0 \\ 0 & 0 & 0 & -u_{3\,av} & 0 & 0 \\ u_{1\,av} & u_{2\,av} & u_{3\,av} & 0 & -1 & 0 \\ 0 & 0 & 0 & 1 & 0 & -\gamma \\ 0 & 0 & 0 & 0 & \gamma & 0 \end{bmatrix}$$

7.5.2 The Exact Tracking Error Dynamics Passive Output Feedback Controller

Let $z^*(\tau)$ and $u^*_{av}(\tau)$ be the nominal state and average control input trajectories. The tracking error system for $e = z - z^*(\tau)$ is given by

$$\mathcal{A}\dot{e} = \mathcal{J}(u_{av})e - \mathcal{R}e + \mathcal{B}^*(\tau)e_u$$

where $e_u = u - u^*(\tau)$.

Let Γ be a diagonal matrix $\Gamma = \text{diag}[\gamma_1, \gamma_2, \gamma_3]$. The dissipation matching condition is readily verified to be uniformly satisfied in accordance with:

$$\mathcal{R} + \mathcal{B}^*(t)\Gamma[\mathcal{B}^*]^T(t) = \begin{bmatrix} Q + \gamma_1[x_4^*]^2 & 0 & 0 & -\gamma_1 x_1^* x_4^* & 0 & 0 \\ 0 & Q + \gamma_2[x_4^*]^2 & 0 & -\gamma_2 x_2^* x_4^* & 0 & 0 \\ 0 & 0 & Q + \gamma_3[x_3^*]^2 & \gamma_3 x_3^* x_4^* & 0 & 0 \\ -\gamma_1 x_1^* x_4^* & -\gamma_2 x_2^* x_4^* & -\gamma_3 x_3^* x_4^* & \frac{1}{Q_L} + \sum_{i=1}^{3} \gamma_i[x_i^*]^2 & 0 & 0 \\ 0 & 0 & 0 & 0 & Q_a & 0 \\ 0 & 0 & 0 & 0 & 0 & Q_b \end{bmatrix} > 0$$

$$\mathcal{A}\dot{e} = \mathcal{J}(u_{av})e - \mathcal{R}e + \mathcal{B}^*(\tau)e_u$$

with

$$\mathcal{B}^*(\tau) = \begin{bmatrix} -z_4^*(\tau) & 0 & 0 \\ 0 & -z_4^*(\tau) & 0 \\ 0 & 0 & -z_4^*(\tau) \\ z_1^*(\tau) & z_2^*(\tau) & z_3^*(\tau) \\ 0 & 0 & 0 \\ 0 & 0 & 0 \end{bmatrix}$$

Let $\Gamma = \text{diag}[\gamma_1, \gamma_2, \gamma_3]$ with $\gamma_i > 0$, $i = 1, 2, 3$. The average feedback control law, based on linear time-varying passive outputs feedback, is given by

$$u_{av} = u_{av}^*(\tau) - \Gamma[B^*(\tau)]^T(z - z^*(\tau))$$

i.e., for $j = 1, 2, 3$.

$$u_{j\,av}(\tau) = u_{j\,av}^*(\tau) - \gamma_j[-z_4^*(\tau)z_j + z_j^*(\tau)z_4]$$

7.5.3 Trajectory Generation

The system is differentially flat, with the three flat outputs given by either one of the following sets of variables

$$\{z_1, z_2, z_6\}, \quad \{z_1, z_3, z_6\}, \quad \{z_2, z_3, z_6\}$$

We choose as nominal average trajectories the ones resulting from a balanced unit factor voltage and current trajectories based on a trajectory planning for the nominal motor angular velocity.

According to the control objectives, we specify, thanks to the flatness of the system, a rest to rest, or equilibrium to equilibrium nominal average trajectory for z_6 as $z_6^*(\tau)$, and a set of balanced, unit factor currents $z_1^*(\tau)$ and $z_2^*(\tau)$.

We set $z_6^*(\tau)$ as a Bézier polynomial smoothly interpolating between two normalized average velocity equilibria. We obtain

7.5 A Three Phase Rectifier-DC Motor System

$$z_5^*(\tau) = \frac{1}{\gamma}(\beta \dot{z}_6^*(\tau) + Q_B z_6^*(\tau))$$

$$z_4^*(\tau) = \frac{\alpha\beta}{\gamma}\ddot{z}_6^*(\tau) + \frac{1}{\gamma}(Q_a\beta + \alpha Q_B)\dot{z}_6^*(\tau) + \left(\gamma + \frac{Q_a Q_B}{\gamma}\right)z_6^*(\tau)$$

Choosing the average normalized flat outputs $z_1^*(\tau)$ and $z_2^*(\tau)$ as balanced unit factor currents with amplitudes $a(\tau)$, yet to be determined

$$z_1^*(\tau) = a(\tau)\cos(\omega_n \tau)$$
$$z_2^*(\tau) = a(\tau)\cos(\omega_n \tau - 2\pi/3)$$
$$z_3^*(\tau) = -(z_1^*(\tau) + z_2^*(\tau)) = a(\tau)\cos(\omega_n \tau + 2\pi/3)$$

The nominal average control inputs are obtained from the current equations and the balanced condition as:

$$u_{1\,av}^*(\tau) = \frac{-\dot{z}_1^*(\tau) - Qz_1^*(\tau) + \cos(\omega_n \tau)}{z_4^*(\tau)},$$

$$u_{2\,av}^*(\tau) = \frac{-\dot{z}_2^*(\tau) - Qz_2^*(\tau) + \cos(\omega_n \tau - 2\pi/3)}{z_4^*(\tau)},$$

$$u_{3\,av}^*(\tau) = -(u_1^*(\tau) + u_2^*(\tau)) = \frac{-\dot{z}_3^*(\tau) - Qz_3^*(\tau) + \cos(\omega_n \tau + 2\pi/3)}{z_4^*(\tau)}$$

In steady state equilibrium conditions we take the amplitude of the balanced currents $a(\tau)$ to be constant of value a. We get the following identity:

$$u_{1\,av}^*(\tau)z_1^*(\tau) + u_{2\,av}^*(\tau)z_2^*(\tau) + u_{3\,av}^*(\tau)z_3^*(\tau) = \frac{3a}{2z_4^*(\tau)}(1 - aQ)$$

The differential equation for the average normalized armature voltage $z_4^*(\tau)$ is given by,

$$\frac{d}{d\tau}z_4^*(\tau) = \frac{3a}{2z_4^*(\tau)}(1 - aQ) - \frac{z_4^*(\tau)}{Q_L} - z_5^*(\tau)$$

The equilibrium average normalized solution for the rectifier output voltage is given by

$$\overline{z}_4^* = -\frac{\overline{z}_5^*}{2} + \sqrt{\frac{[\overline{z}_5^*]^2}{4} + \frac{3}{2}a(1-aQ)Q_L}$$

which, in terms of the steady state equilibrium value of the angular velocity $\overline{z}_6^*$, is expressed as

$$\overline{z}_4^* = -\frac{Q_B}{2\gamma}\overline{z}_6^* + \sqrt{\frac{Q_B^2}{4\gamma^2}[\overline{z}_6^*]^2 + \frac{3}{2}a(1-aQ)Q_L}$$

Since, from the differential parametrization we have:

$$\overline{z}_4^* = \left(\gamma + \frac{Q_a Q_B}{\gamma}\right) \overline{z}_6^*$$

The steady state line current constant voltage amplitude a may be expressed in terms of the equilibrium value for the motor angular velocity $\overline{z}_6^*$, by solving for a from the previous equations. We obtain:

$$a = \frac{1}{2Q} - \sqrt{\frac{1}{4Q^2} - \frac{2}{3QQ_L}\left[\left(\gamma + \frac{2Q_a Q_B + Q_B}{2\gamma}\right)^2 - \frac{Q_B^2}{4\gamma^2}\right][\overline{z}_6^*]^2}$$

The minus sign being chosen just to get smaller line currents.

A solvability condition which represents a natural limitation for the system equilibrium motor angular velocity is given by an imposed real nature on the voltage amplitude.

$$\overline{z}_6^* < \sqrt{\frac{3}{8}\left(\frac{Q_L}{Q}\right)\frac{4\gamma^2}{(2\gamma^2 + 2Q_B(1+Q_a))(2\gamma^2 + 2Q_a Q_B)}}$$

Note that the relation:

$$a = \frac{1}{2Q} - \sqrt{\frac{1}{4Q^2} - \frac{2}{3QQ_L}\left[\left(\gamma + \frac{2Q_a Q_B + Q_B}{2\gamma}\right)^2 - \frac{Q_B^2}{4\gamma^2}\right][\overline{z}_6^*]^2}$$

allows us to carry out a trajectory planning for the balanced reference current common amplitude $a(\tau)$ for a rest to rest maneuver of the motor angular velocity.

Indeed, we may specify,

$$a(\tau) = \frac{1}{2Q} - \sqrt{\frac{1}{4Q^2} - \frac{2}{3QQ_L}\left[\left(\gamma + \frac{2Q_a Q_B + Q_B}{2\gamma}\right)^2 - \frac{Q_B^2}{4\gamma^2}\right][\overline{z}_6^*(\tau)]^2}$$

and thus plan the nominal normalized angular velocity reference trajectory $z_6^*(\tau)$ so that the solvability condition is uniformly satisfied.

7.5.4 Simulations

We performed simulations on a rectifier and a DC motor with the following parameter data:

Motor Parameters

$J = 1.625 \times 10^{-4}$ N.m.rad/s^2, $\quad L_m = 25$ mH, $\quad R_a = 6$ Ω, $\quad K = 9$ V.s/rad,

$B = 1.2 \times 10^{-5}$ N.m.s/rad

Rectifier Parameters

$$L = 2 \text{ mH}, \quad C = 2 \text{ mF}, \quad R_L = 5.9 \text{ }\Omega,$$
$$V = 230\sqrt{2} \text{ V}, \quad R = 0.037 \text{ }\Omega$$

Figure 7.11 depicts a smooth rest-to-rest maneuver for the angular velocity of the described DC motor controlled by the AC Boost rectifier circuit.

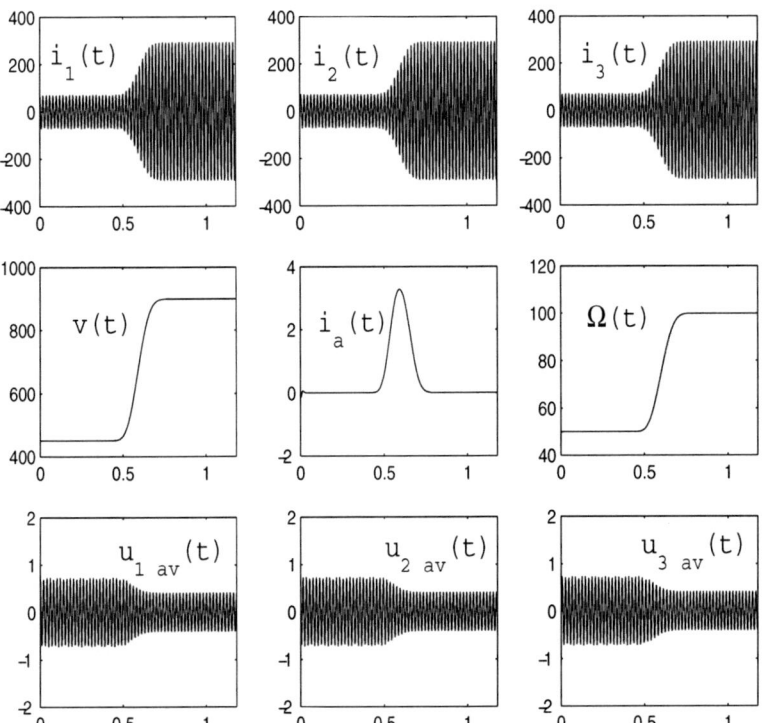

Fig. 7.11. A rest-to-rest angular velocity maneuver for a Boost rectifier-DC motor system.

References

1. Batarseh I (2004) Power electronics circuits. John Wiley & Sons, New York
2. Benvenuti L, Di Benedetto MD, Grizzle JW (1993) Trajectory control of an aircraft using approximate output tracking, 2nd European Control Conference, 1638-1643, Groningen, The Netherlands
3. Blasko V, Kaura K (2000) A new mathematical model and control of a three phase ac-dc voltage source converter. IEEE Trans Power Electr 15:953–959
4. Bose BK (1992) Modern power electronics: evolution, technology and applications. IEEE Press, New York
5. Brockett RW (1964) Finite dimensional linear systems. Addison-Wesely, Boston
6. Cáceres R, Barbi I (1996) Sliding mode controller for the boost inverter. *IEEE International Power Electronics Congress*, Cuernavaca, México, October 14-17, 1996. pp. 247-252
7. Coughlin RF, Driscoll FF (1982) Operational amplifiers and linear integrated circuits. Prentice Hall, Englewood Cliffs, New Jersey
8. Czaki F, Ganszky K, Ipsits I, Marti S (1980) Power electronics. Akadémiai Kiadó, Budapest
9. Edwards C, Spurgeon SK (1998) Sliding mode control. Taylor and Francis, London
10. Emelyanov SV (1967) Variable structure control Systems. Nauka, Moscow
11. Escobar G (1999) On nonlinear control of switching power electronics systems. PhD Thesis, Université de Paris-Süd, Orsay
12. Escobar G, Chevreau D, Ortega R, Mendes E (2001) An adaptive passivity-based controller for a unity power factor rectifier. IEEE Trans on Control Syst Tech 9:637-644
13. Escobar G, Van der Schaft A, Ortega R (1999) A Hamiltonian viewpoint in the modeling of switching power converters. Automatica 35:445-452
14. Escobar G, Chevreau D, Ortega R, Mendes E (2001) An adaptive passivity based controller for a unity power factor rectifier. IEEE Trans on Contr Syst Tech 9:637-644
15. Fliess M (1990) Generalized controller canonical form for linear and nonlinear dynamics. IEEE Trans on Auto Control 35:994-1001
16. Fliess M (2000) Sur des Pensers Nouveaux Faisons des Vers Anciens. Actes Conférence Internationale Francophone d'Automatique (CIFA-2000), Lille. France

17. Fliess M, Lévine J, Martín P, Rouchon P (1996) A Lie-Bäcklund approach to dynamic feedback equivalence and flatness (Lecture Notes in Control and Information Sciences Vol. 247) pp. 247-268, Garofalo F and Glielmo L (Eds.), Springer-Verlag, London
18. Fliess M, Levine J, Martin P, Rouchon P (1995) Flatness and defect of nonlinear systems: introductory theory and examples. Int J of Control 61:1327-1361
19. Fliess M, Márquez R (2000) Continuous time linear predictive control and flatness: a module theoretic setting with examples. Int J of Control 73:606-623
20. Fliess M, Márquez R, Delaleau E (2000) State feedbacks without asymptotic observers and generalized pid regulators, in Nonlinear Control in the Year 2000, Isidori A, Lamnabhi-Lagarrigue F, Respondek W (Eds.) (Lecture Notes in Control and Information Sciences), Springer, London
21. Fliess M, Marquez R, Delaleau E, Sira-Ramírez H (2002) Correcteurs proportionels.intégraux généralisés. ESAIM Control Opt and Calc of Variat 7:23-41
22. Fliess M, Sira-Ramírez H, Márquez R (1998) Regulation of non-minimum phase outputs: a flatness based approach in Perspectives in Control: theory and applications, Normand-Cyrot D (Ed.), Springer London
23. Fliess M, Sira-Ramírez H (2003) An algebraic framework for linear identification. ESAIM, Control Opt and Calc of Variat 9:151-168
24. Fliess M, Sira-Ramírez H (2004) Reconstructeurs d'etat. C.R. Acad Sci de Paris Série I 338:91–96
25. Fliess M, Sira-Ramírez H (2004) Control via state estimations of some nonlinear systems. *IFAC Symposium on Nonlinear Control Systems* (NOLCOS-2004). Stuttgart, Germany, September 1-3
26. Furuta K, Sano A, Atherton D (1988) State variable methods in automatic control. John Wiley and Sons, Chichester, England
27. García PD, Barbi I (1990) A family of resonant dc link voltage source inverters. *IEEE International Conference on Industrial Electronics* pp. 844-849
28. Gauthier JP, Kupka A (1994) Observability and observers for non-linear systems. *SIAM J. on Control and Opt.* 32:975-994.
29. Glad T, Ljung L (2000) Control theory. Taylor and Francis, London
30. Hsieh GC, Lin CH, Li JM, Hsu YC (1996) A Study of series resonant dc/ac inverter. IEEE Trans on Power Electr 11:641-652
31. Isidori A (1995) Nonlinear control systems. Springer-Verlag, London
32. Jung SL, Tzou YY (1996) Discrete sliding-mode control of a pwm inverter for sinusoidal output waveform synthesis with optimal sliding curve. IEEE Trans on Power Electr 11:567-577
33. Kailath T (1980) Linear systems. Prentice Hall, Englewood Cliffs, New Jersey
34. Karagiannis D, Mendes E, Astolfi A, Ortega R (2003) An experimental comparison of several PWM controllers for a single-phase AC-DC converter. IEEE Trans on Control Syst Techn, 11:940-947
35. Kassakian J, Schlecht M, Verghese G (1991) Principles of power electronics. Addison-Wesley, Reading, Massachusetts
36. Kazimierczuk M, Czarkowski D (1993) Application of the principle of energy conservation to modeling of the pwm converters. *Proc. 2nd IEEE Conf. Contr. Appl.* Vol. 1, 291-296, Vancouver, B.C.
37. Khalil H (2000) Nonlinear systems. Prentice Hall, Upper Saddle River, New Jersey
38. Kwatny H, Blankenship G (2000) Nonlinear control and analytical mechanics: a computational Approach. Birkhäuser, Boston

39. Lee TS (2003) Input-output linearization and zero-dynamics control of three-phase ac/dc voltage-source converters. IEEE Trans on Power Electr 18:11-22
40. Lee DC, Lee GM, Lee KD (2000) DC-bus voltage control of three phase ac/dc PWM converters using feedback linearization. IEEE Trans Ind Applact 36:826-833
41. Marino R, Tomei P (1995) Nonlinear control design. Prentice Hall International, London
42. Mendes de Seixas F (1993) Analysis of the zvm-pwm commutation cell and its applications to the dc-dc and dc-ac converters. MS Thesis, Federal University of Santa Clara, Florianópolis
43. Mendes de Seixas F, Cruz-Martins D (1997) The ZVS-PWM commutation cell applied to the DC-AC converter. IEEE Trans on Power Electr 12:726-733
44. Mohan M, Undeland T, Robbins W (1989) Power electronics: converters, applications and design. John Wiley and Sons, New York
45. Morici R, Rossi C, Tonieli A (1994) Variable structure controller of ac/dc boost converter *20th Int. Conf. Ind. Electron., Contr. Instrument.* Vol. 1, pp 1449-1454.
46. Nijmeijer H, Fossen T (Eds.) (1999) New directions in nonlinear observer design. (Lecture Notes in Control and Information Sciences, Vol. 244) Springer, London
47. Norsworthy S, Schreier R, Temes G (1997) Delta-sigma data converters: theory, design, and simulation. IEEE Press, Piscataway, New Jersey
48. Ortega R, Loria A, Nicklasson H, Sira-Ramírez H (1998) Passivity based control of Euler-Lagrange systems: mechanical, electrical and electromechanical applications. Springer, London
49. Perruquetti W, Barbot J (Eds.) (2002) Sliding mode control in Engineering. Marcel Dekker, New York
50. Rashid M (1993) Power electronics: circuits, devices, and applications. Prentice Hall International, Englewood Cliffs, New Jersey
51. Rugh W (1996) Linear system theory. Prentice Hall Information and System Sciences Series, Upper Saddle River, New Jersey
52. Sabanovoic A, Fridman L, Spurgeon S (Eds.) (2004) Variable structure systems: from principles to implementation. IEE Press, London
53. Sastry S (1999) Nonlinear systems: analysis, stability and control. Springer, New York
54. Severns R, Bloom G (1983) Modern dc-to-dc switchmode power converter circuits. Van Nostrand Reinhold, New York
55. Silva, JF (1999) Sliding mode control of boost-type unity-power-factor PWM rectifiers. IEEE Trans Ind Electron 46:594-603.
56. Sira-Ramírez H (1987) Sliding Regimes in analog signal encoding and delta modulation circuits. *Proc. 25th Annual Allerton Conference on Communications, Control and Computing.* Vol. 1, pp. 78-87. University of Illinois at Urbana-Champaign, Ill.
57. Sira-Ramírez H (1988) Differential geometric methods in variable structure systems. Int J of Control 48:1359-1391
58. Sira-Ramírez H (1998) A general canonical form for feedback passivity of nonlinear systems. Int J of Control 71:891-905
59. Sira-Ramírez H (1999) A general canonical form for sliding mode control of nonlinear systems in variable structure systems, sliding mode and nonlinear control, Young KKD and Özgüner Ü (Eds.) (Lecture Notes in Control and Information Sciences Vol. 247) pp. 123-142, Springer-Verlag

60. Sira-Ramírez H (2001) Dc to ac power conversion on a "boost" converter. Int J of Robust and Nonl Control 11:589-600
61. Sira-Ramírez H (2003) On the generalized pi sliding mode control of dc-to-dc power converters: a tutorial. Int J of Control 76:1018-1033
62. Sira-Ramírez H (2003) Sliding mode Δ modulators and generalized proportional integral control of linear systems. Asian J of Control 5:467-475
63. Sira-Ramírez H, Agrawal S (2004) Differentially Flat Systems. Marcel Dekker, New York
64. Sira-Ramírez H, Fliess M (2004) On the output feedback control of a synchronous generator. *43rd IEEE Conference on Decision and Control*, Bahamas
65. Sira-Ramírez H, Fossas E, Fliess M (2002) Output trajectory tracking in an uncertain double bridge "buck" dc-to-dc power converter: an algebraic, on-line, parameter identification approach. *41st IEEE Conference on Decision and Control*, Las Vegas, Nevada, December 10-13
66. Sira-Ramírez H, Ilic-Spong M (1987) Sliding motions in bilinear switched networks. IEEE Trans on Circ and Syst 34:919-933
67. Sira-Ramírez H, Ilic-Spong M (1989) Exact linearization in switch mode DC-to-DC power converters, Int. J. of Control. 50:511-524.
68. Sira-Ramírez H, Lischinsky-Arenas P (1991) The differential algebraic approach in nonlinear dynamical compensator design for DC-to-DC power converters. Int J of Control 54:111-134
69. Sira-Ramírez H, Marquez R, Fliess M (2002) Sliding mode control of dc-to-dc power converters using integral reconstructors. Int J of Robust and Nonl Contr 12:1173-1186
70. Sira-Ramírez H, Ortega R, García-Esteban M (1998) Adaptive passivity-based control of average dc-to-dc power converter models. Int J of Adapt Contr and Signal Proc 12:63-80
71. Sira-Ramírez H, Ortega R, Pérez-Moreno R, García–Esteban M (1995) A sliding mode controller-observer for DC–to–DC power converters: a passivity approach. *34th IEEE Conference on Decision and Control*. New Orleans, Louisiana, USA, December 13-15
72. Sira-Ramírez H, Silva-Navarro G (2002) Regulation and tracking for the average boost converter circuit: a generalized proportional integral approach. Int J of Control 75:988-1001
73. Slotine J, Li W (1991) Applied nonlinear control. Prentice Hall, Englewood Cliffs, New Jersey
74. Steele R (1975) Delta modulation systems. Pentech Press, London
75. Utkin V (1978) Sliding modes and their applications in variable structure systems, Mir Publishers, Moscow
76. Utkin V (1992) Sliding modes in control and optimization, Springer-Verlag, New York
77. Utkin V, Guldner J, Shi J (1999) Sliding mode control in electromechanical systems. Taylor and Francis, London
78. Wong, MC, Zhao, ZY, Han YD, Zhao LB (2001) Three dimensional pulse-width-modulation technique in three-level power inverters for the three phase four wired system. IEEE Trans Power Electronics 16:418-427.
79. Wood P (1981) Switching power Converters. Van Nostrand Reinhold, New York
80. Wu R, Dewan SB, Slemon GR (1991) Analysis of an ac-to-dc source converter using pwm with phase and amplitude control. IEEE Trans on Ind Appl 27:353-368

81. Yacoubi L, Al-Haddad K, Dessaint LA, Fnaiech F (2005) A DSP based implementation of a nonlinear model reference adaptive control for a three phase three level NPC boost rectifier prototype. IEEE Trans on Power Electr 20:1084-1092
82. Yeung KS, Cheng CC, Kwan CM (1993) A unified design of sliding mode and classical controllers. IEEE Trans on Auto Contr 38:1422-1427
83. Young K, Özgüner Ü (Eds.) (1999) Variable structure systems, sliding mode and nonlinear control. (Lecture Notes in Control and Information Sciences Vol. 247) Springer, London
84. Żak S (2002) Systems and control. Oxford University Press, Oxford
85. Zinober A, Fossas-Colet E, Scarrat J, Biel D (1998) Two sliding mode approaches to the control of a buck-boost system. *The 5th International Workshop on Variable Structure Systems*. Longboat Key, Florida, December 11-13

Index

AC rectifiers
  Monophasic Boost
    average model, 386
    DC motor loads, 400
    ETEDPOF controller, 388
    Flatness, 389
    steady state, 387
  Three phase Boost
    average model, 393
    DC motor loads, 408
    ETEDPOF control, 395
    Flatness, 395
    Sigma-Delta control implementation, 398
Ackermann's formula, 224, 225
Amplitude limiter circuit, 164

Bezier polynomials, 167, 314, 316, 410
Boost converter, **20**
  experimental prototype, 25
  model, 22
    alternative, 24
    non-ideal, 53
  normalization, 23
  sliding mode control, 71
  static transfer function, 23
Boost-Boost converter, **46**
  experimental prototype, 50
  equilibrium point, 47
  Experimental prototype, 105
  model, 47
    alternative, 49
  sliding mode control, 102

  static transfer function, 47
Buck Boost converter
  sliding mode control, 78
Buck converter, **13**
  experimental prototype, 18
  model, 14
  normalization, 15
  static transfer function, 16
Buck-Boost converter, **27**
  experimental prototype, 30
  model, 27
    alternative, 29
  normalization, 28
  static transfer function, 29

Cúk converter, **34**
  equilibrium point, 37
  model, 35
  normalization, 36
  sliding mode control, 82
  static transfer function, 37
Cayley-Hamilton theorem, 131
Comparator circuit, 76
Control via approximate linearization, **123**
Controllability, 130

DC-to-AC Conversion, **361**
DC-to-AC conversion
  Approximate linearization approach Boost, 371
  approximate linearization approach Buck-Boost, 373
  ETEDPOF approach

Boost, 381
Buck, 380
Buck-Boost, 383
flatness approach
 Boost, 375
 Buck, 374
 Buck-Boost, 376
sliding mode approach
 Boost, 378
Dissipation matching condition, 138, 142, 187, 262, 287
Double Buck-Boost converter, **50**
 equilibrium point, 51
 model, 51
 sliding mode control, 108
 static transfer function, 51
Dual dissipation matching condition, 332
Dynamic PI feedback control
 Buck-Boost, 195

Equivalent control
 MIMO case, 99
 SISO case, 65

Feedback linearization, **236**
 Brunovsky's canonical form, 242
 Frobenius's theorem, 241
 Input-output linearization, 238
 Involutivity condition, 252
 Isidori's canonical form, 236
 State feedback linearization, 240
Flatness, **130**
 Buck-Boost converter, 191
 Linearized Boost, 170
 quadratic converter, 222
Fliess' generalized canonical forms, **150**
 Controller design, 316
 Buck-Boost converter, 199
 linearized Boost, 181
Full state feedback
 observer design, 126
 Pole placement, 124

Generalized Proportional Integral control, **133**
 Buck converter, 154
 GPI-Sliding mode control, 343

robustness to load variations, 348, 355
linearized Boost converter, 183
linearized Buck-Boost converter, 202
State structural reconstruction, 136

Hamiltonian representation
 Boost converter, 187
 Boost-Boost converter, 233
 Buck converter, 159
 Buck-Boost converter, 205
 Cúk converter, 213
 quadratic Buck converter, 227
 the Zeta converter, 218
Hamiltonian systems, 139

Iterative solutions, 368

LaSalle's theorem, 252, 260, 273, 277, 280, 294, 297, 300, 303, 312

Minimum phase behavior
 Boost-Boost converter, 229
 Buck converter, 143
 Buck-Boost converter, 191
 Cúk converter, 212
 the Zeta converter, 218

Nominal trajectories in DC-to-AC conversion
 Boost, 366
 Buck, 363
 Buck-Boost, 370
Non-inverting Buck-Boost converter, **31**
 model, 31
 normalization, 32
 static transfer function, 33
Non-Minimum phase behavior
 the Zeta converter, 218
Non-minimum phase behavior
 Boost, 170
 Boost-Boost converter, 229
 Cúk converter, 212
Nonlinear Control Methods, 235
Nonlinear observer design, **331**
 Boost converter, 333
 Buck-Boost converter, 335
 Full order observers, 331
 Reduced order observers, 337

Boost converter, 337
Buck-Boost converter, 341

Passivity based control, **136**
  Buck converter, 156
    experimental results, 162
  Buck-Boost converter, 204
    experimental results, 207
  Energy shaping plus damping injection, 136
  Exogenous system, 137
  linearized Boost, 185
  Projection operator, 138
Perturbation matching condition, 70, 100
Proportional derivative control
  Buck converter, 145
  linearized Boost, 174

Quadratic Buck converter, **44**
  equilibrium point, 45
  model, 44
  sliding mode control, 91
  static transfer function, 46

Reduced order observers, 128
Routh-Hurwitz array, 217, 256

Separation principle, 127
Sepic converter, **38**
  equilibrium point, 40
  model, 39
  normalization, 39

static transfer function, 41
Sigma-Delta modulation, **112**
  average feedbacks, 115
  hardware implementation, 118
  theorem, 117
  two sided, 365
Sigma-Delta modulator, 20, 113
Sliding mode control
  MIMO systems, 95
  SISO systems, 62
Sliding modes
  existence, 69
  experimental implementation, 75, 77
  invariance conditions, 69
Sliding surface
  accessibility
    MIMO case, 101
    SISO case, 67
  MIMO case , 97
  SISO case, 64

Trajectory tracking
  linearized Boost, 176

Variable Structure Systems, 62

Zeta converter, **41**
  equilibrium point, 43
  model, 41
  normalization, 43
  sliding mode control, 87
  static transfer function, 44

Printed in the United States
54155LVS00001B/91-120